CO_2 Emissions from Vehicles
(Volume II)

CO_2 Emissions from Vehicles (Volume II)

Guest Editors

Maksymilian Mądziel
Kazimierz Lejda
Artur Jaworski

 Basel • Beijing • Wuhan • Barcelona • Belgrade • Novi Sad • Cluj • Manchester

Guest Editors

Maksymilian Mądziel
Department of Automotive
Vehicles and Transport
Engineering
Rzeszow University
of Technology
Rzeszow
Poland

Kazimierz Lejda
Department of Automotive
Vehicles and Transport
Engineering
Rzeszow University
of Technology
Rzeszow
Poland

Artur Jaworski
Department of Automotive
Vehicles and Transport
Engineering
Rzeszow University
of Technology
Rzeszow
Poland

Editorial Office
MDPI AG
Grosspeteranlage 5
4052 Basel, Switzerland

This is a reprint of the Special Issue, published open access by the journal *Energies* (ISSN 1996-1073), freely accessible at: www.mdpi.com/journal/energies/special_issues/W461DQFHHQ.

For citation purposes, cite each article independently as indicated on the article page online and using the guide below:

Lastname, A.A.; Lastname, B.B. Article Title. *Journal Name* **Year**, *Volume Number*, Page Range.

ISBN 978-3-7258-3438-9 (Hbk)
ISBN 978-3-7258-3437-2 (PDF)
https://doi.org/10.3390/books978-3-7258-3437-2

Cover image courtesy of Maksymilian Mądziel

© 2025 by the authors. Articles in this book are Open Access and distributed under the Creative Commons Attribution (CC BY) license. The book as a whole is distributed by MDPI under the terms and conditions of the Creative Commons Attribution-NonCommercial-NoDerivs (CC BY-NC-ND) license (https://creativecommons.org/licenses/by-nc-nd/4.0/).

Contents

About the Editors vii

Preface ix

Piotr Pryciński, Jacek Pielecha, Jarosław Korzeb, Roland Jachimowski and Piotr Pielecha
Impact of Vehicle Aging and Mileage on Air Pollution Emissions
Reprinted from: *Energies* **2025**, *18*, 939, https://doi.org/10.3390/en18040939 1

Lingyao Wang, Taofeng Wu and Fangrong Ren
Analysis of Spatial Differences and Influencing Factors of Carbon-Emission Reduction Efficiency of New-Energy Vehicles in China
Reprinted from: *Energies* **2025**, *18*, 635, https://doi.org/10.3390/en18030635 21

Monika Ziemska-Osuch and Dawid Osuch
Analysis of the Impact of Turn Signal Usage at Roundabouts on CO Emissions and Traffic Flows
Reprinted from: *Energies* **2024**, *17*, 6145, https://doi.org/10.3390/en17236145 48

Monika Bortnowska and Arkadiusz Zmuda
The Possibility of Using Hydrogen as a Green Alternative to Traditional Marine Fuels on an Offshore Vessel Serving Wind Farms
Reprinted from: *Energies* **2024**, *17*, 5915, https://doi.org/10.3390/en17235915 59

Wojciech Lewicki, Milena Bera and Monika Śpiewak-Szyjka
The Correlation of the Smart City Concept with the Costs of Toxic Exhaust Gas Emissions Based on the Analysis of a Selected Population of Motor Vehicles in Urban Traffic
Reprinted from: *Energies* **2024**, *17*, 5375, https://doi.org/10.3390/en17215375 94

Tomasz Cepowski
Utilizing Artificial Neural Network Ensembles for Ship Design Optimization to Reduce Added Wave Resistance and CO_2 Emissions
Reprinted from: *Energies* **2024**, *17*, 5326, https://doi.org/10.3390/en17215326 113

Fredy Rosero, Carlos Xavier Rosero and Carlos Segovia
Towards Simpler Approaches for Assessing Fuel Efficiency and CO_2 Emissions of Vehicle Engines in Real Traffic Conditions Using On-Board Diagnostic Data
Reprinted from: *Energies* **2024**, *17*, 4814, https://doi.org/10.3390/en17194814 134

Natalia Szymlet, Michalina Kamińska, Andrzej Ziółkowski and Jakub Sobczak
Analysis of Non-Road Mobile Machinery Homologation Standards in Relation to Actual Exhaust Emissions
Reprinted from: *Energies* **2024**, *17*, 3624, https://doi.org/10.3390/en17153624 152

Tomasz Cepowski and Paweł Kacprzak
Reducing CO_2 Emissions through the Strategic Optimization of a Bulk Carrier Fleet for Loading and Transporting Polymetallic Nodules from the Clarion-Clipperton Zone
Reprinted from: *Energies* **2024**, *17*, 3383, https://doi.org/10.3390/en17143383 172

Artur Krzemiński and Adam Ustrzycki
Visualisation Testing of the Vertex Angle of the Spray Formed by Injected Diesel–Ethanol Fuel Blends
Reprinted from: *Energies* **2024**, *17*, 3012, https://doi.org/10.3390/en17123012 202

Diming Lou, Guofu Song, Kaiwen Xu, Yunhua Zhang and Kan Zhu
The Oxidation Performance of a Carbon Soot Catalyst Based on the Pt-Pd Synergy Effect
Reprinted from: *Energies* 2024, 17, 1737, https://doi.org/10.3390/en17071737 217

Francesca Maria Grimaldi and Pietro Capaldi
The Effectiveness of HEVs Phase-Out by 2035 in Favor of BEVs with Respect to the Production of CO_2 Emissions: The Italian Case
Reprinted from: *Energies* 2024, 17, 961, https://doi.org/10.3390/en17040961 228

Andrzej Ziółkowski, Paweł Fuć, Piotr Lijewski, Maciej Bednarek, Aleks Jagielski and Władysław Kusiak et al.
The Influence of the Type and Condition of Road Surfaces on the Exhaust Emissions and Fuel Consumption in the Transport of Timber
Reprinted from: *Energies* 2023, 16, 7257, https://doi.org/10.3390/en16217257 248

Jan Monieta and Lech Kasyk
Application of Machine Learning to Classify the Technical Condition of Marine Engine Injectors Based on Experimental Vibration Displacement Parameters
Reprinted from: *Energies* 2023, 16, 6898, https://doi.org/10.3390/en16196898 265

Peter Tapak, Michal Kocur and Juraj Matej
On-Board On-Board Fuel Consumption Meter Field Testing Results
Reprinted from: *Energies* 2023, 16, 6861, https://doi.org/10.3390/en16196861 286

Maksymilian Madziel
Vehicle Emission Models and Traffic Simulators: A Review
Reprinted from: *Energies* 2023, 16, 3941, https://doi.org/10.3390/en16093941 300

About the Editors

Maksymilian Mądziel

Dr. Maksymilian Mądziel is a specialist in vehicle emissions, air quality, and sustainable transport solutions. His research focuses on modeling CO_2 emissions, real-world vehicle emission analysis, and the energy efficiency of hybrid and electric vehicles. He has made significant contributions to the study of emission reduction strategies and sustainable mobility, integrating advanced data analysis and artificial intelligence techniques.

Dr. Mądziel has authored and co-authored numerous scientific publications in high-impact journals such as *Energies, Sustainability, Environmental Science and Pollution Research*, and *Fuel*. His work leverages portable emission measurement systems (PEMSs), machine learning models, and traffic simulation tools to assess the environmental impact of transportation. His studies cover a broad range of topics, including the effects of road infrastructure design on vehicle emissions, fuel efficiency assessments, and alternative fuel applications.

Dr. Mądziel actively participates in international research projects aimed at optimizing transport systems and implementing emission reduction technologies. He collaborates with leading experts from various countries, contributing to the development of sustainable mobility solutions in urban and intercity transport. His research plays a vital role in shaping policies for low-emission zones, green vehicle adoption, and future transportation strategies.

Kazimierz Lejda

Prof. Kazimierz Lejda is a respected researcher specializing in fuels, emissions, and their environmental impact. His research focuses on fuel quality analysis, combustion process optimization, and methods for reducing pollutant emissions.

His publications frequently address the influence of driving resistance on energy consumption and exhaust emissions in motor vehicles. His work includes both laboratory tests and real-world studies, providing a comprehensive evaluation of various fuels and propulsion technologies.

Prof. Lejda is also actively involved in the development of alternative fuels and their modification to enhance environmental and operational performance. He collaborates with scientists from different countries, participating in international research projects and scientific conferences.

His extensive research contributes to the search for innovative solutions in fuels and propulsion technologies, playing a crucial role in sustainable development and environmental protection.

Artur Jaworski

Dr. Artur Jaworski is a specialist in internal combustion engines and alternative fuels, focusing on emission reduction and energy efficiency in transport. His research covers the impact of fuel blends, exhaust emissions analysis, and vehicle performance assessment under real-world conditions.

He has authored and co-authored numerous scientific publications in journals such as *Energies, Sustainability, Fuel, Applied Sciences*, and *Combustion Engines*. His work involves experimental studies, portable emissions measurement systems (PEMSs), and simulation techniques to optimize engine performance and reduce environmental impact.

Preface

Today, one of the most pressing issues of society is the challenge of reducing greenhouse gas emissions, particularly carbon dioxide (CO_2). As the global community continues to seek sustainable solutions to mitigate climate change, the transportation sector remains a key area due to its significant contribution to anthropogenic emissions. The need for innovative and effective approaches to reducing transport-related emissions is more urgent than ever.

This Special Issue, titled CO_2 Emissions from Vehicles (Volume II), aims to provide a platform for researchers and industry experts to share their latest findings on mitigating CO_2 emissions from transportation sources. The contributions included in this issue explore a diverse range of solutions, from advancements in internal combustion engines and after-treatment systems to the electrification of transport and the application of artificial intelligence in emission modeling. By addressing both local and global perspectives, the research presented here provides valuable insights toward a cleaner and more sustainable future.

The papers in this issue cover various aspects of emission reduction strategies, including the impact of vehicle aging on air pollution, the effectiveness of new-energy vehicles in different regions, and the role of hydrogen as a potential alternative fuel. Furthermore, studies on machine learning applications in emission modeling, and fuel efficiency assessments highlight the growing importance of digital technologies in tackling environmental challenges. By incorporating legislative reviews and comprehensive analyses of emission data, this Special Issue also sheds light on the regulatory frameworks that shape the future of transportation sustainability.

We hope that the findings presented in this Special Issue inspire further research and policy development, ultimately leading to meaningful advancements in reducing CO_2 emissions from vehicles. The collaboration between academia, industry, and policymakers is essential in achieving our collective goal of decarbonizing transportation and protecting our environment for future generations.

We extend our sincere gratitude to all the authors, reviewers, and contributors who have made this Special Issue possible. Their dedication to advancing scientific knowledge and providing practical solutions for emission reduction is invaluable in the fight against climate change.

Maksymilian Mądziel, Kazimierz Lejda, and Artur Jaworski
Guest Editors

Article

Impact of Vehicle Aging and Mileage on Air Pollution Emissions

Piotr Pryciński [1,*], Jacek Pielecha [2], Jarosław Korzeb [1], Roland Jachimowski [1] and Piotr Pielecha [3]

[1] Faculty of Transport, Warsaw University of Technology, 00-661 Warsaw, Poland; jaroslaw.korzeb@pw.edu.pl (J.K.); roland.jachimowski@pw.edu.pl (R.J.)
[2] Faculty of Civil Engineering and Transport, Poznan University of Technology, 60-965 Poznan, Poland; jacek.pielecha@put.poznan.pl
[3] Faculty of Civil Engineering and Transport, Doctoral School, Poznan University of Technology, 60-965 Poznan, Poland; piotr.pielecha@doctorate.put.poznan.pl
* Correspondence: piotr.prycinski@pw.edu.pl

Abstract: The research described in this paper aimed to verify the impact of road vehicle aging on air pollutant emissions. The problem of vehicle aging and the resulting changing air pollutant emissions was identified with the operational mileage of passenger cars. The validity of such an approach to research on air pollutant emissions changing over time was confirmed by a preliminary review of publications in scientific databases such as Scopus and Web of Science. The research problem presented in this paper was to assess the impact of vehicle aging on essential air pollutant emissions (CO, CO_2, NO_X). The research method included measuring the actual RDE air pollutant emissions using research equipment, i.e., the SEMTECH DS gaseous exhaust gas component analyzer. This study was conducted on vehicles with diesel engines, different operating ages, different mileages, and engines with similar displacement, mostly 1.9–2.0 dm^3, and equipped with manual transmissions. The tests were conducted in the Poznań agglomeration. The results of measured air pollutant emissions in the RDE mode allowed for mapping the changes in air pollutant emissions for diesel engine vehicles with similar displacement as a function of operating age (mileage) and also for collecting preliminary data for analyses in the field of modeling road air pollutant emissions within the vehicle aging phenomenon.

Keywords: aging of vehicles; emission of pollutants; environmental impact

Academic Editor: Roberto Finesso

Received: 10 December 2024
Revised: 7 February 2025
Accepted: 10 February 2025
Published: 16 February 2025

Citation: Pryciński, P.; Pielecha, J.; Korzeb, J.; Jachimowski, R.; Pielecha, P. Impact of Vehicle Aging and Mileage on Air Pollution Emissions. *Energies* **2025**, *18*, 939. https://doi.org/10.3390/en18040939

Copyright: © 2025 by the authors. Licensee MDPI, Basel, Switzerland. This article is an open access article distributed under the terms and conditions of the Creative Commons Attribution (CC BY) license (https://creativecommons.org/licenses/by/4.0/).

1. Introduction

Intensive legislative actions taken in the European Union, aimed at reducing the negative environmental impact of transport, necessitate the exploration of new areas and solutions to decrease the emissions associated with transport activities. This paper addresses the issue of examining the impact of vehicle aging on air pollutant emissions. It reviews existing research focused on studying the vehicle aging phenomenon and its influence on air pollutant emissions. Those resulting from fuel combustion are a dominant source of environmental pollution. This paper contains an assessment of the changes in air pollutant emissions from road vehicles in relation to their increasing operational age. This research problem is justified in the context of implementing low-emission zones in Poland and Europe. The legal basis for establishing low-emission zones in Poland is the Act on Electromobility and Alternative Fuels [1]. This act regulates issues related to introducing low-emission zones in urban areas. In terms of implementing low-emission zones in Poland, it is crucial to verify actual air pollutant emissions and determine how these emissions change with the operational age of vehicles using data from real-world emissions testing conducted with the Real Driving Emission (RDE) method.

Over the past decade, Poland has experienced rapid and unprecedented changes regarding vehicle ownership on a European scale [2,3]. Statistical data on the number of motor vehicles operating on Polish roads shows that the average number of vehicles was over 32.4 million between 2017 and 2022, including more than 24.5 million passenger cars. Passenger cars thus accounted for more than 76% of all vehicles in Poland. However, according to the Polish Automotive Industry Association's data, the number of passenger cars registered in Poland should be adjusted for vehicles older than 10 years whose data have not been updated in the central vehicle database during this period. A lack of updates in the central vehicle database may indicate that periodic technical inspections of these vehicles were not carried out and, consequently, that these vehicles may not be in actual use. After adjusting the data for such vehicles, an average of approximately 18.5 million passenger cars were registered annually between 2018 and 2022.

This highlights the need for methods to improve the verification of actual air pollution emissions resulting from the operation of numerous passenger vehicles. The average age of a passenger car in Poland in 2022, among the population of vehicles with updated data, was 14.9 years, an increase of 0.4 years compared to 2021, while the median age remained at 15 years, the same as in 2021 [2,3]. Passenger cars up to 4 years old accounted for only 11.4% of the passenger car fleet at the end of 2022, a decrease of 0.9% compared to the end of 2021. Passenger cars aged between 5 and 10 years constituted 17.0% of vehicles, i.e., the same as in 2021. The largest group of vehicles were those aged between 11 and 20 years, representing 49% of vehicles. The oldest vehicles in Poland, i.e., those over 20 years old, accounted for 23% of all passenger cars. The number of the oldest vehicles was nearly twice as low as the number of vehicles with an operational age of up to 4 years [2,3].

Verifying and assessing air pollution emissions for vehicles operating in Poland remains a relevant and justified issue given these observations. This phenomenon is primarily caused by the mass import of used vehicles from Western Europe. Solutions aimed at reducing transport-related emissions, such as the implementation of clean transport zones, are being sought and implemented. However, it can be assumed that air pollution emissions vary with the operational age and mileage of passenger vehicles. To verify this and determine compliance with access requirements to selected clean transport zones, it is necessary to conduct on-road emissions testing.

Real Driving Emission (RDE) testing for road vehicles is the most accurate method for determining the release of harmful substances into the environment [4]. On-road air pollution emissions testing is standardized and conducted under real-world driving conditions. The research methodology used in this paper involves conducting on-road air pollution emissions tests on eight vehicles with mileage ranging from 1000 to over 300,000 km, a maximum operational age close to the average vehicle age in Poland, i.e., over 14 years. The test vehicles were equipped with diesel engines. For precise air pollution emissions measurements, SEMTECH DS testing equipment was used as part of the research methodology.

The air pollution emissions measured for the eight vehicles during on-road tests were compared with the data on the vehicles' operational age and mileage. The vehicle with the highest mileage was additionally characterized in terms of repairs and technical inspections performed.

2. Aging of Vehicles and Its Impact on the Environment—The State of the Art and Literature Review

Passenger cars have the largest share in air pollution emissions in Poland. Their share over the period from 2017 to 2023 is presented in Figure 1.

Figure 1. Motor vehicles and passenger cars in Poland in the years 2017–2023 (thousands of units) [3].

After updating the data in the Central Motor Vehicle Database, the actual number of passenger cars registered in Poland between 2018 and 2022 significantly dropped to an average of 18.5 million passenger cars per year. The number of passenger cars after the update is presented in Figure 2. The amount of passenger cars in 2018–2022 shows a clear upward trend, with estimates suggesting that the number of passenger cars registered in Poland and regularly used—linked to mandatory technical inspections recorded in the central vehicle database—exceeds 19.5 million vehicles. Considering substantial imports of vehicles from Western Europe, the number of passenger cars surpassed 20 million in 2024. Nearly 400,000 new passenger cars were registered by the end of the third quarter of 2024 alone [5].

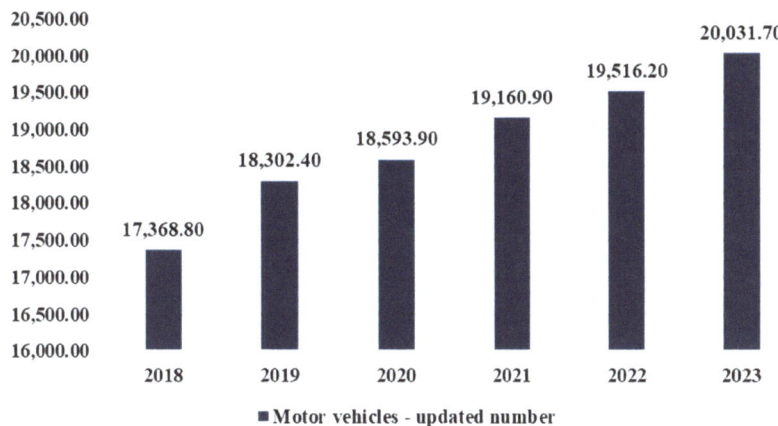

Figure 2. Updated numbers of passenger cars in Poland in the years 2018–2022 (thousands of units) [3].

For many years, the aging fleet of motor vehicles in Poland has been primarily fueled by imports from the secondary markets of other countries, where the majority of vehicles are over 10 years old. Figure 3 presents the age structure of passenger cars in Poland [3]. The consequences of maintaining an aging fleet of passenger cars are borne not only by car owners, often in the form of higher repair bills, but also by society through external costs. Older passenger cars are less safe and less environmentally friendly. Worn-out engines emit significantly more toxic exhaust gases than new internal combustion engines. In 2022, despite ambitious plans to achieve zero-emission transport in the European Union, passenger cars powered by fossil fuels continued to dominate in Poland (gasoline—45%, diesel—39%, LPG—13%) [3].

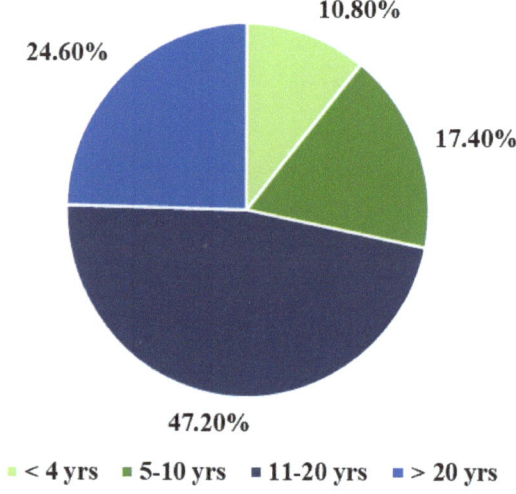

Figure 3. Updated age structure of passenger cars in Poland in the years 2018–2022 (thousands of units) [3].

Zero- or low-emission vehicles accounted for only 3% of passenger cars. Over 72% of vehicles in Poland are older than 11 years, while nearly one in ten passenger cars is younger than 4 years. The environmental impact of passenger cars is therefore closely related to their technical condition and the emission standard they comply with.

Given the data above, the aging of mechanical systems is an inevitable result of the operation of any equipment, including passenger cars. In transportation, this issue is addressed during the vehicles' design phase. In terms of environmental impact, the focus is typically on the type of transport mode and operating conditions, with less emphasis on the technical condition and age of vehicles, as well as their use and maintenance. Air pollution caused by human activities related to the resource extraction, production, operation, and decommissioning of vehicles represents a significant share of anthropogenic emissions into the environment.

For many years, and continuing into the future, legislative actions aimed at reducing emissions from fuel combustion have been planned and implemented across the European Union, which assumed the role of a global leader in environmental protection with the Lisbon Strategy in 2000. Notable initiatives include the Green Deal [6] and the Fit for 55 package [7], which set ambitious plans to reduce air pollution and shift the structure of transport by increasing the use of rail for long-distance freight. Solutions such as restricting access to selected functional urban zones for vehicles not meeting emission standards are becoming increasingly common. However, emission standards are established during the homologation phase based on initial emissions testing without verification of the actual vehicle emissions during their operational lifetime.

The issue of aging mechanical systems and devices is well documented and represents a crucial stage in the operation of any equipment. This stage is considered during the design phase of transportation systems. Depending on usage, periodic inspections and major repairs, the lifespan and reliability of vehicles change over time. Similarly, the aging of road vehicles impacts the level of air pollution emissions. In this context, environmental impact is measured as the number of emissions generated and expressed in grams per unit of vehicle mileage.

In the discussion on the environmental impact of technical transport systems, attention is usually directed to fuel types, external conditions of transport operations, and vehicle types. However, more precise verification should focus on the technical condition, operational age, and maintenance of vehicles. Enhanced verification of a vehicle's technical condition would allow for an analysis of how vehicle wear affects emissions released into the environment by passenger cars.

In the literature, the aging of vehicles is often associated with the evaluation of mechanical system wear (e.g., catalytic converters) in internal combustion vehicles responsible for reducing air pollution emissions. Publications rarely directly verify changes in air pollution emissions (and compliance with emission standards) as a function of vehicle mileage. Emission standards define the maximum permissible levels of harmful substances emitted by motor vehicles and other fuel-burning devices. These standards typically regulate substances such as nitrogen oxides (NO_X), carbon monoxide (CO), particulate matter (PM), hydrocarbons (HCs), and other chemical compounds emitted by combustion engines and devices.

Emission standards vary depending on the vehicle type, fuel type, and geographic location. The regulations aim to control and reduce harmful substance emissions to protect public health and the environment. Key legislative frameworks related to air emission standards include the following:

- Environmental protection laws that include regulations on harmful substance emissions into the atmosphere. These laws often define air quality targets and emission reduction requirements for various industrial sectors, vehicles and other emission sources.
- International treaties and agreements on environmental protection, such as the Kyoto protocol or the Paris agreement, which may impose emission reduction obligations on individual countries or regions.
- Regulatory agencies responsible for monitoring air quality and enforcing emission regulations. Examples include the Environmental Protection Agency (EPA) in the United States and the European Environment Agency (EEA).
- Technical standards that set out permissible emission levels for various types of vehicles, industrial equipment, and other emission sources. These standards are often developed in collaboration with industry, scientists, and other stakeholders.

As presented in paper [8], to meet increasingly stringent emission standards for conventional vehicles, it is essential to employ a variety of technologies, often complementary, to purify the substances contained in automotive exhaust gases. Among these, the most effective are exhaust purification devices, including exhaust gas cleaning systems such as the Gasoline Particulate Filter (GPF) for spark-ignition vehicles, the Diesel Particulate Filter (DPF) for compression-ignition vehicles, and Exhaust Gas Recirculation (EGR) systems supported by the use of ammonia or urea. The technical solutions currently employed in vehicles for exhaust gas purification ensure compliance with strict Euro emission standards. However, air pollution emission standards primarily address greenhouse gas emissions. Unfortunately, exhaust emission standards for road transport do not account for a crucial aspect of exhaust impact on human health, namely acetaldehyde, benzene, and benzo(a)pyrene [9]. These compounds are not monitored or regulated, yet they significantly contribute to the toxicity of exhaust gases.

Analyses of exhaust gas compositions conducted by the authors of paper [9] point to the presence of highly carcinogenic compounds emitted by vehicles despite meeting strict air pollution emission standards. Current emission control methods used to certify vehicles for road use are not aimed at eliminating highly carcinogenic chemical compounds [10]. It would therefore be advisable to supplement measurement methods with more detailed analyses of exhaust gas mixtures, especially hydrocarbons, to unambiguously determine their actual harmful effects on human health. Consequently, an increase in carcinogenic compounds emitted by vehicles can be expected as the operational age of road transport vehicles increases.

It is important to emphasize that the efficiency of exhaust gas purification devices inevitably deteriorates with prolonged vehicle use. Several mechanisms can be utilized to explain, study, and simulate aging phenomena. As noted in paper [8], thermal aging and mechanical aging are the most significant causes of performance degradation over time in the case of catalytic devices. Regarding Diesel Particulate Filters (DPFs), clogging with so-called ash is a critical issue for their stable operation. To more effectively and rapidly develop and test exhaust gas purification devices, accelerated thermal aging methods should be investigated and applied. The so-called small-sample aging methods allow for accelerated catalyst aging tests at very low costs, although the accuracy of these methods may not always be sufficient.

The impact of external forces acting on exhaust gas purification systems during typical vehicle operation is not considered. The results of methods simulating the aging of entire vehicles and engine aging conducted in laboratory conditions align more closely with real-world usage patterns, yet they are too time-consuming and costly for widespread application [8]. Nevertheless, real-world studies are the most reliable in determining the aging of key components responsible for purifying exhaust gases generated by conventional vehicles.

For vehicles with diesel engines, particulate matter from exhaust gases in the form of ash primarily originates from fuel additives, lubricant additives, engine component wear, and exhaust system corrosion [10]. It can be inferred that as vehicles age, the emission of pollutants, including particulate matter $PM_{2.5}$ and PM_{10}, increases. Some studies aim to establish a relationship between laboratory testing and the actual effects of exhaust system aging. As indicated in paper [11], operating a catalyst at 800 °C for 50 h yields exhaust system aging results similar to the effects observed in vehicles with a mileage of 150,000 km by applying a rapid aging method to catalysts. Similarly, the authors of paper [12] conducted catalyst aging tests under various conditions, finding that 16 h of catalyst operation at 800 °C approximates the aging effects of a vehicle with a mileage of 135,000.

The age distribution of vehicles is therefore a significant factor to consider when assessing the impact of air pollution emissions. These studies focused on components responsible for exhaust gas purification. However, large-scale real-world emission studies have not been widely conducted. Such research is described only in paper [13], which examined the impact of vehicle aging on emissions over a sample of 600 vehicles. A strong correlation between vehicle age and emissions was observed. The study's conclusions emphasized the need for proper engine maintenance and stricter air pollution emission controls. Based on paper [13], under Israeli emission standards, older vehicles (over 12 years of age) exceeded air pollution emission limits by as much as 50%. This observation highlights the importance of reassessing policies such as vehicle access restrictions to specific urban zones based on vehicle age.

Publication [13] provides comprehensive results from mass studies on the impact of vehicle age on air pollution emissions. However, these studies did not employ modern Real Driving Emissions (RDE) testing methods. Key conclusions from this research underline the necessity of improving vehicle engine maintenance processes and continuing efforts to monitor and control vehicle emissions. Additionally, air pollution emissions from vehicles are influenced by external factors such as the quality of fuel materials [14] and standards governing their production.

The effects of aging and the technological substitution of motor vehicles and their impact on air pollution emissions are discussed in paper [15]. The authors utilize a model simulating the internal dynamics of vehicle population growth and air pollution emissions. They highlight the issue of older vehicles originating from the secondary market. It is possible to model the vehicle renewal rate, allowing for sensitivity analysis of the vehicle fleet to specific age and technological parameters. This enables the study of emission deterioration, evaluation of the effects of implementing inspection and maintenance programs, and the introduction of cleaner fuels. Vehicle aging modeling in [15] is conducted using the Weibull function. The authors also identify various parameters and factors influencing air pollution emissions, which are categorized into technological aspects and vehicle age-related aspects.

The degradation of exhaust purification systems and vehicle age are mentioned among the age-related parameters. These parameters contribute to increased air pollutant emissions per unit of pollutant generated by transportation. However, the decreasing intensity of vehicle use with advancing vehicle age acts as a mitigating factor for total emissions. Technological parameters focus on the development of new exhaust purification technologies and the implementation of advanced vehicle control systems. These factors include the introduction of new fuels and modeling the aging processes of critical components involved in vehicle exhaust purification.

It is worth noting that determining the emissions inventory in road transport is challenging due to the variability in technologies and operating conditions encountered on the roads. An important variable is the annual mileage of vehicles. In paper [16], the authors collected extensive data on vehicle mileage from used car sellers in Italy and analyzed them to understand the impact of vehicle age on actual mileage. The data enabled the development of a function describing annual mileage as a function of vehicle age. It was found that the average mileage of 10-year-old vehicles is only about 40% of the vehicle's mileage in the first year, while for 20-year-old vehicles, this percentage drops to just 10%.

These findings are significant for environmental calculations, as nitrogen oxide and particulate matter emissions in road transport decrease by over 20% when corrected vehicle age functions are applied compared to a constant mileage model. The reduced intensity of vehicle use with age diminishes the relevance of accelerated scrappage programs and low-emission zones for air quality improvement. A similar analysis of vehicle fleets registered in selected areas is described in paper [17]. This analysis involved registration data and assessed the impact of socio-economic conditions on air pollution emissions.

The results revealed several interesting trends in vehicle types and ages compared to national statistics and average resident incomes. The average vehicle age strongly correlated with income levels. Vehicle fleets were oldest in regions with the lowest incomes and newest in regions with the highest incomes. The average vehicle age was 10.8 years in the former case and only 5.9 years in the latter. This disparity in vehicle age results in average emission factors from transport sources being 63% higher for nitrogen oxides, 73% higher for carbon monoxide, and 104% higher for volatile organic compounds in low-income regions compared to high-income regions.

Methods for determining air pollution emissions based on statistical and real-world research results are presented in papers [18,19]. However, these methods do not account for the impact of passenger car age on air pollution emissions.

As part of the research, a literature review was conducted to identify publications dealing with reliability that align with the environmental impact of transportation. Topics related to the technical condition and reliability of vehicles have been analyzed by many authors. For instance, in paper [20], the authors analyzed the reliability of passenger cars based on periodic technical inspections of nearly 2 million vehicles using data from a central vehicle registry. However, the research methodology did not include measurements of air pollution emissions at specific periodic inspection points. The results indicated that the frequency of defects and failures increases with a car's age and mileage. The correlation between these variables significantly strengthens with the size of the analyzed sample, expressed as the number of vehicles undergoing diagnostic tests.

Regarding the intensity of defects, it was found to peak for vehicles with annual mileages between 15,000 and 20,000 km. Gasoline vehicles were found to be more prone to failure. The issue of reducing the negative environmental impact of transportation was addressed in paper [21]. The authors focused on reducing particulate emissions and air pollution by employing advanced tribological piston/cylinder systems in diesel engines. In terms of optimizing the performance of road vehicle engines, the study investigated materials that enhanced the sliding surface of pistons and cylinders made of aluminum alloys. Over time, diesel engine emissions have been increasingly restricted by stringent emission regulations. Consequently, advanced technologies such as high-pressure direct fuel injection, turbocharging, exhaust gas recirculation, variable valve timing, downsizing, catalytic treatment of combustion products, and particulate filters are being developed.

The operational age of vehicles has also been analyzed in the literature beyond the context of transportation ecology. In paper [22], the author examined the impact of vehicle age on road safety. While this approach is relevant to the external costs of transportation, it does not propose new solutions or concepts concerning air pollution emissions or the impact of vehicle age on emissions. According to the author of [22], 75% of road accidents caused by vehicle technical conditions are due to lighting and tire failures.

An important topic related to potential future air pollution emissions is the forecasting of vehicle technical conditions [23,24]. Conducting appropriate testing procedures and monitoring the technical condition of vehicles in terms of air pollution emissions could be valuable. This requires developing proper diagnostic procedures, which are currently not in place. Real Driving Emissions (RDE) testing could serve as a data source for such research and analysis. Air pollution emissions could thus be treated as a diagnostic signal. Forecasting involves determining changes in the diagnostic parameters of transportation means, which characterize the deterioration of their condition over time.

The literature includes research on the operational efficiency of vehicles at various stages of their lifecycle. Paper [25] examined technical readiness, repair costs, and revenues from the operation of road transport vehicles. The research determined the operational efficiency characteristics for buses, showing that vehicle age significantly impacts operational efficiency. However, these analyses do not address air pollution emissions.

Vehicle operation system modeling is often explored in terms of operation and reliability. Semi-Markov processes, based on three operational states (usage, downtime, and repair), are commonly used for this purpose [26]. Due to the stochastic nature of vehicle failures, understanding stochastic processes is crucial for maintaining effective and safe operation. For example, paper [27] used a Poisson process to model the corrective maintenance of specific vehicle parts, demonstrating the practical applications and limitations of this approach for assessing the risk of future costs in continued operation.

Air pollution emission studies are widely known [28], but large-scale research on emissions in terms of the changing operational age of transport vehicles is not commonly conducted. Among the less frequently applied studies examining the impact of vehicle age and technical condition on air pollution emissions are those mentioned in [13–17]. The distribution of vehicle ages appears to be a significant factor that should be considered when assessing the environmental impact of air pollution emissions.

3. Research Problem and Methodology

The research methodology employed in this paper involved measuring real-world emissions using advanced research equipment, specifically the SEMTECH DS gas analyzer (Sensors Inc., Saline, MI, USA) and a particulate matter analyzer. The test equipment was used to conduct the RDE (Real Driving Emission) test. Vehicles manufactured after 2017 must meet the requirements of the RDE test. During the test, it is necessary to select a route that meets the test's conditions. The RDE test should be conducted on paved roads during working days in urban, suburban, and motorway areas. The regulations prohibit the excessive use of neutral gear in the initial phase after starting the engine. The route should be at least 16 km long in urban, rural, and highway conditions. The speed achieved in urban areas should not exceed 60 km/h, and the average speed should be between 15 and 45 km/h. The share of each test section should be about one-third of the total distance. In addition, stops, i.e., when the car does not move faster than 1 km/h, should not constitute more than 30% of the total travel distance in the test's urban section. In the extra-urban section, the car should travel at least 16 km between 60 and 90 km/h. On the highway section, speeds of more than 90 km/h must be observed, and the maximum speed should not exceed 145 km/h. An essential aspect of RDE tests is that an elevation difference of 100 m above sea level between the starting and ending points must be observed. This eliminates the problem of repeatability of results in different parts of the world due to increased emissions during uphill climbs.

This study analyzed eight diesel engine vehicles with varying operational ages, ranging from manufacturing years of 2006 to 2018. The vehicles had mileage spanning from 20,000 km to 317,000 km. All vehicles were equipped with engines of similar displacement, between 1.9 dm^3 and 2.0 dm^3, and featured manual transmissions.

Before testing, all vehicles and the measurement equipment were stationed in garage conditions at a temperature of approximately 20 °C. The exhaust gases entered the analyzer through a special filter that separated particulate matter. Subsequently, the gases were directed to a Flame Ionization Detector (FID), which measured hydrocarbon concentrations. The exhaust gases were then cooled to 4 °C and directed to a Non-Dispersive Ultraviolet (NDUV) analyzer to determine nitrogen oxide (NO_X) concentrations. Finally, the exhaust gas mixture was analyzed using a Non-Dispersive Infrared (NDIR) analyzer, which measured carbon monoxide (CO) and carbon dioxide (CO_2) concentrations. The SEMTECH DS device also employed an electrochemical analyzer to measure oxygen (O_2) concentrations.

The gaseous components of exhaust gases were measured using the SEMTECH DS analyzer, which comprehensively measures the CO, CO_2, and NO_X content in exhaust gases. An exhaust gas collection system was mounted between the exhaust gas analyzer and the vehicle's exhaust system, in which the exhaust gases moved at a temperature of 191 °C to prevent water condensation. After introducing the exhaust gases into the analyzer through a special filter, solid particles were separated, and the exhaust gases went to the FID, where the hydrocarbon concentration was measured. Then, the exhaust gases were cooled to a temperature of 4 °C and moved to a non-dispersive analyzer using ultraviolet radiation (NDUV), where the concentration of nitrogen oxides was determined. The exhaust gas mixture was then directed to the NDIR analyzer, which also used the non-dispersive method. However, the carbon monoxide and carbon dioxide concentrations were measured using infrared radiation. Finally, the SEMTECH DS device measured the oxygen concentration using an electrochemical analyzer. The technical data of the SEMTECH DS are presented in Table 1.

Table 1. Technical data of the Semtech DS analyzer.

Parameter	Measurement Method	Accuracy
CO	NDIR, 0–10%	±3% range
HC	FID, 0–10,000 ppm	±2.5% range
NO_X (NO + NO_2)	NDUV, 0–3000 ppm	±3% range
CO_2	NDIR, 0–20%	±3% range
O_2	paramagnetic, 0–20%	±1% range

The separated solid particles were also sent to the second TSI analyzer (the solid particle counter). The exhaust gas flow rate was measured using a 2″ measurement probe. Due to road conditions, it was necessary to attach a probe to ensure tightness with the vehicle's exhaust system. The TSI Incorporated—EEPS 3090 (Engine Exhaust Particle Sizer™ Spectrometer) analyzer (TSI Incorporated, Aachen, Germany) was used to measure particle diameters ranging from 5.6 nm to 560 nm. The technical data of the TSI 3090 EEPS analyzer are presented in Table 2.

Table 2. Technical data of the EEPS particulate matter analyzer.

Operating Features	Value
Particle size range	5.6 nm to 560 nm
Particle size resolution	16 channels per decade (32 in total)
Electrometer channels	22
Time resolution	10 size distributions/s
Sample flow	10 dm^3/min
Sheath air	40 dm^3/min
Operating temperature	0 °C to 40 °C

A schematic of the research methodology is presented in Figure 4. Additionally, an On-Board Diagnostics (OBD) system was used to record basic driving parameters and geographic coordinates in a spreadsheet, enabling the creation of a route map.

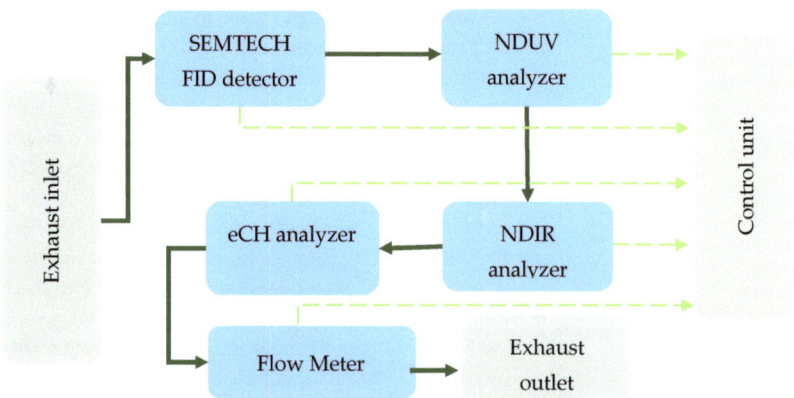

Figure 4. Diagram of the measurement method using the SEMTECH DS and flow meter (own study).

Research using the presented testing system has also been conducted previously for other vehicle types [18,19,28], including non-passenger vehicles. Earlier studies addressed vehicles such as commercial vehicles [29], specialized vehicles [30], hybrid and electric vehicles [31], and delivery vehicles [32].

Passenger cars homologated since 2017 must comply with the requirements of the Real Driving Emissions (RDE) test [28]. During the RDE test, emissions of harmful compounds in exhaust gases are measured under urban, rural, and highway driving conditions. The test route must be appropriately designed to ensure measurement process continuity, requiring the development of a reliable and repeatable driving route. During the tests, data must be recorded without any interruptions.

The RDE tests for vehicles, as outlined in the methodology adopted in this paper, were conducted on business days on paved roads within the Poznań metropolitan area and nearby regions. The test route included at least 16 km each in urban, rural, and highway conditions. In urban areas, speeds during testing did not exceed 60 km/h, with an average speed range between 15 and 45 km/h. Each segment constituted approximately one-third of the total route length. Additionally, stationary periods, during which the vehicle moved slower than 1 km/h, could not exceed 30% of the total urban driving time.

In the rural segment, vehicles covered at least 16 km at speeds ranging from 60 to 90 km/h. On highways, speeds ranged from 90 to 145 km/h. An important aspect of RDE testing is ensuring that the difference in elevation between the start and end points of the test does not exceed 100 m. These conditions were met for all vehicles subjected to the tests.

RDE measurement methods described in the literature [33] include analyses of factors influencing air pollution emissions. Such factors include vehicle load, terrain topography, traffic congestion [34,35], and other variables like fuel type, driving conditions, vehicle type, and the location of operation [36,37]. However, these studies do not examine the impact of vehicle operational age and aging on air pollution emissions, which is the focus of this article.

The operational age of passenger cars is a factor that can significantly influence air pollution emissions, making it a crucial issue in terms of the sustainable development of transport systems. This topic is also addressed in the literature [38–40].

4. Research Experiment

As part of a computational example, air pollution emission tests were conducted for seven passenger vehicles. The vehicles underwent the Real Driving Emissions (RDE) test. The analysis utilized vehicles with characteristics outlined in Table 3. The RDE tests were performed in the Poznań metropolitan area, adhering to the guidelines specified for RDE test conditions.

Table 3. Parameters of vehicles subjected to the RDE test (own data).

Vehicle Feature	Vehicle V1	Vehicle V2	Vehicle V3	Vehicle V4	Vehicle V5	Vehicle V6	Vehicle V7
Mileage	20,000	25,000	86,000	120,000	140,000	160,000	317,000
Fuel type	Diesel	Diesel	Diesel	Diesel	Diesel	Diesel	Diesel
Engine capacity [dm^3]	2.0	1.9	2.0	1.9	2.0	2.0	1.9
Year of manufacture	2018	2006	2010	2008	2008	2011	2010
Engine power [kW]	75	85	81	110	115	163	77
Emission standard	Euro 6	Euro 4	Euro 5	Euro 4	Euro 4	Euro 5	Euro 4
Manufacturer/Model	VW Caddy	Audi A4	Skoda Yeti	Opel Vectra	VW Transporter	BMW 320d	VW Touran

The urban portion of the test passed through the main arterial roads of Poznań and ended in the city's northeastern part at the intersection with National Road No. 92. Here, the rural section of the RDE test began, comprising a segment of National Road No. 92 and the S5 expressway. At the Wierzyce interchange, the S5 expressway looped back, transitioning the test into the highway section. In this section, the vehicle traveled at the highest permissible speed on the S5 segment, and upon entering the A2 highway, the speed was maintained at up to 130 km/h.

The route transitions were also illustrated on area maps, including annotations for changes in vehicle speed. The urban route, with speed variations between 0 and 50 km/h, is marked in green; the rural route, with speed variations in the range of 60–90 km/h, is marked in blue; and the highway section, with speeds exceeding 90 km/h, is marked in purple. The route, which was consistent for all eight tested vehicles during the RDE test, is presented in Figure 5.

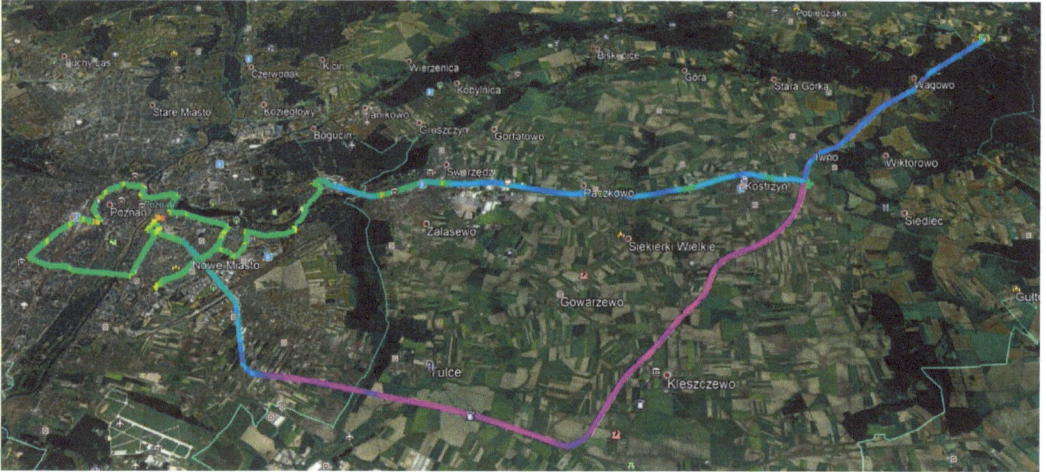

Figure 5. Route of vehicles V1–V7 in the RDE tests (own study). Different colors in this figure mean different profiles of the average speeds of vehicles during the RDE test.

Separate colors in Figure 5 show the vehicle speed ranges in the individual sections of the RDE test. Yellow and green mark the urban route (up to 60 km/h), light blue and blue mark the suburban test stage (up to 90 km/h), and navy blue and purple mark the highway test stage (up to 130 km/h).

In each section of the RDE test, the vehicles traveled approximately 30 km, nearly double the required distance. This also resulted in a total driving time that was twice as long. Vehicles V1–V7 covered nearly 100 km in less than 120 min during the RDE tests. Each section of the test accounted for about 33% of the total route, aligning with the test requirements. The share of stops and driving time at speeds exceeding 100 km/h were very similar across all RDE tests. Photographic illustrations of the field tests carried out for the selected vehicles are presented in Figure 6.

Although all seven vehicles were intended to follow the same route, unplanned roadworks, traffic congestion, or detours caused slight variations in the total distance covered by the vehicles. Nevertheless, all the trips met the RDE test requirements, and the minor differences in distance and duration did not affect the emission measurement results. The emissions measured during the RDE tests are presented in Figures 7–9.

The NO$_X$ emissions were lowest for vehicles with mileage up to 86,000 km, specifically for vehicles V1, V2, and V3. Among these, only vehicle V1 met the NO$_X$ emission standards, with a very low mileage of just 20,000 km. For vehicles V2 and V4, the NO$_X$ emission profiles were similar across urban, suburban, and highway driving conditions.

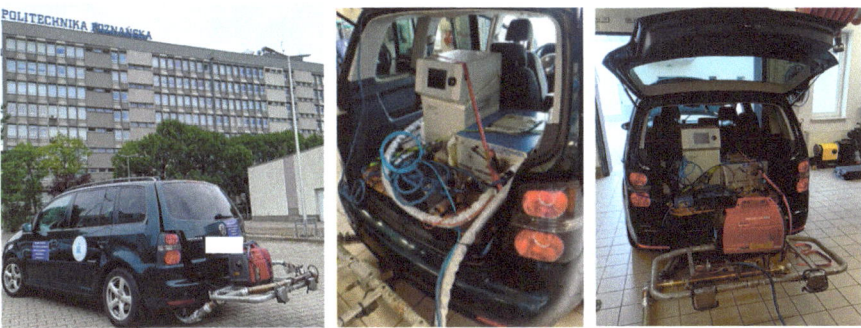

Figure 6. RDE test equipment installation in vehicle V7 (own study).

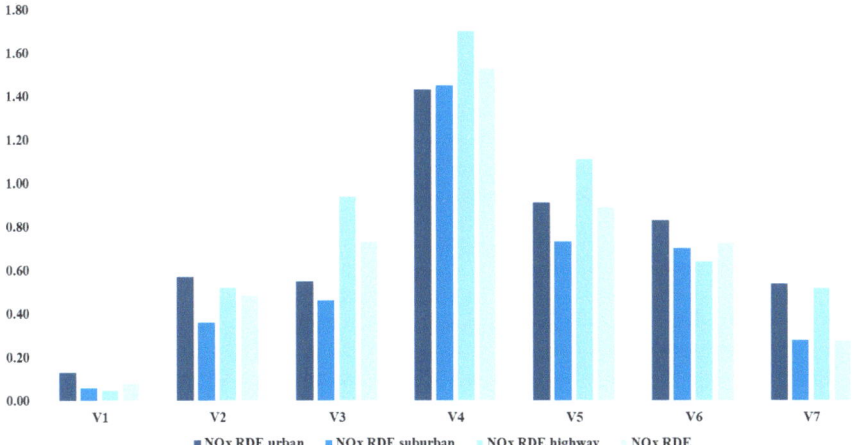

Figure 7. NO$_X$ emissions for vehicles V1–V7 in the RDE tests [g].

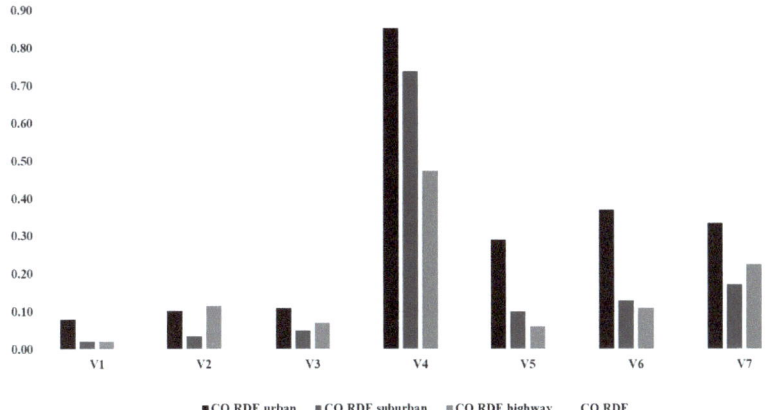

Figure 8. CO emissions for vehicles V1–V7 in the RDE tests [g].

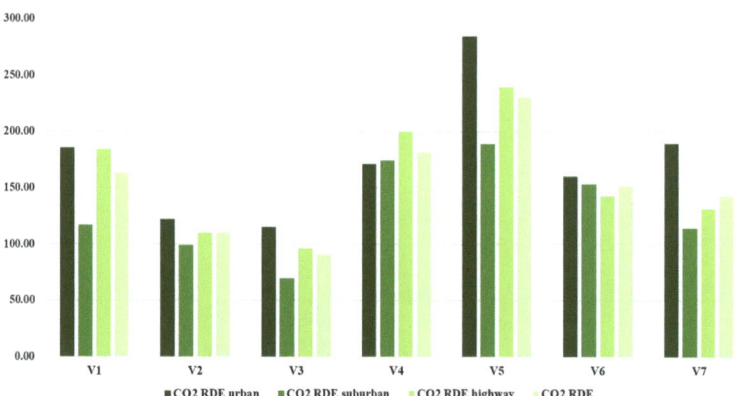

Figure 9. CO$_2$ emissions for vehicles V1–V7 in the RDE tests [g].

NO$_X$ emissions increased significantly for vehicles with mileage of 86,001–160,000 km. The average NOx emissions in the RDE test for vehicles with mileage between 86,001 km and 160,000 km (V4, V5, V6) are 70% higher, rising from 1.29 g/km to 3.14 g/km. The average NOx emissions in the RDE test for vehicles with mileage between 86,001 km and 317,000 km (V4, V5, V6 and V7) are 61% higher than in the case of vehicles with mileage of up to 86,000 km.

In the case of vehicle V7, equipped with a 1.9 TDI engine, its NO$_X$ emissions, despite its high mileage and operational age, were comparable to vehicles with mileage of up to 86,000 km. This could be attributed to major repairs to the engine's turbocharging system, including turbine regeneration and the repair of injector seals, performed at 302,000 km. The NO$_X$ emission results indicated that the exhaust after-treatment system was functioning correctly.

The emission profile under urban, suburban, and highway conditions is similar for all vehicles. However, a value analysis shows that CO emissions are lowest for vehicles with mileage of up to 86,000 km, i.e., vehicles V1, V2, and V3. The average CO emission for vehicles with mileage of up to 86,000 km is more than 23% lower compared to the average emission generated by vehicles with mileage of 86,001–160,000 km and is more than 19% lower compared to the average emission generated by vehicles with mileage of 86,001–317,000 km. The highest CO emissions were recorded for vehicles with mileage in the range of 86,001 km to 317,000 km.

In vehicle V4, equipped with a 1.9 CDTI engine compliant with the Euro 4 standard, a significant increase in CO and NO$_X$ emissions was observed during the RDE test. This vehicle was the only one to exceed the CO emission standard. This phenomenon occurred in the urban, suburban, and highway sections. The results suggest that the combustion process and the emission control system in vehicle V4 were not functioning properly.

Several potential causes could lead to such symptoms. High NO$_X$ emissions point to excessively high temperatures in the combustion chamber or a malfunctioning Exhaust Gas Recirculation (EGR) system. High CO emissions could suggest incomplete fuel combustion, which might derive from an improper air–fuel mixture, a faulty catalytic converter, or clogged injectors.

The EGR system issues are particularly notable, as this system plays a key role in reducing NO$_X$ emissions by recirculating part of the exhaust gases back into the cylinders. The EGR valve could have become stuck in a closed position or restricted, preventing proper recirculation. The lack of recirculated exhaust gases might have caused cylinder temperatures, leading to incomplete combustion and, consequently, higher CO emissions.

A malfunctioning catalytic converter explains the high CO levels, which are also associated with increased exhaust gas temperatures that further raise NO_X emissions. Another issue could involve the fuel injection system, such as operational failures in the injectors, which could disrupt the mixture's uniformity in the combustion chamber. This could result in incomplete combustion, increasing CO, and uneven combustion, raising temperatures and, consequently, NO_X emissions.

Other potential causes for increased emissions could include the contamination of mass air flow (MAF) sensors, which might provide incorrect data on the air volume and mixture composition, leading to suboptimal combustion and a malfunctioning NO_X sensor reporting incorrect data to the engine control unit. Another possible scenario could be turbocharger damage, reduced boost or intake leaks, leading to insufficient air supply, a rich mixture, incomplete combustion (increasing CO), and uneven combustion (increasing NO_X).

Finally, the soot regeneration process in the Diesel Particulate Filter (DPF) could cause backpressure in the exhaust system, disrupting the combustion process, during which CO and NO_X emissions could temporarily increase.

A value analysis reveals that CO_2 emissions are lowest for vehicles with mileage of up to 86,000 km, specifically for vehicles V1, V2, and V2. The average CO_2 emissions for vehicles with mileage of up to 86,000 km are over 180% lower compared to the average emissions generated by vehicles with mileage in the range of 86,001–160,000 km, and are over 106% by vehicles with mileage of 86,001–317,000 km. The highest CO_2 emissions were recorded for vehicles V4 and V5, with mileages of 120,000 km and 140,000 km, respectively.

Notably, the CO_2 emissions of vehicle V7 deserve attention, as its operational mileage might suggest a significant deviation from the observed results. The recorded CO_2 values for V7 can be attributed to diligent engine maintenance, including regular replacement of filters, oil, and fuel, as well as consistent monitoring of the vehicle's technical condition. Despite having the highest mileage of over 317,000 km, vehicle V7 met the EURO 4 emission standard for NO_X and CO, along with vehicle V1.

Regular technical inspections and major repairs have ensured that V1, despite intensive use during the first three years of operation, does not exhibit elevated levels of air pollutant emissions. The characteristics of V7's recorded annual mileage are presented in Figure 10.

Vehicle V7 underwent periodic oil changes during its operation. The oil was changed every 30,000 km up to a mileage of 200,000 km and every 15,000 km from 200,000 km to 317,000 km. The scope of inspection and repair activities for vehicle V7 are presented in Table 4.

Table 4. Types of maintenance activities performed on vehicle V7 (own data).

Maintenance Type	Overview of Vehicle V7 Maintenance
TYPE 0	Vehicle registration and general inspection
TYPE 1	Oil change, replacement of filters: oil, air, fuel, cabin
TYPE 2	Timing system replacement
TYPE 3	Coolant, brake fluid, and gearbox oil replacement
TYPE 4	Air conditioning service
TYPE 5	Shock absorber replacement
TYPE 6	Brake repair–brake pads
TYPE 7	Brake repair–brake disks
TYPE 8	Clutch and dual-mass flywheel replacement, repair of injector seals
TYPE 9	Battery replacement
TYPE 10	Engine turbocharger regeneration
TYPE 11	Glow plug sensor replacement
TYPE 12	Mass airflow sensor replacement

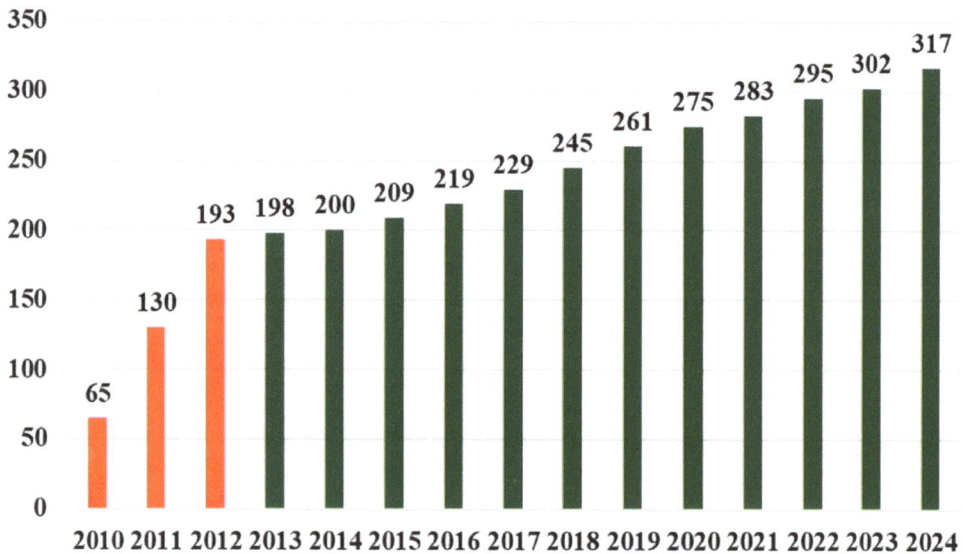

Figure 10. Annual mileage of vehicle V7 [thousand km].

The frequency of technical inspections related to the replacement of operating fluids increased upon reaching a mileage of 200,000 km. The types of repairs and inspections as a function of the operational mileage of vehicle V7 are presented in Table 5. These inspections and repairs significantly contributed to the low level of air pollutant emissions, despite vehicle V7 having the highest mileage compared to the other vehicles tested.

Table 5. Maintenance history of vehicle V7 (own data).

Date	Mileage	Maintenance Scope	Date	Mileage	Maintenance Scope
05.2010	0	TYPE 0	06.2018	244,000	TYPE 4
08.2010	33,469	TYPE 1	08.2018	245,000	TYPE 1
12.2010	65,138	TYPE 1	01.2019	250,500	TYPE 2
01.2011	97,595	TYPE 1	10.2019	260,502	TYPE 1
09.2011	129,689	TYPE 1	12.2020	275,000	TYPE 1 and TYPE 6
04.2012	163,079	TYPE 1 and TYPE 2	08.2021	282,620	TYPE 6 and TYPE 7
10.2012	193,080	TYPE 1	02.2022	288,000	TYPE 1
09.2014	200,000	TYPE 3 and TYPE 4	08.2022	291,500	TYPE 8
06.2015	209,000	TYPE 5	09.2022	295,000	TYPE 9
06.2016	219,000	TYPE 4	08.2023	302,000	TYPE 1 and TYPE 10
06.2017	229,000	TYPE 1	04.2024	309,000	TYPE 4 and TYPE 11
02.2018	240,000	TYPE 6	08.2024	317,000	TYPE 12

5. Conclusions

The primary objective of planned actions in the coming years will be to identify solutions that contribute to reducing air pollution emissions. In Poland, efforts to eliminate vehicles that do not meet exhaust emission standards are likely to continue. Changes in passenger transportation and freight movement methods are anticipated, including the integration of various transport modes for single transits, the promotion of eco-friendly solutions for businesses, increased funding for investments, fleet modernization, and infrastructure upgrades. Additionally, initiatives to enhance resilience to future economic shocks will be promoted along with the establishment of clean transport zones in urban areas. The research results presented in the paper may serve as recommendations for modifying the rules of access to clean transport zones.

The literature lacks research results verifying the impact of vehicle aging, expressed as operational mileage, on air pollutant emissions. There is also a shortage of data to develop a method for assessing the actual changes in air pollution emissions during the progressive operation of transport modes. This issue is particularly important in terms of the future implementation of clean transport zones, where compliance with air pollution emission standards set at the time of a vehicle's manufacture will determine access to selected urban areas.

As part of the conducted research using specialized equipment, key air pollutant emissions—namely carbon monoxide, carbon dioxide, and nitrogen oxides—were measured and correlated with vehicle operational age. The research was conducted on vehicles with engines of comparable displacement volumes. The results point to significant differences in the emissions of selected chemical compounds depending on the vehicle's mileage and operational age, specifically:

- NO_X emissions increased significantly for vehicles with a mileage of 86,001–160,000 km. The average NO_X emissions in the RDE test for vehicles with mileage between 86,001 km and 160,000 km (V4, V5, V6) is 70% higher, rising from 1.29 g/km to 3.14 g/km. The average NO_X emissions in the RDE test for vehicles with mileage between 86,001 km and 317,000 km (V4, V5, V6, and V7) is 61% higher than for vehicles with mileage of up to 86,000 km. In the case of vehicle V7 with a 1.9 TDI engine, despite its high mileage and operational age, NO_X emissions were comparable to those of vehicles with mileage of up to 86,000 km.
- Average CO emissions for vehicles with mileage of up to 86,000 km were more than 23% lower compared to the average emissions generated by vehicles with mileage of 86,001–160,000 km and were more than 19% lower compared to the average emissions generated by vehicles with mileage of 86,001–317,000 km. The highest CO emissions were recorded for vehicles with mileage in the range of 86,001 km to 317,000 km.
- Average CO_2 emissions for vehicles with mileage of up to 86,000 km were more than 182% lower compared to the average emissions generated by vehicles with mileage of 86,001–160,000 km and were more than 106% lower compared to the average emissions generated by vehicles with mileage of 86,001–317,000 km.

Based on the research conducted, it can be indicated that the air pollutant emissions tested at the homologation stage are subject to change over the vehicle's service lifetime. The changes in air pollutant emissions are different for individual chemical compounds, as indicated above.

The vehicle's operational age, mileage and technical condition are factors affecting air pollutant emissions. Long-term research must be conducted on identical vehicles, including repetitive transportation tasks, using comparable techniques and driving styles, with regular maintenance, to accurately study the impact of vehicle aging on emissions. Such an approach would provide more precise insights into the variability of air pollutant emissions as the vehicle's operational age increases. Such studies are planned for the future.

In this study, the analyzed vehicles had similar but not identical diesel engine displacements. Additionally, the vehicles' technical conditions varied, as evidenced by the air pollutant emission results of vehicles V4 and V7. Driving style also affects emissions, and it is clear that the driving styles for the tested vehicles were not comparable.

Considering the observed results and changes in pollutant emissions for vehicles with mileages of up to 86,000 km, between 86,001 km and 160,000 km, and above 160,000 km, it is recommended to conduct real-world emission studies using RDE tests. These tests should focus on vehicles with uniform engine configurations, varying mileages, and comparable driving styles and habits. This would enable the development of correctional indicator methods to adjust vehicle emissions, accounting for the operational age factor as represented by mileage.

The conducted studies constitute a novelty because similar results have not been published before. The results were obtained using the most accurate evaluation of the air pollutant emission measurement method in road conditions (RDE). The limitation of this method is the need to perform time-consuming and cost-intensive tests on a large population of vehicles. To achieve more accurate results, it is necessary to conduct tests on a larger sample of vehicles, considering the testing of emissions of identical vehicle types over a long period of operation.

Author Contributions: Conceptualization, P.P. (Piotr Pryciński), J.K. and P.P. (Piotr Pielecha); Methodology, P.P. (Piotr Pryciński) and J.P.; Validation, P.P. (Piotr Pryciński); Formal analysis, P.P. (Piotr Pryciński) and R.J.; Investigation, P.P. (Piotr Pryciński) and J.K.; Resources, P.P. (Piotr Pryciński), J.P. and P.P. (Piotr Pielecha); Data curation, P.P. (Piotr Pryciński), J.P. and P.P. (Piotr Pielecha); Writing—original draft, P.P. (Piotr Pryciński) and R.J.; Writing—review & editing, J.P., J.K. and P.P. (Piotr Pielecha); Visualization, P.P. (Piotr Pryciński) and P.P. (Piotr Pielecha); Supervision, J.P., J.K. and R.J. All authors have read and agreed to the published version of the manuscript.

Funding: The article was developed as part of a research project supporting scientific activities in the discipline of Civil Engineering and Transport, titled "Study of the Aging Phenomenon of Transport Means in the Context of Vehicle Impact on the Environment".

Data Availability Statement: The original contributions presented in this study are included in the article. Further inquiries can be directed to the corresponding author.

Conflicts of Interest: The authors declare no conflict of interest.

References

1. Act of January 11, 2018 on Electromobility and Alternative Fuels. Available online: https://isap.sejm.gov.pl/isap.nsf/download.xsp/WDU20230000875/U/D20230875Lj.pdf (accessed on 3 December 2024).
2. Polish Automotive Industry Association. Automotive Industry Report, 2023/2024. Available online: https://www.pzpm.org.pl/pl/Publikacje/Raporty/Roczny-Raport-Branzy-Motoryzacyjnej-PZPM-2023-2024 (accessed on 3 December 2024).
3. Polish Automotive Industry Association. Automotive Industry Report, 2024/2025. Available online: https://www.pzpm.org.pl/en/Automotive-market/Reports/PZPM-Automotive-Industry-Report-2024-2025 (accessed on 3 December 2024).
4. Commission Regulation (EU) 2016/427 of 10 March 2016 amending Regulation (EC) No 692/2008 as regards emissions from light passenger and commercial vehicles (Euro 6) (Text with EEA relevance). *Off. J. Eur. Union* **2016**, *82*, 1–98.
5. Polish Automotive Industry Association. Automotive Industry, Edition Q1–Q3 2024, Evolution of Alternative Fuels on the Path to Sustainability. Available online: https://www.pzpm.org.pl/en/Automotive-market/Reports/Automotive-Industry-Quarterly-report-prepared-by-PZPM-and-KPMG-edition-Q1-Q3-2024 (accessed on 3 December 2024).
6. Available online: https://commission.europa.eu/strategy-and-policy/priorities-2019-2024/european-green-deal_pl (accessed on 3 December 2024).
7. Available online: https://www.consilium.europa.eu/en/policies/fit-for-55/ (accessed on 3 December 2024).
8. Yu, T.; Li, K.; Wu, Q.; Yao, P.; Ke, J.; Wang, B.; Wang, Y. Diesel Engine Emission Aftertreatment Device Aging Mechanism and Durability Assessment Methods: A Review. *Atmosphere* **2023**, *14*, 314. [CrossRef]
9. Kęska, A. Analiza i ocena składu toksycznych grup związków emitowanych w spalinach pojazdów zgodnych z normą Euro 6 [Analysis and evaluation of the composition of toxic groups of compounds emitted in the exhaust of Euro 6 vehicles]. *Przemysł Chem.* **2022**, *101*, 606–610.
10. Bodek, K.M.; Wong, V.V. The Effects of Sulfated Ash, Phosphorus and Sulfur on Diesel Aftertreatment Systems—A Review. In Proceedings of the JSAE/SAE International Fuels & Lubricants Meeting, Kyoto, Japan, 23 July 2007.
11. Kunwar, D.; Carrillo, C.; Xiong, H.; Peterson, E.; DeLaRiva, A.; Ghosh, A.; Datye, A.K. Investigating anomalous growth of platinum particles during accelerated aging of diesel oxidation catalysts. *Appl. Catal. B Environ.* **2020**, *266*, 118598. [CrossRef]

12. Schmieg, S.J.; Oh, S.H.; Kim, C.H.; Brown, D.B.; Lee, J.H.; Peden, C.H.; Kim, D.H. Thermal durability of Cu-CHA NH3-SCR catalysts for diesel NOx reduction. *Catal. Today* **2012**, *184*, 252–261. [CrossRef]
13. Anilovich, I.; Hakkert, A.S. Survey of vehicle emissions in Israel related to vehicle age and periodic inspection. *Sci. Total Environ.* **1996**, *189*, 197–203. [CrossRef]
14. Liu, H.; Qi, L.; Liang, C.; Deng, F.; Man, H.; He, K. How aging process changes characteristics of vehicle emissions? A review. *Crit. Rev. Environ. Sci. Technol.* **2020**, *50*, 1796–1828. [CrossRef]
15. Zachariadis, T.; Ntziachristos, L.; Samaras, Z. The effect of age and technological change on motor vehicle emissions. *Transp. Res. Part D Transp. Environ.* **2001**, *6*, 221–227. [CrossRef]
16. Caserini, S.; Pastorello, C.; Gaifami, P.; Ntziachristos, L. Impact of the dropping activity with vehicle age on air pollutant emissions. *Atmos. Pollut. Res.* **2013**, *4*, 282–289. [CrossRef]
17. Miller, T.L.; Davis, W.T.; Reed, G.D.; Doraiswamy, P.; Tang, A. Effect of county-level income on vehicle age distribution and emissions. *Transp. Res. Rec.* **2002**, *1815*, 47–53. [CrossRef]
18. Pryciński, P.; Wawryszczuk, R.; Korzeb, J.; Pielecha, P. Indicator Method for Determining the Emissivity of Road Transport Means from the Point of Supplied Energy. *Energies* **2023**, *16*, 4541. [CrossRef]
19. Pryciński, P.; Wawryszczuk, R.; Korzeb, J.; Pielecha, P.; Murawski, J. Selected vehicle emission assessment issues in passenger transport services. *Combust. Engines* **2023**, *195*, 14–22.
20. Dziedziak, P.; Szczepański, T.; Niewczas, A.; Ślęzak, M. Car reliability analysis based on periodic technical tests. *Open Eng.* **2021**, *11*, 630–638. [CrossRef]
21. Milojević, S.; Glišović, J.; Savić, S.; Bošković, G.; Bukvić, M.; Stojanović, B. Particulate Matter Emission and Air Pollution Reduction by Applying Variable Systems in Tribologically Optimized Diesel Engines for Vehicles in Road Traffic. *Atmosphere* **2024**, *15*, 184. [CrossRef]
22. Jędra, I. Wpływ wieku samochodów na bezpieczeństwo w transporcie drogowym. [Impact of vehicle age on road transport safety]. *Autobusy Tech. Eksploat. Syst. Transp. [Buses Technol. Oper. Transp. Syst.]* **2017**, *18*, 178–181.
23. Tylicki, H.F. Prognozowanie stanu technicznego środków transportu. [Forecasting the technical condition of transport means]. *Autobusy Tech. Eksploat. Syst. Transp. [Buses Technol. Oper. Transp. Syst.]* **2016**, *17*, 1189–1192.
24. Oprzędkiewicz, J.; Młynarski, S. *Problemy Predykcji Jakości i Niezawodności Pojazdów Samochodowych [Quality and Reliability Prediction Problems for Motor Vehicles]*; Cracow University of Technology: Kraków, Polansd, 2000.
25. Niewczas, A.; Rymarz, J.; Dębicka, E. Stages of operating vehicles with respect to operational efficiency using city buses as an example. *Eksploat. I Niezawodn.—Maint. Reliab.* **2019**, *21*, 21–27. [CrossRef]
26. Borucka, A.; Niewczas, A.; Hasilova, K. Forecasting the readiness of special vehicles using the semi-Markov model. *Eksploat. I Niezawodn.—Maint. Reliab.* **2019**, *21*, 662–669. [CrossRef]
27. Andrzejczak, K.; Młyńczak, M.; Selech, J. Poisson-distributed failures in the predicting of the cost of corrective maintenance. *Eksploat. I Niezawodn.—Maint. Reliab.* **2018**, *20*, 602–609. [CrossRef]
28. Pryciński, P.; Pielecha, P.; Korzeb, J.; Pielecha, J.; Kostrzewski, M.; Eliwa, A. Air Pollutant Emissions of Passenger Cars in Poland in Terms of Their Environmental Impact and Type of Energy Consumption. *Energies* **2024**, *17*, 5357. [CrossRef]
29. Siedlecki, M.; Szymlet, N.; Fuć, P.; Kurc, B. Analysis of the Possibilities of Reduction of Exhaust Emissions from a Farm Tractor by Retrofitting Exhaust Aftertreatment. *Energies* **2022**, *15*, 7963. [CrossRef]
30. Lijewski, P.; Merkisz, J.; Fuć, P.; Ziółkowski, A.; Rymaniak, Ł.; Kusiak, W. Fuel consumption and exhaust emissions in the process of mechanized timber extraction and transport. *Eur. J. For. Res.* **2017**, *136*, 153–160. [CrossRef]
31. Pielecha, J.; Skobiej, K.; Kurtyka, K. Exhaust emissions and energy consumption analysis of conventional, hybrid, and electric vehicles in real driving cycles. *Energies* **2020**, *13*, 6423. [CrossRef]
32. Nowak, M.; Rymaniak, Ł.; Fuć, P.; Andrzejewski, M.; Daszkiewicz, P. Testing the emissions of gaseous components and particulate matter by a light-duty commercial vehicle in real operating conditions. *Buses Technol. Oper. Transp. Syst.* **2017**, *18*, 327–331.
33. Agarwal, A.K.; Mustafi, N.N. Real-world automotive emissions: Monitoring methodologies, and control measures. *Renew. Sustain. Energy Rev.* **2021**, *137*, 110624. [CrossRef]
34. Suarez-Bertoa, R.; Valverde, V.; Clairotte, M.; Pavlovic, J.; Giechaskiel, B.; Franco, V.; Astorga, C. On-road emissions of passenger cars beyond the boundary conditions of the real-driving emissions test. *Environ. Res.* **2019**, *176*, 108572. [CrossRef] [PubMed]
35. Rosero, F.; Fonseca, N.; López, J.M.; Casanova, J. Effects of passenger load, road grade, and congestion level on real-world fuel consumption and emissions from compressed natural gas and diesel urban buses. *Appl. Energy* **2021**, *282*, 116195. [CrossRef]
36. Skobiej, K.; Pielecha, J. Analysis of the Exhaust Emissions of Hybrid Vehicles for the Current and Future RDE Driving Cycle. *Energies* **2022**, *15*, 8691. [CrossRef]
37. Zhai, Z.; Xu, J.; Zhang, M.; Wang, A.; Hatzopoulou, M. Quantifying start emissions and impact of reducing cold and warm starts for gasoline and hybrid vehicles. *Atmos. Pollut. Res.* **2023**, *14*, 101646. [CrossRef]
38. Jacyna, M.; Żochowska, R.; Sobota, A.; Wasiak, M. Scenario analyses of exhaust emissions reduction through the introduction of electric vehicles into the city. *Energies* **2021**, *14*, 2030. [CrossRef]

39. Jacyna, M.; Wasiak, M.; Lewczuk, K.; Kłodawski, M. Simulation model of transport system of Poland as a tool for developing sustainable transport. *Arch. Transp.* **2014**, *31*, 23–35. [CrossRef]
40. Gołda, I.J.; Gołębiowski, P.; Izdebski, M.; Kłodawski, M.; Jachimowski, R.; Szczepański, E. The evaluation of the sustainable transport system development with the scenario analyses procedure. *J. Vibroeng.* **2017**, *19*, 5627–5638. [CrossRef]

Disclaimer/Publisher's Note: The statements, opinions and data contained in all publications are solely those of the individual author(s) and contributor(s) and not of MDPI and/or the editor(s). MDPI and/or the editor(s) disclaim responsibility for any injury to people or property resulting from any ideas, methods, instructions or products referred to in the content.

Article

Analysis of Spatial Differences and Influencing Factors of Carbon-Emission Reduction Efficiency of New-Energy Vehicles in China

Lingyao Wang [1,†], **Taofeng Wu [2,*,†]** and **Fangrong Ren [2,†]**

1. School of Internet of Things, Nanjing University of Posts and Telecommunications, Nanjing 210003, China; b22100201@njupt.edu.cn
2. College of Economics and Management, Nanjing Forestry University, Nanjing 210037, China; 180213120008@hhu.edu.cn
* Correspondence: wuu@njfu.edu.cn
† These authors have contributed equally to this work and share first authorship.

Abstract: As new-energy vehicles (NEVs) gradually gain public attention, their carbon-reduction issues have become a focal point in academia. This study evaluates the carbon-reduction efficiency of NEVs in 21 Chinese provinces using an improved three-stage DEA model, analyzes spatial disparities with the Dagum Gini coefficient, and decomposes carbon-emission factors using the LMDI method. Results show that the overall carbon-reduction efficiency is low, with an average value of only 0.266. Significant differences exist in production- and consumption-stage efficiencies across regions. Shanxi Province performed the best, with efficiency scores of 1 in both stages, while the carbon-reduction stage showed the lowest efficiency, ranging between 0.2 and 0.3 in most regions. The central region exhibited the highest carbon-reduction efficiency, followed by the western and eastern regions, primarily influenced by intra-regional disparities. Energy intensity significantly suppresses carbon emissions, followed by energy structure, while economic development and population size positively contribute to carbon emissions. This study provides theoretical support for regional governments to formulate policies related to the NEV industry and offers practical guidance for its further development.

Keywords: NEVs; carbon-reduction efficiency; spatial differences; DEA; LMDI

Academic Editors: Maksymilian Mądziel, Kazimierz Lejda and Artur Jaworski

Received: 11 January 2025
Revised: 29 January 2025
Accepted: 29 January 2025
Published: 30 January 2025

Citation: Wang, L.; Wu, T.; Ren, F. Analysis of Spatial Differences and Influencing Factors of Carbon-Emission Reduction Efficiency of New-Energy Vehicles in China. *Energies* **2025**, *18*, 635. https://doi.org/10.3390/en18030635

Copyright: © 2025 by the authors. Licensee MDPI, Basel, Switzerland. This article is an open access article distributed under the terms and conditions of the Creative Commons Attribution (CC BY) license (https://creativecommons.org/licenses/by/4.0/).

1. Introduction

The development of new energy sources is an important part of the response to global climate change, including the promotion of new-energy vehicles [1,2]. China has been actively taking measures to achieve carbon-emission reduction. China started developing NEVs in 2009, mainly in the form of pilot policies in different cities. Since then, China's new-energy vehicle industry has been booming. According to relevant data from the World Resources Institute, the transportation sector accounts for about 10 to 15% of carbon emissions, with small cars being the largest carbon-emission tool in transportation, accounting for about 45%. According to China Energy News, while global transportation carbon emissions rose by 254 million tons from 2021 to 2022, China's transportation emissions dropped by 3.1% during the same period. There is no doubt that NEVs have made a huge contribution to this, especially pure electric vehicles. Due to their high energy efficiency, pure electric vehicles are able to reduce the amount of oil consumed per kilometer by about 98% [3].

Currently, there is still academic controversy over whether NEVs are conducive to low-carbon development, and the main reason for the controversy is the different lifecycles in focus. From a full lifecycle perspective, NEVs have higher carbon emissions in the vehicle cycle. However, from the perspective of the fuel cycle, NEVs still have an obvious emission reduction effect compared with ordinary fuel vehicles. Therefore, it is one of the focuses of this study to assess the carbon-reduction efficiency and analyze the influencing factors of NEVs by starting from their whole lifecycle and including their production and use stages in the research framework at the same time.

In terms of the coordination of regional development, there is a huge difference in regional development between NEV markets, with a serious imbalance in the distribution of production and sales. Most NEV production capacity is concentrated in the Yangtze River Delta (YRD), Chengdu, and Chongqing, while the penetration level is higher in the YRD and the Pearl River Delta (PRD). Therefore, a deeper reading of the development of NEVs in each region and the carbon-reduction efficiency during this period is also one of the focuses of this paper.

This study focuses on the NEV industry in 21 provinces of China and evaluates its carbon-reduction efficiency using a three-stage DEA model that accounts for undesirable outputs. Additionally, regional differences are analyzed using the Dagum Gini coefficient, and the LMDI method is employed to decompose the carbon emissions of each province into four aspects. The findings provide theoretical guidance for the further development of the NEV industry in various provinces and serve as a reference for the achievement of China's "dual carbon" goals. The contributions of this study are as follows:

First, in assessing the carbon-reduction efficiency of NEVs, this paper comprehensively considers the stages of production, use, and low-carbon development and applies a three-stage DEA model that considers non-desired outputs to construct a comprehensive evaluation index system. In the selection of input indicators, the carbon emissions of NEVs are comprehensively considered in relation to their actual development. Secondly, the variability of regional power structure will lead to the different carbon-reduction efficiency of NEVs in each region, so this paper starts from the provincial level to compare and analyze the differences in carbon-reduction efficiency in each region so as to provide a theoretical basis for the local government's differentiated management of carbon reduction in the whole lifecycle of the new-energy vehicle industry. Third, in order to dig deeper into the carbon-emission factors driving the new-energy automobile industry, this paper applies the LMDI to quantitatively analyze from multiple dimensions and quantify the contribution value of relevant influencing factors.

2. Literature Review

2.1. New-Energy Vehicles and Carbon Emissions

Conventional fuel vehicles bring great convenience to society and, at the same time, cause serious environmental pollution. NEVs are considered by most scholars as one of the important ways to alleviate environmental pressure [4,5]. However, "clean and environmentally friendly" is only a part of the fact of NEVs and a relative concept. As an indispensable part of environmental performance assessment, existing studies mainly focus on how to accurately calculate carbon-emission efficiency and deeply explore the various influencing factors behind it. Evaluating the carbon emissions of NEVs needs to be approached from the perspective of their full lifecycle [6]. Life cycle assessment has been widely used by scholars in environmental impact related analysis by evaluating the carbon emissions of NEVs at various stages.

Currently, there are still controversies about the environmental benefits of NEVs. On the one hand, some scholars believe that NEVs are conducive to reducing environmental pollution [7–10]. Elshurafa and Peerbocus (2020) deeply analyzed carbon-emission reduction from NEVs in Saudi Arabia by constructing a model of the power system in the region [10]. The results show that if 1% of NEVs are deployed, it is expected to reduce carbon emissions by at least 0.5%, and under optimal conditions, this reduction may even be enhanced to 0.9%. The carbon-reduction effect of NEVs has been verified in the same way in most countries, such as Ghana [11], Brazil [12], South Korea [13], the United States [14] and China [15], among other regions. Therefore, NEVs have great potential to become cleaner vehicles with low operating costs as long as the related technical issues are effectively addressed [16].

In addition, supply-chain optimization of new-energy vehicles [17,18] and green path transport [19–21] are also the focus of carbon emissions. Some scholars have constructed multi-objective optimization models around new-energy vehicles and carbon emissions. These studies, ranging from supply-chain management to vehicle path optimization, have proposed an inventory transport model considering dynamic carbon-emission factors and product characteristics, a multi-objective supplier selection model, and a green path model based on vehicle load changes, respectively. This not only verifies the effectiveness of high-standard vehicle use and optimization algorithms in reducing carbon emissions and costs but also provides theoretical support for the low-carbon development of new-energy vehicles.

On the other hand, some researchers have questioned the environmental benefits of NEVs [22,23], mainly focusing on the power battery production process and the current electricity mix. A 2020 report by Greenpeace and CEPF projects that global passenger EV batteries will require over 2.05 million tons of cobalt from 2021 to 2030. This value accounts for almost 30% of the total extractable amount of known cobalt ore resources in the world.

Some scholars find that NEVs increase particulate matter (PM2.5), which is not conducive to the improvement of environmental quality, which may be due to the fact that the popularity of NEVs in China is not high and the environmental benefits are not obvious enough [23]. In addition, pollutants generated during the production and manufacturing process of automobiles can cause pollution elsewhere through exhaust pipes [22,24]. In terms of the electricity structure, the main source of electricity for EVs is the grid, so the cleanliness of electricity is largely determined by the carbon intensity of the technology used to generate it. For example, in Southwest China, where hydropower is the most prominent part of the power mix, NEVs have shown significant GHG reduction benefits. In the north, high coal-fired power use and climate factors result in NEV carbon intensity exceeding that of conventional fuel vehicles [25]. Currently, China's power structure is still dominated by coal, and the electricity required by NEVs in the power generation process causes serious environmental pollution [22], so assessing the carbon-emission efficiency of NEVs during the full period needs to be emphasized.

2.2. Analysis of Factors Affecting Carbon Emissions

As the issue of climate change caused by carbon emissions has attracted great global attention, carbon peaking, and carbon neutrality have become a topic of focus for countries around the world. At present, the studies of factors influencing carbon emissions are mainly divided into two categories, namely single-factor and multifactor impact studies.

Scholars vary in their perspectives from single-factor cuts. From the perspective of urban expansion, some scholars have used the Support Vector Machine (SVM) algorithm to reveal the growth trend of carbon emissions in Bangladesh and the significant impact of urban expansion on carbon-emission rates [26]. Some scholars used LMDI and PDA methods to analyze regional carbon-emission efficiency differences in China, finding that energy efficiency, scale, and technological progress are key factors, while regional, industry, and management effects are less significant [27]. In terms of green technological innovation, some researchers apply the Super-SBM and fixed-effects models, discovering that such innovation enhances manufacturing carbon efficiency in both quality and quantity, with a more pronounced impact in the central region [28]. In addition, other scholars have made in-depth analyses of carbon emissions from NEVs from the perspectives of capital allocation efficiency [29], policy subsidies [30], and carbon tax conditions [31].

Regarding the multifactor decomposition of carbon emissions, existing studies commonly use the SDA and the IDA. The calculation process of the IDA method is simpler than that of the SDA method, which does not need to analyze the input (output) data but only needs to integrate the temporal data of the various production organizations, and the required data are easier to collect. Among IDA, LMDI is the most popular among scholars. Zhang et al. (2019) [32] concluded that economic development and energy intensity are the key contributors influencing carbon emissions. On this basis, Li et al. (2024) [33] combined the LMDI and STIRPAT models to study regional differences in carbon output across Chinese provinces, revealing significant disparity, with economic factors as the main contributor. This confirms that economic growth significantly drives China's carbon emissions, a conclusion supported by most scholars [34–37].

In developing countries, rapid economic growth typically brings accelerated industrialization, urbanization, and enhanced transport systems. These stages of development require the support of large amounts of energy, but since their energy mix is often dominated by non-renewable energy sources, economic growth is an important source of carbon emissions for developing countries. This view is also shared by numerous scholars. The opposite is true for developed countries, where carbon emissions are not strongly associated with economic growth [38].

Summarizing the above analysis, although the promotion and application of NEVs is considered a feasible transition strategy, scholars are still controversial about whether it can reduce carbon emissions. Whether NEVs really contribute to reducing carbon emissions and whether there are regional differences still need to be further verified. Therefore, it is crucial to evaluate the carbon reduction of NEVs in different regions from a multidimensional perspective to explore the intrinsic causal relationship between NEV promotion and carbon reduction. On this basis, carbon-emission decomposition is further carried out to illustrate the carbon-reduction effect of NEVs from different channels.

3. Research Method

To comprehensively assess the carbon-reduction efficiency and its influencing factors during the operation of new-energy vehicles in the sample regions, this paper combines the DEA model with the LMDI method. Specifically, this paper assesses the carbon-reduction efficiency of 21 provinces (cities) during the new-energy operation from 2018 to 2022 based on a three-stage DEA model considering non-expected outputs. Meanwhile, in order to promote the improvement of carbon-reduction efficiency in each region, this paper applies the LMDI method on the basis of the assessment results and digs deeper into the specific impacts of energy structure, energy intensity, economic development, and population size on carbon emissions. The specific framework structure is shown in Figure 1.

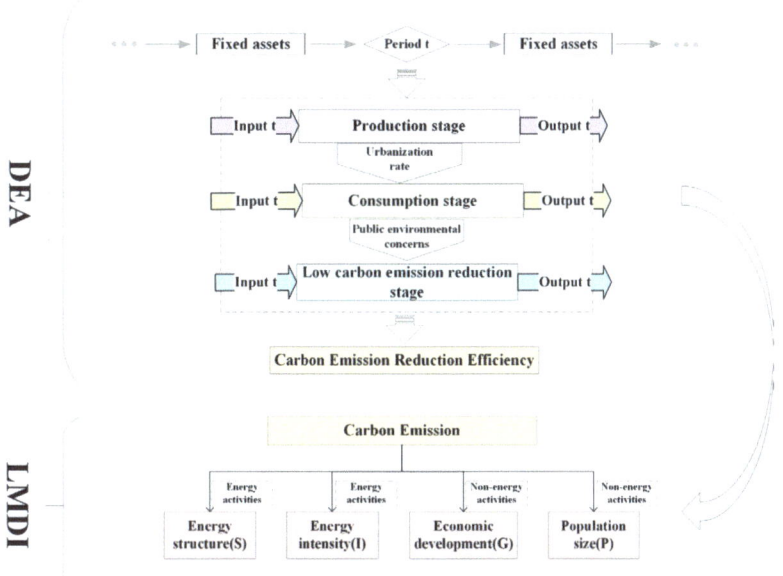

Figure 1. Structure of the research methodology.

In the assessment of carbon reduction efficiency during the operation of new-energy vehicles, they are divided into a production phase, a consumption phase, and a low-carbon reduction phase. The production phase is mainly used to analyze the efficiency of resource use during the manufacturing process of new-energy vehicles, especially the cleanliness and low-carbon nature of energy use. The level of infrastructure, such as power grid facilities and modernization of production equipment in industrial parks, will directly affect the carbon-emission reduction efficiency of new-energy vehicle production. The consumption phase is used to assess the market efficiency of the promotion and use of new-energy vehicles, focusing on user acceptance and the usage environment. The low-carbon emission reduction stage focuses on analyzing the contribution of new-energy vehicles to overall carbon emissions during operation, as well as the regional emission reduction capacity. Regions powered by clean energy have a more significant carbon-reduction effect, while optimization of the energy mix is the key to efficiency improvement.

The LMDI method further explains inter-regional differences in energy structure and infrastructure, etc., by decomposing the drivers of carbon emissions. The LMDI method quantitatively decomposes the contribution of changes in energy and non-energy activities to carbon emissions and reveals the extent to which energy use is optimized in the sample regions. Therefore, based on the LMDI method, this paper proposes the following hypotheses:

Hypothesis 1: *Optimization of energy structure and energy intensity can reduce regional carbon emissions.*

Hypothesis 2: *The increase in economic development level and population size can increase regional carbon emissions.*

In the following, this paper will explain in detail the specific principles of the DEA model and the LMDI method.

3.1. DEA Model

The traditional Data Envelopment Analysis (DEA) neglects the non-expected outputs, so this paper improves on this basis and adopts a three-stage DEA model that considers the non-expected outputs. The specific process framework is shown in Figure 1 and defined as follows.

Suppose there are m $DMU_j (j = 1, 2 \ldots, M)$, each DMU containing t cycles ($t = 1, 2 \ldots, T$), and s stages ($s = 1, 2 \ldots, S$). Each stage contains an input x_{ijt} and an output y_{pjt}^L. Each stage is connected to the other by z_{sjt}^L (Link), and each period is connected to each other by z_{rit}^c (Carry-over).

Where the input x_{ijt}^1 of $DMU_j (j = 1, 2 \ldots, m)$ in the production stage is the number of employees in the automobile industry, the fixed capital stock in the automobile industry, and the output y_{Pjt}^1 is the production quantity of NEVs. In the consumption stage, the input x_{ijt}^2 of $DMU_j (j = 1, 2 \ldots, m)$ is the per capita consumption expenditure of residents, the desired output y_{Pjt}^2 is the gross regional product and the non-desired output y_{Pjt}^{2u} is the total energy consumption of the automobile industry. In the low-carbon emission reduction stage, the input x_{ijt}^3 of $DMU_j (j = 1, 2 \ldots, m)$ is the number of effective patents and the undesired output y_{Pjt}^{3u} is the carbon emission during the operation of new energy. z_{sjt}^{L1} between the production stage and the consumption stage is the urbanization rate and z_{sjt}^{L2} between the consumption stage and the low-carbon emission reduction stage is a public environmental concern. z_{rjt}^c between period t and t + 1 is the fixed-asset investment of each province and city.

Referring to Lu et al. (2022) [39] and Zhang et al. (2021) [40], the total efficiency of DMU_j is:

$$\max \sum_{t=1}^{T} \varphi_t \left(r_t^1 \lambda_t^1 + r_t^2 \lambda_t^2 + r_t^3 \lambda_t^3 \right) \tag{1}$$

where $\sum_{t=1}^{T} \varphi_t = 1$, r_t^1, r_t^2, r_t^3 are the baseline weights for the production phase, consumption phase, and low carbon emission reduction phase of cycle t, respectively. $r_t^1 + r_t^2 + r_t^3 = 1$.

S.T.

Stage 1 (Production stage)	Stage 2 (Consumption stage)	Stage 3 (Low carbon emission reduction stage)
$\sum_{j=1}^{n} x_{ijt}^1 \lambda_{jt}^1 \leq x_{ijt}^1$ $\sum_{j=1}^{n} y_{pjt}^1 \lambda_{jt}^1 \geq y_{pjt}^1$ $\sum_{j=1}^{n} \lambda_{jt}^1 = 1$	$\sum_{j=1}^{n} x_{ijt}^2 \lambda_{jt}^2 \leq x_{ijt}^2$ $\sum_{j=1}^{n} y_{pjt}^2 \lambda_{jt}^2 \geq y_{pjt}^2$ $\sum_{j=1}^{n} y_{pjt}^{2u} \lambda_{jt}^2 \leq y_{pjt}^{2u}$ $\sum_{j=1}^{n} \lambda_{jt}^2 = 1$	$\sum_{j=1}^{n} x_{ijt}^3 \lambda_{jt}^3 \leq x_{ijt}^3$ $\sum_{j=1}^{n} y_{pjt}^{3u} \lambda_{jt}^3 \leq y_{pjt}^{3u}$ $\sum_{j=1}^{n} \lambda_{jt}^3 = 1$
Stage link		Period link
$\sum_{j=1}^{n} \lambda_{jt} z_{sjt}^{L1} = \sum_{j=1}^{n} \mu_{jt} z_{sjt}^{L1}$ $\sum_{j=1}^{n} \mu_{jt} z_{sjt}^{L2} = \sum_{j=1}^{n} \eta_{jt} z_{sjt}^{L2}$		$\sum_{j=1}^{n} \mu_{j,t-1} z_{rjt}^c = \sum_{j=1}^{n} \mu_{j,t} z_{rjt}^c$

For the measurement of the efficiency of each indicator in this paper, refer to Hu and Wang (2006) [41] as follows.

Input:

$$TFE = \frac{\text{Target Input}}{\text{Actual Input}} \tag{2}$$

Desired outputs:
$$\text{TFE} = \frac{\text{Actual Desirable Input}}{\text{Target Desirable Input}} \tag{3}$$

Undesired outputs:
$$\text{TFE} = \frac{\text{Target Undesirable Input}}{\text{Actual Undesirable Input}} \tag{4}$$

3.2. LMDI Approach

The factor decomposition method aims to consider the basic index factors of the research object in depth and split these factors in the form of a product to ensure that the results are not distorted due to multiple covariance among the factors in the analysis process. The complex multi-parameter coupled system problem is decomposed into several influencing factors by the factor decomposition method.

The Logarithmic Mean Divisia Index (LMDI) method, as a technique based on exponential decomposition analysis, has now become the main empirical research method in the field of studying carbon emissions due to its characteristic of not including residual variables [27,42].

Referring to Yang et al. (2020) [42], this paper utilizes the LMDI model combined with Kaya's constant equation to decompose the carbon-emission influences of China's provinces during the operation of the new energy source in terms of four aspects: energy structure, energy intensity, economic development, and population size.

$$C = \sum_i C_i = \sum_i \frac{C_i}{E_i} \times \frac{E_i}{E} \times \frac{E}{Y} \times \frac{Y}{P} \times P = \sum_i F_i \times S_i \times I \times G \times P \tag{5}$$

Under the LMDI method, applying the multiplicative decomposition and summation decomposition, the change in carbon emissions ΔC_t from 2011 (base year) to year t can be expressed as the following equation.

$$\Delta C_t = \Delta C_F + \Delta C_S + \Delta C_I + \Delta C_G + \Delta C_P \tag{6}$$

The contribution of each influencing factor to carbon emissions can be expressed as:

$$\eta_1 = \frac{\Delta C_S}{\Delta C_t},\ \eta_2 = \frac{\Delta C_I}{\Delta C_t},\ \eta_3 = \frac{\Delta C_G}{\Delta C_t},\ \eta_4 = \frac{\Delta C_P}{\Delta C_t} \tag{7}$$

If $\eta_i > 0$, then it means that the factor i has a positive driving effect on carbon emissions, and vice versa, a negative inhibiting effect.

4. Empirical Analysis

4.1. Data Description

In this study, 21 provinces in China from 2011 to 2022 were used as the study sample, and the data used were from the China Energy Statistical Yearbook, China Statistical Yearbook, and Statistical Bulletin of National Economic and Social Development of Chinese Cities. The selection of only 21 sample provinces (municipalities) in this study is due to data availability. Regions with significant data gaps were excluded after manual screening. Including regions with extremely low carbon reduction efficiency or incomplete data might dilute the overall research results and obscure the focus of regional efficiency evaluation.

Regarding the public environmental concern indicators, reference is made to Zheng et al. (2012) [43], which utilized the Baidu index of residents' searches for specific environmental keywords ("environmental pollution") to construct public environmental concern indicators for each province and city. Due to the missing data in some cities, this paper adopts the interpolation method to deal with the missing data. Table 1 shows the three-stage DEA model and the related indicators used in this paper, taking into account the non-desired outputs.

Table 1. Input and output variables meaning.

Stage	Character	Variant	Unit
Production stage	Input	Number of employees (automotive industry)	Persons
		Fixed capital stock (automotive industry)	Million RMB
	Output	Number of NEVs produced	Number
	Link	Urbanization rate	%
Consumption stage	Input	Consumption expenditure per inhabitant	RMB
	Output	Total energy consumption (automotive industry)	Million tons of standard coal
		Gross regional product (GDP)	100 million RMB
	Link	Public environmental concerns	-
Low carbon emission reduction stage	Input	Number of active patents	Pieces
	Output	Carbon emissions during operation of new energy sources	Tons
Carry-over		fixed-asset investment	100 million RMB

4.2. Regional Efficiency Analysis

4.2.1. Analysis of the Total Efficiency of the Region

Figure 2 reports the high and low carbon reduction efficiency of NEVs in each province and city. The specific values are shown in Appendix A. Overall, the carbon-reduction efficiency of NEVs in each province and city is at a low level, with the value of the sample areas being below 0.7, and the overall average value is only 0.266, which is still a large space for development. First, the low carbon reduction efficiency is influenced by the power structure. The carbon emission of NEVs mainly comes from the power production link. Despite the low carbon emissions in the use phase of NEVs, the source of electricity is mainly coal-fired power generation, as China is still dominated by high-carbon fossil energy. The cleanliness of electricity is largely determined by the level of purity of the technology used to generate it, so the transformation of China's power structure is the key to realizing the carbon-reduction efficiency of NEVs. Second, the limitation of battery technology is also one of the keys to the difficulty of improving the carbon-reduction efficiency of NEVs. The energy density and cycle utilization rate of power batteries will directly affect the carbon-emission efficiency of vehicles. Therefore, manufacturers need to optimize the production process, improve energy efficiency, and use green energy to achieve decarbonization in the battery production process. Finally, external environments such as policy support, market acceptance, and infrastructure conditions will also affect the carbon-reduction efficiency of NEVs.

Looking at the differences between provinces and cities, Shanxi Province has the highest carbon-reduction efficiency for NEVs, at 0.678. Shanxi Province emphasizes the promotion of NEVs in the public service sector. Shanxi Province has put forward a policy for the promotion of NEVs in the public service sector, in which NEVs are not less than a certain percentage of the new or renewed vehicles in the public service sector, such as urban passenger transportation, sanitation, airport commuting, and public security patrol. Among the sample regions, Tianjin has the lowest carbon-reduction efficiency for NEVs, at 0.039, less than 20% of the overall level. The average annual travel carbon reduction for new-energy vehicles in Tianjin in 2022 is 9% lower compared to the national level. This is mainly due to the incomplete uploading of monitoring data and the low mileage of some new-energy vehicles.

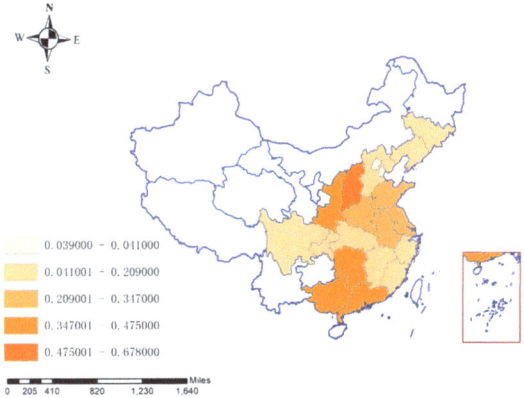

Figure 2. Carbon-reduction efficiency of NEVs in various provinces.

4.2.2. Stage Efficiency Analysis of Areas

Figure 3 reports the changes in the annual average values of efficiency in the production stage (S1), the consumption stage (S2), and the low-carbon emission reduction stage (S3) for each province and city. Figure 4 reports the average value of efficiency for each province and city in the three stages.

From the time dimension, the efficiencies of the production stage, consumption stage, and low-carbon emission reduction stage of NEVs do not show a significant upward trend from 2018 to 2022. First, the efficiency of the production stage of NEVs in all provinces and cities shows a slight increase from 2018 to 2021, from 0.333 to 0.517. On the one hand, this is due to the continuous progress of NEV technology. Both the breakthroughs in core technologies, such as safety performance and service life, as well as cost reduction, have been developed rapidly. On the other hand, the national subsidy policy for the NEVs industry is also one of the key reasons to promote the efficiency of the production stage. The subsidy policy reduces the purchase cost of NEVs through financial incentives, which improves consumers' motivation to purchase vehicles and can also promote the improvement of efficiency in the production stage. However, the efficiency of this stage has dropped significantly in 2022, by 0.139, mainly due to the adjustment of the national subsidy policy for NEVs. The relevant government authorities in China issued a notice at the end of 2021, announcing that the subsidy rate for NEVs in 2022 would be rolled back by 30% from the 2021 level. In addition, the 2022 NEVs purchase subsidy policy was terminated on 31 December 2022. The gradual withdrawal of national policy subsidies will lead to more serious cost pressure on some enterprises, which in turn will lead to a significant decline in the efficiency of the production phase.

Second, the consumption stage in all provinces and cities mainly fluctuates around 0.5 and reaches its peak of 0.519 in 2020, which is still quite a distance away from playing an effective role. On the one hand, the technical maturity of NEVs itself and the issue of range is one of the key points to be overcome in its consumption stage. Currently, NEV technology is not yet fully mature, and consumers have limited awareness and trust in NEVs. Inadequate charging facilities and consumer concerns about insufficient power for long-distance travel can lead to poor acceptance of NEVs. On the other hand, changes in the government's subsidy policy for NEVs can also cause uncertainty in consumers' vehicle purchases, which in turn can lead to inefficiencies in the consumption phase.

Finally, compared with the production and consumption phases, NEVs have the lowest efficiency in the low-carbon emission reduction phase, floating around 0.25 overall, about half the efficiency of the consumption phase. This is mainly related to the current coal-based energy structure system in China, the high carbon of power battery production, and green technology. Therefore, NEVs should pay more attention to the whole lifecycle of carbon reduction and increase the proportion of green power usage.

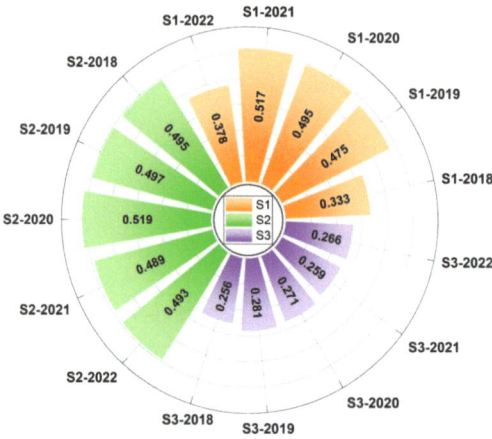

Figure 3. Annual average of three stages of efficiency of new-energy vehicles.

The production-stage efficiency and consumption-stage efficiency of NEVs vary greatly between regions. For example, in Shanxi Province, the efficiencies of NEVs in the production and consumption stages are both 1, while in Tianjin, the efficiencies of the two stages are as low as 0.194 and 0.133, respectively, which may be related to the imperfections of the charging facilities and the imbalanced distribution. Except for Shanxi, the efficiencies of NEVs at the low-carbon emission reduction stage in all regions are below 0.4.

At the production stage, there are only four provinces and cities with optimal efficiency, accounting for 19.05%, namely Hunan, Shanghai, Shanxi, and Shaanxi. The efficiency of NEVs in Sichuan Province in the production stage is the lowest, only 0.149. On the one hand, the scale of NEVs in Sichuan Province is relatively small and lacks leading enterprises, especially in key products such as cars. On the other hand, during the industrialization of NEVs in Sichuan Province, there is a lack of core collaborative platforms among various links such as materials, systems, vehicles, and energy storage applications, resulting in large differences in the level of technology and the size of product scale. Therefore, the improvement of efficiency at the production stage is the current bottleneck to be broken in Sichuan Province.

In the consumption stage of NEVs, most provinces and cities have efficiencies below 0.5, and only five have efficiency values above 0.7, namely Hebei, Guangdong, Jiangsu, Shandong, and Shanxi. The efficiency of the consumption stage of NEVs in Beijing is only 0.059, which is a big gap with other provinces and cities and needs to be emphasized more.

In the low-carbon emission reduction stage, the efficiency values of provinces and cities are mostly between 0.2 and 0.3. Beijing and Tianjin NEVs have the lowest efficiency in the low-carbon emission reduction stage, at 0.119 and 0.102, respectively. Guangxi province has the highest efficiency in this stage, at 0.439, although it still fails to play an effective role. Therefore, regions should boost green energy use and integrate it with NEVs, promoting industry-wide green development.

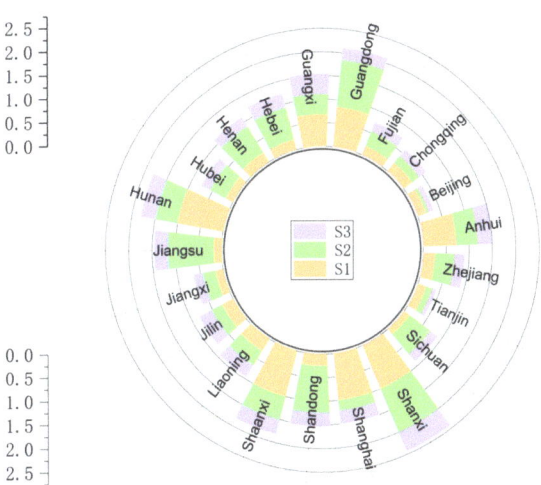

Figure 4. Average value of the three stages of efficiency in the provinces.

4.2.3. Regional Differences in the Efficiency of Key Indicators

Figure 5 reports the efficiency of key indicators in the carbon-reduction system for NEVs in each province and city, namely carbon emissions, number of valid patents, and number of NEVs produced. The horizontal coordinates represent the number of each province and city, and the number, as well as the specific efficiency values are shown in Appendix B.

From the efficiency value of each indicator, the production quantity efficiency of NEVs far exceeds that of other variables. Among them, there are 8 provinces and cities with NEV production quantity efficiency above 0.8, accounting for 38.1% of the research sample, and those reaching the optimal efficiency are Hunan, Shanxi, Shaanxi, and Shanghai. Shanghai has a strong automotive industry foundation and a perfect industrial chain layout. For example, Shanghai has gathered a number of whole-vehicle eco-enterprises and parts producers above the scale, forming core industrial clusters such as Jiading Anting, Pudong Jinqiao, and Lingang. In addition, many well-known automotive companies, such as SAIC and Tesla, have set up R&D centers or production bases in Shanghai, and all of these initiatives have greatly improved the efficiency of the production quantity of NEVs in Shanghai. The production quantity efficiency of NEVs in Sichuan Province is the lowest, only 0.3, and there is still a need for continuous improvement in this area.

The efficiency gap between the number of effective patents in each province and city is large, and the overall level is low, with seven of them having an efficiency value below 0.1, accounting for 33.3% of the total. This is mainly due to the low share of invention patents with the highest gold content in each region of China, and the key technologies and core fields often encounter bottlenecks. In addition, the low percentage of patents in key industries and core technology fields and the short duration of patent maintenance are also important reasons for the inefficiency of the number of effective patents in each region of China. Guangxi Province has the highest efficiency in the number of effective patents, at 0.614, but there is still a long way to go before it can play an effective role.

From the perspective of carbon-emission efficiency, all regions performed poorly, with 90.5% having an efficiency value below 0.3. Among them, Liaoning Province ranked first with a carbon-emission efficiency of 0.529. The carbon-emission efficiency of Jilin Province is 0.464, ranking second. In addition to the constraints on the electricity supply structure, NEVs also generate a large amount of carbon emissions in some parts of the production process, such as the collection stage of raw materials, the production and manufacturing of core components, and the subsequent end-of-life recycling stage of the vehicle.

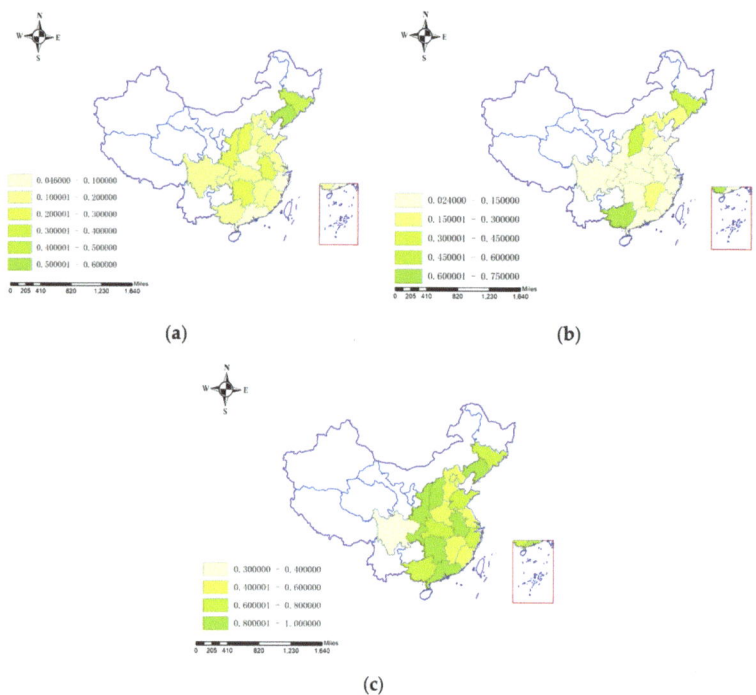

Figure 5. Key indicator efficiency: (**a**) Carbon Emissions; (**b**) Number of active patents; (**c**) Number of new-energy vehicles produced.

4.3. Analysis of Spatial Variation

4.3.1. Analysis of the Efficiency of Economic Zones

The previous analysis examined efficiency differences among various provinces. This section will focus on analyzing regional differences among the three major economic zones.

Figure 6 illustrates the regional distribution of carbon-reduction efficiency during the operation of new-energy vehicles (NEVs). The average carbon-reduction efficiency of NEVs in the central region is significantly higher than that in the eastern and western regions. This can be attributed to several factors.

On the one hand, the adoption rate and vehicle type structure of NEVs across different regions affect carbon-reduction efficiency. In the central region, effective NEV promotion policies, combined with relatively lower vehicle purchase costs, have boosted NEV adoption in the household vehicle market, thereby significantly enhancing carbon-reduction efficiency. In contrast, the western region struggles with insufficient infrastructure, resulting in a lower penetration rate of NEVs and limited carbon-reduction effects.

On the other hand, economic development levels and urban layouts also play a role. The central region features medium-sized cities with stable urbanization levels, and NEV usage is concentrated in short- and medium-distance commuting scenarios, which have a more pronounced effect on improving carbon-reduction efficiency. Conversely, in the eastern region, most large cities are densely populated with longer commuting distances. As a result, NEVs operating on congested roads consume more electricity, limiting their carbon-reduction efficiency.

Contrary to popular perception, the carbon-reduction efficiency of new-energy vehicles (NEVs) in the eastern region ranks lowest among the three major economic zones. On one hand, the eastern region's reliance on high-carbon electricity is a significant factor. As the economic center of China, the eastern region has a massive demand for industrial and residential electricity, making it difficult to reduce its dependence on traditional coal-fired power. Although NEVs do not produce direct emissions during operation, the high-carbon intensity of electricity generation results in lower overall carbon-reduction efficiency. Additionally, the eastern region has limited land resources, restricting the development of renewable energy sources such as wind and solar power. Consequently, the carbon-emission intensity of the electricity used by NEVs in this region is higher than that in the central and some western regions.

On the other hand, the higher proportion of premium NEVs in the eastern region contributes to greater carbon emissions during their production and throughout their lifecycle. With stronger purchasing power, consumers in the eastern region tend to favor high-performance NEVs with large-capacity batteries. These models have higher carbon emissions during the production phase, especially in battery manufacturing, which undermines overall carbon-reduction efficiency. Furthermore, the NEV industry in the eastern region heavily relies on imported components, such as high-performance battery materials. The cross-border transportation of these components adds to hidden carbon emissions.

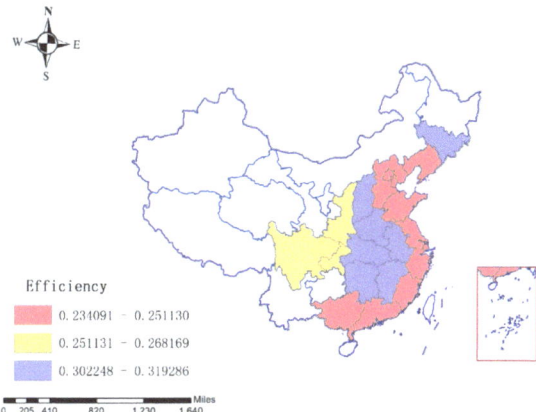

Figure 6. Regional efficiency of city clusters.

4.3.2. Decomposition of Spatial Differences

This study analyzes the spatial distribution differences in low-carbon reduction efficiency across the sample regions by utilizing the Dagum Gini coefficient. As shown in Table 2, the overall Gini coefficient of low-carbon reduction efficiency exhibits a declining trend with fluctuations during the study period, decreasing from 0.207 in 2018 to 0.157 in 2022. This indicates a significant reduction in overall disparities in low-carbon reduction efficiency.

From the perspective of contribution rates to disparities, both the intra-regional contribution rate and the hypervariable density contribution rate remained around 40% throughout the study period, approximately double that of the inter-regional contribution rate. This suggests that intra-regional imbalances are the primary factor influencing the spatial differences in low-carbon reduction efficiency during the operational phase of NEVs. This may be closely related to significant disparities within provinces or cities in terms of NEV penetration rates, usage scenarios, infrastructure development, and policy support levels.

Further analysis reveals that the hypervariable density contribution rate showed an overall trend of rising initially and then declining during the study period, with a continuous decrease after 2019. This indicates that the heterogeneity in low-carbon reduction efficiency during the NEV operational phase across different regions initially expanded but gradually diminished over time. The sharing and diffusion of core technologies and management expertise for NEVs among regions have further promoted balanced efficiency development across regions.

In contrast, the inter-regional contribution rate exhibited a trend of first declining and then rising, increasing continuously from its lowest value of 14.028% in 2019 to 27.215% in 2022. This trend suggests a certain degree of fluctuation in inter-regional disparities, particularly as some regions, driven by policies and technological advancements, may further widen the gap with less developed regions. Therefore, government departments need to place greater emphasis on promoting coordinated inter-regional development when formulating policies.

It is worth noting that 2019 marked a critical turning point for intra-regional, inter-regional, and hypervariable density contribution rates. This may be attributed to China's introduction of more systematic NEV development strategies in 2019, such as strengthening the NEV credit policy and optimizing subsidies, which provided greater incentives for different regions to develop NEVs. Additionally, around 2019, the nationwide efforts to expand charging station and power grid infrastructure intensified, enabling regions that lagged in NEV development to gradually address their infrastructure shortcomings for NEV promotion.

Table 2. Spatial differences in efficiency at the low-carbon reduction stage.

Year	Overall	Intra-Regional Contribution	Inter-Regional Contribution	Hypervariable Density Contribution Rate
2018	0.207	39.853%	19.624%	40.523%
2019	0.161	41.949%	14.028%	44.022%
2020	0.187	39.204%	21.248%	39.547%
2021	0.148	40.400%	24.442%	35.157%
2022	0.157	37.408%	27.215%	35.377%

Table 3 reports the intra-regional Gini coefficient for the low-carbon reduction stage efficiency. The intra-regional Gini coefficient of the Eastern region is higher compared to the Central and Western regions. Specifically, the intra-regional Gini coefficient in the Eastern region shows a continuous downward trend. This indicates that the efficiency differences in the low-carbon reduction phase in the Eastern region gradually narrowed during the study period. In the Central region, the intra-regional Gini coefficient fluctuated downward, from 0.167 in 2018 to 0.12 in 2022. Therefore, the low-carbon transition in the Central region still faces significant internal challenges, requiring more precise support tailored to regional issues, such as resource allocation and policy coverage. In the Western region, the intra-regional Gini coefficient for low-carbon reduction efficiency was highly unstable, experiencing multiple sharp increases and decreases. The spiral decline in the Western region may be attributed to the phased implementation of policies. The region likely experienced waves of concentrated policy support and investment, such as special funds for renewable energy. These measures drove efficiency improvements in the short term, but after the policies ended, efficiency declined again.

Table 3. The intra-regional Gini coefficient for low-carbon reduction stage efficiency.

Year	East	Central	West
2018	0.217	0.167	0.197
2019	0.187	0.14	0.082
2020	0.188	0.161	0.163
2021	0.169	0.115	0.078
2022	0.156	0.12	0.157

Figure 7 shows the trend of the inter-regional Gini coefficient. For example, in 2022, regional differences exhibited an imbalanced spatial pattern of "East and West > Central and West > East and Central". Further analysis reveals that during the study period, the Gini coefficients between the East and West regions, as well as between the Central and West regions, showed a clear "W" shape. This also reflects the significant stage changes in the regional differences in low-carbon reduction phase efficiency for new-energy vehicles, and this fluctuating trend can be attributed to the non-sustained impact of policies and markets. Meanwhile, the inter-regional Gini coefficient between the East and Central regions showed a noticeable decline, from 0.201 in 2018 to 0.154 in 2022. This indicates that the gap in low-carbon reduction phase efficiency between the East and Central regions has been narrowing. Therefore, the inter-regional low-carbon transition is gradually showing an equilibrating trend, but further efforts are needed to deepen the application of green technologies and promote sustainable development in the central region.

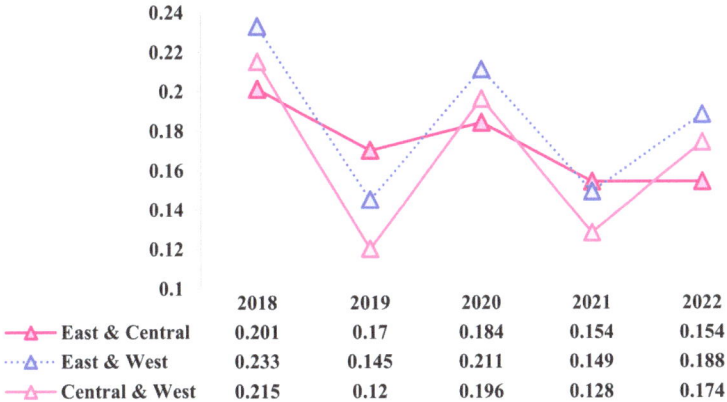

Figure 7. Inter-regional trends in the Gini coefficient.

4.4. Analysis of Influencing Factors

4.4.1. Main Factors Influencing Carbon Emissions

Using the LMDI decomposition method, this paper investigates the degree of contribution of four influencing factors, namely energy structure effect, energy intensity effect, economic development effect, and population size effect, for the carbon emissions during the new energy operation period from 2011 to 2022, with 2011 as the base period. Furthermore, the corresponding contribution rate of each driving factor was further obtained. The specific results are shown in Table 4 and Figure 8. Among them, the positive contribution value indicates that a factor has a positive influence on carbon emissions, and the negative contribution value indicates the existence of a negative influence, while the size of the value reflects the degree of influence on carbon emissions.

Overall, energy structure and energy intensity significantly reduce NEV operational carbon emissions, with cumulative contributions of −12.703% and −234.535%. This indicates that as the contribution values of these factors increase, carbon emissions decrease accordingly. The level of economic development and population size effect generally positively influence the carbon emissions during the operation of NEVs, with cumulative contribution values of 331.231% and 16.008%, respectively.

The change in energy structure led to a reduction of 89,825,100 tons of carbon emissions during the operation of NEVs during the period from 2011 to 2022, with a cumulative contribution of −12.703%, which provides a significant inhibition of carbon emissions. However, the degree of influence of energy structure is relatively weak compared with the energy intensity effect, and it is still necessary to continue to promote the evolution. Within the current electricity supply system of NEVs, the main source of energy consumption is still fossil fuels due to China's unique energy structure.

Table 4. Decomposition of factors affecting carbon emissions (tons) (2011–2022).

Year	Energy Structure Effect	Energy Intensity Effect	Economic Development Effects	Population Size Effect	Total Effect
2011–2012	−888.3154	−17,803.6200	25,060.2100	2288.0050	8656.2796
2012–2013	−845.2513	−8877.4170	24,114.1500	1482.4460	15,873.9277
2013–2014	−1011.5360	−13,119.5000	17,404.1800	1677.2620	4950.4060
2014–2015	−1575.2650	−13,211.6100	12,303.6600	680.0912	−1803.1238
2015–2016	−1014.9570	−12,963.6500	15,407.9100	2106.0030	3535.3060
2016–2017	−1435.0720	−22,764.4400	27,936.1300	1260.7220	4997.3400
2017–2018	−964.3012	−16,861.1000	22,679.4000	878.7462	5732.7450
2018–2019	−979.7866	−3839.8030	11,462.3000	825.5944	7468.3048
2019–2020	112.0494	−2693.0100	8402.5920	419.3560	6240.9874
2020–2021	−991.1002	−32,674.5500	46,128.8400	−21.7310	12,441.4588
2021–2022	611.0225	−21,031.7600	23,314.6800	−277.2934	2616.6491
Total	−8982.51 (−12.703%)	−165,840.46 (−234.535%)	234,214.05 (331.231%)	11,319.20 (16.008%)	70,710.2806

Figure 8. Decomposition of the contribution of factors influencing carbon emissions (%).

From the perspective of energy intensity, it can indicate the energy utilization rate, which is inherently volatile as a complex indicator involving many factors such as policy, market and economy. In addition, the energy intensity effect can also reflect the technological progress effect of NEVs side by side. During the period from 2011 to 2022, the change in energy intensity led to a reduction in carbon emissions by 165,840,400 tons, with a cumulative contribution rate of −234.535%, which has a significant carbon-reduction effect.

Changes in economic development from 2011 to 2022 led to an increase in carbon emissions by 234,210,140,500 tons, with a cumulative contribution of 331.231%, which is the most important factor affecting carbon emissions. In addition, the positive contribution of the economic development factor to carbon emissions shows an increasing trend. The contribution of this factor increased from 289.503 to 891.013% during the study period, and although it experienced a significant drop during the period, it has continued to rise sharply since then. The expansion of economic scale is usually accompanied by an increase in national consumption capacity. Given the central role that energy plays in economic development, this increased consumption capacity will ultimately lead to increasing consumption of natural resources for social production and living, which in turn will lead to a rise in total carbon emissions.

The population size increase from 2011 to 2022 contributed 113.192 million tons to carbon emissions, accounting for 16.008% of the total. Population growth creates a demand for livelihoods, and increased demand for land increases the ecological burden, which in turn leads to higher carbon emissions. However, its positive effect on emissions during NEVs' operation is weakening, even turning negative in 2020. With the increase in population, especially the rapid expansion of urban populations, the demand for public charging facilities has grown significantly. To meet the expanding demand, the construction of charging piles and switching stations has been accelerated around the world, improving the convenience of using new-energy vehicles. This not only promotes the popularity of new-energy vehicles but also promotes the decarbonization of the transport energy mix, allowing the emission reduction potential of new-energy vehicles to be gradually released. In addition, population growth has been accompanied by a trend of rejuvenation and increased environmental awareness, with the younger generation of consumers more inclined to choose low-carbon modes of transport, such as shared mobility and public transport with new-energy vehicles. This trend further strengthens the market demand for new-energy vehicles and drives down carbon emissions. Most importantly, population growth is not only reflected in increased numbers but is also accompanied by higher levels of urbanization and shifts in consumer attitudes. Denser urban populations create higher market demand and drive the popularity of new-energy vehicles. With policies such as license restrictions, new-energy vehicles have accelerated the replacement of fuel vehicles with lower costs and environmental advantages, thus weakening the positive impact of population size on carbon emissions.

4.4.2. Energy Activities and Carbon-Reduction Efficiency

In the previous section, this study revealed the factors affecting carbon emissions and the extent of their contribution based on LMDI. However, the LMDI method focuses on quantitative decomposition and does not directly involve the causal relationship between variables. Its decomposition results may be affected by factors such as data quality and model assumptions. Therefore, this section uses a fixed-effects model to verify whether the decomposed factors related to energy activities (energy intensity and energy structure) have a statistically significant effect on the carbon-emission reduction efficiency of new-energy vehicles so as to enhance the reliability and robustness of the conclusions. The reason for testing only the effect of energy activities is that energy intensity and energy structure are relatively more susceptible to the influence of policy and technological means. For example, carbon-emission levels can be directly improved by improving energy use efficiency or accelerating clean energy substitution. Although population size and economic development are important factors affecting carbon emissions, they tend to be macro background variables with a strong exogenous and uncontrollable nature. Energy intensity and energy structure, on the other hand, are endogenous energy activity variables, and their impacts are closer to the specific mechanisms of carbon emissions from new-energy vehicles, making them more suitable for research focus.

Based on the above analyses, this paper tests the specific effects of energy structure and energy intensity on the carbon-reduction efficiency of new-energy vehicles. The regression results are shown in Table 5. The results show that the regression coefficients of energy structure and energy intensity are both significant at the 1% level. Therefore, the improvement of both energy structure and energy intensity can promote the improvement of carbon-reduction efficiency, and thus achieve the reduction of carbon emissions. This is also related to the increasing use of clean energy in recent years. Figure 9 shows the changes in selected energy consumption from 2011 to 2022.

Table 5. Energy structure, energy intensity, and carbon-reduction efficiency.

	Carbon-Emission Reduction Efficiency	Carbon-Emission Reduction Efficiency
Energy structure	1.136 ***	
	(8.565)	
Energy intensity		0.056 ***
		(6.301)
_cons	−0.515 ***	0.218 ***
	(−5.645)	(12.671)
Year_FE	YES	YES
N	105	105
R^2	0.432	0.294
Adj.R^2	0.403	0.258

Note: t statistics in parentheses, *** $p < 0.01$.

As shown in Figure 9, natural gas consumption increases every year. Until 2022, natural gas consumption ranks first among the selected energy sources. With the rising use of natural gas, the energy structure is gradually changing from high-carbon energy dominated by coal and oil to low-carbon and clean energy. As a low-carbon energy source, natural gas has significantly lower carbon emissions per unit of electricity generated or per unit of energy released than coal and oil. In addition, natural gas is not only a clean energy source, its application efficiency is also higher. In recent years, the application of natural gas in power generation, industrial energy supply, and residential energy use has gradually expanded. At the same time, the widespread application of modern natural gas utilization technologies, such as combined cycle power generation, has further improved energy conversion efficiency.

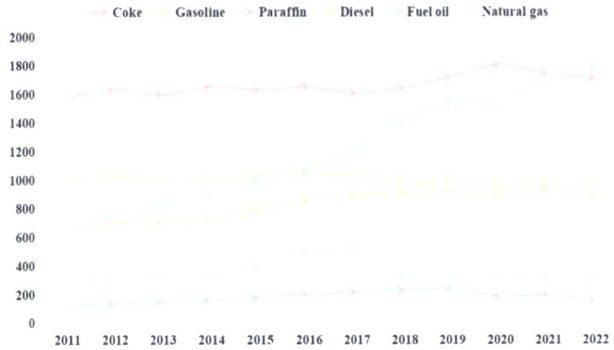

Figure 9. Changes in consumption of selected energy sources (10,000 tons of standard coal).

Figure 10 shows the structure of selected energy shares for 21 provinces and cities. Looking at the provinces and cities, the coke share is dominant among the selected energy sources, especially in Hebei Province, with a share of more than 70%. The coke share of Shanxi Province is second only to Hebei Province, with about 55%. Theoretically, although new-energy vehicles achieve zero emissions in the end-use phase, the electricity sources they use lead to high-carbon indirect emissions, which in turn undermine the overall emission reduction effect. However, according to the assessment results of this paper, the carbon-reduction efficiency of new-energy vehicles in Shanxi Province ranks first among the sample regions. This suggests that the carbon-reduction efficiency of new-energy vehicles is not only affected by the energy mix but is also closely related to local policies, charging infrastructure, grid cleanliness, technical performance, and regional economic structure. This can also be seen in the higher share of natural gas in Beijing. Among the sample regions, Beijing ranks first in terms of natural gas share, but its carbon-reduction efficiency for new-energy vehicles is at a low level. Despite the high share of natural gas, policies to promote new-energy vehicles may be more biased towards meeting short-term goals, such as purchase subsidies and target allocation, and fail to effectively optimize overall energy efficiency. In addition, Beijing's electricity demand is dense, and some of its electricity may rely on coal power input from external regions, thus indirectly driving up carbon emissions.

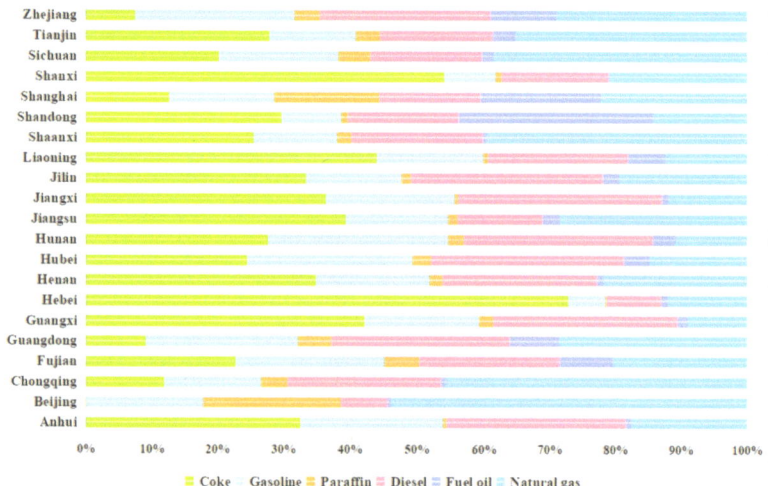

Figure 10. Share of energy consumption by province.

5. Conclusions and Suggestions

5.1. Discussion

This paper evaluates the carbon-reduction efficiency during the operation of NEVs in 21 provinces in China, and analyzes the influencing factors of carbon emissions in four dimensions.

On the one hand, this paper assesses the carbon-reduction efficiency of NEVs based on a three-stage DEA model that considers undesired outputs. The assessment results are in line with Su et al. (2021) [23]. The overall carbon-reduction efficiency of NEVs is not satisfactory, and there is still a considerable distance from playing an effective role. This may be due to the fact that the popularity of NEVs is not high enough, and therefore, their green effect is not obvious [23]. In addition, the high pollution in the production and manufacturing process of automobiles is also an important reason for their low carbon reduction efficiency [22,24]. In the comparative analysis of economic zone efficiencies, the conclusions of this study differ significantly from the existing literature. Unlike most studies that suggest the East region has the highest carbon-reduction efficiency, this study finds that the carbon-reduction efficiency of new-energy vehicles in the East region is relatively low, leaving substantial room for improvement. Furthermore, the primary reason for the significant differences in low-carbon reduction stage efficiency among the three major economic zones lies in intra-regional disparities.

On the other hand, this paper applies the LMDI method to decompose the carbon emissions during the period of China's new energy operation from 2011 to 2022. This paper's findings align with Yang et al. (2020) [42], showing that economic development is the primary driver of increased carbon emissions [44,45], while energy intensity plays a key role in reducing them. In line with Quan et al. (2020) [46], the population size effect is second only to economic development in driving carbon emissions. However, in terms of energy structure, the study in this paper finds differences with Quan et al. (2020) [46]. Unlike Quan et al. (2020) [46], which argues that energy structure contributes to carbon emissions, this study argues that energy structure instead suppresses carbon emissions, which also matches the findings of He et al. (2022) [47]. The reason for the difference in some studies may be due to the selection of the study sample and the study period.

This study evaluates the carbon-emission reduction efficiency during the operation of new-energy vehicles in each province, but due to the availability of data, there are still some limitations in this study, with a regional sample of only 21. Some regions lack continuous, stable, and reliable statistical data, and the statistical standards are not uniform, making it difficult to make a direct comparison with other regions. To ensure data quality and consistency, this study sacrificed the comprehensiveness of the sample size. The narrow sample coverage may lead to limitations in the applicability of the study's conclusions on a national scale. There are large differences in the level of economic development, energy structure, technological level, and policy environment between regions, while regions not included in the sample may present different carbon-emission reduction efficiency characteristics, which cannot comprehensively reflect the overall situation of new-energy vehicle operation in the country. Therefore, future research will aim to broaden the sample scope to encompass additional provinces, cities, and specific regions, such as economically developed and less developed areas, while accounting for the impact of inter-regional heterogeneity on the findings. For instance, subregional studies or stratified analytical methods will be employed to uncover the unique characteristics of each region. Moreover, to address challenges related to data continuity and reliability, efforts will focus on establishing a more comprehensive and standardized database. Future studies may integrate big data technologies and employ modern tools such as remote sensing and satellite monitoring to acquire more precise energy and emission data.

The factor decomposition analysis in this study focuses on carbon-emission drivers over the period 2011–2022. Variations in carbon-emission reduction efficiency often exhibit long-term and phased patterns, making short-term data insufficient to capture the complete dynamic trends. Consequently, future research will extend the temporal scope to incorporate long-term data, enabling a more comprehensive understanding of policy effects and technological advancements. Furthermore, predictive models may be employed to simulate the potential impacts of policies and technologies on future carbon-emission reduction efficiency.

5.2. Conclusions

This study draws the findings by evaluating the carbon-reduction efficiency during the operation of NEVs in 21 provinces in China and analyzing the factor decomposition of carbon emissions from four aspects.

1. From the perspective of efficiency assessment, the carbon-reduction efficiencies of NEVs in the sample areas during the operation period are all lower than 0.7, and the overall average value is only 0.266, which still has much room for development. Among them, Tianjin has the lowest carbon-reduction efficiency of NEVs, which is only 0.039. This

is also related to the high carbon emission during the production of automobile power and the imperfection of the current power structure. There is a large gap between the production-stage efficiency and the consumption-stage efficiency in each province and city. Among them, Shanxi Province has the best performance in both stages, with both stages reaching optimal efficiency 1. Some cities performed poorly, such as Chongqing and Beijing, with efficiencies below 0.2 for both stages. The efficiency of the low-carbon emission reduction stage is at the lowest level among the three stages, with most regions having an efficiency between 0.2 and 0.3. Guangxi province has the highest efficiency in this stage, but at only 0.439, it still does not play an effective role. The efficiency of the quantity of NEVs produced is at the highest level compared to the efficiency of the other key indicators. A total of eight provinces and cities, accounting for 38.1% of the study sample, had an efficiency above 0.8.

2. From the perspective of efficiency differences among economic regions, the central region demonstrates the highest low-carbon reduction efficiency, followed by the western and eastern regions. The primary reason for this phenomenon lies in intra-regional disparities, while the proportion of inter-regional disparities has also gradually increased in recent years.

3. From the perspective of the decomposition of carbon-emission influencing factors, the carbon emissions are dampened by the energy structure and energy intensity, with cumulative contributions of -12.703% and -234.535%, respectively. Although NEVs have some advantages in fuel cycle emissions, their charging link has a high proportion of carbon-emission intensity. Economic development is the main driver of increased carbon emissions, followed by population size.

This study reveals regional differences in carbon-emission reduction efficiency of new-energy vehicles, stage imbalances, and the significant impact of energy structure on carbon emissions. It suggests that policies should focus more on targeting, precision, and whole-life management. Regional differences in efficiency highlight the importance of tailoring to local conditions. Inefficiencies in the production, consumption, and emission reduction stages reflect the need for the industry to deeply optimize technological innovation, green power supply, and supply-chain management. The findings provide a clear direction for policymakers to promote a shift from single-link subsidies to greening the entire industry chain and for the industry to find more sustainable solutions in technology R&D, resource integration, and regional synergy, thus fully unleashing the potential of new-energy vehicles in carbon-emission reduction.

5.3. Suggestions

Suggestions for new-energy vehicle enterprises. On the one hand, new-energy vehicle enterprises need to strengthen low-carbon power battery research and development by shifting the focus of R&D and fostering cross-disciplinary technological synergies. Enterprises can focus on core technologies such as solid-state batteries, high-energy-density batteries (e.g., lithium-sulfur batteries), and low-cost electrolytes to reduce the carbon footprint of battery production and usage. The path of technological innovation can draw inspiration from the "Kirin Battery" R&D model of CATL, establishing a three-tier innovation system of "basic research–pilot bases–industrial incubation". Additionally, they can learn from Toyota's approach to solid-state battery R&D, which integrates materials science and chemical engineering, to foster deeper collaboration between domestic universities and enterprises and establish a low-carbon battery technology R&D alliance. On the other hand, new-energy vehicle enterprises can create a green supply chain with regional characteristics. This involves launching differentiated supply-chain transformation pilots tailored to regions with varying resource endowments and industrial foundations. For instance, in Shanxi Province, enterprises can integrate coal-based hydrogen energy with the new-energy vehicle industry chain, while in Yunnan Province, they can promote synergy with the clean aluminum industry to establish an aluminum lightweight body supply chain. The implementation framework for the green supply chain can be divided into three phases: in the short term (1–2 years), enterprises can implement the "Guidelines for the Construction and Operation of Power Battery Recycling Service Networks for New Energy Vehicles" and require first-tier suppliers to achieve [ISO 50001] [48] certification. In the medium term (3–5 years), they can establish a regional supply-chain carbon footprint platform and create a supplier carbon account system, taking inspiration from BYD's "Cloud Rail" project. In the long term (more than 5 years), they can enforce whole-chain carbon labeling management, such as the closed-loop material traceability system implemented for BMW iX3 models.

Recommendations for government departments. On the one hand, government departments can implement a 'two-dimensional' regulatory system that addresses both spatial and temporal dimensions. In the spatial dimension, the atmospheric hotspot grid technology developed by the Ministry of Ecology and Environment of China, which employs high-resolution remote sensing satellites and UAV technology, can be utilized to monitor, in real time, key enterprises involved in the production processes of new-energy vehicles, such as battery manufacturing. National industrial parks and key new-energy vehicle enterprises can be classified into high-, medium-, and low-emission zones while simultaneously establishing a dynamic emission database. Concurrently, an enterprise-level carbon-emission warning mechanism should be established, which sets thresholds based on remote sensing data and emission intensity of enterprises, thereby triggering early warnings and implementing corrective actions. In the temporal dimension, government departments can implement a battery carbon passport system. Specifically, carbon footprint data covering the entire lifecycle of the power battery—including raw material extraction, manufacturing, transportation, usage, recycling, and other stages—should be integrated into the national carbon emissions monitoring platform to create a comprehensive 'battery carbon passport.' Power battery manufacturers should be mandated to deploy carbon footprint tracking systems throughout their production processes and upload real-time data to the national platform. Furthermore, independent third-party organizations, such as the China Quality Certification Centre, should be engaged to verify carbon footprint data, ensuring both its transparency and accuracy. On the other hand, government departments should innovate models for the construction of low-carbon charging stations. Based on regional economic development levels, underdeveloped areas can be categorized into regions with weak infrastructure and those with limited economic resources, with differentiated PPP models tailored to each region's characteristics. In regions with weak infrastructure, the government will take the initiative to introduce social capital for the construction and operation of charging stations, for instance, using the BOT (Build-Operate-Transfer) model, which will be transferred to local governments when the project matures. In regions with limited economic resources, the government's underwriting role will be enhanced, and the EPC + F (design-procurement-construction and financing) model will be adopted to accelerate project implementation and mitigate long-term risks to social capital.

Author Contributions: L.W.: formal analysis and writing—original draft preparation. T.W.: data curation, methodology and software, writing—review and editing. F.R.: conceptualization, resources, and project administration. All authors read and contributed to the manuscript. All authors have read and agreed to the published version of the manuscript.

Funding: This research was funded by the National Social Science Foundation Project, established, and supported under the National Office for Philosophy and Social Science Work, grant number 24BJY142; Jiangsu Province Social Science Foundation Project, established, and supported under Jiangsu Province Philosophy and Social Science Planning Office, grant number 22GLD019; Major Project of Philosophy and Social Science Research in Universities of Jiangsu Province, established, and supported under Jiangsu Province Education Department, grant number 2022SJZD053; Postgraduate Research & Practice Innovation Program of Jiangsu Province, established, and supported under Jiangsu Province Education Department, SJCX24_0372.

Data Availability Statement: The data are contained within Section 4.4 of this article.

Conflicts of Interest: The authors declare that the research was conducted in the absence of any commercial or financial relationships that could be construed as a potential conflict of interest.

Appendix A

Table A1. Carbon-reduction efficiency of NEVs by Province.

NO.	DMU	S	S1	S2	S3
1	Anhui	0.347	0.719	0.466	0.309
2	Beijing	0.041	0.188	0.059	0.119
3	Chongqing	0.136	0.232	0.174	0.199
4	Fujian	0.176	0.200	0.331	0.226
5	Guangdong	0.475	0.846	1.000	0.265
6	Guangxi	0.418	0.672	0.435	0.439
7	Hebei	0.209	0.220	0.701	0.314
8	Henan	0.252	0.235	0.656	0.238
9	Hubei	0.175	0.172	0.393	0.212
10	Hunan	0.430	1.000	0.500	0.322
11	Jiangsu	0.271	0.169	1.000	0.291
12	Jiangxi	0.161	0.167	0.288	0.207
13	Jilin	0.192	0.266	0.266	0.302
14	Liaoning	0.199	0.301	0.365	0.287
15	Shaanxi	0.456	1.000	0.514	0.314
16	Shandong	0.316	0.249	1.000	0.296
17	Shanghai	0.254	1.000	0.222	0.284
18	Shanxi	0.678	1.000	1.000	0.438
19	Sichuan	0.175	0.149	0.547	0.208
20	Tianjin	0.039	0.194	0.133	0.102
21	Zhejiang	0.177	0.254	0.418	0.230
Average	-	0.266	0.440	0.498	0.267

Appendix B

Table A2. Efficiency of key indicators in the carbon-reduction system for NEVs.

DMU	Carbon Emissions	Number of Active Patents	Number of New-Energy Vehicles Produced
Anhui	0.226	0.097	0.817
Beijing	0.057	0.024	0.518
Chongqing	0.195	0.132	0.943
Fujian	0.168	0.123	0.541
Guangdong	0.056	0.031	0.983
Guangxi	0.169	0.614	0.739
Hebei	0.194	0.263	0.419
Henan	0.096	0.140	0.547
Hubei	0.171	0.100	0.745
Hunan	0.238	0.134	1.000
Jiangsu	0.188	0.053	0.570
Jiangxi	0.148	0.218	0.441
Jilin	0.464	0.464	0.719
Liaoning	0.529	0.200	0.805
Shaanxi	0.256	0.149	1.000
Shandong	0.111	0.117	0.676
Shanghai	0.106	0.077	1.000
Shanxi	0.220	0.562	1.000
Sichuan	0.165	0.126	0.300
Tianjin	0.046	0.081	0.559
Zhejiang	0.083	0.088	0.758

References

1. Zhang, X.; Li, Z.; Luo, L.; Fan, Y.; Du, Z. A review on thermal management of lithium-ion batteries for electric vehicles. *Energy* **2022**, *238*, 121652. [CrossRef]
2. Onat, N.; Kucukvar, M. A systematic review on sustainability assessment of electric vehicles: Knowledge gaps and future perspectives. *Environ. Impact Assess. Rev.* **2022**, *97*, 106867. [CrossRef]
3. Yuan, H.; Ma, M.; Zhou, N.; Xie, H.; Ma, Z.; Xiang, X.; Ma, X. Battery electric vehicle charging in China: Energy demand and emissions trends in the 2020s. *Appl. Energy* **2024**, *365*, 123153. [CrossRef]
4. Zhai, H.; Christopher Frey, H. Development of a modal emissions model for a hybrid electric vehicle. *Transport. Res. Transport Environ.* **2011**, *16*, 444–450. [CrossRef]
5. de Rubens, G.Z. Who will buy electric vehicles after early adopters? Using machine learning to identify the electric vehicle mainstream market. *Energy* **2019**, *172*, 243–254. [CrossRef]
6. Zahoor, A.; Yu, Y.; Zhang, H.; Nihed, B.; Afrane, S.; Peng, S.; S'api, A.; Lin, C.J.; Mao, G. Can the new energy vehicles (NEVs) and power battery industry help China to meet the carbon neutrality goal before 2060? *J. Environ. Manag.* **2023**, *336*, 117663. [CrossRef]
7. Shi, H.X.; Wang, S.Y.; Zhao, D.T. Exploring urban resident's vehicular PM2.5 reduction behavior intention: An application of the extended theory of planned behavior. *J. Clean. Prod.* **2017**, *147*, 603–613. [CrossRef]
8. Wang, L.; Yu, Y.J.; Huang, K.; Zhang, Z.Q.; Li, X. The inharmonious mechanism of CO_2, NO_X, SO_2, and PM2.5 electric vehicle emission reductions in Northern China. *J. Environ. Manag.* **2020**, *274*, 111236. [CrossRef]
9. Zheng, J.; Sun, X.; Jia, L.; Zhou, Y. Electric passenger vehicles sales and carbon dioxide emission reduction potential in China's leading markets. *J. Clean. Prod.* **2020**, *243*, 118607. [CrossRef]

10. Elshurafa, A.M.; Peerbocus, N. Electric vehicle deployment and carbon emissions in Saudi Arabia: A power system perspective. *Electr. J.* **2020**, *33*, 106774. [CrossRef]
11. Ayetor, G.K.; Quansah, D.A.; Adjei, E.A. Towards zero vehicle emissions in Africa: A case study of Ghana. *Energy Pol.* **2020**, *143*, 111606. [CrossRef]
12. Teixeira, A.C.R.; Sodré, J.R. Simulation of the impacts on carbon dioxide emissions from replacement of a conventional Brazilian taxi fleet by electric vehicles. *Energy* **2016**, *115*, 1617–1622. [CrossRef]
13. Choi, W.; Yoo, E.; Seol, E.; Kim, M.; Song, H.H. Greenhouse gas emissions of conventional and alternative vehicles: Predictions based on energy policy analysis in South Korea. *Appl. Energy* **2020**, *265*, 114754. [CrossRef]
14. Jenn, A.; Azevedo, I.L.; Michalek, J.J. Alternative-fuel-vehicle policy interactions increase U.S. greenhouse gas emissions. *Transport. Res. Pol. Pract.* **2019**, *124*, 396–407. [CrossRef]
15. Li, N.; Chen, J.; Tsai, I.; He, Q.; Chi, S.; Lin, Y.; Fu, T. Potential impacts of electric vehicles on air quality in Taiwan. *Sci. Total Environ.* **2016**, *566*, 919–928. [CrossRef]
16. Lévay, P.Z.; Drossinos, Y.; Thiel, C. The effect of fiscal incentives on market penetration of electric vehicles: A pairwise comparison of total cost of ownership. *Energy Pol.* **2017**, *105*, 524–533. [CrossRef]
17. Xu, Z.; Li, Y.; Li, F. Electric vehicle supply chain under dual-credit and subsidy policies: Technology innovation, infrastructure construction and coordination. *Energy Pol.* **2024**, *195*, 114339. [CrossRef]
18. Xing, P.; Wang, M. The interplay of recycling channel selection and blockchain adoption in the new energy vehicle supply chain under the government reward-penalty scheme. *J. Clean. Prod.* **2025**, *487*, 144384. [CrossRef]
19. Eslamipoor, R. An optimization model for green supply chain by regarding emission tax rate in incongruous vehicles. *Model. Earth Syst. Environ.* **2023**, *9*, 227–238. [CrossRef]
20. Eslamipoor, R.A. fuzzy multi-objective model for supplier selection to mitigate the impact of vehicle transportation gases and delivery time. *J. Data Inf. Manag.* **2022**, *4*, 231–241. [CrossRef]
21. Eslamipoor, R. Direct and indirect emissions: A bi-objective model for hybrid vehicle routing problem. *J. Bus. Econ.* **2024**, *94*, 413–436. [CrossRef]
22. Zhao, J.X.; Xi, X.; Na, Q.; Wang, S.S.; Kadry, S.N.; Kumar, P.M. The technological innovation of hybrid and plug-in electric vehicles for environment carbon pollution control. *Environ. Impact Assess. Rev.* **2021**, *86*, 106506. [CrossRef]
23. Su, C.W.; Yuan, X.; Tao, R.; Umar, M. Can new energy vehicles help to achieve carbon neutrality targets? *J. Environ. Manag.* **2021**, *297*, 113348. [CrossRef] [PubMed]
24. Lin, B.Q.; Tan, R.P. Estimation of the environmental values of electric vehicles in Chinese cities. *Energy Pol.* **2017**, *104*, 221–229. [CrossRef]
25. Gan, Y.; Lu, Z.; He, X.; Hao, C.; Wang, Y.; Cai, H.; Wang, M.; Elgowainy, A.; Przesmitzki, S.; Bouchard, J. Provincial greenhouse gas emissions of gasoline and plug-in electric vehicles in China: Comparison from the consumption-based electricity perspective. *Environ. Sci. Technol.* **2021**, *55*, 6944–6956. [CrossRef]
26. Rahman, M.N.; Akter, K.S.; Faridatul, M.I. Assessing the Impact of Urban Expansion on Carbon Emission. *Environ. Sustain. Ind.* **2024**, *23*, 100416. [CrossRef]
27. Li, R.; Han, X.; Wang, Q. Do technical differences lead to a widening gap in China's regional carbon emissions efficiency? Evidence from a combination of LMDI and PDA approach. *Renew. Sustain. Energy Rev.* **2023**, *182*, 113361. [CrossRef]
28. Miao, C.; Chen, Z.; Zhang, A. Green technology innovation and carbon emission efficiency: The moderating role of environmental uncertainty. *Sci. Total Environ.* **2024**, *938*, 173551. [CrossRef]
29. Zhao, M.; Sun, T.; Feng, Q. Capital allocation efficiency, technological innovation and vehicle carbon emissions: Evidence from a panel threshold model of Chinese new energy vehicles enterprises. *Sci. Total Environ.* **2021**, *784*, 147104. [CrossRef]
30. Li, J.; Jiang, M.; Li, G. Does the new energy vehicles subsidy policy decrease the carbon emissions of the urban transport industry? Evidence from Chinese cities in Yangtze River Delta. *Energy* **2024**, *298*, 131322. [CrossRef]
31. Zheng, P.; Pei, W.; Pan, W. Impact of different carbon tax conditions on the behavioral strategies of new energy vehicle manufacturers and governments-A dynamic analysis and simulation based on prospect theory. *J. Clean. Prod.* **2023**, *407*, 137132. [CrossRef]
32. Zhang, C.; Su, B.; Zhou, K.; Yang, S. Decomposition analysis of China's CO_2 emissions (2000–2016) and scenario analysis of its carbon intensity targets in 2020 and 2030. *Sci. Total Environ.* **2019**, *668*, 432–442. [CrossRef]
33. Li, S.; Yao, L.; Zhang, Y.; Zhao, Y.; Sun, L. China's provincial carbon emission driving factors analysis and scenario forecasting. *Environ. Sustain. Ind.* **2024**, *22*, 100390. [CrossRef]
34. Lin, B.; Long, H. Emissions reduction in China's chemical industry–Based on LMDI. *Renew. Sustain. Energy Rev.* **2016**, *53*, 1348–1355. [CrossRef]
35. Wang, M.; Feng, C. Decomposition of energy-related CO_2 emissions in China: An empirical analysis based on provincial panel data of three sectors. *Appl. Energy* **2017**, *190*, 772–787. [CrossRef]

36. Jin, Y.; Zhang, K.; Li, D.; Wang, S.; Liu, W. Analysis of the spatial–temporal evolution and driving factors of carbon emission efficiency in the Yangtze River economic Belt. *Ecol. Indic.* **2024**, *165*, 112092. [CrossRef]
37. Gao, Z.; Zhang, Q.; Liu, B.; Liu, J.; Wang, G.; Ni, R.; Yang, K. The driving factors and mitigation strategy of CO_2 emissions from China's passenger vehicle sector towards carbon neutrality. *Energy* **2024**, *294*, 130830. [CrossRef]
38. Waheed, R.; Sarwar, S.; Wei, C. The survey of economic growth, energy consumption and carbon emission. *Energy Rep.* **2019**, *5*, 1103–1115. [CrossRef]
39. Lu, L.C.; Chiu, S.Y.; Chiu, Y.H.; Chang, T.H. Three-stage circular efficiency evaluation of agricultural food production, food consumption, and food waste recycling in EU countries. *J. Clean. Prod.* **2022**, *343*, 130870. [CrossRef]
40. Zhang, L.; Zhao, L.; Zha, Y. Efficiency evaluation of Chinese regional industrial systems using a dynamic two-stage DEA approach. *Socio-Econ. Plan. Sci.* **2021**, *77*, 101031. [CrossRef]
41. Hu, J.L.; Wang, S.C. Total-factor energy efficiency of regions in China. *Energy Pol.* **2006**, *34*, 3206–3217. [CrossRef]
42. Yang, J.; Cai, W.; Ma, M.; Liu, C.; Ma, X.; Li, L.; Chen, X. Driving forces of China's CO_2 emissions from energy consumption based on Kaya-LMDI methods. *Sci. Total Environ.* **2020**, *711*, 134569. [CrossRef] [PubMed]
43. Zheng, S.; Wu, J.; Kahn, M.E.; Deng, Y. The nascent market for "green" real estate in Beijing. *Eur. Econ. Rev.* **2012**, *56*, 974–984. [CrossRef]
44. Zhang, S.; Zhao, T. Identifying major influencing factors of CO_2 emissions in China: Regional disparities analysis based on STIRPAT model from 1996 to 2015. *Atmos. Environ.* **2019**, *207*, 136–147. [CrossRef]
45. Wang, Q.; Chiu, Y.H.; Chiu, C.R. Driving factors behind carbon dioxide emissions in China: A modified production-theoretical decomposition analysis. *Energy Econ.* **2015**, *51*, 252–260. [CrossRef]
46. Quan, C.; Cheng, X.; Yu, S.; Ye, X. Analysis on the influencing factors of carbon emission in China's logistics industry based on LMDI method. *Sci. Total Environ.* **2020**, *734*, 138473. [CrossRef]
47. He, Y.; Xing, Y.; Zeng, X.; Ji, Y.; Hou, H.; Zhang, Y.; Zhu, Z. Factors influencing carbon emissions from China's electricity industry: Analysis using the combination of LMDI and K-means clustering. *Environ. Impact Assess. Rev.* **2022**, *93*, 106724. [CrossRef]
48. Available online: https://www.iso.org/iso-50001-energy-management.html (accessed on 28 December 2024).

Disclaimer/Publisher's Note: The statements, opinions and data contained in all publications are solely those of the individual author(s) and contributor(s) and not of MDPI and/or the editor(s). MDPI and/or the editor(s) disclaim responsibility for any injury to people or property resulting from any ideas, methods, instructions or products referred to in the content.

Article

Analysis of the Impact of Turn Signal Usage at Roundabouts on CO Emissions and Traffic Flows

Monika Ziemska-Osuch *[ ] and Dawid Osuch

Department of Modeling and Mathematical Methods in Transportation, Gdynia Maritime University, 81-225 Gdynia, Poland
* Correspondence: m.ziemska@wn.umg.edu.pl

Abstract: In contemporary times, one of the challenges in road traffic is the failure of drivers to adhere to traffic regulations. While the use of turn signals may seem trivial, the studies presented in this article demonstrate the significant impact that this practice can have on road traffic, both in terms of urban network capacity and fuel consumption, which consequently affects the emission of exhaust gases into the natural environment. A common example of the failure to use turn signals is when drivers navigate roundabouts. The example presented here analyzes an existing road network configuration consisting of two roundabouts and an intersection between them. The method of microsimulation was employed using the PTV Vissim 25 software. This study examined a scenario where the percentage of drivers adhering to the use of turn signals increased by 10% in each successive simulation version. The results clearly indicate that the capacity of the network, as well as the emissions of road pollutants, depend not only on traffic volume but also, most importantly, on traffic flow efficiency.

Keywords: sustainable transport; fuel consumption; transport microsimulation; transport modeling for a cleaner environment

Citation: Ziemska-Osuch, M.; Osuch, D. Analysis of the Impact of Turn Signal Usage at Roundabouts on CO Emissions and Traffic Flows. *Energies* **2024**, *17*, 6145. https://doi.org/10.3390/en17236145

Academic Editors: Kazimierz Lejda, Artur Jaworski and Maksymilian Mądziel

Received: 11 November 2024
Revised: 2 December 2024
Accepted: 3 December 2024
Published: 6 December 2024

Copyright: © 2024 by the authors. Licensee MDPI, Basel, Switzerland. This article is an open access article distributed under the terms and conditions of the Creative Commons Attribution (CC BY) license (https://creativecommons.org/licenses/by/4.0/).

1. Introduction

In contemporary times, one of the most crucial aspects of road networks is adherence to traffic regulations. These rules are established in specific cases to ensure that every road user can feel safe while traveling to their destination [1]. Therefore, these regulations are of utmost importance, especially for ensuring the safety of the most vulnerable road users, such as pedestrians and cyclists [2]. Nowadays, many studies explore the huge problem of driving behavior [3]. What was studied most often is aggressive driving [4], seat belt use [5], and using mobile phones during driving [6]. It has been observed that, in modern times, a significant issue is the failure of drivers to use turn signals when they are exiting a roundabout. Not many studies have investigated the rates of turn signal usage at multiline urban roundabouts, but one of them is an investigation about the factors affecting turn signal usage [7]. The reasons behind omitting turn signal usage remain largely unknown. Drivers are most likely engaging in this practice due to inattention or negligence. However, even such a seemingly minor issue can have a significant impact on urban traffic flow [8]. It is particularly frustrating when, as a driver, we wish to enter traffic on a roundabout, but the driver exiting the roundabout fails to signal their intention to leave, preventing us from completing our maneuver sooner despite being able to do so safely. Roundabouts are among the safest types of intersections in road traffic [9]; therefore, other factors are directly dependent on the actions of the drivers themselves. Fuel consumption [10] by vehicles is the result of numerous factors, including the driver's style of driving, their skills, and their experience. Sudden acceleration and braking contribute to increased fuel consumption and, consequently, a rise in exhaust emissions [11,12] into the environment [13]. Road conditions [14] also directly affect the number of pollutants released; thus, in cities striving for sustainable transport [15], one of the key objectives of traffic management is often the improvement of traffic flow [16]. Additional factors that influence fuel consumption are closely related to the vehicle itself, such as its weight, technical condition, and the type of engine. It is well known that a heavy-duty vehicle will consume more fuel than a small passenger car [17]. The main objective of this study is to show how a parameter such as measuring drivers using turn signals at a roundabout can affect the accuracy of microsimulation measurement results in road traffic.

2. Materials and Methods

In order to address the question of the actual impact of not using turn signals when exiting a roundabout to the right, a model was developed using PTV VISSIM software [18]. Vissim's traffic flow model is a stochastic, time step-based, microscopic model that treats driver–vehicle units as basic entities [19]. As the model accounts for psychological aspects as well as for physiological restrictions of drivers' perception, it is called the psycho-physical car-following model. It contains a psycho-physical car following model for longitudinal vehicle movement and a rule-based algorithm for lateral vehicle movement. This software is designed for microsimulation of road traffic. It employs a leader–following traffic model, specifically the Wiedemann model [20]. Microsimulation of road traffic is crucial in studies focused on specific cases within the road network. It utilizes very fine data aggregation and high model detail, as each road user plays an essential role in the scenarios under investigation. Microsimulation is applied for various aspects, such as assessing road capacity, verifying the functionality of traffic signals, considering changes to urban infrastructure, or examining the impact of different transport modes or road users on traffic flow. All of these factors can be tested in a properly constructed model to prevent incorrect implementations in the urban network. In the case under study, the model was developed based on sections of the road network, with traffic volume data added according to measurements taken by the Municipal Roads Authority of Warsaw, valid as of the specific date 24 October 2023 [21]. The data used are real and reflect the challenging traffic conditions during the morning peak hours. Subsequently, the type and directional structure of traffic within the network were incorporated. Due to the detailed nature of the study, data regarding bicycle crossings as well as pedestrian volumes at specific pedestrian crossings within the network were also included. The studied section of the road network does not feature traffic signals; thus, this element was excluded from the constructed model. In the prepared network, the traffic regulations applicable in Poland were implemented. On the studied section of the road network, these regulations follow the right-hand driving system. A parameter, referred to as Anticipate Routes, was introduced to distinguish the results, and it will be described in greater detail later in this chapter. This parameter was directly related to the "conflict areas" function within PTV VISSIM, which allowed us to determine which vehicle had priority at a conflict point. The driving behavior of vehicles approaching a conflict area shall produce the maximum capacity for a minor flow, without affecting vehicles in the major flow. Vehicles in the major flow might be hindered by vehicles on the merge lane, and the smaller the user-defined safety distance factor is, the greater the hindrance [19].

The model was then calibrated and validated. To ensure the accuracy of the number of vehicles exiting the network, the GEH statistic [22] was applied, and the results are shown in Table 1. GEH is a metric used for comparing two values of traffic flow—measured and modeled. The desired result should be below 5, which indicates that the GEH suggests the model accurately reflects the actual traffic volume.

The model was created using PTV VISSIM version 25. The model was built using real traffic measurement data from the city of Warsaw. Warsaw, the capital of Poland, is located in Central Europe. Figure 1 below shows the layout of the studied road segment. The studied road network consists of a roundabout, an intersection, and another roundabout. The morning peak period, characterized by exceptionally high traffic volume due to the proximity of residential areas, was examined. Some evaluation was conducted via node evaluation. Node evaluation was especially used to determine specific data from intersections without first having to define all sections manually in order to determine the data. In Figure 1, the measurement area of the node is highlighted by a black line. Table 2 shows the distance from the stop line to the edge of the node.

Table 1. Traffic Flow—GEH statistic results.

	Turn	Measurements (7:00–8:00 a.m.) [veh/h]	Measurements (8:00–9:00 a.m.) [veh/h]	Model (7:00–8:00 a.m.) [veh/h]	Model (8:00–9:00 a.m.) [veh/h]	GEH (7:00–8:00 a.m.) [-]	GEH (8:00–9:00 a.m.) [-]
46_NW	R	66	75	68	76	0.24	0.12
	T	45	83	40	84	0.77	0.11
	L	26	34	28	35	0.38	0.17
	B	0	0	0	0	0.00	0.00
46_SW	R	309	429	313	428	0.23	0.05
	T	384	375	387	377	0.15	0.10
	L	19	22	20	24	0.23	0.42
	B	19	15	17	11	0.47	1.11
46_SE	R	80	18	77	16	0.34	0.49
	T	83	99	81	103	0.22	0.40
	L	444	564	449	562	0.24	0.08
	B	3	4	6	1	1.41	1.90
46_NE	R	27	68	32	69	0.92	0.12
	T	493	406	491	404	0.09	0.10
	L	26	9	31	9	0.94	0.00
	B	0	0	0	0	0.00	0.00
47_NW	R	84	109	80	109	0.44	0.00
	T	219	276	214	280	0.34	0.24
	L	135	177	134	177	0.09	0.00
	B	8	13	11	18	0.97	1.27
47_SW	R	39	46	36	43	0.49	0.45
	T	375	330	379	331	0.21	0.06
	L	75	53	76	57	0.12	0.54
	B	2	0	3	0	0.63	0.00
47_SE	R	92	74	89	73	0.32	0.12
	T	579	572	576	576	0.12	0.17
	L	119	80	114	79	0.46	0.11
	B	4	3	8	2	1.63	0.63
47_NE	R	444	483	449	485	0.24	0.09
	T	342	292	342	296	0.00	0.23
	L	58	70	61	74	0.39	0.47
	B	5	5	8	1	1.18	2.31

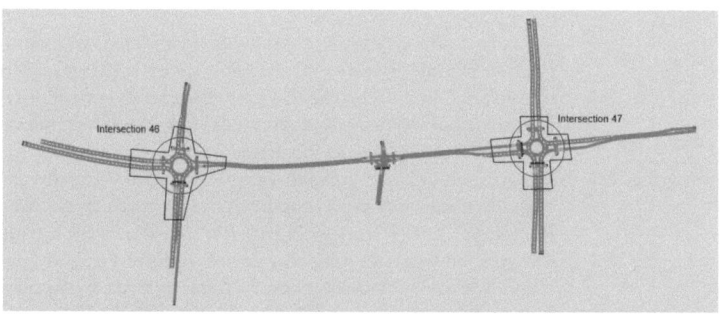

Figure 1. Modeled road network.

Table 2. Node areas.

Inlet	46_NW	46_SW	46_SE	46_NE	47_NW	47_SW	47_SE	47_NE
distance [m]	40	62	66	54	34	54	50	33

Roundabout 1 (52.145412, 21.051792) was designated as Intersection 47 for the purposes of this study. This roundabout consists of the following: NW (Komisji Edukacji Narodowej Avenue)—two traffic lanes on the approach, including one lane for straight and right turns, and one lane for straight and left turns; NE (Płaskowickiej Street)—two traffic lanes on the approach, including an additional, dedicated lane for right turns, and one lane for straight, right, and left turns; SE (Komisji Edukacji Narodowej Avenue)—two traffic lanes on the approach, including one lane for straight and right turns, and one lane for straight and left turns; SW (Płaskowickiej Street)—two traffic lanes on the approach, including an additional, dedicated lane for straight, right, and left turns, and one lane for right turns. Pedestrian crossings are present at each approach. Bicycle crossings are located at the SW (Płaskowickiej Street) and SE (Komisji Edukacji Narodowej Avenue) approaches. A visual representation of the roundabout is shown in the diagram below—Figure 2.

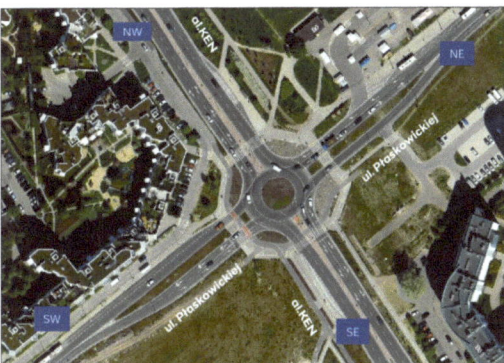

Figure 2. Intersection 47 [21].

The traffic composition at Intersection 46 was as follows: 1.1% motorcycles, 91.9% passenger vehicles, 4.5% delivery vehicles, 1% buses, 1.4% trucks, and 0.1% articulated vehicles (trucks with trailers/semi-trailers).

Roundabout 2 (52.142270, 21.044947) in the model was designed as Intersection 46. This roundabout consists of four entry points: North-West (Dereniowa Street)—Two lanes on the entry, including a dedicated right-turn lane, and one lane for straight-through and left turns. North-East (Płaskowickiej Street)—One lane on the entry. South-East (Stryjeńskich Street)—Two lanes on the entry, including a dedicated left-turn lane and one lane for straight-through, right, and left turns. South-West (Płaskowickiej Street)—Two lanes on the entry, including one lane for right turns and one lane for straight-through, right, and left turns. Pedestrian crossings are present at each entry point. There are no bicycle crossings. A visual representation of the roundabout is shown in the figure below—Figure 3.

The vehicle composition was as follows: motorcycles, 0.6%; passenger cars, 91.8%; delivery vehicles, 4.6%; buses, 1.2%; trucks, 1.6%; and articulated trucks with trailers/semi-trailers, 0.2%.

Between the roundabouts, there is an intersection that is primarily transit-oriented. The traffic flow from the subordinate entries is low, as these are mainly access points to local small parking areas.

Figure 3. Intersection 46 [21].

The model utilized the Scenario Management option. Each scenario included an increment of 10% in the Anticipate Routes parameter. The accompanying software manual explains this attribute as follows: "Percentage of vehicles required to yield that account for the routes of vehicles with the right of way. These vehicles approach with the major flow and will turn further upstream, meaning they will not reach the conflict area. For example, this can be used to determine the percentage of vehicles entering a roundabout because they trust that priority vehicles will exit the roundabout before reaching the conflict area, based on the turn signal of the priority vehicle. The value range is from 0 to 100%, with the default set at 0%. This attribute does not affect pedestrians who do not have to wait. For pedestrians who are required to wait, the following applies: 100% means all pedestrians required to wait to observe the routes of approaching vehicles with the right of way in the major flow. Less than 100% means pedestrians required to wait do not observe the routes of approaching vehicles with the right of way in the major flow". Thus, starting with a value of 0, representing 0% of drivers paying attention to the turn signals of vehicles turning right at the roundabout, this scenario reflects the worst traffic conditions. In this case, the driver entering the roundabout would be forced to stop before making the entry maneuver until ideal traffic conditions are achieved. At 100%, where 100% of drivers pay attention to the turn signals of other drivers, theoretical conditions would result in the best possible traffic flow. In this scenario, the movement would be smooth. A total of 11 variants of the microsimulation were tested. For each scenario, the traffic intensity was identical and matched the measured actual traffic intensities. In these prepared scenarios, fuel consumption was measured at each of the studied roundabouts. It should be noted that the PTV Vissim software measures this in gallons. Since in Poland, as the unit of measurement for fuel is liters, the results were converted from gallons to liters. Simultaneously, the total carbon monoxide emissions (CO) [23–25] were measured at each of the studied roundabouts.

3. Results

The measurement results for the morning peak traffic intensity were obtained by recording parameters during the period from 7:30 a.m. to 8:30 a.m. Additionally, the model employed a network filling method for the first 15 min, with traffic intensity consistent with real-world conditions. This was followed by two hours representing the morning peak and an additional 15 min of simulation for a cool-down phase to stabilize the network. In total, the simulation lasted 2.5 h, lasting from 7:15 a.m. to 9:00 a.m.

CO Emissions

In the first part of the microsimulation analysis, CO emissions produced by vehicles in the studied network are compared to the number of vehicles. It is important to note that the number of vehicles, or traffic intensity, is consistent across all tested scenarios. The discrepancies observed concerning the percentage of anticipated routes are due to the fact that this analysis was conducted during the morning peak hours. As a result, varying numbers of vehicles were able to reach the control points, where traffic counters were placed. It is crucial to understand that the number of vehicles entering the network is directly influenced by several factors, such as traffic flow, intensity, and the ability to perform certain maneuvers, such as turning or going straight.

At Intersection 46, the total number of vehicles at the measurement control points ranges from a minimum of 2141 to a maximum of 2190 vehicles, resulting in a difference of nearly 50 vehicles per hour. Similarly, the total CO emissions fluctuate within a range of 3764 g (minimum) to 4559 g (maximum). CO emission results are directly related to traffic intensity. Therefore, such a comparative analysis is presented in Figure 4. At Intersection 47, the discrepancies in the results are even more pronounced. The total number of vehicles at the measurement control points ranges from a minimum of 2568 to a maximum of 2625 vehicles, leading to a difference of 57 vehicles per hour. Similarly, the total CO emissions range from a minimum of 3747 g to a maximum of 8276 g. As with Intersection 46, at Intersection 47, the results show a decreasing trend with each subsequent scenario (Figure 5).

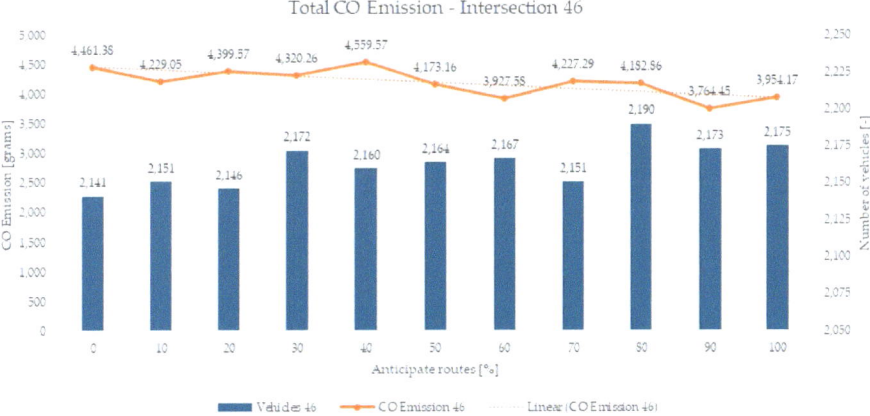

Figure 4. Analysis of all CO emission variants according to the number of vehicles—Intersection 46.

Since the comparative results of the number of vehicles and CO emissions were not conclusive, the microsimulation model also analyzed the total number of vehicle stops at the studied roundabouts. This indicator, along with the number of vehicles identified in the previous study, revealed that at Intersection 46, the difference between the highest and lowest number of stops was nearly 950 (Figure 6). On the other hand, for Junction 47, these differences were much more pronounced and began to drastically decrease only when the share of vehicles in the Anticipate Routes indicator reached 80%. The difference between the highest and lowest number of stops at this junction was almost 6500 (Figure 7). Further detailed analysis in the microsimulation model was also conducted with a breakdown by individual entries. Since the entries to the roundabouts significantly differed in terms of both geometry and traffic intensity during the morning peak hours, the CO emissions at these entries also varied considerably.

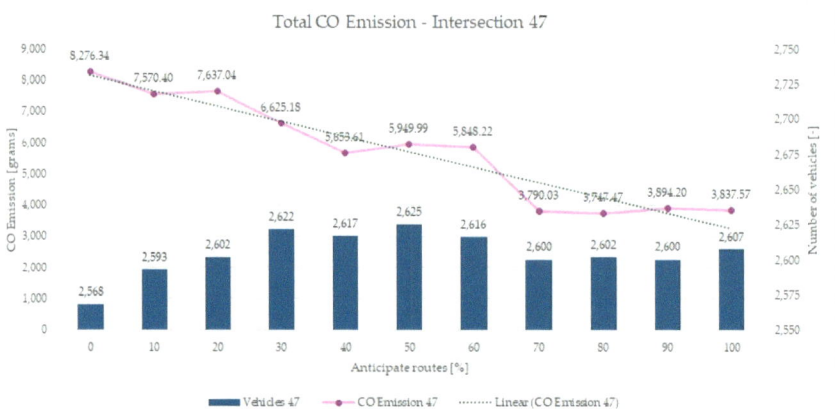

Figure 5. Analysis of all CO emission variants according to the number of vehicles—Intersection 47.

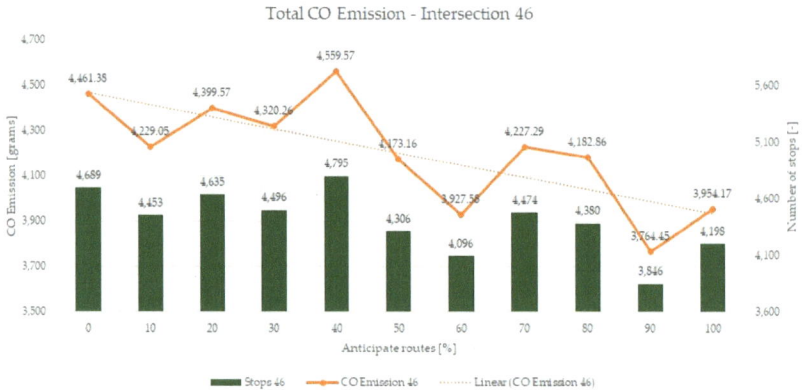

Figure 6. Analysis of all CO emission variants according to the number of stops—Intersection 46.

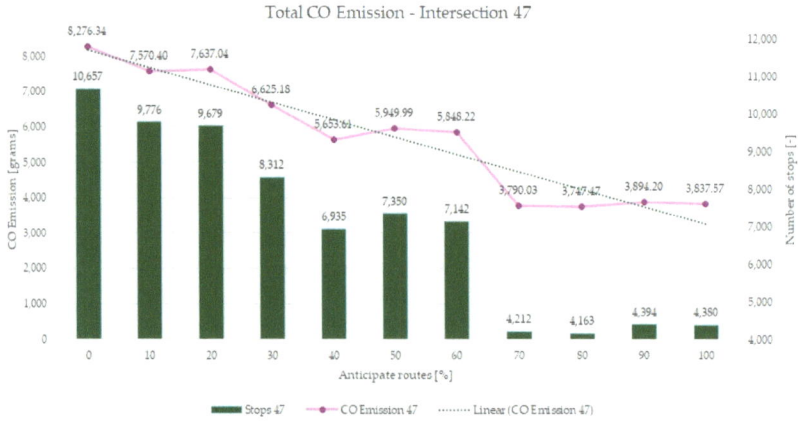

Figure 7. Analysis of all CO emission variants according to the number of stops—Intersection 47.

Factors such as traffic volume, the share of turning movements, and the influence of other entries and their turning movements also contribute to these results. The entry to a roundabout is not independent in itself, as is the case with traffic signals and directional signals. When comparing the entries at Intersection 46, it is evident that only the SE entry is dominant, and it is primarily responsible for the majority of CO emissions at this studied roundabout (Figure 8). A similar situation occurs at Intersection 47, where the SE entry also has the greatest impact on the emission of pollutants. Additionally, at this entry, it is observable that as the number of drivers using turn signals increases, the emission of pollutants related to individual transportation decreases (Figure 9).

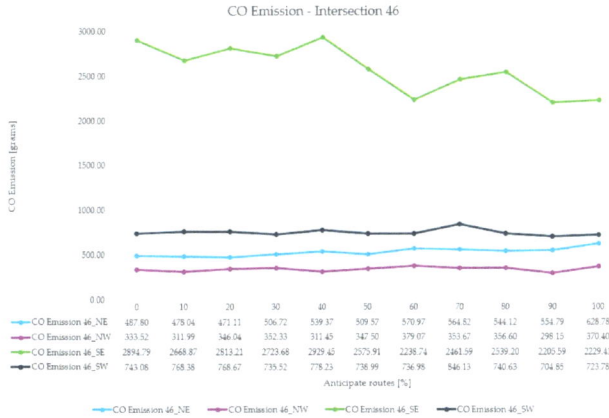

Figure 8. CO Emission—Intersection 46 inlets comparison.

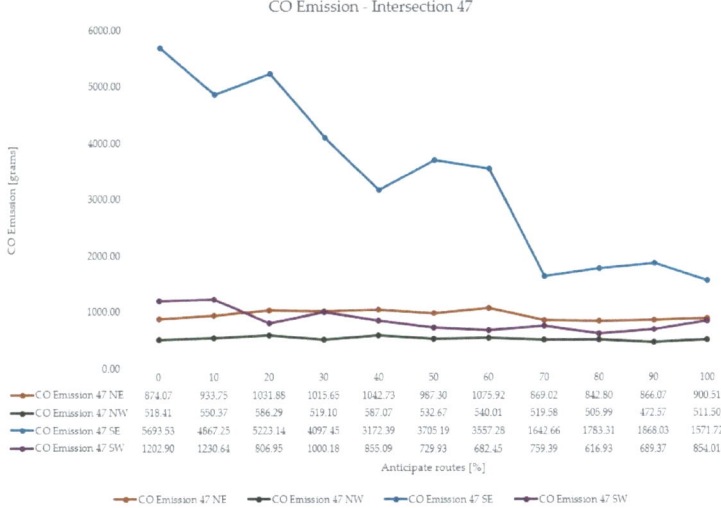

Figure 9. CO Emission—Intersection 47 inlets comparison.

The results are directly correlated with the traffic volume and directional structure at the roundabout. It is well established that vehicles intending to exit to the right from their entry point spend the least time on the roundabout, as they have the shortest distance to cover. These vehicles also have the least impact on other directions and entry points. In scenarios such as the one analyzed in this study, where the highest traffic volume occurs at the SE entry point and the majority of vehicles make a left turn, this significantly influences all measured traffic parameters. In a microscopic simulation model, where the research methodology is based on the follow-the-leader approach, these factors are of critical importance. In cases similar to the one presented, it is not feasible to expect linear increases or decreases in the results due to the numerous variables present in the modeled road network. The findings highlight the significance of considering the percentage of drivers using turn signals when studying CO emissions.

4. Conclusions

The presented study illustrates how a seemingly simple issue, such as the use of turn signals when exiting a roundabout, can significantly affect traffic flow, and consequently the environmental pollution associated with vehicle emissions. The use of turn signals plays a crucial role in facilitating smooth traffic movement by improving communication between road users, reducing uncertainty, and preventing abrupt maneuvers. As a result, this leads to more efficient traffic circulation, reducing congestion and the need for sudden acceleration or braking, which in turn decreases fuel consumption and lowers the levels of harmful exhaust emissions. The results presented reveal a clear correlation: when more than 30% of drivers cease to use their turn signals while exiting a roundabout, particularly when combined with an entry point experiencing the highest traffic volume, there is a significant impact on emissions and air pollution. This study shows that this behavior leads to an increase in the average number of stops, which in turn raises fuel consumption. With each additional percentage of drivers failing to comply with the regulations in this scenario, the conditions worsen further, amplifying the negative effects on both traffic flow and environmental pollution. The results available in the work refer to those in which the speed of vehicles with the parameters of the infrastructure itself was tested [26]. A similar convergence of results can also be seen in [27], where the authors wrote "emission data obtained by crossing the roundabout, it can be noticed an increased emission of harmful exhaust components, at the moment of acceleration at entries, and the driving around the island, as well as at the exit points, wherein particular the highest CO_2 emission occurs". The next similarity in results can be found in publication [28], where authors measured quantities of traffic and emission impacts of single-lane roundabouts in urban corridors.

The most unlikely scenario, where no one adheres to traffic regulations, is unlikely to ever occur in reality. However, it serves as a reference point for comparing the remaining measurement data. By considering this extreme case, it becomes possible to better understand the impact of varying degrees of non-compliance with traffic laws on traffic flow, fuel consumption, and environmental factors.

The presented microsimulation study highlights another crucial relationship. Specifically, it suggests that future general measurement studies should consider incorporating the tracking of drivers who fail to use their turn signals. Due to significant variations in vehicle flow within the network during peak traffic periods (in this case, the morning rush), it becomes apparent how this factor influences the measured data. A well-calibrated and validated model is essential to obtain reliable and meaningful simulation results. Such an approach ensures that the data reflect real-world conditions more accurately, allowing for better insights into the impacts of driver behavior on traffic flow and environmental factors. This underscores the importance of incorporating behavioral elements, such as the use of turn signals, into traffic simulation models to enhance the quality and applicability of the results. The presented study also demonstrates the validity of implementing public awareness campaigns highlighting the negative consequences of failing to use turn signals when exiting a roundabout. Such campaigns would help to inform drivers about how this simple yet crucial action can significantly improve traffic flow, reduce fuel consumption, and lower emissions. By addressing the behavioral aspects of road safety, these campaigns could contribute to safer, more efficient driving practices, ultimately benefiting both traffic efficiency and environmental sustainability.

Further research in this area is needed to be able to enter the comparison page of the capabilities of the PTV Vissim software for determining priority using priority rules, where without the use of additional external programmable parameters, it is not possible to adjust the number of drivers complying with the use of the turn signal compared to the method used in this article.

Author Contributions: Conceptualization, M.Z.-O.; Methodology, M.Z.-O.; Software, D.O.; Validation, D.O.; Formal analysis, M.Z.-O. and D.O.; Investigation, M.Z.-O.; Resources, D.O.; Data curation, D.O.; Writing—original draft, M.Z.-O.; Writing—review & editing, M.Z.-O. All authors have read and agreed to the published version of the manuscript.

Funding: This research was funded by the statutory activities of Gdynia Maritime University, grant number WN/2024/PZ/06.

Data Availability Statement: Data available in a publicly accessible repository that does not issue DOIs. Publicly available datasets were analyzed in this study. This data can be found here: https://zdm.waw.pl/dzialania/badania-i-analizy (accessed on 10 November 2024).

Conflicts of Interest: The author declares no conflicts of interest.

Abbreviations

CO	Carbon monoxide
E	East
N	North
S	South
W	West
R	Right (turn)
T	Through movement
L	Left (turn)
B	Back (turn)

References

1. Alkaabi, K. Identification of Hotspot Areas for Traffic Accidents and Analyzing Drivers' Behaviors and Road Accidents. *Transp. Res. Interdiscip. Perspect.* **2023**, *22*, 100929. [CrossRef]
2. Alyamani, H.; Alharbi, N.; Robooey, A.; Kavakli, M. The Impact of Gamifications and Serious Games on Driving under Unfamiliar Traffic Regulations. *Appl. Sci.* **2023**, *13*, 3262. [CrossRef]
3. Mikoski, P.; Zlupko, G.; Owens, D.A. Drivers' Assessments of the Risks of Distraction, Poor Visibility at Night, and Safety-Related Behaviors of Themselves and Other Drivers. *Transp. Res. Part F Traffic Psychol. Behav.* **2019**, *62*, 416–434. [CrossRef]
4. Luo, X.; Ge, Y.; Qu, W. The Association between the Big Five Personality Traits and Driving Behaviors: A Systematic Review and Meta-Analysis. *Accid. Anal. Prev.* **2023**, *183*, 106968. [CrossRef]
5. Taylor, N.L.; Daily, M. Self-Reported Factors That Influence Rear Seat Belt Use among Adults. *J. Saf. Res.* **2019**, *70*, 25–31. [CrossRef]
6. Li, W.; Huang, J.; Xie, G.; Karray, F.; Li, R. A Survey on Vision-Based Driver Distraction Analysis. *J. Syst. Arch.* **2021**, *121*, 102319. [CrossRef]
7. Muley, D. Investigation of Factors Affecting Turn Signal Usage at Modern Roundabouts in State of Qatar. *Urban Plan. Transp. Res.* **2023**, *11*, 2234974. [CrossRef]
8. Thaker, P.; Gokhale, S. The Impact of Traffic-Flow Patterns on Air Quality in Urban Street Canyons. *Environ. Pollut.* **2016**, *208*, 161–169. [CrossRef]
9. Li, L.; Zhang, Z.; Xu, Z.G.; Yang, W.C.; Lu, Q.C. The Role of Traffic Conflicts in Roundabout Safety Evaluation: A Review. *Accid. Anal. Prev.* **2024**, *196*, 107430. [CrossRef]
10. Ziemska, M. Exhaust Emissions and Fuel Consumption Analysis on the Example of an Increasing Number of HGVs in the Port City. *Sustainability* **2021**, *13*, 7428. [CrossRef]
11. Ou, Y.; West, J.J.; Smith, S.J.; Nolte, C.G.; Loughlin, D.H. Air Pollution Control Strategies Directly Limiting National Health Damages in the US. *Nat. Commun.* **2020**, *11*, 957. [CrossRef] [PubMed]
12. Zhang, X.; Wang, Q.; Qin, W.; Guo, L. Sustainable Policy Evaluation of Vehicle Exhaust Control—Empirical Data from China's Air Pollution Control. *Sustainability* **2019**, *12*, 125. [CrossRef]
13. Jacyna, M.; Żochowska, R.; Sobota, A.; Wasiak, M. Scenario Analyses of Exhaust Emissions Reduction through the Introduction of Electric Vehicles into the City. *Energies* **2021**, *14*, 2030. [CrossRef]

14. Bergel-Hayat, R.; Zukowska, J. Road Safety Trends at National Level in Europe: A Review of Time-Series Analysis Performed during the Period 2000–12. *Transp. Rev.* **2015**, *35*, 650–671. [CrossRef]
15. Pojani, D.; Stead, D. Sustainable Urban Transport in the Developing World: Beyond Megacities. *Sustainability* **2015**, *7*, 7784–7805. [CrossRef]
16. Lee, S.; Oh, J.; Kim, M.; Lim, M.; Yun, K.; Yun, H.; Kim, C.; Lee, J. A Study on Reducing Traffic Congestion in the Roadside Unit for Autonomous Vehicles Using BSM and PVD. *World Electr. Veh. J.* **2024**, *15*, 117. [CrossRef]
17. Ziemska-Osuch, M.; Guze, S. Analysis of the Impact of Road Traffic Generated by Port Areas on the Urban Transport Network—Case Study of the Port of Gdynia. *Appl. Sci.* **2022**, *13*, 200. [CrossRef]
18. Gunarathne, D.; Amarasingha, N.; Wickramasighe, V. Traffic Signal Controller Optimization Through VISSIM to Minimize Traffic Congestion, CO and NOx Emissions, and Fuel Consumption. *Sci. Eng. Technol.* **2023**, *3*, 9–21. [CrossRef]
19. PTV Vissim and PTV Viswalk Help. Available online: https://cgi.ptvgroup.com/vision-help/VISSIM_2022_ENG/Content/11_Auswertungen/Ausw_a_ausfuehren.htm (accessed on 17 June 2022).
20. Chaudhari, A.A.; Srinivasan, K.K.; Chilukuri, B.R.; Treiber, M.; Okhrin, O. Calibrating Wiedemann-99 Model Parameters to Trajectory Data of Mixed Vehicular Traffic. *Transp. Res. Rec. J. Transp. Res. Board* **2021**, *2676*, 718–735. [CrossRef]
21. Warszawa, Z.D.M. Badania i Analizy. Available online: https://zdm.waw.pl/dzialania/badania-i-analizy/ (accessed on 11 November 2024).
22. de Villa, A.R.; Casas, J.; Breen, M.; Perarnau, J. Static OD Estimation Minimizing the Relative Error and the GEH Index. *Procedia Soc. Behav. Sci.* **2014**, *111*, 810–818. [CrossRef]
23. Cobley, L.A.E.; Pataki, D.E. Vehicle Emissions and Fertilizer Impact the Leaf Chemistry of Urban Trees in Salt Lake Valley, UT. *Environ. Pollut.* **2019**, *254*, 112984. [CrossRef] [PubMed]
24. Ravi, S.S.; Osipov, S.; Turner, J.W.G. Impact of Modern Vehicular Technologies and Emission Regulations on Improving Global Air Quality. *Atmosphere* **2023**, *14*, 1164. [CrossRef]
25. Barth, M.; Boriboonsomsin, K. Real-World Carbon Dioxide Impacts of Traffic Congestion. *Transp. Res. Rec.* **2008**, *2058*, 163–171. [CrossRef]
26. Fernandes, P.; Tomás, R.; Acuto, F.; Pascale, A.; Bahmankhah, B.; Guarnaccia, C.; Granà, A.; Coelho, M.C. Impacts of Roundabouts in Suburban Areas on Congestion-Specific Vehicle Speed Profiles, Pollutant and Noise Emissions: An Empirical Analysis. *Sustain. Cities Soc.* **2020**, *62*, 102386. [CrossRef]
27. Jaworski, A.; Mądziel, M.; Lejda, K. Creating an Emission Model Based on Portable Emission Measurement System for the Purpose of a Roundabout. *Environ. Sci. Pollut. Res.* **2019**, *26*, 21641–21654. [CrossRef]
28. Coelho, M.C.; Farias, T.L.; Rouphail, N.M. Effect of Roundabout Operations on Pollutant Emissions. *Transp. Res. Part D Transp. Environ.* **2006**, *11*, 333–343. [CrossRef]

Disclaimer/Publisher's Note: The statements, opinions and data contained in all publications are solely those of the individual author(s) and contributor(s) and not of MDPI and/or the editor(s). MDPI and/or the editor(s) disclaim responsibility for any injury to people or property resulting from any ideas, methods, instructions or products referred to in the content.

Article

The Possibility of Using Hydrogen as a Green Alternative to Traditional Marine Fuels on an Offshore Vessel Serving Wind Farms

Monika Bortnowska * and Arkadiusz Zmuda *

Department of Naval Architecture and Shipbuilding, Faculty of Navigation, Maritime University of Szczecin, 70-500 Szczecin, Poland
* Correspondence: m.bortnowska@pm.szczecin.pl (M.B.); arkadiusz.zmuda@pm.szczecin.pl (A.Z.)

Abstract: Achieving the required decarbonisation targets by the shipping industry requires a transition to technologies with zero or near-zero greenhouse gas (GHG) emissions. One promising shipping fuel with zero emission of exhaust gases (including CO_2) is green hydrogen. This type of fuel, recognised as a 100% clean solution, is being investigated for feasible use on a service offshore vessel (SOV) working for offshore wind farms. This study aims to examine whether hydrogen may be used on an SOV in terms of the technical and economic challenges associated with the design process and other factors. In the analyses, a reference has been made to the current International Maritime Organization (IMO) guidelines and regulations. In this study, it was assumed that hydrogen would be directly combusted in a reciprocating internal combustion engine. This engine type was reviewed. In further research, hydrogen fuel cell propulsion systems will also be considered. The hydrogen demand was calculated for the assumed data of the SOV, and then the volume and number of high-pressure tanks were estimated. The analyses revealed that the SOV cannot undertake 14-day missions using hydrogen fuel stored in cylinders on board. These cylinders occupy 66% of the ship's current volume, and their weight, including the modular system, accounts for 62% of its deadweight. The costs are over 100% higher compared to MDO and LNG fuels and 30% higher than methanol. The actual autonomy of the SOV with hydrogen fuel is 3 days.

Keywords: hydrogen fuel; hydrogen-powered marine engine; composite tanks; high-pressure type IV; IGF code; SOV vessel; hydrogen fuel costs

Citation: Bortnowska, M.; Zmuda, A. The Possibility of Using Hydrogen as a Green Alternative to Traditional Marine Fuels on an Offshore Vessel Serving Wind Farms. *Energies* **2024**, *17*, 5915. https://doi.org/10.3390/en17235915

Academic Editors: Kazimierz Lejda, Artur Jaworski and Maksymilian Mądziel

Received: 1 November 2024
Revised: 21 November 2024
Accepted: 22 November 2024
Published: 25 November 2024

Copyright: © 2024 by the authors. Licensee MDPI, Basel, Switzerland. This article is an open access article distributed under the terms and conditions of the Creative Commons Attribution (CC BY) license (https://creativecommons.org/licenses/by/4.0/).

1. Introduction

1.1. Research Background

Nearly 60–70% of global CO_2 emissions come from the combustion of fuels in various economic sectors, with maritime transport accounting for about 3.5% of all greenhouse gas emissions. According to the IMO [1], maritime transport emissions will continue to increase by up to 50% by 2050 if the current 99% of transport shipping fuel demand (covered by fossil fuels) is not drastically reduced in favour of alternative fuels. Without extensive action, emissions could reach up to 130% of 2008 levels by 2050 [2]. Figure 1 shows CO_2 emissions from 2012 to 2023, expressed in tonnes, associated with maritime transport by major ship types [2].

If maritime transport is to meet decarbonisation targets—whether set by the IMO [3], the EU or other national regulations—through a transition to technologies with zero or near-zero greenhouse gas (GHG) emissions (i.e., electric, hybrid or fuel cell drives), the requirements include using clean fuels (i.e., methanol, ammonia, hydrogen or biofuels) and/or alternative energy sources (i.e., solar and wind power or nuclear power). Figure 2 depicts the IMO's current strategy for reducing GHGs from maritime transport, which uses 2008 as the baseline and should achieve net zero by 2050.

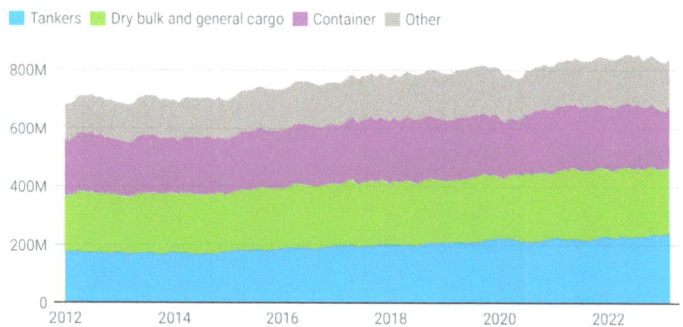

Figure 1. Carbon dioxide emissions by main vessel types, tons, 2012–2023, acc. [2]. Note: The group 'other' includes vehicles and roll-on/roll-off ships, passenger ships, offshore ships and service and miscellaneous ships.

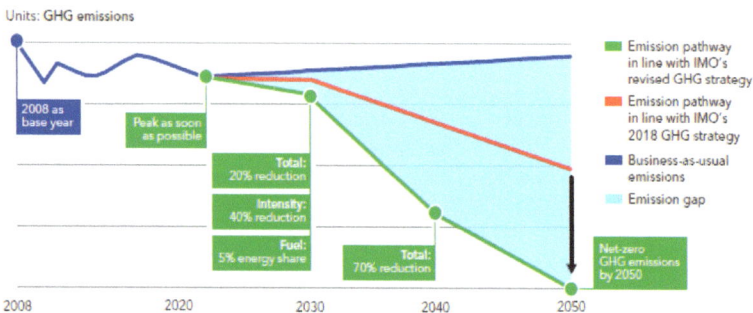

Figure 2. Outline of ambitions and minimum indicative checkpoints in the revised IMO GHG strategy, acc. [3].

According to [4], hydrogen is currently becoming increasingly important in the shipping sector, although implementation is still at an early stage. The current fleet of hydrogen-powered vessels is relatively small, and most projects are in the demonstration or development phase [5]. However, orders for vessels with alternative propulsion systems represent an increasing share, with around 49% of new orders involving vessels using various forms of alternative fuels [4]. According to [6], 135 LNG-capable vessels have been ordered, and current orders also include 249 vessels adapted to use ammonia as fuel, 247 vessels adapted to use methanol and 14 vessels adapted to run on hydrogen.

Orders for new ships definitely show a clear trend towards alternative fuels, and the most promising zero-emission marine fuels include green hydrogen. For this reason, the authors have chosen to examine this new fuel, with great potential for SOV propulsion. Hydrogen can be produced from renewable energy sources, so the offshore wind farms in the Baltic Sea assumed in this study as the working area of the vessel could become a potential source of hydrogen, produced by electrolysis using the electricity generated by the offshore wind farms. This hydrogen could be used to power service vessels operating in the vicinity of these farms.

Recent studies indicate that global hydrogen demand has reached 97 Mt in 2023, an increase of 2.5% compared to 2022 [4], with the majority being met by hydrogen produced from fossil fuels. Green hydrogen is currently less available (its production was less than 1 Mt in 2023). Although the production costs are high relative to grey and black hydrogen, it is expected to be commercialised and have competitive prices [7].

According to these sources [8,9], the world's first zero-emission, small passenger ferry using hydrogen fuel cells, batteries and electric propulsion, is the Switch Maritime-owned catamaran 'Sea Change', which began operating in the San Francisco Bay area in July 2024.

It has 10 high-pressure (250 bar), onboard hydrogen tanks (from Hexagon) that store 246 kg of compressed hydrogen.

In Europe, the first certified vessel was a 14-metre-long passenger ship 'Hydroville', powered by hydrogen and diesel oil. Launched in 2017 for CMB Technologies, the ferry is equipped with two internal combustion engines running on hydrogen and diesel, with a total output of 441 kW. The hydrogen fuel is transported in 12,205-litre tanks at 200 bar pressure and two diesel tanks of 265 litres each [10].

According to [7,11], in recent years, several new and considerably larger hydrogen-powered vessels have entered the market, and a group of other vessels are at the design and research stage. One example, the passenger/car ferry MF 'Hydra', over 80 m long, made its first trip off the coast of Norway on liquid hydrogen fuel in March 2023. The vessel can accommodate up to 300 passengers and 80 vehicles. It carries 4000 kg of hydrogen fuel and is equipped with two 200 kW fuel cells, two 440 kW generating sets and two Shottel thrusters [12].

In addition, a new SOV concept has emerged, featuring a liquid hydrogen power plant, jointly developed by shipowner Louis Dreyfus Armateurs (LDA) and Norwegian company Salt Ship Design. It is estimated [13,14] that the ship will be able to operate 95% of the time with zero CO_2 emissions and will only release water during standard operations. This would have a positive impact on the emissions associated with the operation of offshore wind farms, preventing the release of around 4000 tonnes of CO_2 per year.

All the hydrogen-powered vessels built to date have resulted from new projects. There is no case in the literature of a vessel conversion or upgrade to run on hydrogen apart from this year's publication by Melideo D. et al. [15]. The authors examined the possibilities of converting ro–ro and ro–pax ferries operating on a short route (about 27 km) off the Italian coast. The scenarios included the replacement of the main and auxiliary engines with fuel cells, together with an analysis of the hydrogen demand and other parameters associated with the use of this fuel. However, the authors did not attempt to explore feasible layouts of the ferry interior as required by the use of fuel cells or hydrogen tanks and their impact on the mass parameters of the entire vessel.

The publication [16] presents a draft scenario for the integrated use of offshore hydrogen-powered vessels with polymer membrane fuel cells (PEMFCs) and offshore wind energy and provides a technical and economic feasibility analysis. This project addresses problems associated with offshore hydrogen transport and offshore wind energy development, including hydrogen storage. Another publication [17] presents an overview of recent research on hydrogen ship propulsion systems and various aspects of the production, transport, storage and use of liquid/gas H_2 and its derivatives as fuel in the shipping industry. The authors state that hydrogen propulsion in maritime transport is still in the experimental phase, and in most cases, these experiments serve as a kind of platform to evaluate the feasibility of different technological solutions. Panagiotis Karvounis et al. [18] made a comparative assessment of the potential for the use of alternative fuels for marine engines, in particular hydrogen, methanol and ammonia, as well as the associated environmental benefits. In addition, they assessed the required storage conditions, space and related costs, paying much attention to the safety issues and requirements for each alternative fuel. The results of this study show that the environmental benefits of alternative fuels are only visible when renewable energy is mainly used for fuel production, while the applicability of these fuels depends on the type of vessel and the relevant storage limitations. The article [19] analyses the possibility of replacing a conventional diesel generator installed on a small hybrid ferry with an innovative system based on PEMFC technology. To this end, the total energy/power demand of the vessel was considered using a typical operational profile, and a preliminary reconfiguration of the propulsion system was proposed along with an energy management strategy. The daily hydrogen consumption was determined. In addition, different storage technologies for compressed and liquefied phases were considered and compared to determine the ship's mass and space requirements.

The overview of current research shows that assessments made to assess the required storage space on ships for alternative fuels used in their energy systems mainly relate to hydrogen fuel cell energy systems, to a lesser extent addressing internal combustion engines. This is probably due to the fact that the advantages of using hydrogen in shipping were only confirmed a few years ago, so there are more questions than clear answers when it comes to the design and construction of ships adapted to this type of fuel. International or class regulations are still in the process of creation. Not surprisingly, there is a lack of information in the literature on the impact of the use of hydrogen fuel on this important issue:

- Design and construction changes to the ship's hull;
- Weight and volume of fuel tanks and their impact on the ship layout (engine room and adjacent compartments), as well as the possibility of installing them in the ship's hull;
- Solutions and configuration of the ship's energy system;
- Cargo carrying capacity and displacement of the vessel;
- Operational parameters, including vessel speed and range.

Over the last 10 years, considerations of alternative fuels in maritime transport, in terms of economic or environmental performance and the level of exhaust gas reduction, have been the subject of analyses presented in a number of publications [20–25]. However, these analyses are mainly concerned with transport vessels operating on established shipping routes. The use of hydrogen as a marine fuel for ships has only gained prominence in the past few years resulting in an increasing number of hydrogen-related topics in scientific publications as well as in the related industries.

One such example is the research [26] carried out on the experimental catamaran 'Energy Observer', which provided valuable information on the use of hydrogen fuel cells to propel this unit using renewable energy sources to produce hydrogen through electrolysis of seawater. On the other hand, the authors [27], using a coastal ferry (12,000 GT) as an example, presented the impact of several alternative fuels: LNG, MGO and hydrogen, by comparing them using a life cycle assessment (LCA) analysis. In the paper, they also made suggestions for a plan to use hydrogen for small coastal ships and to actually achieve a zero-emission ship. Similarly, in publication [28], the authors, using the LCA method, assessed the environmental impact of marine propulsion systems using hydrogen and ammonia as marine fuels. Selected for the study were two-stroke and four-stroke engines of tankers powered by conventional fuels and, for comparison, powered by 'green' and 'blue' hydrogen and ammonia with the support of pilot fuel. Whereas in [29], the authors broadly considered the safety of hydrogen storage. Discussing the advantages and disadvantages of hydrogen production technologies, the authors also highlighted the challenges that need to be overcome to safely use hydrogen as a marine fuel.

The key findings from the literature overview are as follows:

- Issues of using alternative fuels such as LNG or methanol (MeOH) are discussed, but the use of hydrogen fuel is still in an experimental phase, and research is limited to smaller coastal vessels;
- No information is available on the impact of fuel type on the design process and ship construction and performance, i.e., the main dimensions, functional–spatial layout of the hull, deadweight or total displacement of the ship;
- For special vessels, i.e., service vessels (apart from a few examples with design plans), there is a lack of scientific publications on the effects and possibilities of using hydrogen fuel with strict safety rules;
- There are no data on the use of hydrogen for direct combustion in reciprocating internal combustion engines used in ship energy systems and thus on the impact of such a solution on the functional and spatial parameters of the ship.

1.2. Hydrogen Properties, Storage and Maritime Transport

Hydrogen (H_2) is the lightest and most common chemical element in the universe. It also has the lowest density of all elements, which depends on its state of aggregation; i.e., in the gaseous phase, it is only 0.00009 g/cm^3; in the liquefied state, it is approx. 0.07 g/cm^3 and in the solid state, 0.08 g/cm^3 [30]. Under normal conditions (0 °C and 1013 hPa), hydrogen is an odourless and colourless gas. It is flammable and, with oxygen, forms an explosive mixture. However, its lightness means that rapid vaporisation occurs if the hydrogen tank is damaged—without posing a risk to humans. When burned in internal combustion engines, it does not pollute the air, as water is the only by-product when powering fuel cells.

The degree of purity of hydrogen, and thus its low or zero emissions, depends on the method of its production. In this research carried out, its purest form was considered, i.e., produced by electrolysis of water using electricity, which can be a 'green source' (when it comes, for example, from wind energy generated by an offshore wind farm). Other methods of production include processes based on steam reforming of natural gas or petrol reforming. These, however, result in by-products in the form of CO_2 [31]. Methods of hydrogen production are discussed in many research publications; hence, this study does not discuss this subject. The method of hydrogen production corresponds to terms associated with its colour: green is the purest hydrogen derived from renewable sources, grey is hydrogen derived from fossil fuels, with high CO_2 emissions, while blue denotes hydrogen derived like grey, with the process being combined with CO_2 capture and storage technology.

Hydrogen storage and mode of transport are key issues related to hydrogen production. On ships, hydrogen as a fuel is most often used in the liquid or gaseous phase:

- Liquefied hydrogen transported in cryogenic tanks (−253 °C);
- Compressed hydrogen transported in metal, high-pressure tanks (older solution) and in composite tanks or high-pressure vessels, i.e., tanks (recent solution).

In this article, the authors focus on an analysis of the use of hydrogen as a gaseous propellant, as hydrogen compression technology has been known for a long time and is well-developed (widely used in both industrial and automotive applications). This technology is simpler, requiring only special pressure vessels, without the use of a complex cooling system during hydrogen storage and transport—as is the case with liquefied hydrogen. The whole process of compression of hydrogen counts as an energy-intensive and costly process—one of the disadvantages of this storage method. From an environmental point of view, the preparation for transport and the transport of compressed hydrogen itself consumes less energy (6–25% for a pressure range of 250–700 bar), as the preparation of liquefied hydrogen requires about 21–30% of the total energy contained in the fuel [31,32].

Of course, in the case of liquefied hydrogen, the hydrogen still needs to be kept in a liquefied state, which will increase the energy consumption by a few percent in relation to the total energy contained in the fuel.

According to literature sources, both types of hydrogen storage systems are crucial to its future as a marine fuel, with liquefied hydrogen being more efficient on long shipping routes and compressed hydrogen offering greater flexibility for smaller, short-range ships. According to [31], it is assumed that this form of hydrogen can be most competitive in coastal region markets and in inter-regional areas where pipelines are technically difficult or not economically justified.

The low density of the gaseous form of hydrogen (under normal conditions) means that it requires a large volume for storage [33]. Compared to natural gas, hydrogen has as much as a three times lower energy content in the same volume. Therefore, when using hydrogen as an energy carrier in gaseous form, it is necessary to compress it in order to obtain a sufficiently large mass volume of hydrogen stored in a tank. Gaseous hydrogen is stored at room temperature (about 298 K) in the pressure range from 150 to 800 bar [34]. Increased pressures allow its energy density to be significantly raised, and thus, more fuel can be carried in a smaller volume, which in turn benefits the limited volume of the ship's hull. Figure 3 shows the hydrogen storage conditions as a function of temperature and pressure [35]. Unfortunately, even so, the low density of hydrogen contributes to the fact that, even at high pressure, little useful energy is stored—which necessitates very large tank volumes and increased storage costs. Consequently, to increase the operating range of seagoing vessels, tanks are required to withstand such high pressures while not increasing the ship's deadweight.

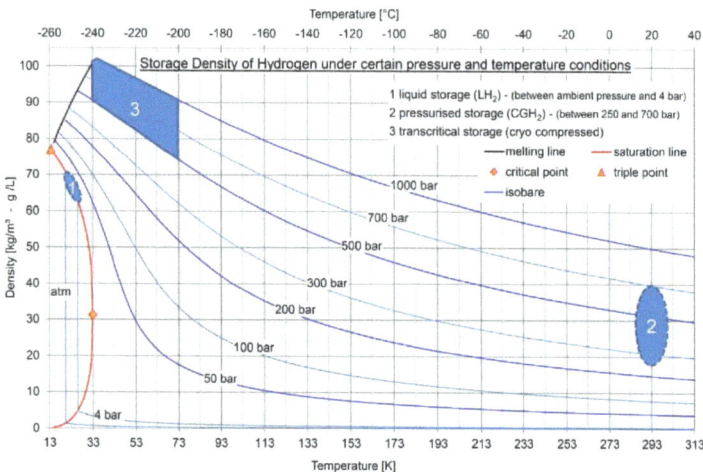

Figure 3. Hydrogen storage conditions as a function of temperature and pressure [35].

For the storage and transport of pressurised hydrogen fuel by ship, containerised sets or independent cylindrical tanks are used, mounted in special racks/frames. When a ship requires larger volumes of hydrogen, companies offer a modular system, both in terms of the diameter and length of such tanks. The choice of how to store the transported hydrogen fuel on the SOV and a broader discussion of the above technologies will be presented further in this study.

There are currently four categories of standard cylinders (types I–IV) for the storage of compressed hydrogen. The fifth generation is in the experimental and research phase (intended mainly for space applications) [36].

The first three categories are tanks made mainly of steel, relatively heavy and not suitable for marine transport use. Type II has composite components to reduce their weight, and type III tanks have a metal liner (usually aluminium) inside, surrounded by full composite cover (e.g., carbon fibre). This offers greater weight reduction compared to type II tanks, which have a metal liner and only partial composite reinforcement. This group of tanks is mainly used in the transport and automotive sectors.

The latest developments, the fourth generation of pressure vessels, are constructed entirely of carbon composite materials. These materials provide high strength and resistance to rusting and fatigue, form a protective barrier to hydrogen, are much lighter than steel tanks and retain a correspondingly large gravimetric volume of fuel [35]. According to [37,38], they have up to 70% less weight compared to type I tanks, so this type of tank is particularly desirable in reducing the cargo carrying capacity as little as possible in the ship under study.

The type IV tank is made by winding or braiding carbon fibres in an epoxy resin matrix over a plastic liner. It allows hydrogen to be stored at a pressure of 700 bar, and the amount of gas stored is equal to 12% of the tank weight. When the gas is compressed to a pressure of 700 bar, it achieves a volumetric volume in the tank that reaches 40.2 g/L [39]. Typical structural elements of a composite hydrogen tank are shown in Figure 4 [37].

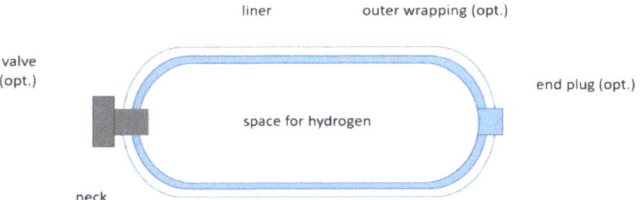

Figure 4. Typical components of a hydrogen tank—source: [37].

Methods of manufacturing this type of vessel vary, and many companies already offer innovative solutions for the production lines. Of the global manufacturers of composite pressure vessels, only a few offer products for the maritime industry. This article focuses on two manufacturers with extensive experience and advanced technological solutions:

- NPROXX, based in the Netherlands;
- Hexagon Purus, with a seat in Ålesund, Norway.

In Poland, research work is also in progress to develop composite tanks for high-pressure hydrogen storage. For example, the Amargo company, in collaboration with the Warsaw University of Technology and the Wrocław University of Technology, is developing pressurised hydrogen tanks for industrial applications [40]. The result of this work is expected to be high-capacity composite hydrogen tanks with a liner of HDPE and PA plastic. Amargo tries to obtain the maximum achievable lifetime (service life) of its tanks. The firm uses original tank condition monitoring elements permanently installed in manufactured tanks and works on implementing a real-time tank condition monitoring system. All these activities are aimed at extending the life cycle of the tanks or predicting tank failures that could compromise the integrity of the chemical-resistant liner [40].

Carrying high-pressure hydrogen cylinders on a ship requires a number of strict safety requirements. For this reason, special structures in the form of containers or racks are used. Hexagon Purus already employs such comprehensive solutions. Table 1 shows examples from the specifications offered by the above-mentioned manufacturer of lightweight composite high-pressure containers for the carriage of hydrogen.

Table 1. Type IV tank specification (nominal working pressure: 700 bar, 15 °C)—data from Hexagon Purus [41].

Reference	Outside Diameter [mm]	Overall Length [mm]	Cylinder Weight [kg]	Water Volume [L]	Hydrogen Capacity [kg]	Weight Ratio (Hydrogen Weight/Cylinder Weight) [%]
H2-70-440X1050	440	1050	59	76	3.1	5.3
H2-70-530X2154	530	2154	188	244	9.8	5.2
H2-70-610X2060	600	2060	208	360	14.5	7.0
H2-70-705X2078	705	2078	264	458	18.4	7.0

Table 2 presents some specifications for composite tanks (type IV) from NPROXX, the company that also specialises in high-pressure hydrogen tanks for the marine market. Data of the two largest tanks offered by the company are shown, currently undergoing certification (anticipated release in the last quarter of 2024) [42].

Table 2. Specifications of selected hydrogen cylinders from NPROXX, source: [43].

Parameter	Specifications	
	AH 710-70	AH 620-70
Volume [L]	553	350
Weight [kg]	338	211
Usable H_2 mass [kg]	22.2	12.4
Outer diameter [mm]	700	610
Length incl. valve [mm]	2454	1917
Mounting	Neck mounting	Neck mounting
Weight ratio (hydrogen weight/cylinder weight) [%]	6.6	5.9

Due to the greater availability of technical data on how to transport and store compressed hydrogen, this study used cylindrical cylinders from the manufacturer Hexagon Purus and a method of modular installation of the cylinders with frame protection on the service vessel.

1.3. The Use of Hydrogen in Internal Combustion Engines

Today, combustion engine technologies dominate the field of converting chemical energy into electricity. This may change when fuel cell technology achieves low overall costs, low recycling costs and high efficiency over a wide range of outputs. The achievement of these goals by fuel cells will make them truly competitive with reciprocating internal combustion engines, which are currently the most efficient devices for converting chemical energy into mechanical energy and indirectly into electrical energy. Considering the above and the well-developed technology of internal combustion engines, the author of the work [44] believes that reciprocating internal combustion engines will be widely used to generate electricity by burning hydrogen.

Hydrogen is more often used in four-stroke (medium- and high-speed) reciprocating engines used in ships operating over short distances (and in trucks) than in two-stroke (slow-speed) high-power piston engines used in deep-sea vessels [45].

Hydrogen can be burned directly in reciprocating internal combustion engines or burned in mixtures of hydrogen with other fuels to generate propulsion power by main engines and to generate electricity by generating sets. In addition, hydrogen can be used in hydrogen fuel cells to generate electricity for ship propulsion. The authors' current research addresses the former solution, i.e., direct combustion of hydrogen in reciprocating internal combustion engines, while the use of hydrogen fuel cells on SOV vessels will be analysed in further stages of this research, including comparative studies of the two solutions from technical and economic points of view.

The market introduction of hydrogen-powered engines is currently underway. For example, the largest marine engine manufacturers, Wärtsilä Corporation (Helsinki, Finland) and MAN Energy Solutions SE (Augsburg, Germany), report that their hydrogen-powered engines will be launched in 2026, but orders may already be placed in early 2025 [45,46].

Wärtsilä Corporation (Helsinki, Finland) will initially offer two solutions for its engines [46]:

- The 31H2 dual-fuel engine, which can run on natural gas or hydrogen, as well as a mixture of these fuels;
- The 31SG-H2 engine, which can run on natural gas and a mixture of natural gas and hydrogen up to 25 per cent by volume; can also be adapted to run on hydrogen alone.

MAN Energy Solutions SE (Augsburg, Germany) is currently working on a hydrogen-powered dual-fuel H_2 engine that will be fully flexible and require only pilot fuel, which can be derived from green synthetic diesel. MAN representatives believe that hydrogen-powered reciprocating internal combustion engines are an extremely attractive option compared to fuel cells due to the lower cost, power density, flexibility of fuel application and high reliability of these engines. Based on their long-term experience, they estimate a service life of more than 30 years for this type of engine (with regular maintenance), which far exceeds that of fuel cells. At the same time, they point out that fuel cells are a good solution in the lower power ranges but not as the main propulsion system solution for coastal and offshore vessels [45].

For its part, WinGD Ltd. (Winterthur, Switzerland), a manufacturer of slow-speed, high-power marine engines, is now betting on ammonia as a zero-emission hydrogen-based fuel that can be produced without greenhouse gas emissions using electricity generated from renewable energy sources. The first WinGD engine that can run on ammonia should be available in 2025. An ongoing project is developing a concept for the use of ammonia for both diesel-powered WinGD X-type engines and dual-fuel LNG X-DF engines [47].

To date, few hydrogen-fuelled engines are available. For example, the company BeHydro (Gent, Belgium) offers an off-the-shelf solution, the $BEH_2YDRO\ DZ\ H_2$ family of hydrogen-only engines (in which combustion of the air–hydrogen mixture is initiated by a spark, as in spark ignition engines). In addition, its $BEH_2YDRO\ DZD\ H_2$ dual-fuel engines are powered by diesel fuel and a fuel composed of 85% gaseous hydrogen and 15% liquid fuel (where the combustion of the air–hydrogen mixture is initiated by injection of liquid fuel, as in compression ignition engines) [48,49].

The company BeHydro (Gent, Belgium) offers the $DZ\ H_2$ and $DZD\ H_2$ engines with in-line cylinder layouts as 6- and 8-cylinder engines and in V-cylinder layouts as 12- and 16-cylinder engines [48,49].

Polish scientists are also involved in research into the use of hydrogen to power internal combustion engines. Researchers from the Krakow University of Technology presented a piston internal combustion engine adapted to run on hydrogen in January 2024. This is a Scania (Södertälje, Sweden) five-cylinder high-speed engine type DI09 074M, with 269 kW power, adapted for marine applications [50,51].

1.4. Research Aim

As a result of the literature review that revealed the lack of published data on analyses of using zero-emission alternative fuels on offshore vessels from the design-construction perspective, the aim of the current research is to analyse the feasibility of using compressed hydrogen fuel to propel an SOV vessel in terms of design and economic challenges. To achieve such an aim, the authors examined the existing design concept of the SOV, together with the ship's operational profile—as well as the operating regime guidelines and corresponding power requirements. These were presented in detail in a publication by Bortnowska M. [52].

To achieve the main objective of this research, the following actions are needed:

- Analyse the physical and chemical properties of hydrogen and the challenges posed by the use of hydrogen on board as a fuel;
- Apply the technical requirements for the safe operation of hydrogen fuel (storage and transport) in accordance with current IMO guidelines;
- Carry out a power demand analysis for the main SOV operating regimes in the area of wind farms located in the Baltic Sea;
- Select the number and power of commercially available compressed hydrogen engines and analyse their configurations;
- Calculate fuel consumption for different ship operating regimes, select the storage and transport system and deploy their elements in the hull, meeting safety requirements;
- Analyse the impact of the use of hydrogen fuel on the design and structural elements and spatial layout of the ship;
- Identify what benefits will be achieved with the different fuels, taking into account the hull volume used, resulting from the volume of the fuel tanks selected;
- Conduct an economic analysis in terms of the cost of the hydrogen fuel used and make a rough comparison with other accepted fuels such as MDO, LNG and MeOH.

2. Initial Assumption and Methodology

2.1. Specification of the Service Vessel and Its Area of Work

Table 3 contains the key data from the ship's specifications that are important for this study, and Figure 5 depicts the sketch profile view of the ship.

Table 3. Technical specifications of service vessel, adopted for further analysis [53].

No.	Parameter	Value	Symbol	Unit
1.	Length between perpendiculars	72.0	L_{bp}	m
2.	Breadth	18.0	B	m
3.	Hull depth	8.9	H	m
4.	Draught	5.3	T	m
5.	Displacement	5004	D	t
6.	Deadweight	1005	DWT	t
7.	Cargo weather deck area	280	A_{wd}	m^2
8.	Technical crew number	60	n	people
9.	Dynamic positioning system	-	DP2	-
10.	Endurance	14	A	days

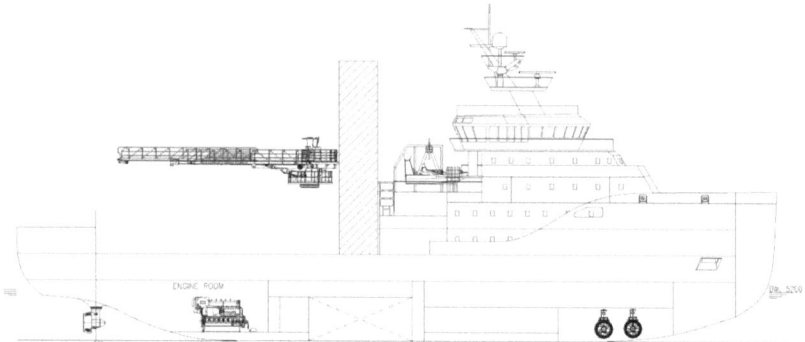

Figure 5. Sketch profile view of the SOV, source: [52].

A diesel–electric propulsion system was adopted for this study based on a database of similar vessels developed by the authors, in which four variants of reciprocating internal combustion engines powered by different types of fuel, i.e., diesel, LNG, methanol and hydrogen, drive generating sets to produce electricity for shipboard needs, including propulsion.

The following engine types used in the selected gensets were used:

- Wärtsilä 20 four-stroke diesel engine (Wärtsilä Corporation, Helsinki, Finland), which can run on marine diesel oil (MDO) and light and heavy fuel oil (LFO, HFO);
- Wärtsilä 20DF four-stroke dual-fuel diesel engine (Wärtsilä Corporation, Helsinki, Finland), that can run on liquified natural gas (LNG), MDO and HFO;
- Wärtsilä 20 methanol-fuelled four-stroke diesel engine (Wärtsilä Corporation, Helsinki, Finland), which, in addition to methanol, can run on MDO, LFO and HFO;
- BEH$_2$YDRO DZD H$_2$ four-stroke hydrogen-fuelled diesel engine (BeHydro, Gent, Belgium), which can run on MDO and LFO in addition to hydrogen.

The use of four engine power variants will allow a ship design analysis to be carried out for the estimation of the volume and weight of fuel tanks relative to the type of fuel used and the area occupied in the ship's hull space. Thus, the analysis will allow an appropriate spatial layout of the ship's power system (gensets and fuel tanks), taking into account the area of the engine room location, and an assessment of the impact of a given variant on the space occupied, displacement or deadweight of the ship.

The computational analysis was carried out for a vessel (Table 3) that was hypothetically planned to operate in a wind farm area in the Baltic Sea at a distance of approximately 80 km from the coastline. For the analysis, the vessel was assumed to operate for average weather conditions in that area: average significant wave height H_s = 2.5 m, and average wind speed v_w = 10 m/s [53].

2.2. Hydrogen Fuel Storage and Transport on a Service Offshore Vessel

Based on the literature review, market analysis and current technologies in maritime transport of hydrogen fuel, the authors decided to use the following in the project:

- Technology for transporting hydrogen in compressed form at high pressure—700 bar;
- A cylindrical tank was selected—composite cylinder of type 4, made by Hexagon Purus, pressure 700 bar and ambient temperature;
- Technical data—H2-70-705X2078:
    - Outside diameter D_{H2} = 705 mm;
    - Overall length L_{H2} = 2078 mm;
    - Cylinder weight W_{H2} = 264 kg;
    - Volume V = 457 litres;
    - Hydrogen capacity V_{H2} = 18.4 kg;
- Hydrogen fuel storage system on board the ship: the cylinders to be mounted on board in modular racks; each rack contains nine cylindrical cylinders, which can easily be combined; due to the limited manufacturer and regulatory guidelines on this issue, the authors chose the above concept as the preliminary one, which may be modified in the future.

The number of cylinders used, and thus the number of racks, will be determined later in this work, after an analysis of the mass and volume quantities of hydrogen required for the application.

2.3. Operational Profile of Service Vessel Work

The service vessel, in performing its tasks, is in various instantaneous operating states, during which it exhibits energy requirements in a variable range [52,53]. The specific nature of the SOV operation consists of several stages, which is why the entire cycle of its operation has been referred to as a mission. A single-ship mission consists of the following stages [52,53]:

1. Normal operation: shipping to the working area;
2. Service work in the offshore wind farms in the Baltic Sea;
3. Normal operation: returning to port;
4. Stopover in port.

The most common missions for a ship last 14 days. Due to the physical and chemical properties of compressed hydrogen fuel, it was found that a 14-day service period at sea would be feasible. It was therefore assumed that the maximum mission time T_s = 14 days.

Of all the mission phases, the main tasks of the vessel include service work around offshore wind farms, and, for this, the following was adopted: five regimes of ship operation marked (I–V) together with their duration in hours [52,53]:

- I—Operation with active working platform and DP2 dynamic positioning system (3.5 h);
- II—Manoeuvring (1.5 h);
- III—Operation with an inactive working platform and DP2 dynamic positioning system (3.5 h);
- IV—Low operation (1.0 h), v_s = 6 knots;
- V—Night break using DP1 (14.5 h).

Each of the above service work phases is characterised by varying power demand. Table 4, based on [52,53], shows the input energy demand values for the particular service work regimes and for the round trip (to the offshore work area and back to port).

Table 4. Power required for individual service work phases.

	Service Work Phases	Power Requirement P_R' [kW] **	Time T [h per Day]	Energy Required	
				E_R [kWh per Day]	E_{MIS} * [kWh per Mission]
		Service Work in the Offshore Wind Farm			
I	Operation with active working platform and DP2 dynamic positioning system	2851	3.5	9978.5	139,699
II	Manoeuvring	1526	1.5	2289	32,046
III	Operation with DP2 dynamic positioning system	1976	3.5	6916	96,826
IV	Low operation—shipping $v_s = 6$ kn	773	1.0	773	10,822
V	Night break using DP1	1426	14.5	20,677	289,478
		Shipping to/from the Offshore Wind Farm			
VI	Normal operation $v = 12$ kn	2190	2 × 4	n/a	17,523 only round trip

* Mission—14 days, ** acc. [52,53], P_R'—power requirement for individual phases.

It follows from the data in Table 4 that the highest power demand occurs in operating regime (I)—with the DP2 system and working active platform, during normal ship operation, i.e., sailing to and from the wind farm, and in operating regime (III)—during operations using the dynamic positioning system.

2.4. Research Method

The new fuel type involves examining the effect of its use for service vessel propulsion and checking its feasibility in terms of safety criteria and volume design in the hull to provide storage space. These tasks can be carried out by analytical and computer-aided methods using CAD-type tools.

The computational research method was used to assess the feasibility of storing, transporting and bunkering hydrogen fuel on the vessel under study. The assessment also covered capabilities to provide demanded power under varying propulsion system load conditions during service work and estimation of the new fuel costs. The method was based on the following:

- Technical and operational data of the service vessel analysed in previous work—acc. [52];
- Latest IMO guidelines relating to hydrogen fuel;
- Available engine solutions adapted to run on this new fuel;
- A storage system that allows hydrogen to be transported under high-pressure conditions.

Using Autodesk Inventor Professional 2025 and AutoCAD Mechanical 2025, a 3D model of cylindrical tanks along with a transport module was utilised for research purposes. The model took into account parameters such as cylinder dimensions, the weight of the cylinder including the gas, and the volume occupied by the module with the cylinders. Thanks to the parameterization of the 3D models, modifications can be made more quickly and with precision. An additional advantage of CAD software is the visualisation of each element and the ability to verify the compatibility of individual modules with the cylinders on the ship, as well as the arrangement of power generators within the hull of the SOV.

However, the computational model is based on general formulas (presented later in this article) related to hourly and specific fuel consumption, enabling the determination of the required fuel reserves (in terms of mass and volume). It also relies on general formulas for calculating fuel consumption costs and engine loads under various operational conditions of the SOV.

Figure 6 hierarchically depicts the main stages of the computational analyses, resulting from the initial assumptions and research background.

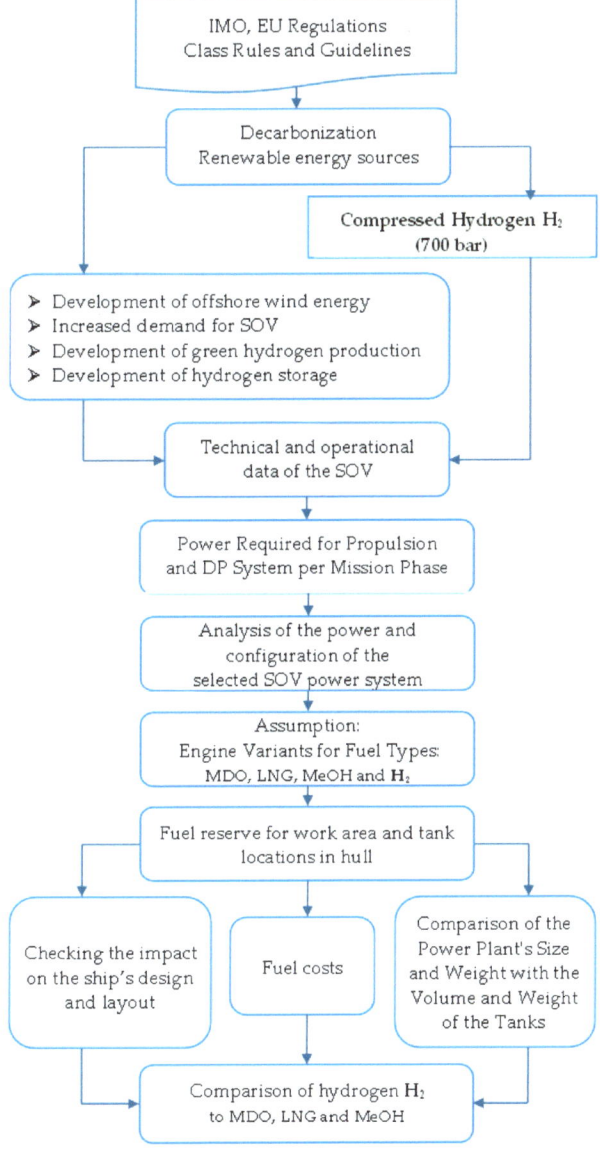

Figure 6. Block diagram of the analysis of the size and mass of the power plant with fuel tanks and fuel costs for the SOV.

3. Design Analysis of the SOV for the Use of Alternative Propulsion Fuels

Taking into account the technical data of the analysed SOV, presented in Section 2, the vessel's design analysis was carried out to estimate the volume and weight of the fuel tanks, depending on the type of fuel used, and to estimate the area occupied in the ship's hull space. For comparison, methanol, LNG and conventional MDO fuel were used in addition to compressed hydrogen.

3.1. Selection of the Energy System Solution

3.1.1. Energy System Solutions for Service Offshore Vessels

From the SOV, CSOV (Construction Service Operation Vessel) and WSV (Windfarm Support Vessel) database of 20 similar vessels (authors' collection used in this review), it can be seen that all vessels are equipped with diesel–electric propulsion or hybrid diesel-electric propulsion supported by a battery system.

In a diesel–electric propulsion system, electricity is produced by one common power station consisting of two, three or four gensets. The electric energy is then supplied to a propulsion system consisting of two electric motors driving two azimuthal thrusters and to the ship's other loads such as the thrusters and all the equipment and facilities in the engine room, on deck, the bridge and accommodation.

In hybrid propulsion, two to four gensets that produce electricity for electric power receivers are supported by a battery bank system.

These considerations refer to combustion–electric propulsion, with reciprocating internal combustion engines powered by a variety of fuels: hydrogen, diesel, LNG and methanol.

SOV power system solutions are dominated by a configuration of three gensets (12 vessels out of 20), with 11 vessels using gensets of the same power and only 1 using a configuration of two generating sets of higher power and one of lower power. For this reason, the authors adopt a configuration of three gensets of the same power for further study.

Considering the operational states of the service vessel (Table 3), we can conclude that the highest power demand occurs in operation mode (I)—with active DP2 system and active working platform. Taking into account the power demand in the different operating states of the examined SOV and the power values occurring on similar vessels (according to the database of service vessels developed by the authors), the total power of the power system was estimated at 5400 kW; i.e., each of the three gensets should have a power of 1800 kW. A diagram of the power system of the analysed ship is shown in Figure 7.

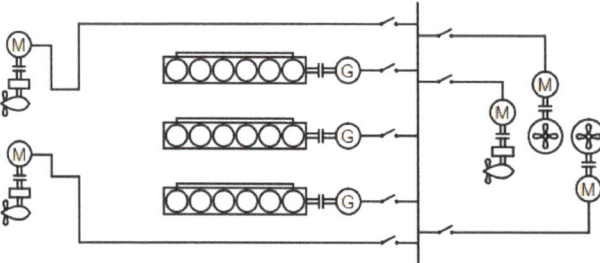

Figure 7. The energy system of the SOV under analysis [authors' drawing].

The database of similar vessels shows that two azimuth thrusters with an estimated power of 1000–1800 kW each and two bow thrusters with an estimated power of 750–1500 kW each are most commonly used for propulsion and manoeuvring on SOVs. In many cases, one retractable azimuth thruster with an estimated power of 800 kW is also used forward in addition to two thrusters.

3.1.2. DP2 Dynamic Positioning System

The high power output of the power system and appropriate propulsion and steering equipment provide the vessel with very good manoeuvring characteristics, typical of vessels equipped with the DP2 dynamic positioning system. Such vessels should meet two basic conditions [54,55]:

(1) Maintain the position throughout the range of variations in hydrometeorological conditions specified in the design guidelines. This condition is met by installing generating sets and thrusters of sufficiently high power.
(2) Maintain position even when a failure of any component occurs. This condition is met by adopting the redundancy concept, which is to maintain the dynamic positioning capability of the vessel in the event of a single failure of any component of the system.

In the current study, condition (1) is met, as the selected power system of 5400 kW is sufficient to drive two azimuth thrusters of approx. 1400 kW each and two bow thrusters of approx. 1200 kW each and one retractable azimuth bow thruster of about 800 kW.

In addition, the ship's energy system can be equipped with an energy storage system consisting of, e.g., two battery units with a capacity of 300–600 kWh each and a maximum power of 800–1600 kW. This system will be able to support the ship's energy system in situations of high demand for electricity, while at low demand the excess electricity from the generating sets will be able to charge battery banks. In addition, the battery banks can supply power to the ship's systems when the gensets are not running, for instance when the ship stays in port.

In the present study, condition (2) is also met, as the power system consists of three gensets and two azimuth thrusters and three bow thrusters, guaranteeing that the dynamic positioning capability of the vessel is maintained even in the event of a single failure of any system component. The limit operating conditions of the SOV are determined by the power (generators, azimuth thrusters and bow thrusters) available once a failure of any system component occurs and not by the nominal power of the installed equipment. Thus, taking into account the power requirements for the engines in different operating states of the vessel estimated in Table 3, a failure of one of the gensets, azimuth thruster or one of the bow thrusters will provide the vessel with the required power to maintain position.

DP systems are equipped with a load-limiting function related to position maintenance priority. In certain situations, e.g., in severe weather conditions, the operation of the thrusters can consume the entire power reserve, thus limiting the power transmitted to other lower priority equipment, such as lifting equipment [55].

3.1.3. Selection of Generating Sets

Several variants of the gensets were adopted for this study. These machines, supplying electricity for the propulsion power of the ship, will be the study focus and will be fed with different fuels: hydrogen and, for comparison, methanol, LNG and conventional MDO fuel. The four gensets are composed of the following four-stroke reciprocating internal combustion engines to drive the generators:

- Wärtsilä 20 (Wärtsilä Corporation, Helsinki, Finland) powered by conventional MDO fuel;
- Wärtsilä 20 dual-fuel (Wärtsilä Corporation, Helsinki, Finland) fed with LNG preceded by a pilot dose of conventional MDO fuel;
- Wärtsilä 20 (Wärtsilä Corporation, Helsinki, Finland) fuelled with methanol (M);
- BEH$_2$YDRO DZD H$_2$ (BeHydro, Gent, Belgium) fuelled with a mixture of hydrogen (H$_2$) and MDO in a ratio of 85/15.

A hydrogen-powered engine from BeHydro (Gent, Belgium) was included alongside the engines from Wärtsilä Corporation (Helsinki, Finland) because hydrogen-fuelled engines from the latter are expected on the market in 2026 [46]. MAN Energy Solutions SE (Augsburg, Germany) also plans to launch hydrogen-powered marine engines in 2026 [45].

Table 5 summarises the technical data of the gensets adopted for the analysis. The data relevant for design purposes include dimensions and mass. The space occupied by the gensets should account for the dimensions of the gensets themselves, as well as all the associated equipment and servicing (as defined by the regulations and the manufacturer's requirements), which obviously increases the footprint. Assuming the ship breadth $B = 18$ m (Table 3), the gensets can be positioned side-by-side along the ship's centre line. In this study, the approximate service space around the gensets was assumed to be 1 m around, with a 1.5 metre spacing between two gensets. Such space ensures convenient day-to-day servicing of the auxiliary engines and provides adequate service space during overhaul. For comparison, for the Wärtsilä 8L20 engine (Wärtsilä Corporation, Helsinki, Finland), the manufacturer recommends [56]:

- Service space for generating set—500 mm;
- Distance needed to dismantle the pump cover with fitted pumps—650 mm;
- Width for dismantling lubricating oil module with lifting tool—approximately 500 mm.

The determined indicative space occupied by the gensets is summarised in Table 5.

Table 5. Main technical data of auxiliary sets [48,49,57–59].

No.	Parameter	Engine Type			
		Diesel Engine (Wärtsilä 8L20) *	Dual-Fuel Engine (Wärtsilä 9L20DF) *	Methanol Engine (Wärtsilä 9L20M) *	Hydrogen Engine (BEH$_2$YDRO 12DZD H$_2$) **
1.	Rated power P_R [kW]	1760	1755	1800	1800
2.	Speed [rpm]	1200	1200	1200	900
3.	Cylinder bore/Piston stroke (mm)	200/280	200/280	200/280	256/310
4.	Dimensions L × B × H [mm]	6700 × 2010 × 2547	6700 × 2010 × 2831	6700 × 2010 × 2831 *	6667 × 1850 × 3131
5.	Area required (engines + service area) [m^2] acc. to design analysis	61	61	61	59,4
6.	Weight M_E [t]	23	25	25 ***	33.7
7.	Fuel type	MDO, LFO, HFO	LNG, MDO, HFO	Methanol (M), MDO, LFO, HFO	Hydrogen (H$_2$), MDO, LFO
8.	IMO	Tier II or III	Tier II or III	Tier II or III	Tier III

* Wärtsilä Corporation (Helsinki, Finland); ** BeHydro (Gent, Belgium); *** Approximate data due to the lack of catalogue data.

A closer look at the data summarised in Table 5 shows that the main dimensions of the hydrogen-fuelled genset are similar to those of the other three Wärtsilä gensets (Wärtsilä Corporation, Helsinki, Finland) running on different fuels. Only the height is larger than the other machines (by 300 mm compared to the 9L20DF and 9L20M engines and 584 mm compared to the 9L20 engine). This will affect the need for more space above the hydrogen genset, while the footprint of all the gensets remains similar. In addition, the BEH$_2$YDRO 12DZD H$_2$ genset (BeHydro, Gent, Belgium) weighs 9–11 tons more than the other gensets, which, for three gensets, will increase the weight by approximately 30 t.

This research focuses on the BEH$_2$YDRO 12DZD H$_2$ hydrogen-fuelled engine solution (BeHydro, Gent, Belgium), a 12-cylinder engine in a V-cylinder arrangement. It is a four-stroke, medium-speed, dual-fuel H$_2$ engine that is turbocharged, intercooled and built for multiple fuel execution; i.e., diesel, marine diesel oil (MDO), gasoil (GO), dual fuel (diesel and hydrogen), biofuel, vegetable oil, fuel cell grade or less purified H$_2$ are possible [48]. According to the manufacturer [49], the engine is characterised by its ability to run on less purified hydrogen, its fast response to varying loads and an 85% CO$_2$ reduction. It complies with EU Stage V requirements when combined with selective catalytic reduction, where a chemical reaction converts nitrogen oxides into nitrogen and oxygen, and a diesel particulate filter (DPF) removes particulate matter or soot from the exhaust gas. Other characteristics include a long life, easy maintenance and the availability of spare parts (rare materials not used). The engine runs on a fuel composed of 85% gaseous hydrogen and 15% liquid fuel, with an option for liquid fuel only. Combustion of the air–hydrogen mixture is initiated by liquid fuel injection, similar to compression–ignition engines.

In order to maintain the safety of the entire hydrogen combustion process, the engine is fitted with equipment for fuel treatment before it is delivered to the engine combustion chamber. The fuel preparation and delivery process are not relevant to the research discussed herein, so its description has been omitted.

3.1.4. Parameters of the Fuels Used

In order to dimension the fuel tanks, it is necessary to characterise the parameters of the fuels used in the selected engines (Table 5). The most relevant physical and chemical properties of the selected fuels are shown in Table 6, with one type of liquid fuel (MDO) selected for comparison, as this fuel does not require the treatment system nor the construction of the tank heating system using insulated fuel pipes routed to the engines, as is required for heavy fuel oil [60]. This will make the installation of liquid fuel less complex and less costly during the construction and operation of the vessel.

Table 6. Comparison of the physical and chemical properties of hydrogen in relation to MDO, LNG fuel and methanol [31,61–63].

| No. | Property | Fuels ||||
		MDO	LNG (Liquid −163 °C)	Methanol (65 °C)	Hydrogen (Compressed 700 bar)
1.	Density (kg/m^3)	830–850	410–500	787–792	40.2
2.	Emergency content LHV (MJ/kg)	42.0–43.0	50.0	19.9–20	120.0
3.	Boiling point (°C)	150–370	−162.0	65.0	−252.77
4.	Flash point (°C)	min. 60	−188.0	9–11.0	−259.20
5.	Auto ignition (°C)	240.0	537.0	385.0–464.0	585.0
6.	Fuel tank size relative to MGO	1.0	1.7	2.3	7.4

As low-carbon fuels, LNG and methanol have a lower carbon emission intensity compared to conventional petroleum-based marine fuels, while hydrogen has virtually zero carbon intensity. Since this study assumed a fuel consisting of 85% gaseous hydrogen and 15% liquid fuel as the fuel for the hydrogen-fuelled engine, carbon dioxide emissions will therefore occur but in much lower quantities.

For easier storage and use, LNG is liquefied by cooling to −163 °C, while hydrogen is compressed to a pressure of 700 bar. Like LNG-fuelled vessels, hydrogen-fuelled vessels use their own engine architecture and require specialised fuel storage systems. Hydrogen fuel has the lowest volumetric energy density, yet the highest specific density MJ/kg compared to other fuels used in vessel propulsion, including SOVs (Table 3), requiring larger tank volumes that will take up significantly more space on board.

3.2. Technical Requirements for Fuel Tanks on Ships

3.2.1. Current Regulations

The use of hydrogen as a fuel in maritime transport has gained prominence in recent decades, but only in 2019 did the International Maritime Organisation (IMO) approve and recognise hydrogen as a potentially safe and environmentally friendly fuel for ship propulsion. Still, hydrogen as an alternative fuel does not yet have full regulations. Work is still in progress, and the regulations are expected to be finalised by 2025. The IMO interim guidelines complement the most important regulation of the International Code of Safety for Ships Using Gases or Other Low-flashpoint Fuels (IGF Code)—concerning safety related to hydrogen storage, fuel systems and propulsion systems on board [64,65].

These regulations aim to provide an international standard for ships operating with gas or low flashpoint liquid. They set out criteria for the arrangement of equipment, machinery and installations on ships using low flashpoint liquids or gases for propulsion, such as hydrogen. There are also national (flagship) regulations and rules of classification societies that contain regulations for the use of hydrogen fuel in ships. Figure 8 illustrates the key rules and regulations relating to the use of hydrogen fuel in maritime transport [1,66].

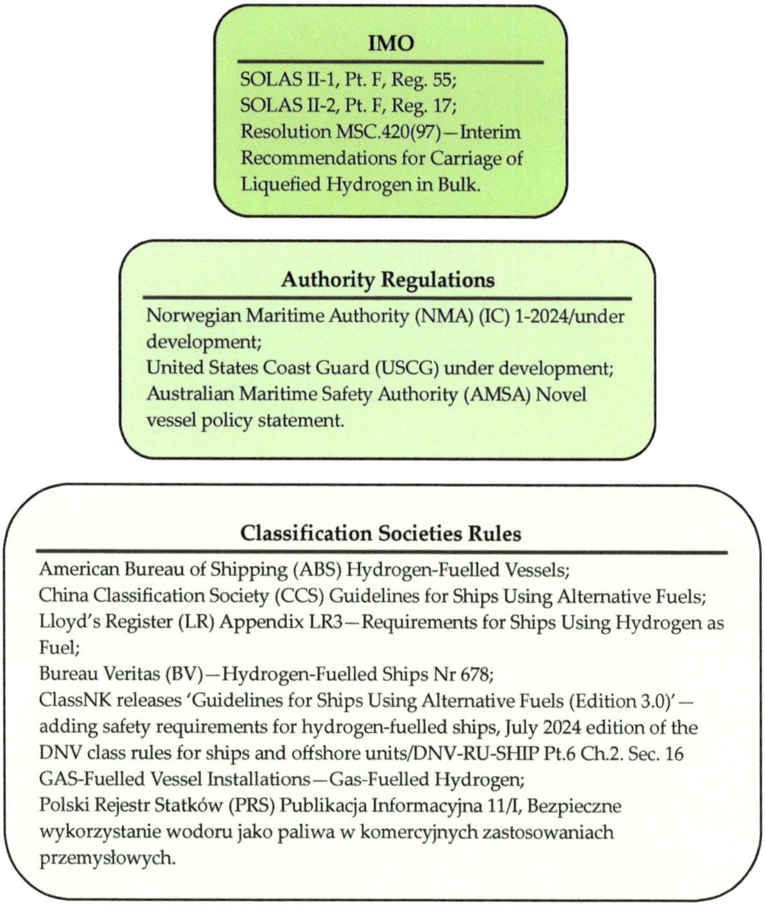

IMO

SOLAS II-1, Pt. F, Reg. 55;
SOLAS II-2, Pt. F, Reg. 17;
Resolution MSC.420(97)—Interim Recommendations for Carriage of Liquefied Hydrogen in Bulk.

Authority Regulations

Norwegian Maritime Authority (NMA) (IC) 1-2024/under development;
United States Coast Guard (USCG) under development;
Australian Maritime Safety Authority (AMSA) Novel vessel policy statement.

Classification Societies Rules

American Bureau of Shipping (ABS) Hydrogen-Fuelled Vessels;
China Classification Society (CCS) Guidelines for Ships Using Alternative Fuels;
Lloyd's Register (LR) Appendix LR3—Requirements for Ships Using Hydrogen as Fuel;
Bureau Veritas (BV)—Hydrogen-Fuelled Ships Nr 678;
ClassNK releases 'Guidelines for Ships Using Alternative Fuels (Edition 3.0)'—adding safety requirements for hydrogen-fuelled ships, July 2024 edition of the DNV class rules for ships and offshore units/DNV-RU-SHIP Pt.6 Ch.2. Sec. 16 GAS-Fuelled Vessel Installations—Gas-Fuelled Hydrogen;
Polski Rejestr Statków (PRS) Publikacja Informacyjna 11/I, Bezpieczne wykorzystanie wodoru jako paliwa w komercyjnych zastosowaniach przemysłowych.

Figure 8. Current regulations for hydrogen fuel.

Many countries are strongly committed to the implementation of hydrogen as a fuel for marine use, which is evident in the development of the infrastructure and projects under development. Notably, at the IMO symposium [67], hydrogen fuel was announced as the long-term future, so it can be concluded that hydrogen has significant potential for the future energy economy.

3.2.2. Fuel Tanks with Compressed Hydrogen CH_2 in the Ship's Hull

Transporting hydrogen as a fuel on ships requires compliance with a number of stringent safety and design requirements due to its unique properties. Due to its characteristics, it increases the risk of leakage or penetration through the tank material, which can make it brittle, which in turn makes it difficult to store and transport. This article presents some of the most relevant design-related guidelines from DNV-RU-SHIP Pt.6 Ch.2. Sec. 16 Gas-Fuelled Vessel Installations—Gas-Fuelled Hydrogen:

- Use high-strength materials that are resistant to hydrogen permeability, hydrogen embrittlement and corrosion (this study uses composite containers with a lower weight and higher resistance to hydrogen diffusion compared to steel). Data on hydrogen embrittlement susceptibility of commonly used metallic materials are given in ISO/TR 15916: 2015, ANSI/AIAA G-095A-2017 or NASA TM-2016-218602;
- Locate the hydrogen fuel tanks in the ship's hull and design the ship's engine room along with the layout of other rooms containing the hydrogen fuel systems;
- Containers must be equipped with safety valves to discharge hydrogen in a controlled manner if the permitted pressure is exceeded (e.g., to the atmosphere via ventilation ducts);
- Hydrogen cylinders must be placed in separate technical rooms, away from heat sources (engines, boilers) and electrical equipment that could be a potential source of ignition;
- Tanks are to be separated from crew spaces by fire barriers (Class A-60);
- Tanks should be above the waterline (as high up in the hull as possible), and if not, very strict safety and access regulations must be met;
- Spaces where tanks are located must be equipped with ventilation systems with continuous air exchange to prevent the accumulation of hydrogen, which can be highly explosive, even at low concentrations, and equipped with hydrogen detector systems to detect its presence at very low concentrations and immediately raise an alarm if a leak is detected. The regulations require tanks to be located in open spaces or in enclosed spaces (equipped with an efficient ventilation system);
- Vessels inside the hull must be protected from mechanical damage, e.g., impacts or shocks, by additional protective covers or barriers;
- Bulkheads must be fire resistant and meet high fire insulation standards (Class A-60 according to IGF Code);
- Spaces with hydrogen tanks must be equipped with automatic fire detection systems and fixed fire extinguishing systems, i.e., water spray or powder extinguishing systems or gas extinguishing systems.

3.3. Calculation of Fuel Supply and Arrangement of Tanks in the Ship's Hull

3.3.1. Estimation of Fuel Quantity

In order to estimate the required fuel supply in store for the analysed SOV, it is necessary to determine the overall efficiency of a piston internal combustion engine, which for modern medium-speed marine engines is 0.38–0.50. The overall efficiency is a measure of the engine's utilisation of the heat flux supplied in the form of fuel and is the basic energy and economic indicator of internal combustion engines. It depends on a number of design and operating factors, as well as the engine's condition and load level. It should also be taken into account that the highest overall efficiency is achieved by internal combustion engines in the load range of 70–90% of rated power [68]. Taking these factors into account, we can assume that the overall efficiency of the reciprocating internal combustion engines is η_{DE} = 0.44, with an average load of 80% of rated power. For simplicity, the efficiency of all engines was assumed to be the same regardless of the type of fuel burned.

In addition, the estimated power demand for the assumed SOV operating states in Table 4, including the duration of each operating state, was also taken into account to determine the required fuel supply.

The following relationships were used to calculate the maximum fuel supply required:

- Fuel consumption per hour G_{fuel} [kg/h]:

$$G_{fuel} = \frac{P_R}{LHV \cdot \eta_{DE}} \cdot 3600 \quad (1)$$

where P_R—rated power [kW], LHV—emergency content [kJ/kg], and η_{DE}—engine efficiency [-];

- Total fuel reserve M_{Tfuel} [t]:

$$M_{Tfuel} = G_{fuel} \cdot T_s \cdot 10^{-3} \quad (2)$$

where G_{fuel}—the fuel consumption per hour [kg/h], and T_S—mission duration [h];

- Fuel volume V_{fuel} [m^3]:

$$V_{fuel} = \frac{M_{Tfuel}}{\rho_{fuel}} \cdot 1000 \quad (3)$$

where M_{Tfuel}—total fuel reserve [t], and ρ_{fuel}—fuel density [kg/m^3];

- Specific fuel consumption g_{fuel} [g/kWh]:

$$g_{fuel} = \frac{G_{fuel}}{P_R} \cdot 1000 \quad (4)$$

where G_{fuel}—the fuel consumption per hour [kg/h], and P_R—rated power [kW].

The hydrogen-fuelled engine runs on a fuel composed of 85% gaseous hydrogen and 15% liquid fuel, with the combustion of the air–hydrogen mixture initiated by injection of MDO liquid fuel. Similarly, during LNG-fuelled dual-fuel engine operation, combustion of the air–gas mixture is initiated with a small amount of MDO pilot fuel, less than 1% of full-load fuel consumption [69]. The amount of this pilot fuel is optimised for optimum combustion using an onboard engine speed and load control and monitoring system [59,69]. The proportions of these fuels were further taken into account in determining the required maximum fuel reserve for the SOV (Table 7).

Table 7. Estimated total fuel reserve and volume for a mission.

No.	Parameter	Fuels			
		MDO	LNG	Methanol	Hydrogen
1.	Total fuel reserve M_{Tfuel} [t]	115.85	99.35	254.74	42.46
	The ratio of hydrogen to:	0.37	0.43	0.17	-
2.	Fuel volume V_{fuel} [m^3]	137.92	220.77	322.45	1056.13
	The ratio of hydrogen to:	7.66	4.78	3.28	-
3.	Fuel volume V_{fuel} [m^3]	137.92	218.57 LNG / 1.38 MDO	322.45	897.71 H$_2$ / 20.69 MDO
4.	Emergency content LHV [MJ/kg]	43	50	20	120
5.	Fuel consumption per hour G_{fuel} [kg/h]	266.94	228.91	586.96	97.83
6.	Specific fuel consumption g_{fuel} [g/kWh]	189.59	163.04 *	407.61 *	67.93 *
7.	Area required (tanks + compartments required) [m^2] acc. to design analysis	230	264	360	1539 impossible

* Due to a lack of published data (due to commercial confidentiality) on specific fuel consumption, this parameter was estimated on the basis of a Formula (4).

The data contained in Table 7 show that a very large volume of hydrogen is required to power the engines, more than the tripled amount of methanol required and more than seven times the amount of MDO fuel required. The use of liquid MDO in the BEH$_2$YDRO 12DZD H$_2$ engine (BeHydro, Gent, Belgium) makes it possible to reduce the required hydrogen supply by 158.42 m^3, with the need to store 20.69 m^3 of MDO fuel.

Also evident is the significantly lower weight of stored hydrogen fuel, six times less than that of methanol and more than twice less than that of LNG fuel. A comparison of the so-called 'energy density' of compressed hydrogen with the other fuels considered in this study is shown in Figure 9.

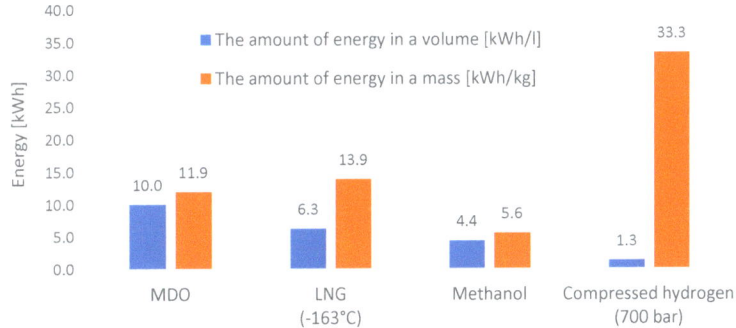

Figure 9. Comparison of the so-called energy density for compressed hydrogen with respect to the other fuels considered in this study [authors' research].

A more detailed picture of the variation in hydrogen fuel demand and volume occupied, depending on the possible variable service time, is presented in Table 8. These results provide the basis for conducting further design analyses regarding the layout of the hydrogen tanks in the SOV hull.

Table 8. Calculated fuel reserve and hydrogen tank volume for ship autonomy (1–14 days).

No.	Endurance	Hydrogen Mass [t]	Hydrogen Volume [m^2]	Hydrogen Volume Minus MDO [m^3]	Number of Cylinders [pcs].
1.	1 day	3.03	75.44	65.68	150
2.	3 days	9.10	226.31	205.75	449
3.	5 days	15.16	377.19	342.91	749
4.	7 days	21.23	528.07	480.08	1048
5.	9 days	27.29	678.94	617.24	1348
6.	11 days	33.36	829.82	754.40	1647
7.	13 days	39.42	980.69	891.57	1947
8.	14 days	42.46	1056.13	960.15	2096

3.3.2. Load Factor Analysis of Reciprocating Internal Combustion Engines

Table 9 shows the loads of the selected reciprocating internal combustion engines depending on the value of the power demand in the assumed operating states of the SOV (Table 4). As mentioned earlier, the highest overall efficiency of reciprocating internal combustion engines, and consequently the lowest specific fuel consumption, is achieved in the load range of 70–90% of rated power [68]. For this reason, the number and power of selected gensets should be such that, as far as possible, they will operate in these rated power ranges in particular operating conditions. This is particularly important in operating states that cover a longer period of time, such as states I, III and V, as the greatest amount of fuel is consumed during these operating states.

The following relationships were used to calculate the engine load:

- Engine load E_{load} [%]:

$$E_{load} = \frac{P_R{'}}{P_R} \cdot 100 \; [\%] \tag{5}$$

where $P_R{'}$—power requirement [kW], and P_R—rated power [kW].

Table 9. Degree of loading of the selected two combustion engine configurations depending on the value of the power demand in the assumed operating states SOV.

No.	Parameter	Operational Profile and Phases of SOV Service Work					
		I	II	III	IV	V	Normal Operation
1.	Power requirement $P_R{'}$ [kW]	2851	1526	1976	773	1426	2190
2.	Time operating T [hours/day]	3.5	1.5	3.5	1.0	14.5	8 total
	Configuration of 3 generating sets with the same power						
3.	Number of working Generators	2	1	2	1	1	2
4.	Rated power P_R [kW]	1800 1800	1800	1800 1800	1800	1800	1800 1800
5.	Engine load E_{load} (%)	79	85	55	43	79	61
	Configuration of 3 generating sets with different power						
6.	Number of working generators	2	1	1	1	1	2
7.	Rated power P_R [kW]	2400 1200	1800	2400	1200	1800	1800 1200
8.	Engine load E_{load} (%)	79	85	82	64	79	73

The data presented in Table 9 show that for the assumed operating states, the use of three gensets of equal engine power may not be economical, as in operating states III, IV and normal operation, the engines will consume more fuel due to their operation below the recommended loads. This should be taken into account in the calculation of the required fuel supply by increasing the amount accordingly.

For this reason, for the same operating conditions, three gensets of different powers of their four-stroke hydrogen-fuelled diesel engines BEH$_2$YDRO DZD H$_2$ (BeHydro—Gent, Belgium) [48] were used for the comparative analysis:

- The 8 DZD H$_2$ with a power of 1200 kW;
- The 12 DZD H$_2$ with a power of 1800 kW;
- The 16 DZD H$_2$ with a power of 2400 kW.

The total power of these engines, 5400 kW, is the same as that of the three pre-selected 12 DZD H$_2$ engines (BeHydro, Gent, Belgium). Similarly, the mass of the three gensets with different powers is practically the same as that of those with the same powers, as it is only 2.6 t, making them 2.5% heavier. However, this configuration of gensets will be troublesome with regard to their siting in the power plant due to their different dimensions and unit masses.

The use of a new configuration of three gensets with different powers improves the load factor of the engines for state III from 0.55 to 0.82, for state IV from 0.43 to 0.64 and for normal operation from 0.61 to 0.73. In the remaining states, i.e., I, II and V, the load factor of the engines remained unchanged. The apparent improvement in the degree of loading of the engines as a result power changes will certainly improve fuel consumption, but an accurate determination of this parameter for the purpose of comparing the two genset configurations requires the exact operating characteristics of the engines.

However, the use of the new configuration of three gensets of varied power output increases the hydrogen reserve by 5.91 t or 124.88 m^3 (14%). This is a significant amount of hydrogen, but given the unfavourable load factor of the engines with the same power (Table 9), it can be assumed that the required hydrogen reserve will also increase in this configuration due to the increased fuel consumption in the three operating states.

A precise determination of the difference in the required hydrogen reserves for the two genset configurations is not possible at this stage of this study due to the lack of load characteristics or universal propulsion engines, so an average value from both configurations, i.e., 960 m^3, was used for further analysis of the required hydrogen volume.

3.3.3. Hydrogen Fuel Volume and Arrangement of Cylinders in the Ship's Hull

In order to store 960 m^3 of hydrogen, 2096 Hydrogen high-pressure Type 4 cylinders (Table 1) from Hexagon Purus will be required, the parameters of which are outlined in Section 2. For the transport of hydrogen fuel cylinders in the ship's hull, a preliminary initial rack structure was provided for the installation of nine cylinders—as illustrated in Figure 10.

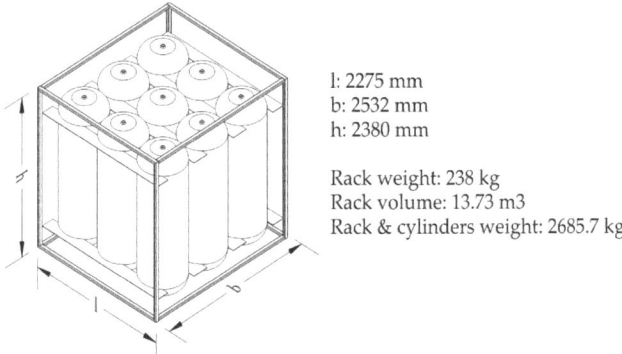

Figure 10. Modular rack for carrying hydrogen cylinders on an SOV [authors' work].

The calculated weight of all the hydrogen-filled tanks was 592.19 t, and after adding the weight of the module frames, the figure rose to 625.6 t, a very high value considering the ship's carrying capacity of 1005 t. It can therefore be seen that storing this amount of hydrogen significantly reduces the available deadweight of the vessel, and thus limits the transport of technical and service personnel and spare parts to offshore wind farms—the ship's primary function.

In this situation, to use hydrogen as a fuel for the SOV, the reserve of hydrogen carried on the SOV must be limited. This can be achieved by the following:

- Reducing the service life of the vessel;
- Creating opportunities for offshore hydrogen refuelling, e.g., at an offshore hydrogen plant, which could ultimately be connected to an offshore wind farm;
- Employing an additional battery system for electricity storage;
- Replacing gensets with reciprocating internal combustion engine generators by hydrogen fuel cells supported by an electricity storage system.

In the latter two solutions, great emphasis is put on the energy efficiency of the vessel, which can be equipped with dual-fuel hydrogen-diesel engines, electric propulsion and a supporting energy storage system or with hydrogen fuel cells, electric propulsion and a supporting energy storage system. However, such SOV energy system solutions will be the subject of future research by the authors, as designs for such vessels are currently being developed.

One solution is based on setting up green hydrogen production facilities at sea in the form of electrolyser stations working in conjunction with offshore wind farms and offshore green hydrogen storage facilities. The availability of such solutions and the creation of an efficient hydrogen refuelling system can provide service vessels with increased service life using a dual-fuel hydrogen-diesel energy system.

In the current study, the authors focused on the analysis of an SOV energy system with hydrogen-fuelled dual-fuel reciprocating engines. To select and deploy the fuel tanks, it was necessary to limit the service life of the SOV analysed. Assuming its daily operation with the same profiles (Table 4), a linear increase in fuel demand was obtained from one operating day to another. Figure 11 shows the calculated demand for hydrogen cylinders (with the assumed parameters), the number of modules with cylinders, together with their total weight to ensure the ship's endurance within the assumed range (1–14 days). At this point, it can be concluded that on an SOV with the assumed parameters, it will not be possible to transport a hydrogen supply providing a 14-day mission—contrary to the case of the same vessel with the other types of fuel (methanol, LNG and MDO).

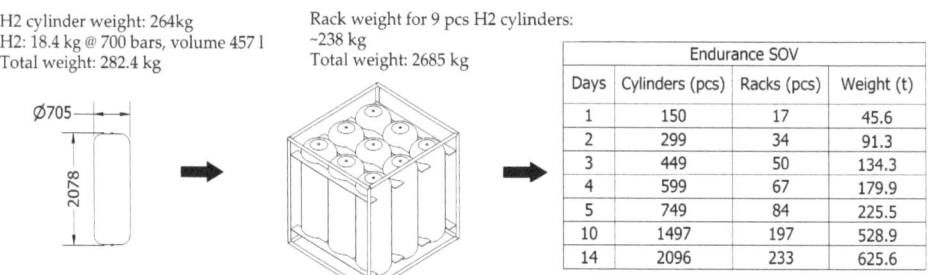

Figure 11. Diagram of the required number of hydrogen cylinders and the modules depending on the endurance of the SOV [authors' work].

By calculating the required fuel mass (Table 7) and the resulting tank volumes for the four fuel types, it was possible to determine and compare the necessary footprint and volume in the ship's hull. Table 10 shows the results of the above volumes for hydrogen fuel, transported in cylinders, over its full range of endurance (1–14 days).

Table 10. Required area and volume for hydrogen cylinder modules in the SOV hull.

No.	Endurance	Number of Cylinder Modules [pcs]	Area for Cylinder Modules in the Hull [m²]	Volume of Cylinder Modules in the Hull [m³]	Percentage Share of Underwater Hull Volume
1.	1 day	17	110	228	5.0
2.	3 days	50	330	685	9.0
3.	5 days	83	550	1142	23.0
4.	7 days	116	769	1599	33.0
5.	9 days	150	989	2056	42.0
6.	11 days	183	1209	2513	51.0
7.	13 days	216	1429	2970	61.0
8.	14 days	233	1539	3198	66.0

By analysing the results of the calculations (Table 10), we conclude that, in terms of footprint and volume, the service vessel will not have the technical capacity to store hydrogen fuel for a 14-day mission. The number of days for which the vessel can remain operational at sea running on hydrogen fuel is marked in green.

3.4. Fuel Costs

For estimated comparisons of the cost-effectiveness of the fuels, this study uses unit prices of the fuels based on price levels as of October 2024. Fuel costs related to the 14-day mission of the vessel, while hydrogen fuel calculations covered the time span of 1–14 days. In addition, annual fuel costs were calculated taking into account the number of voyages performed for the assumed T_R = 250 operating days.

It is important to note that marine fuel prices have fluctuated significantly over the past year depending on the type of fuel and global economic factors. Comparing them to last year's prices—adopted in this study, the prices of methanol fuel and LNG gas have fallen slightly, while traditional fuel has seen an increase of more than 30%—which are now relatively expensive fuel, dependent on global oil prices and emissions regulations. Hydrogen fuel is one of the most expensive marine fuels, with green hydrogen reaching a few thousand dollars per tonne. In addition, expensive high-pressure composite tanks and a hull-mounted system must be added as hydrogen-related costs, further increasing the ultimate hull value.

According to [70], the price of green hydrogen can vary from only EUR 1.12/kg to as much as EUR 16.06/kg depending on geographical location or the technology used. It varies widely; therefore, the authors assumed a value range of EUR 3.80–9.40/kg for the areas of Norway, the UK and Germany—European countries located close to the Baltic Sea. The authors of [70] suggest that by 2050, hydrogen may cost around 1 EUR/kg. The challenge in achieving this competitive price of green hydrogen calls for further technological and infrastructural advances, comprising hydrogen storage, transport and distribution, major factors affecting the cost of the entire supply chain.

This study is based on the price of P_{fuel} as of 14 October 2024, published in [71]:

- P_{MDO} = 650.50 [USD/mt]—the unit price of MDO;
- P_{LNG} = 806 [USD/mt]—the unit price of LNG;
- $P_{methanol}$ = 454 [USD/mt]—the unit price of methanol;
- P_{H2} = 3.80–9.40 [EUR/kg]—the unit price of hydrogen (at an average exchange rate of EUR 1 = USD 1.05, P_{H2} = 3990–9870 [USD/mt]).

The following relationships were used to calculate the fuel costs required:

- Fuel costs K_{fuel} [USD/mission]:

$$K_{fuel} = M_{fuel} \cdot P_{fuel} \qquad (6)$$

where M_{Tfuel}—total fuel reserve [mt], and P_{fuel}—fuel unit price (depending on the type of fuel assumed) [USD/mt];

- Number of voyages per year n_M [-]:

$$n_M = \frac{T_R}{T_S} \qquad (7)$$

where T_R—operating days [day], and T_S—mission duration [day];

- Annual fuel costs K_{Tfuel} [USD/year]:

$$K_{Tfuel} = n_M \cdot K_{fuel} \qquad (8)$$

where n_M—number of voyages per year [-], and K_{fuel}—fuel costs [USD/mission].

4. Results

4.1. Arrangement of Hydrogen Cylinders on the SOV

As a result of the calculation analyses, the vessel's possible operating time is limited to about 3–4 days. Given the current limited regulations (Section 3.2) and the lack of clear guidance on the possibility of installing cylinders inside the ship's hull, the authors have decided to locate hydrogen cylinders on the SOV working deck (significantly limiting the main working area of the ship) and in the afterpeak.

Figures 12 and 13 show variants of the hypothetical layout of hydrogen cylinder modules:

- Variant 1—150 cylinders deployed in 17 modules (red)—the case for a one-day mission. This option was considered a good reference for other cases considered;
- Variant 2—deployment of 449 cylinders in 50 modules (blue)—for a 3-day ship mission—the most likely and feasible option:
  - Total fuel reserve M_{Tfuel}—9.1 t;
  - Fuel volume V_{fuel}—205.75 m^3;
  - Full cylinder weight (with hydrogen)—134.3 t.

The values obtained in the second variant are acceptable given the parameters of the ship under analysis. The weight of the full hydrogen tanks represents 13.4% of the ship's deadweight, and the volume of hydrogen in the modules at 700 bar represents about 14.0% of the volume of the underwater part of the hull.

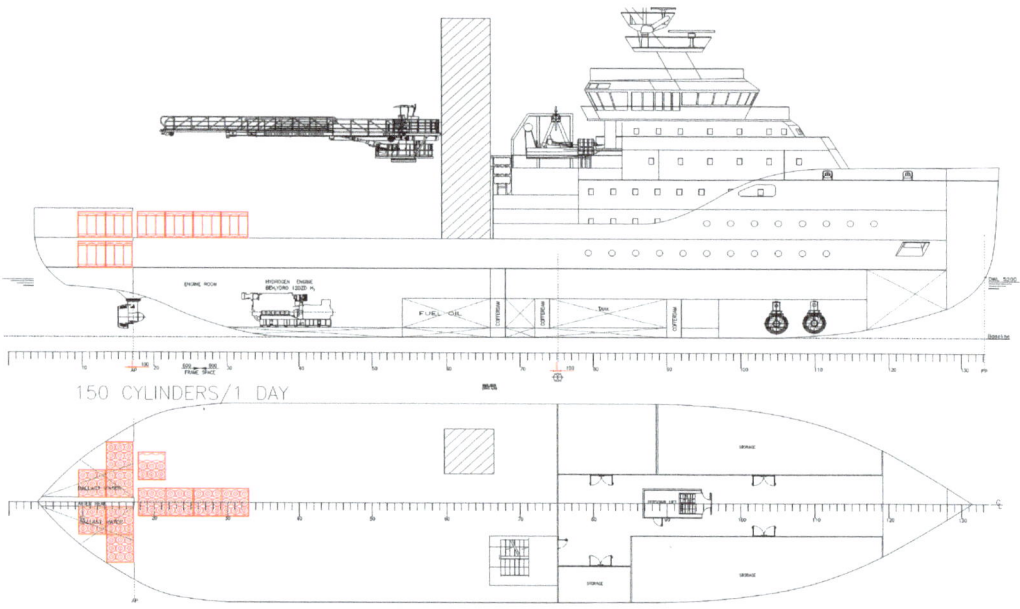

Figure 12. Layout concept of cylindrical tanks, side and deck view—variant 1 (150 cylinders).

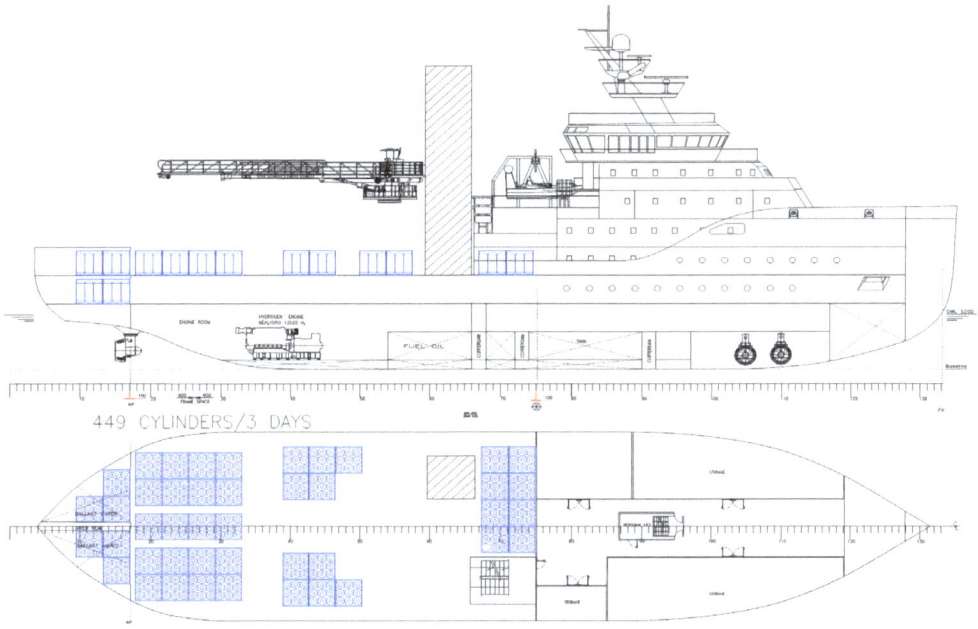

Figure 13. Layout concept of cylindrical tanks, side and deck view—variant 2 (449 cylinders).

Figure 14 shows a section of a view of the power plant with H_2 fuel generating sets.

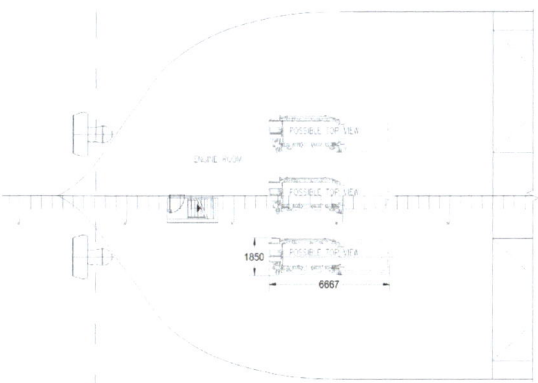

Figure 14. Fragment of a view of the ship's engine room with generating sets.

The estimated floor area required for hydrogen fuel cylinders outside the engine room was 1539 m^2, for methanol fuel and its required compartments 264 m^2, for LNG 230 m^2 and for MDO 360 m^2 (Table 7). Considering that the area occupied by the gensets is very similar for all four of their types, the occupied space in the ship's hull will be mainly determined by the fuel tanks and the additional compartments and facilities required when alternative fuels such as LNG, methanol and hydrogen are used, e.g., regulatory protective compartments, fuel treatment rooms, the so-called preparation room for methanol and the airlocks for both alternative fuels.

4.2. Needed Masses and Volumes of Hydrogen

The results of the analyses are presented graphically in the charts below. Figures 15 and 16 show the masses and occupied volumes of the generating sets, respectively, compared to the masses and volumes of the individual fuels required for a 14-day SOV mission.

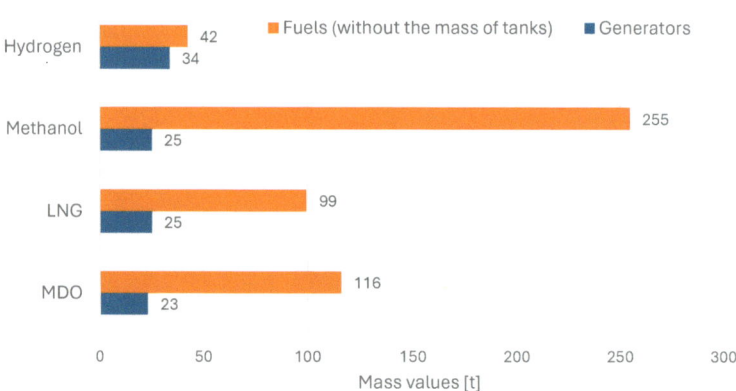

Figure 15. Comparison of the mass of generating sets with the mass of fuel required for a 14-day SOV mission.

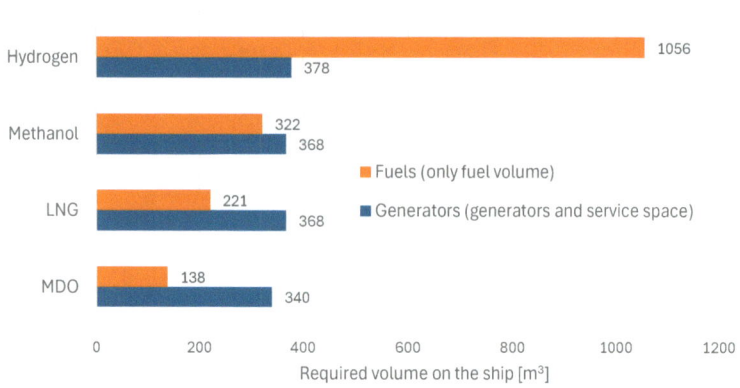

Figure 16. Comparison of the volume occupied by generating sets (including service volume) with the volume of fuel required for a 14-day SOV mission.

In the graphs shown, an inverse correlation between the mass and volume of the fuel itself is apparent: The mass of hydrogen is only slightly greater than that of the generating sets, but its volume is nearly three times larger than the space occupied by the generating sets, including the service bay. The situation is entirely different for the other fuels, as their mass is significantly greater than that of the generating sets, especially in the case of methanol, while the required volume is smaller than the volume occupied by the generating sets, particularly in the case of MDO.

The following chart (Figure 17) provides a summary of the required mass and volume of the various fuels for a 14-day SOV mission.

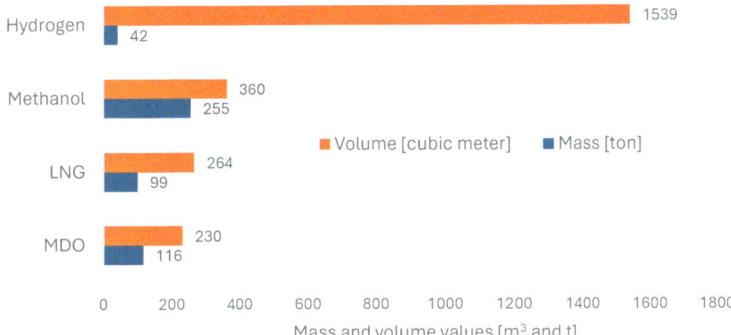

Figure 17. Comparison of mass and volume of fuel required for a 14-day SOV mission.

In the diagram above, it is clear that the required amount of hydrogen has a significantly lower mass but a much greater volume compared to the other fuels. However, due to the very high mass of the compressed hydrogen tanks, its use for SOV propulsion on a 14-day mission is virtually impractical, as its reserves occupy excessive cargo space on the ship, significantly limiting the vessel's carrying capacity.

A comparative overview of the required reserves of various fuels for a 14-day SOV mission is presented in the graph showing the cumulative percentages of mass and volume of these fuels (Figure 18).

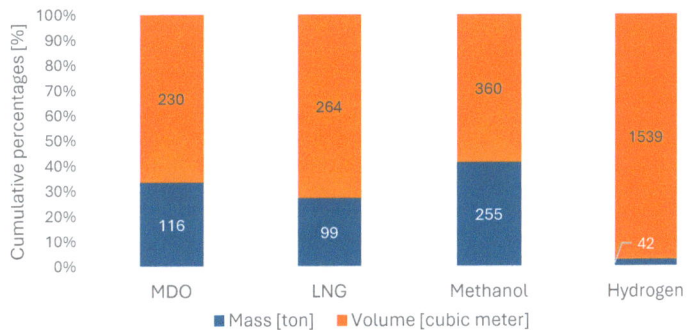

Figure 18. Comparison of the cumulative percentage values of the mass and volume of the fuel required for a 14-day SOV mission.

Figure 19 shows the percentage comparison of the mass of hydrogen-filled tanks in relation to the deadweight of the analysed SOV and, for comparison, the mass of the remaining fuels for a 14-day SOV mission.

The drawing shows the very large mass of the hydrogen tanks, which takes up over 60% of the ship's deadweight.

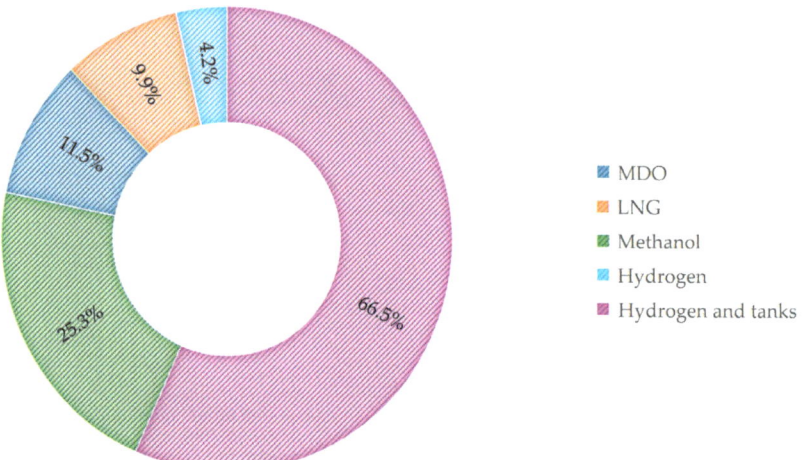

Figure 19. Percentage of mass of the fuel required for a 14-day SOV mission compared to a deadweight of the ship.

4.3. Calculated Fuel Costs

Figure 20 shows the estimated costs of the individual K_{fuel} for a single SOV mission and additionally, for comparison, the costs of hydrogen for a 1-day and a 3-day mission, which are related to the previously presented conclusion that the use of hydrogen for SOV propulsion for a 14-day mission is practically impossible (too large volume and mass of hydrogen tanks).

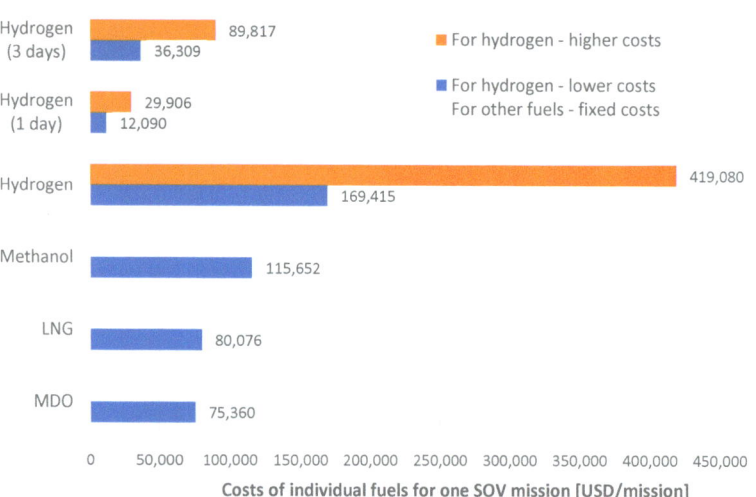

Figure 20. Costs of individual fuels for one SOV mission.

If sailing on 14-day missions, in the assumed period of operations, one vessel may make 18 trips per year, which, after taking into account the fuel costs per mission, gives an estimate of annual fuel costs. For the hydrogen variant, it was assumed for economic justification that the SOV could carry out a 14-day mission, provided that it has the opportunity to replenish hydrogen fuel from hydrogen storage facilities located in the wind farm area to ensure mission continuity. Taking this into account, the significant cost of this fuel is apparent compared to the other fuels included in this study.

5. Summary and Conclusions

The present study appears to be relevant, as it presents the first results on the feasibility of using an innovative hydrogen fuel, transported under high pressure, on a small service vessel for operation in the area of wind farms in the Baltic Sea. For the purpose of this study, the analyses referred to a selected ship type, i.e., a service offshore vessel, and the power demand for the SOV's assumed operating regimes. In addition, a comparison of hydrogen fuel is made to three fuel types, i.e., MDO, LNG and MeOH, which are presented in [52]. In order to achieve the main objective, the state-of-the-art in the application of hydrogen fuel in the marine industry was reviewed, including the capabilities of modern hydrogen storage tank, and marine hydrogen-fuelled engines, as well as the current regulations.

As a result of this research carried out, it has been concluded that the use of hydrogen fuel, carried and stored in the compressed form, burned directly in reciprocating internal combustion engines for propulsion, would not be viable for 14-day ship missions due to the technical problem of storing large volumes of hydrogen, as well as the significant weight of composite cylindrical cylinders because of the following:

- The weight of the hydrogen tanks and the modular system amounted to about 626 t, which is 62% of the ship's current carrying capacity;
- The volume of the hydrogen tanks and the modular system occupies 3198 m^3, which is 66% of the ship's current volume.

Therefore, with regard to the other fuels (MDO, LNG and MeOH), hydrogen fuel and the ship's transfer and storage system require a much larger, if not impossible, weight and volume.

Regarding fuel costs, with the currently estimated and rather complex unit hydrogen prices, the calculated costs of this fuel for SOVs (considering the minimum price) are as follows:

- More than 100% higher than MDO and LNG;
- More than 30% higher than MeOH.

The choice of hydrogen fuel for the service vessel would be justified in the case of a very short ship's autonomy time allowing for up to three-day missions or the use of a solution that fits in with the zero-emission decarbonisation goal, i.e., the use of renewable energy. This means refuelling hydrogen from offshore storage facilities supplied with hydrogen generated from the electrolysis of seawater using electricity generated by an offshore wind farm. For the adopted scenarios, for one-day endurance (recharging from renewable energy at sea on the other mission days), the fuel costs are USD 12,090, and for three-day endurance, the costs are USD 36,309. With such high costs, it is necessary to stabilize the prices of hydrogen, especially its 'green' variety.

The analysis of the SOV design was limited to identifying and comparing the different engine and fuel options in the available hull space and identifying the differences in spatial layout, in line with the current regulations.

The authors' current research has focused on the direct combustion of hydrogen in reciprocating internal combustion engines, while the use of hydrogen fuel cells on an SOV vessel will be analysed in further stages of this research and also by comparing the two solutions from a technical and economic point of view. Moreover, due to the identified problem of storing such large volumes of hydrogen on SOVs and the very large mass of the tanks required for this, the authors will investigate the problem of overcoming the challenges related to storing and refuelling hydrogen at sea in further research.

The conducted research has also demonstrated the potential use of offshore wind farms as bunkering stations for hydrogen fuel obtained from seawater electrolysis. As renewable energy sources such as offshore wind power continue to develop, the price of electricity generated from these sources can be expected to fall. Consequently, the reduced cost of electricity will directly translate into the cost of the entire transport chain—which could make green hydrogen more competitive in the energy market.

With regard to the main objectives of this work, the following conclusions can be made:

1. The analysis of the configuration of the generating sets should be carried out very carefully, as it affects the change in the required hydrogen in stores and may cause problems in the spatial layout of the power plant.
2. The use of compressed hydrogen as fuel on the SOV is possible and has a significant impact on the design of structural elements and spatial layout of the vessel; however, in order to fully exploit the potential of hydrogen, several important problems need to be solved related to hydrogen generation and storage and ship refuelling in offshore areas, such as offshore wind farms.
3. In order to increase the economic viability of operating a hydrogen-powered SOV using reciprocating internal combustion engines in its energy system, hydrogen prices must go down, especially the 'green' variety. With moderate hydrogen prices, the operation of such a vessel will be viable, especially when we appreciate zero CO_2 emissions.

The coming years will show whether hydrogen fuel technology, especially when used on smaller vessels, will become a permanent part of the energy supply of ships and contribute to improved air quality.

Author Contributions: Conceptualization, M.B. and A.Z.; Methodology, M.B. and A.Z.; Formal analysis, M.B. and A.Z.; Resources, M.B. and A.Z.; Data curation, M.B. and A.Z.; Writing—original draft, M.B. and A.Z.; Writing—review & editing, M.B. and A.Z. All authors have read and agreed to the published version of the manuscript.

Funding: This research received no external funding.

Data Availability Statement: The original contributions presented in this study are included in the article. Further inquiries can be directed to the corresponding authors.

Conflicts of Interest: The authors declare no conflicts of interest.

References

1. International Maritime Organization (IMO). Available online: https://www.imo.org/ (accessed on 10 October 2024).
2. Review of Maritime Transport 2023. UN Trade and Development. Available online: https://unctad.org/publication/review-maritime-transport-2023 (accessed on 10 October 2024).
3. Ørbeck-Nilssen, K. CEO Maritime DNV. A Deep Dive into Shipping's Decarbonization Journey. Energy Transition Outlook 2023, Maritime Forecast to 2050. Available online: https://www.dnv.com/publications/maritime-forecast-to-2050-edition-2023/ (accessed on 10 October 2024).
4. International Energy Agency. Global Hydrogen Review 2024. Revised Version October 2024. Available online: https://www.iea.org/reports/global-hydrogen-review-2024 (accessed on 10 October 2024).
5. Maritime Forecast to 2050, Report DNV. Available online: https://www.dnv.com/maritime/publications/maritime-forecast (accessed on 10 October 2024).
6. Mandra, J.O. Clarksons: 45% of Ships Ordered in 2023 Embrace Alternative Fuels, with LNG Still in the Lead. Business Developments & Projects. 3 January 2024. Available online: https://www.offshore-energy.biz/clarksons-45-of-ships-ordered-in-2023-embrace-alternative-fuels-with-lng-still-in-the-lead/ (accessed on 10 October 2024).
7. Potential of Hydrogen as Fuel for Shipping by ABS, CE DELFT & ARCSILEA, European Maritime Safety Agency EMSA/Hydrogen—2022/2023—4837444. 31 August 2023. Available online: https://emsa.europa.eu/publications/reports/item/5062-potential-of-hydrogen-as-fuel-for-shipping.html (accessed on 10 October 2024).
8. E-Magazyny. Available online: https://e-magazyny.pl/aktualnosci/elektromobilnosc/usa-pierwszy-statek-na-wodor-seachange/ (accessed on 3 June 2024).
9. Switch Maritime. Available online: https://www.switchmaritime.com (accessed on 1 August 2024).
10. Ship Technology. Available online: https://www.ship-technology.com/projects/hydroville-passenger-ferry/?cf-view&cf-closed (accessed on 1 August 2024).
11. Ostvik, I. MF Hydrs—World's First LH2 Driven Ship and He Challenges Ahead Towards Zero-Emission Shipping. NORLED, The Greatest Travel Experiences. 2021. Available online: https://www.uib.no/sites/w3.uib.no/files/attachments/norled_mf_hydra_dec_2021.pdf (accessed on 23 August 2024).

12. Berger, M. Statki Napędzane Wodorem. Tak, to Tylko Kwestia Czasu. Bezemisyjny Transport Wodny. Available online: https://elektromobilni.pl/statki-napedzane-wodorem-tak-to-tylko-kwestia-czasu (accessed on 8 September 2024).
13. Grzybowski, M. The Hydrogen SOV Is Sailing Out into the Open Waters. BSSC. 2024. Available online: https://www.bssc.pl/2024/05/16/the-hydrogen-sov-is-sailing-out-into-the-open-waters/ (accessed on 24 September 2024).
14. Offshore WIND. Available online: https://www.offshorewind.biz/2024/03/19/louis-dreyfus-armateurs-unveils-liquid-hydrogen-sov-concept/ (accessed on 24 September 2024).
15. Melideo, D.; Desideri, U. The use of hydrogen as alternative fuel for ship propulsion: A case study of full and partial retrofitting of roll-on/roll-off vessels for short distance routes. *Int. J. Hydrogen Energy* **2024**, *50*, 1045–1055. [CrossRef]
16. Liu, X.; Yang, G.; Sun, B.; Wang, H.; Li, Y.; Wang, R. Assist in GHG Abatement of Offshore Ships: Design and Economic Analysis of an Integrated Utilization Model of hydrogen-powered Ship and Offshore Wind Power. *Sustain. Mar. Struct.* **2023**, *5*, 15–25. [CrossRef]
17. Kołodziejski, M. Review of hydrogen-based propulsion systems in the maritime sector. *Arch. Thermodyn.* **2024**, *44*, 335–380. [CrossRef]
18. Karvounis, P.; Tsoumpris, C.; Boulougouris, E.; Theotokatos, G. Recent advances in sustainable and safe marine engine operation with alternative fuels. *Front. Mech. Eng.* **2022**, *8*, 994942. [CrossRef]
19. Minutillo, M.; Cigolotti, V.; Di Ilio, G.; Bionda, A.; Boonen, E.J.; Wannemacher, T. Hydrogen-based technologies in maritime sector: Technical analysis and prospective. In Proceedings of the European Fuel Cells and Hydrogen Piero Lunghi Conference, Online, 15–17 December 2021; Volume 334, p. 06011.
20. Lee, G.; Kim, J.M.; Yung, K.H.; Park, H.; Jang, H.; Lee, C.S.; Lee, J.W. Environmental Life-Cycle Assessment of Eco-Friendly Alternative Ship Fuels (MGO, LNG, and Hydrogen) for 170 GT Nearshore Ferry. *J. Mar. Sci. Eng.* **2022**, *10*, 755. [CrossRef]
21. Perčić, M.; Vladimir, N.; Koričan, M.; Jovanović, I.; Haramina, T. Alternative Fuels for the Marine Sector and Their Applicability for Purse Seiners in a Life-Cycle Framework. *Appl. Sci.* **2023**, *13*, 13068. [CrossRef]
22. Krantz, G.; Moretti, C.; Brandão, M.; Hedenqvist, M.S.; Nilsson, F. Assessing the Environmental Impact of Eight Alternative Fuels in International Shipping: A Comparison of Marginal vs. Average Emissions. *Environments* **2023**, *10*, 155. [CrossRef]
23. Wang, Q.; Zhang, H.; Huang, J.; Zhang, P. The use of alternative fuels for maritime decarbonization. Special marine environmental risks and solutions from an international law perspective. *Front. Mar. Sci.* **2023**, *9*, 1082453. [CrossRef]
24. dos Santos, V.A.; da Silva, P.P.; Serrano, L. The Maritime Sector and Its Problematic Decarbonization: A Systematic Review of the Contribution of Alternative Fuels. *Energies* **2022**, *15*, 3571. [CrossRef]
25. McKenney, T.A. The impact of maritime decarbonization on ship design: State-of-the-Art Report. In Proceedings of the 15th International Marine Design Conference (IMDC-2024), Delft, The Netherlands, 2–6 June 2024. Available online: https://www.researchgate.net/publication/380892629_The_impact_of_maritime_decarbonization_on_ship_design_State-of-the-Art_Report (accessed on 10 September 2024).
26. Bacquart, T.; Moore, N.; Wilmot, R.; Bartlett, S.; Morris, A.S.O.; Olden, J.; Becker, H.; Aarhaug, T.A.; Germe, S.; Riot, P.; et al. Hydrogen for Maritime Application—Quality of Hydrogen Generated Onboard Ship by Electrolysis of Purified Seawater. *Processes* **2021**, *9*, 1252. [CrossRef]
27. Hwang, S.S.; Gil, S.J.; Lee, G.N.; Lee, J.W.; Park, H.; Jung, K.H.; Suh, S.B. Life cycle assessment of alternative ship fuels for coastal ferry operating in Republic of Korea. *J. Mar. Sci. Eng.* **2020**, *8*, 660. [CrossRef]
28. Dong, D.; Schönborn, A.; Christodoulou, A.; Ölçer, A.; González-Celis, J. Life cycle assessment of ammonia/hydrogen-driven marine propulsion. *Proc. Inst. Mech. Eng. Part M J. Eng. Marit. Environ.* **2024**, *238*, 531–542. [CrossRef]
29. Panić, I.; Cuculić, A.; Ćelić, J. Color-coded hydrogen: Production and storage in maritime sector. *J. Mar. Sci. Eng.* **2022**, *10*, 1995. [CrossRef]
30. Zintegrowana Platforma Edukacyjna Ministerstwa Edukacji Narodowej. Available online: https://zpe.gov.pl/a/wo-dor/DbFQCGctg (accessed on 10 October 2024).
31. Publikacja Informacyjna PRS 11/I. Bezpieczne Wykorzystanie Wodoru Jako Paliwa w Komercyjnych Zastosowaniach Przemysłowych. PRS. Czerwiec 2021. Available online: https://prs.pl/wp-content/uploads/2024/03/p11i_pl.pdf (accessed on 7 July 2024).
32. Züttel, A. Materials for hydrogen storage. *Mater. Today* **2003**, *6*, 24–33. [CrossRef]
33. Ligęza, K.; Narloch, P. Perspektywy magazynowania wodoru z odnawialnych źródeł energii w Polsce cz. 1. *Nowa Energ.* **2023**, *1*, 73–81.
34. Zuttel, A.; Borgschulte, A.; Schlapbach, L. *Hydrogen as a Future Energy Carrier*; Wiley-VCH Verlag GmbH & Co. KGaA: Weinheim, Germany, 2008.
35. Kozikowski, S.; Szymlek, K. Wodór—Zielone Złoto Bezpieczne Magazynowanie—Szanse i Wyzwania. Innowacje w Energetyce 2022. Available online: https://www.cire.pl/filemanager/Materia%C5%82y%20Problemowe%20(Wies%C5%82aw%20Drozdowa)%20/046366fa71c120f78917931799e8aae2abc9495995753756f4091b2bf5a3d369.pdf (accessed on 17 September 2024).
36. CW Composites World. Available online: https://www.compositesworld.com/articles/infinite-composites-type-v-tanks-for-space-hydrogen-automotive-and-more (accessed on 17 September 2024).
37. Hyfindr. Available online: https://hyfindr.com/en/hydrogen-knowledge/hydrogen-tank (accessed on 17 September 2024).
38. Rynek Zbiorników Ciśnieniowych—Materiały Kompozytowe—Composite Marerials (kompozyty.net). Available online: https://kompozyty.net/rynek-zbiornikow-cisnieniowych/ (accessed on 24 September 2024).

39. Magazynowanie Wodoru. Available online: https://wme-z1.pwr.edu.pl/wp-content/uploads/2020/01/3-Magazynowanie-wodoru-2019.pdf (accessed on 24 September 2024).
40. Amargo Tank Think Tank. Available online: https://www.amargo.pl/zbiorniki-kompozytowe-do-magazynowania-wodoru-typ-iv-potencjal-projektu-i-aktualny-poziom-gotowosci-technologicznej/ (accessed on 24 September 2024).
41. Hexagon Purus. Type-4-Infrastructure-and-Mobility-Hexagon-Purus-Datasheet.pdf. Rev. 04/2024. Available online: https://hexagonpurus.com/our-solutions/download-our-brochures/datasheets (accessed on 17 September 2024).
42. NPROXX. Available online: https://www.nproxx.com/hydrogen-powered-heavy-duty-vehicles/shipping-maritime/ (accessed on 17 September 2024).
43. NPROXX. Available online: https://www.nproxx.com/nproxx-introduces-the-new-700-bar-hydrogen-pressure-vessel-ah620-70-2-targeting-heavy-vehicles-long-haul-trucks-and-the-automotive-sector/ (accessed on 17 September 2024).
44. Szwaja, S. *Wodór Jako Paliwo Podstawowe i Dodatkowe do Tłokowego Silnika Spalinowego*; Wydawnictwo Politechniki Częstochowskiej: Częstochowa, Poland, 2019.
45. Katalog Firmy MAN. *Hydrogen in Shipping*; MAN Energy Solutions: Augsburg, Germany, 2020.
46. Portal Internetowy Firmy Wärtsilä—The World's First Large-Scale Hydrogen Engine Power Plant Is Here. Available online: https://www.wartsila.com/energy/sustainable-fuels/hydrogen-power-plant (accessed on 25 September 2024).
47. Portal WinGD. WinGD and Hyundai Heavy Industries Collaborate on Ammonia Two-Stroke Engine Delivery. 2022. Available online: https://www.wingd.com/en/news-media/press-releases/wingd-and-hyundai-heavy-industries-collaborate-on-ammonia-two-stroke-engine-delivery/ (accessed on 25 September 2024).
48. Katalog Silników BEH2YDRO. "Datasheet DZD H_2". Anglo Belgian Corporation. 2022. Available online: https://www.behydro.com/uploads/files/DATASHEETS-Behydro.pdf (accessed on 25 September 2024).
49. *Katalog Zastosowania Silników BEH$_2$YDRO "Information Sheet"*; Anglo Belgian Corporation: Gent, Belgium, 2022.
50. *Wynalazcy z Politechniki Przystosowali Silnik Spalinowy do Zasilania Wodorem*; Politechnika Krakowska: Kraków, Poland. 2024. Available online: https://www.pk.edu.pl/index.php?option=com_content&view=article&id=5193&Ite-mid=1152&lang=pl (accessed on 25 September 2024).
51. Katalog Silnika Scania: Scania Marine Engines DI09 074M. Scania CV AB. 2022. Available online: https://www.scania.com/content/dam/scanianoe/market/master/products-and-services/engines/pdf/specs/marine/DI09074M_269kW.pdf (accessed on 25 September 2024).
52. Bortnowska, M. Projected Reductions in CO_2 Emissions by Using Alternative Methanol Fuel to Power a Service Operation Vessel. *Energies* **2023**, *16*, 7419. [CrossRef]
53. Oleksiuk, J. Projekt Koncepcyjny Statku SOV do Obsługi Farm Wiatrowych. Master's Thesis, Faculty of Navigation, Maritime University in Szczecin, Szczecin, Poland, 2023.
54. Cichocki, M. *Napędy Statków Dynamicznie Pozycjonowanych: Aspekty Eksploatacyjne*; Wydawnictwo Naukowe PWN: Warszawa, Poland, 2018.
55. Clark, I.C.; Hancox, M. *Anchor Handling Tug Operations*; The ABR Company Limited: Trowbridge, UK, 2012.
56. Katalog Silników. *Wärtsilä 20—Product Guide*; Wärtsilä Corporation: Helsinki, Finland, 2020.
57. Katalog Silników. *Wärtsilä 20*; Wärtsilä Corporation: Helsinki, Finland, 2024.
58. Katalog Silników. *Wärtsilä Methanol Engines—The Power to Achieve Carbon Neutrality*; Wärtsilä Corporation: Helsinki, Finland, 2023.
59. Katalog Silników. *Wärtsilä 20DF*; Wärtsilä Corporation: Helsinki, Finland, 2023.
60. Giernalczyk, M.; Górski, Z. *Siłownie Okrętowe—Część II: Instalacje Okrętowe*; Wydawnictwo Akademii Morskiej w Gdyni: Gdynia, Poland, 2014.
61. Nasser, M.A.; Elgohary, M.M.; Abdelnaby, M.; Shouman, M.R. Environmental and economic performance investigation of natural gas and methanol as a marine alternative fuel. *Res. Sq.* **2022**. Available online: https://www.researchgate.net/publication/363371092_Environmental_and_economic_performance_investigation_of_natral_gas_and_-methanol_as_a_marine_alternative_fuel (accessed on 14 September 2024).
62. Methanol Safe Handling Technical Bulletin—Part 1—B: Physical & Chemical Properties of Selected Fuels (1). Methanol Institute. Available online: https://www.methanol.org/wp-content/uploads/2020/04/Part-1-B_-Physical-Chemical-Properties-of-Selected-Fuels-.pdf (accessed on 14 September 2024).
63. The Methanol—Fuelled MAN B&W LGIM Engine. MAN Energy Solutions. Available online: https://www.man-es.com/docs/default-source/document-sync-archive/the-methanol-fuelled-man-b-w-lgim-engine-eng.pdf?sfvrsn=36b925d2_7 (accessed on 14 September 2024).
64. International Maritime Organization (IMO). MSC.1/Circ.1212/Rev.2. 2024. Available online: https://www.skanregistry.com/uploads/download-directory/pdf/357/document.pdf (accessed on 12 October 2024).
65. Hellenic Shipping News. Available online: https://www.hellenicshippingnews.com/imo-ccc-10-interim-guidelines-for-ammonia-and-hydrogen-as-fuel/ (accessed on 12 October 2024).
66. International Maritime Organization (IMO). Available online: https://www.imo.org/en/OurWork/Safety/Pages/IGF-Code.aspx (accessed on 12 October 2024).
67. International Maritime Organization (IMO). Available online: https://www.imo.org/en/MediaCentre/Pages/WhatsNew-1392 (accessed on 12 October 2024).
68. Piotrowski, I.; Witkowski, K. *Okrętowe Silniki Spalinowe*; Trademar: Gdynia, Poland, 2013.

69. Katalog Silników. *Dual-Fuel Engines—WÄRTSILÄ 20DF, 34DF, 46DF and 50DF*; Wärtsilä Corporation: Helsinki, Finland, 2015.
70. Gómez, J.; Castro, R. Green Hydrogen Energy Systems: A Review on Their Contribution to a Renewable Energy System, Energy System. *Energies* **2024**, *17*, 3110. [CrossRef]
71. Ship & Bunker. Available online: https://shipandbunker.com/prices/emea/nwe/nl-rtm-rotterdam (accessed on 10 October 2024).

Disclaimer/Publisher's Note: The statements, opinions and data contained in all publications are solely those of the individual author(s) and contributor(s) and not of MDPI and/or the editor(s). MDPI and/or the editor(s) disclaim responsibility for any injury to people or property resulting from any ideas, methods, instructions or products referred to in the content.

Article

The Correlation of the Smart City Concept with the Costs of Toxic Exhaust Gas Emissions Based on the Analysis of a Selected Population of Motor Vehicles in Urban Traffic

Wojciech Lewicki [1,*], Milena Bera [2,*] and Monika Śpiewak-Szyjka [2]

[1] Department of Regional and European Studies, Faculty of Economics, West Pomeranian University of Technology in Szczecin, Żołnierska 47, 71-210 Szczecin, Poland
[2] Department of Real Estate, Faculty of Economics, West Pomeranian University of Technology in Szczecin, Żołnierska 47, 71-210 Szczecin, Poland; monika.spiewak-szyjka@zut.edu.pl
* Correspondence: wojciech.lewicki@zut.edu.pl (W.L.); milena.bera@zut.edu.pl (M.B.)

Citation: Lewicki, W.; Bera, M.; Śpiewak-Szyjka, M. The Correlation of the Smart City Concept with the Costs of Toxic Exhaust Gas Emissions Based on the Analysis of a Selected Population of Motor Vehicles in Urban Traffic. *Energies* 2024, 17, 5375. https://doi.org/10.3390/en17215375

Academic Editor: Wen-Hsien Tsai

Received: 18 September 2024
Revised: 21 October 2024
Accepted: 22 October 2024
Published: 29 October 2024

Copyright: © 2024 by the authors. Licensee MDPI, Basel, Switzerland. This article is an open access article distributed under the terms and conditions of the Creative Commons Attribution (CC BY) license (https://creativecommons.org/licenses/by/4.0/).

Abstract: The intensive development of road transport has resulted in a significant increase in air pollution. This phenomenon is particularly noticeable in urban areas. This creates the need for analyses and forecasts of the scale and extent of future emissions of harmful substances into the environment. The aim of this study was to estimate the costs of the emission of toxic components of exhaust gases generated by all users of conventionally propelled vehicles travelling on a section of urban road in the next 25 years. The traffic study was carried out on an urban traffic route, playing a key role for road transport in the dimension of a given urban agglomeration. The traffic forecast for the analysed road section was based on the results of our own measurements carried out in April 2023 and external data from the General Directorate for Roads and Motorways. The results of the observations concerned six categories of vehicles for the morning and afternoon rush hours. Based on the data obtained, the generic structure of the vehicle population on the analysed section and the average daily traffic were determined. Using the methodology contained in the Blue Book of Road Infrastructure, parameters were calculated in the form of annual indicators of traffic growth on the analysed section, travel speed, and annual air pollution costs for selected vehicle categories, remembering at the same time that the Blue Book-based methodology does not distinguish between unit costs in relation to the type of emissions. The results of the study confirmed that there was an increase in the cost of toxic emissions for each vehicle category over the projected 25-year period. The largest increases were seen for trucks with trailers and passenger cars. In total, for all vehicle categories, emission costs nearly doubled from 2024 to 2046, from EUR 3,745,229 to EUR 7,443,384, due to the doubling of the number of vehicles resulting from the traffic forecast. The analyses presented here provide an answer to the question of what pollution costs may be faced by cities in which road transport will continue to be based on conventional types of propulsion. In addition, the research presented can be used to develop urban mobility transformation plans for the coming years, within the scope of the widely promoted smart city concept and the idea of electromobility, by pointing out to local authorities the direct economic benefits of these changes.

Keywords: exhaust emission costs; transport; combustion vehicles; urban traffic; smart city; electromobility; exhaust modelling

1. Introduction

As the observations of the socio-economic environment indicate, air pollution is an issue that is extremely topical, especially in the present era of discussions on global warming and its role in this phenomenon of harmful substances [1,2]. The increase in interest in such topics is primarily linked to the increasing use of environmental resources and technological advances, as well as the increasing role of road transport. The high economic level of highly industrialised countries, in particular, is propelling a modest increase in the car population, which contributes to an increase in environmental pollution. The combustion process itself, associated with the generation of energy for automotive purposes, is one of the main sources of emissions, including CO_2, CO, SO_2, CH, and PM, as well as NO_X and many other compounds [3,4]. In addition, it takes a similar amount of air to burn 1 kg of petrol as an average adult uses in a day, and in the case of a passenger car it is a matter of driving several kilometres [5,6]. An analysis of the developments in the automotive market reveals a trend towards environmentally friendly solutions. In terms of technological changes, many researchers point to the development of hydrogen drive technology, emphasising their zero-emissions feature [7–9]. However, it should be emphasised that currently only a few models are available on the passenger car market, such as the GLC F-Cell model produced by a German manufacturer, and two Japanese models, Mirai and Nexo. According to automotive market experts, this type of vehicle may prove useful in the near future in public urban transport, but not in individual transport, due to high purchase costs and limited market availability. The European Union has already noticed this problem, and indicated a second focus related to the development of e-fuel "drop-in" technology, without the need to make any modifications to the drive system of conventional vehicles, following the opinion of vehicle manufacturers that conventional engines will be in use until 2050. In addition, for several decades, there has been a development in legal conditions, tightening the requirements for exhaust emission standards for vehicles equipped with conventionally fuelled engines [10]. Initially, this development was linked to a reduction in fuel consumption and, more recently, to an increase in environmental considerations. All these developments have the fundamental aim of reducing the emissions of harmful substances in urban traffic. This is because cities are areas in which pollution is concentrated. The accumulation of large buildings, increased traffic volumes, and specific street layouts determine the deterioration of air quality [11]. Despite the implementation of increasingly modern technological and structural solutions in road transport, a further increase in traffic is expected, especially in cities. This will result in a further increase in toxic emissions. Analysis and research into the impact of vehicle populations on urban air quality is therefore becoming an extremely critical issue. The research imperative itself focuses on identifying the actual costs as well as the post-intubation costs of the non-transformation of individual road transport modes in relation to urban mobility processes. Due to the expected increase in traffic intensity in many cities, emissions are expected to continue to increase if no action is taken to control them, considering that there are more and more signals that the transformation of motorization towards electromobility will be longer and more difficult than expected. It should therefore be emphasised that studies, such as those presented by the authors on the characteristics of vehicle emission costs in urban areas with the highest traffic intensity, may be of key importance for the systemic implementation of green transport zones in cities and the acceleration of transformation in the domain of urban mobility.

The considerations presented in this paper focus on modelling the costs of the emission of harmful components of exhaust gases on the example of a selected population of motor vehicles in urban traffic. These considerations relate to the extremely topical issue of forecasting the costs of pollutant emissions from road transport, starting a discussion on accelerating the processes of electromobility development, the framework for which has been defined in the promoted smart city concept.

The innovativeness of the presented research lies in the use of a model based on the assumptions of the "Blue Book-Road Infrastructure". This analysis takes into account the dynamics of vehicle traffic in real road conditions in the coming years, considering forecast changes in the number and structure of the vehicle fleet, such as an increase in the number not only of passenger cars but also of buses, delivery trucks, and tractor-trailers with semi-trailers.

At the same time, it introduces a new approach to the issue of research and analysis on the continued operation of combustion-powered vehicles in urban areas. The work complements the research and practice-oriented analysis in the area of measuring toxic emissions from road transport over a longer time horizon. The research presented below therefore has both a theoretical and a practical dimension. It directly fills a research gap in such related areas as: II (emission costs) or II (combustion vehicles), III (urban traffic). IV (exhaust modelling) V (smart city concept).

The article itself is divided into the following sections: Section 1 provides an introduction to the topic. Section 2 reviews the literature on the link between the smart city concept and the reduction in road transport pollution. Section 3 describes the research methodology used to analyse and estimate the cost of toxic exhaust emissions. Section 4 contains the results of the research and numerical experiment and extensive commentary on them. Section 5 contains the research conclusions and perspectives for further development for the analysis and research on this topic.

2. Literature Review: Linking the Smart City Concept to Pollution Reduction

There are many definitions of 'smart city' in the literature, which emphasise different aspects of the issue. Some define a smart city as a city that uses ICT to improve the functioning of urban infrastructure and services, while others emphasise the role of sustainability and public participation [12]. One broader definition of a smart city has been proposed by the authors of the report Mapping Smart Cities in the EU [13]. According to this definition, a smart city is a city in which public issues are addressed using information and communication technologies (ICT), involving different types of stakeholders working in partnership with the city authorities. ICT enables the interconnection of different urban systems and stimulates innovations that facilitate the realisation of urban policy goals. Energy savings on a city scale can be achieved, for example, through the use of intelligent electricity grids that match energy supply to actual demand or by providing appropriate information to individual users so that they take into account not only cost, but also environmental aspects when choosing appliances [14]. One of the researchers presented the key aspects that make up the image of a sustainable city, structured in a contemporary interpretation as a smart city, and its dimensions related to the generation and implementation of innovative solutions in various areas of [15]. In the context of the smart city, there is an increasing emphasis on involving citizens in decision-making processes, leading to the term 'smart citizens'. Citizens are not only users, but also co-creators of city services [16]. The smart city is also a response to sustainability challenges, such as reducing CO_2 emissions, efficient waste and water management, and protecting the environment [17]. The European approach to smart cities focuses primarily on carbon reduction measures and energy efficiency in every area of city functioning, while improving the quality of life of the inhabitants [18]. Smart cities, thanks to information and communication technologies, make more efficient use of available resources to improve the quality of life in a city and ensure its sustainability. An example of such a city is Vienna, which has successfully implemented the concept and is currently leading many rankings on the smart city concept in the area of transport transformation [19]. Other researchers reviewed experiences and solutions of the smart city concept in the field of transport adopted in selected European cities [20]. Voronina R. mentioned positive aspects and trends in the development of Lviv's transport system and innovative solutions. A traffic management system prioritising public transport was presented, which allowed for a better organisation of vehicle traffic, an increase in the quality of tram transport on priority lines, and the monitoring of traffic at intersections with a consequent increase in safety. The author presents proposals for the development of public transport and the promotion of the use of bicycles for travel as a result of the 'Smart city Lviv' project [21].

One of the significant challenges that smart cities are trying to address is the problem of toxic emissions, which have a significant impact on public health and the environment. This topic is widely discussed in the literature, particularly in the context of reducing emissions through innovative technologies and sustainable transport solutions for road transport. It is emphasised that smart cities can significantly contribute to the reduction of toxic emissions through the implementation of intelligent traffic management systems, the development of public transport, and the promotion of low-carbon mobility. These cities can monitor and analyse air quality data, allowing them to take quick and effective action [22]. Some researchers describe the evolution of the smart city concept towards a green smart city, with an emphasis on reducing energy consumption and emissions [23]. Other researchers analysed the smart city concept in one city, with a focus on emission-reducing technologies. The implementation of smart transport systems in this city was aimed at optimising public transport travel times, the fast and efficient transmission of traffic information, improving the quality of public transport, and enhancing the quality of life in the city. The solutions involved the modernisation and creation of new traffic lights linked to systems that take into account more intersections [24]. One of the researchers, in turn, analysed the development of smart cities from the perspective of investment in human and social capital and advanced industrial technologies [25]. Another researcher focused on the role of small and medium-sized enterprises in the smart city space and innovation in business models [26]. Other researchers discussed the use of technology in waste management in smart cities, with an emphasis on minimising resources and reducing emissions of pollutants [27]. An important part of ICT is the increasingly popular Internet of Things (IoT), i.e., networked devices that are connected via the Internet. Through the use of the Internet of Things, e.g., in sensors in bins, the collected data can be generated and analysed, making the waste management process more functional [28].

Recent years have shown that, within the smart city concept, mobility is becoming one of the key elements in addressing the issue of reducing transport pollution. According to one researcher, the processes observed bring about innovations in transport that enable better use of the road infrastructure, thus improving the mobility of residents [29]. Given that road transport is one of the main sources of air pollutant emissions, including nitrogen oxides (NOx), particulate matter (PM) and volatile organic compounds (VOCs). In the smart city itself, efforts are being made to reduce these emissions through the development of intelligent traffic management systems and the promotion of sustainable modes of transport, such as bicycles, electric vehicles, and public transport [30]. According to one researcher, intelligent traffic management systems, such as adaptive traffic lights, car-sharing systems, and dynamic parking management, can significantly reduce emissions by optimising traffic flow and reducing congestion [31]. The role of electromobility in reducing toxic emissions is also widely discussed in the literature. Smart cities promote the development of charging infrastructure for electric vehicles and offer incentives for the use of clean modes of transport [32]. The research postulates that investment in smart city technologies can bring long-term economic benefits by reducing the costs associated with the removal of environmental pollution costs [33]. One researcher has attempted to model the ICE emissions of urban vehicles, highlighting their environmental impact. Simulation studies have been carried out using real data for the intersection area of one city, enabling an objective assessment of the effectiveness of planned traffic organisation changes [34]. Empirical studies of vehicle emissions under urban conditions were also undertaken, with particular emphasis on the variability of results [35]. One of the researchers also analysed the emission characteristics of traction engines, taking into account different urban driving conditions [36]. The literature also raises the issue of identifying sources of pollutant emissions and proposing solutions to reduce them. These measures are key to improving urban air quality. Research often includes an analysis of the impact of alternative fuels such as CNG and optimisation of urban logistics, including so-called 'last mile' solutions. All these approaches aim to reduce the negative impact of transport on the urban environment. Several researchers have analysed the impact of alternative fuels on emissions, particularly in an urban context [37]. A study of transport solutions to reduce urban emissions and an assessment of the impact of logistics solutions to reduce urban emissions has been carried out [38]. Furthermore, others have carried out a statistical analysis of emissions during simulated urban vehicle use [39]. Another paper discusses how the dominance of road transport contributes to air quality, especially in large urban areas [40]. In contrast, one researcher analysed the external costs of passenger transport, taking into account the level of emissions and their impact on public health and the environment in Poland's provincial cities [41]. Examples of research also include an analysis of the impact of traffic calming on emissions and the efficiency of public transport, particularly buses, in terms of air pollution. Another study examines how traffic-calming measures affect pollutant emissions, pointing to the role of driving style in emissions [42]. Other researchers attempted to answer the question of how public transport buses affect air pollution, highlighting the importance of internal combustion vehicle technology for urban air quality [43]. The negative effects of these pollutants on the environment and the health of the population are also indicated [44]. Another researcher focused on modelling road transport emissions and their dispersion in urban space. Other researchers discussed the potential of electric mobility to reduce transport-related emissions. They argued that the larger-scale deployment of electric vehicles in cities could significantly reduce harmful emissions and improve air quality [45]. In addition, another study attempts to answer the question of why, despite the measures introduced, there is no decrease in pollution, but on the contrary, in some cases there is a noticeable increase in pollution [46]. Yet another researcher highlighted the environmental benefits of reducing emissions when using this fuel in cities [47]. The following paper compares emissions from electric and internal combustion vehicles. The differences in emissions under different operating conditions and the impact on air quality were analysed. Issues were raised, remarking that electric vehicles tend to emit less locally than internal combustion vehicles, but the overall environmental benefits depend on the source of energy used to charge the electric vehicles [48]. Others have carried out studies on the impact of low-emission transport, including electric vehicles, on urban emissions. They concluded that emissions are reduced through the use of low-emission vehicles in urban transport [49]. The following paper analyses the impact of different forms of transport on the urban environment and emissions [50]. One researcher presented different strategies for reducing vehicle traffic and their impact on emissions [51]. Other researchers have focused on emission reduction strategies, including the role of electric vehicles in achieving sustainability goals [52].

In terms of pollution forecasting, one researcher used modelling to estimate pollutant emissions for specific streets and the city as a whole. The resulting analyses attempted to identify the levels of pollutant emissions in different parts of the city, enabling the better management of air quality [53]. Subsequent researchers carried out emission measurements on different routes in the city, which allowed for the identification of zones with different levels of pollution. These studies have become important in the context of optimising bus routes to minimise negative environmental impacts [54]. Other researchers like Guo, B., Feng, Y., Lin, and J., Wang, X. analysed the effects of implementing the New Energy Demonstration City (NEDC) programme in the context of reducing urban emissions. This research used a spatial difference model to assess the effectiveness of such initiatives in reducing emission levels. The results suggest that such programmes can contribute to emission reductions, although the effects may vary depending on local circumstances [55]. In contrast, yet another researcher focused on modelling the fuel consumption and emissions of a fleet of city buses in Madrid. The authors analysed the kinetic variables of buses that affect energy consumption and emissions, identifying the most important factors that can be controlled to reduce emissions [56]. Another paper presents analyses of real-world fuel efficiency and emissions from compression–ignition engines of city buses under dynamic conditions, highlighting differences from laboratory tests [57]. Yet another study undertook a comparison of Euro VI diesel and compressed natural gas (CNG) bus emissions, highlighting the benefits of CNG in reducing emissions in urban areas [58]. There are also scientific papers in which studies have been carried out on the fuel consumption and emissions of hybrid buses on the metropolitan road network, indicating the efficiency of hybrid vehicles [59]. Zhang Y. and other researchers, meanwhile, analysed particulate emissions from biodiesel-fuelled city buses, highlighting the potential for emission reductions using after-treatment systems [60]. Research was also undertaken into the impact of ethanol–diesel blends on emissions from different bus fleets, focusing on nitrogen oxides and particulate emissions [61].

The topic of correlating smart city concepts with the cost of chemical emissions of exhaust components, especially in the context of the analysis of motor vehicles in urban traffic, also faces criticism in the literature from different perspectives, including technological, social, environmental, and economic perspectives. Townsend A. warns of risks with system interoperability, data volume management, and high costs. It also touches on social issues, such as the lack of universal access to the technology and its marginalising impact on certain groups [62]. Knox P. and Pinch S. analyse the impact of technology on urban space. They point out that smart city management and emissions systems are often imposed in a public context, which limits their operation. They also focus on the issues surrounding the introduction of new technologies and their transmission to teams [63]. Sheller M. and Urry J. indicate that, despite the benefits of intelligent traffic management, such solutions may have unintended consequences, such as increased emissions. For example, if traffic management systems optimise the flow of vehicles, they may encourage greater use of private cars, resulting in higher toxic emissions [64].

The above considerations provide room for further research, which could focus on analysing the impact of different technological solutions in the context of the smart city on air quality, public health and the mobility behaviour of residents.

Summarising the presented analysis of the literature relevant to the research topic undertaken, it is noticeable that there is a research gap regarding current analyses and studies in the area of modelling the costs of harmful exhaust emissions covering several categories of conventionally fuelled vehicles in the long term.

3. Materials and Methods

As the analysis of the literature indicates, the problem of modelling urban traffic and, at the same time, exhaust emissions, is very complicated [65–67]. Firstly, a very large number of determinants need to be defined, many of which are difficult to measure objectively, making it necessary to introduce certain model simplifications. In addition, there is a lack of unambiguous research to establish fixed relationships. Given that exhaust emissions and fuel consumption are highly dependent on the behaviour of the individual driver, an experimental approach is needed, adopting the research scheme described below. The aim of this study was to estimate the cost of toxic exhaust emissions generated by all road users travelling on the section of road under study in a 25-year perspective.

The research presented in this article is based on the assumptions of the 'Blue Book-Road Infrastructure' developed by JASPERS (Joint Assistance to Support Projects in European Regions). This is a set of methodologies for assessing the economic impact of infrastructure projects in the road transport sector. The document offers a consistent approach to analysing the costs and benefits of road investments, taking into account both technical and environmental and socio-economic aspects. Due to the volume of the article, the basic tool used in the Blue Book, the Cost–Benefit Analysis (CBA), was used for the purpose of this article. This analysis allows for the cost-effectiveness of road projects to be assessed from an economic and social perspective. The model takes into account both the direct costs, such as expenditure on construction and the maintenance and upgrading of infrastructure, and the benefits resulting from improved road quality (e.g., reduced travel times, reduced accidents). In the analysis, unit economic indicators were used to calculate the costs associated with air pollution, covering emissions from different categories of vehicles (cars, trucks, buses), which have a negative impact on human health and the environment. The costs of noise or the costs of traffic accidents were not taken into account.

One of the key assumptions of the Blue Book model is the traffic forecast. This analysis takes into account the dynamics of vehicle traffic in the following years, considering the projected changes in the number and structure of the vehicle fleet, such as an increase in the number of cars or trucks. These forecasts have an important impact on the calculation of the cost of pollutant emissions, which vary according to traffic intensity. All analyses are based on a detailed distinction between vehicle categories (e.g., cars, vans, trucks, buses), as different vehicles emit different amounts of pollution and generate different costs for the infrastructure. Separate unit cost indices are used for each vehicle category, taking into account factors such as speed, road type or terrain conditions. The Blue Book model is based on a long-term time horizon, usually analysed over 20–30 years. This long-term analysis allows for the effects of infrastructure projects to be assessed in the context of changing traffic, technology, and environmental regulations. The book assumes that road investments not only have a direct impact on infrastructure, but also on the environment and public health. Therefore, the costs of pollution, both local (e.g., particulates, nitrogen oxides) and global (e.g., CO_2 emissions), are analysed. Health costs resulting from air pollution are included in the model to assess the overall impact on society.

The Blue Book of Road Infrastructure is a document that acts as a methodological guide for assessing the costs and benefits of road infrastructure. It provides a set of guidelines to support the economic analysis and assessment of the economic and environmental impact of road investments.

The document focuses on many aspects of road infrastructure, such as the costs of road construction, maintenance and operation, but also takes into account factors affecting the environment, including the costs of air pollution. Within the 'Blue Book', there are unit economic indicators to estimate the costs associated with the negative environmental impacts of road transport, including emissions for different vehicle categories.

In summary, the 'Blue Book-Road Infrastructure' is a key reference document in the planning and evaluation of road projects, taking into account both economic and environmental aspects, making it an essential tool in the investment decision-making process in the area of transport.

The traffic study was carried out in a communication route connecting the seaport with the national and international road system on national road No. 10 on the selected section, which constitutes one of the two available road connections between the right-and left-bank parts of the city, playing a key role in local and regional transport the road point covered by the analysis is presented in Figure 1.

Figure 1. The location and view of the urban road section selected for analysis. Source: own study based on Google Maps.

The traffic forecast for the analysed section of the national road was based on the results of our own measurements carried out in April 2023 and external data from the General Traffic Measurement 2023/2024. Traffic measurements were carried out in a selected metropolitan road section of a European city at six stations. Due to the fixed measurement location at pedestrian crossings, public transport buses starting from the bus terminal or entering it from the direction of the city centre were not counted, while those arriving from the right side of the city were counted. The results of the observations were summed up in 15 min intervals, broken down into the six categories of vehicles (without division by type of assembled power unit) characterised below:

- Motorbikes and mopeds [including three-wheeled vehicles]—M;
- Passenger cars [including cars with trailers, vans, SUVs]—SO;
- Vans [no distinction made as to whether they are used for transporting goods or passengers, including minibuses]—SD;
- Buses [vehicles used for the carriage of passengers with more than 20 seats, excluding smaller units which require a licence to drive lorries]—A;
- Heavy goods vehicles [over 3.5 tonnes, including concrete mixers, refuse collection lorries, excavators]—SC;
- Trucks with trailers or semi-trailers [including truck tractors]—SCp.

The measurement was carried out on all weekdays, with data collected from 0^{00}–24^{00}, and additionally, traffic was measured during the morning and midday rush hours from 6^{30}–9^{30} and 14^{30}–17^{30} on weekdays or 10^{00}–13^{00} and 17^{30}–20^{30} on Saturdays and Sundays. Measurement results are presented in the tables below, presenting data for the analysed section during the morning and afternoon peak hours. The data were collected in a 24 h cycle and are presented in synthetic form as average daily vehicle traffic [SDR]. Based on the traffic measurements, the types of vehicles on the analysed section was determined. The dominant share of passenger cars, accounting for as much as 83.43% of all traffic participants, is evident. The smallest share was recorded among motorcycle users, accounting for just 0.18%, and buses, whose share was 1.94%. On the basis of traffic measurements carried out on the analysed section of the national road, SDR was determined—the average daily traffic in a year, i.e., the number of vehicle trips passing through the road section during 24 consecutive hours on an average day in a year. The value of the indicator is expressed in the actual number of vehicles per day and is presented in Table 1.

Table 1. Traffic volume on the analysed road section [for individual vehicles/24 h].

Direction	SDR [Average Daily Traffic]
	[Vehicle/24 h]
from the Pioneer Bridge to the Port Bridge	37214
- cars	30280
- vans	3791
- buses	1081
- trucks	875
- trucks with trailers or semi-trailers	1187
from the Port Bridge to the Pioneer Bridge	40556
- cars	34215
- vans	3796
- buses	352
- trucks	845
- trucks with trailers or semi-trailers	1348
both directions together	77770

Source: own study.

The traffic forecast covered a period of 25 years. In accordance with the methodology contained in sub-chapter 1.6, regarding traffic forecasts, in the document *"Blue Book-Road Infrastructure"*; forecasts of average daily traffic were carried out using a simplified method.

The values of the average daily traffic are a product of the traffic in the base year and the traffic growth indicators for particular vehicle categories, estimated with the use of the method of the General Directorate for National Roads and Motorways. The traffic forecast was prepared according to the scheme contained in the document *"Blue Book-Road Infra-structure"*, shown in Figure 2 below.

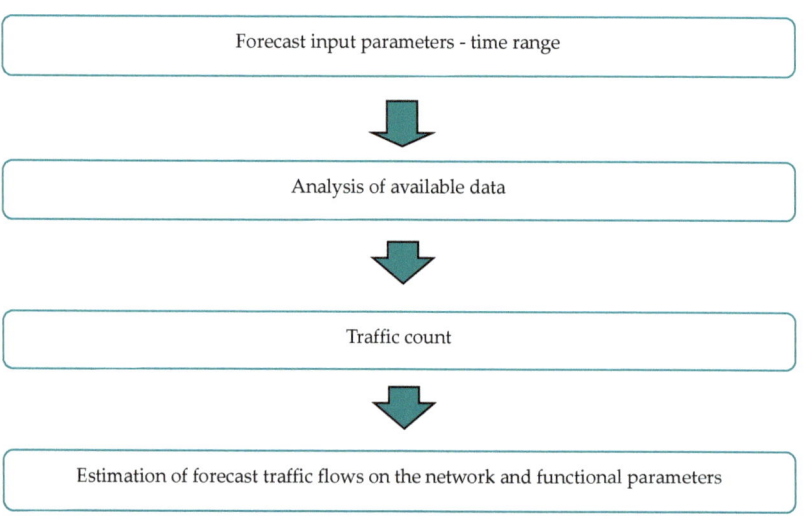

Figure 2. Diagram of traffic forecasting stages. Source: own study based on the methodology in the document "Blue Book-In-road infrastructure".

The traffic forecast has been prepared according to the scheme contained in the document 'Blue Book-Road Infrastructure'. This analysis takes into account the dynamics of vehicle traffic in the following years, taking into account projected changes in the number and structure of the vehicle fleet, such as an increase in the number of cars or trucks. These projections have an important impact on the calculation of the costs of pollutant emissions, noise, and other factors that vary with traffic intensity. All analyses are based on a detailed distinction between vehicle categories (e.g., cars, vans, trucks, buses), as different vehicles emit different amounts of pollution and generate different costs for the infrastructure. Separate unit-cost factors are used for each vehicle category, taking into account factors such as speed, road type, or terrain conditions. The Blue Book model is based on a long-term time horizon, usually analysed over 20–30 years. This long-term analysis allows for the effects of infrastructure projects to be assessed in the context of changing traffic, technology, and environmental regulations. The book assumes that road investments not only have a direct impact on infrastructure, but also on the environment and public health. Therefore, the costs of pollution, both local (e.g., particulates, nitrogen oxides) and global (e.g., CO_2 emissions), are analysed. The health costs of air pollution are included in the model to assess the overall impact on society. Introducing the matter in question could compromise the accepted principles and consistency of the results.

In accordance with the methodology of the "Blue Book-Road Infrastructure", the traffic forecast assumptions developed by the General Directorate for Roads and Motorways were used to calculate the annual percentage increase in traffic. In order to estimate the indicators of the annual percentage increase in traffic [RWR] for a given category of vehicles, the following were used:

Attachment 1 from 14 March 2007—"Projections of the GDP growth rate for the period 2007–2037 for planning and design purposes for national roads";

Attachment 2 from 14 March 2007—"Principles of forecasting the internal traffic growth rates for the period 2007—2013 on the road network for planning and design purposes".

On the basis of the above data, annual traffic growth rates [RWR] were estimated according to the following relation:

$$\text{RWR} = \frac{1 + [W_e * W_{GDP}]}{100} \qquad (1)$$

where

RWR—annual traffic growth rate;
W_e—elasticity index;
W_{GDP}—annual growth rate of GDP.

According to the guidelines of the General Directorate for National Roads and Motorways, the values of the internal traffic growth index for buses for planning purposes were assumed to be at the level of 1.0. The values of the annual traffic growth indexes for the analysed region and city are presented in Table 2. The obtained traffic growth indicators for individual vehicle categories estimated with the use of the method of the General Directorate for National Roads and Motorways were used to calculate the values of the annual traffic growth indicators [RWR] [multiplying the average daily traffic in the base year by the traffic growth indicators for individual vehicle categories for the sub-region of the city under study]. Due to the varying traffic growth dynamics of the individual road categories, elasticity factors [equal to 1 for the national road category for all types of vehicles] were applied to the annual traffic growth rates in accordance with the guidelines of the "Blue Book-Road Infrastructure".

Table 2. Traffic forecast for the analysed road.

Years	SO	SD	SC	SCp	A *
2023	73093	7948	1807	2940	1433
2024	76119	8069	1836	3085	1433
2025	79270	8191	1866	3237	1433
2026	82552	8315	1896	3396	1433
2027	85895	8438	1926	3560	1433
2028	89374	8563	1956	3731	1433
2029	92591	8690	1987	3899	1433
2030	95924	8819	2018	4074	1433
2031	99301	8947	2049	4253	1433
2032	102796	9077	2081	4440	1433
2033	106332	9206	2112	4631	1433
2034	109905	9334	2143	4826	1433
2035	113598	9463	2175	5029	1433
2036	117233	9588	2205	5230	1433
2037	120891	9711	2235	5434	1433
2038	124469	9830	2264	5635	1433
2039	128054	9947	2293	5838	1433
2040	131537	10059	2320	6036	1433
2041	135010	10169	2347	6235	1433
2042	138358	10273	2372	6428	1433
2043	141679	10375	2397	6621	1433
2044	144966	10474	2421	6813	1433
2045	148213	10571	2445	7004	1433
2046	151533	10669	2469	7200	1433
2047	154927	10768	2493	7402	1433

Source: Own calculations. * According to the source methodology Instruction for Assessing the Economic Efficiency of Road and Bridge Projects for Provincial Roads, [A] the annual traffic growth factor is 1000 for average daily bus traffic, i.e., the number of buses is assumed to be constant over the analysis period.

According to the methodology contained in the document "Blue Book-Road Infrastructure", the speeds of travel streams were estimated on the basis of the speed tables from the "Manual for calculating the effectiveness of road and bridge in-vestments" developed by the Road and Bridge Research Institute.

To determine the travel speeds for all vehicle categories on urban roads, the auxiliary hourly traffic volume N_1 was calculated according to the formula:

$$N_1 = 0.5 * [SO + SD + 2 * (SC + SCp + A)] * k \qquad (2)$$

where

N_1—auxiliary hourly traffic volume in [vehicles/hour];
SO—average daily passenger vehicle traffic in [vehicles/day];
SD—average daily commercial vehicle traffic in [vehicles/day];
SC—average daily truck traffic in [vehicles/day];
SC_p—average daily truck traffic with trailers in [vehicles/day];
A—average daily bus traffic in [vehicles/day];
k—conversion factor for hourly traffic [according to the IBDiM manual for economic roads, the value of k factor is 0.095].

For the calculation of the travel speed, the parameters identifying the analysed road section were determined. The technical parameters for determining the travel speed are presented in Table 3.

Table 3. Technical parameters for determining travel speed.

Specification	Unit of Measure	Option
Terrain type: P—flat; F—Rolling; G—Mountainous	-	P
Type of area: M—built; Z—undeveloped	-	M
Character of traffic on the road section: G—Business; T—Tourist; R—Recreational	-	G
Road class [GP, G, Z]	-	G
Number and width of roadways	m	2 × 10.5 m
Sections with overtaking visibility > 450 m	%	100%

Source: Own study.

On the basis of the above data and the tables included in the "Manual for calculating the efficiency of road and bridge investments" prepared by the Road and Bridge Research Institute, travel speeds were calculated for all vehicle categories are presented in Table 4.

Table 4. Hourly traffic intensity and travel speeds of cars, vans, trucks and buses.

The Year	N_1 [Vehicle/Hour]	Travel Speed [SO]; [SD]; [SC]; [SCp]; [A] [km/h]
2023	3964	35
2024	4117	30.6 *
2025	4274	30.6
2026	4437	30.6
2027	4603	30.6
2028	4775	30.6
2029	4955	30.6
2030	5138	30.6
2031	5328	30.6
2032	5506	30.6
2033	5690	30.6
2034	5877	30.6
2035	6070	30.6
2036	6265	30.6

Table 4. Cont.

The Year	N₁ [Vehicle/Hour]	Travel Speed [SO]; [SD]; [SC]; [SCp]; [A] [km/h]
2037	6462	30.6
2038	6666	30.6
2039	6866	30.6
2040	7068	30.6
2041	7266	30.6
2042	7464	30.6
2043	7656	30.6
2044	7847	30.6
2045	8032	30.6
2046	8215	30.6
2047	8397	30.6

Source: Own study. * The methodology for speeds above 4000 vehicles/hour does not foresee lower values than 30.6. At the same time, the value in the table for speeds from 3900 to 3999 is 35.

The analysis carried out points to the need to transform the transport sector, particularly with regard to high-emission vehicles. The implementation of new technologies, such as electric vehicles or hydrogen propulsion, can significantly reduce emission costs and their negative impact on the environment and human health. Emission costs will continue to rise as traffic increases, suggesting that transforming the transport sector should be one of the priorities for environmental action. It is worth noting that emission costs are not just about the economic aspects of transport itself, but also include long-term health and environmental costs, which are important in the context of global sustainable development strategies.

4. Results and Discussion

The pollution costs are assumed to be the total costs generated by all vehicle users travelling on the section of road that is the subject of this analysis. The costs of pollution consist of the costs associated with the environmental impact of transport, including:

- negative impact on human health;
- material damage and environmental damage;
- CO_2 emissions.

The most significant transport-related air pollutants include particulate matter [PM10, PM2.5], nitrogen oxides [NO$_x$], sulphur dioxide [SO$_2$], volatile organic compounds [VOCs], and ozone [O$_3$] as an indirect pollutant. Greenhouse gases [GHGs] are not included in the air pollution cost group, as they have no toxic properties. According to the methodology of the 'Blue Book-Road Infrastructure', the basis for calculating the cost of pollution is the unit economic cost of the environment depending on speed and individual vehicle categories. The unit environmental economic costs contained in Appendix A of the 'Blue Book-Road Infrastructure' were used for the calculations. The pollution cost values were estimated according to the formula for calculating environmental pollution costs.

$$K_Z = 365 * L * \sum_{j=1}^{2} k_{s,j}[V_{pdrtj}, T, S] * SDR_j \qquad (3)$$

where

K_z—annual air pollution costs of motor vehicles [vehicle/hour];
J—number of vehicle categories;
$k_{s,j}$ [V_{pdrtj}, T,S]—unit air pollution costs for vehicle category 'j' as a function of travel speed; $Vpdrt$, j, terrain T and pavement condition [vehicle/km];
SDR_j—annual average daily traffic volume for vehicle category 'j' [vehicles/day];
SC_p—average daily traffic of trucks with trailers [vehicles/day];
L—length of road section [km];

Wkm—transport workload for j category vehicles depending on the length of the road section and in the speed range $Vpdrt, j$, in [vehicle km/day].

The economic costs of air pollution are calculated, taking into account the different vehicle categories, as different types of vehicles emit different amounts and types of pollutants, leading to different environmental and public health impacts. Therefore, each vehicle type is analysed separately in order to accurately estimate the costs resulting from its use. This process is carried out separately for each variant and each year of the economic analysis, which allows for costs to be assessed over the long term, taking into account changing traffic volumes and forecasts of the future load on the road infrastructure. These forecasts take into account growth in the number of vehicles as well as changes in the vehicle fleet structure. In addition, traffic forecasts for individual vehicle categories are a key element in the calculation of pollution costs. This includes estimates for both cars, which dominate urban traffic, and trucks, which account for a larger share of highway and freight emissions.

The toxic emissions costs for the different vehicle categories are shown in Table 5.

Table 5. Costs of pollutant emissions [EUR].

	Passenger Cars [SO]	Vans [SD]	Trucks [SC]	Trucks with Trailers [SCp]	Buses *	All Modes of Transport
2023	961469	139398	530432	1673111	440819	3745229
2024	1001273	141520	538945	1755629	440819	3878186
2025	1042721	143660	547751	1842130	440819	4017081
2026	1085893	145835	556557	1932614	440819	4161718
2027	1129867	147992	565364	2025944	440819	4309986
2028	1175630	150184	574170	2123258	440819	4464061
2029	1217946	152412	583270	2218864	440819	4613311
2030	1261789	154674	592370	2318454	440819	4768106
2031	1306210	156919	601469	2420321	440819	4925738
2032	1352183	159199	610863	2526740	440819	5089804
2033	1398696	161461	619963	2635435	440819	5256374
2034	1445695	163706	629062	2746407	440819	5425689
2035	1494273	165969	638456	2861931	440819	5601448
2036	1542088	168161	647262	2976317	440819	5774647
2037	1590205	170319	656068	3092410	440819	5949821
2038	1637271	172406	664581	3206797	440819	6121874
2039	1684428	174458	673094	3322321	440819	6295120
2040	1730243	176422	681020	3435000	440819	6463504
2041	1775928	178351	688945	3548248	440819	6632291
2042	1819967	180175	696284	3658082	440819	6795327
2043	1863652	181964	703622	3767915	440819	6957972
2044	1906889	183701	710667	3877179	440819	7119255
2045	1949600	185402	717712	3985875	440819	7279408
2046	1993272	187121	724757	4097415	440819	7443384

Source: Own study. * The value is constant as the traffic forecast for this vehicle category is fixed (explanation Table 2).

The analysis of the data clearly shows that for each vehicle category there has been an increase in the cost of toxic emissions over the projected 25-year period. The largest increases were for trucks with trailers and passenger cars. In total, for all vehicle categories, emissions costs nearly doubled from 2023 to 2046, from €3,745,229 to €7,443,384. Figure 3 identifies the vehicle groups that could be prioritised for constraints, thus indicating which of the vehicle groups analysed is in need of transformation in the form of replacing current propulsion systems with low-emission ones.

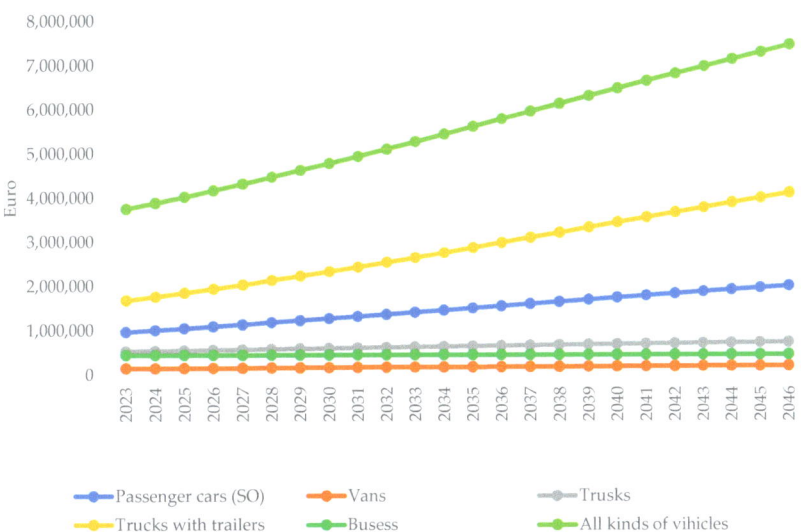

Figure 3. Costs of pollutant emissions for individual vehicle groups and together in [EUR]. Source: own study.

5. Conclusions

There is no longer any doubt in the minds of any researcher that road transport has a negative impact on the environment through, among other things, atmospheric air pollution, thereby generating environmental costs [41,68]. In line with the smart city concept, which is currently being promoted, the aim should be to achieve a way of travelling in cities that not only does not inconvenience the demand side but also reduces the harmful effects of exhaust fumes on the environment. [69]. However, implementing low-carbon transport can be quite a challenge. Given that the average age of the car fleet in the geographical study area is 15.5 years [70], it is certainly reasonable to assume that conventionally powered vehicles will be in use even after their official production ceases after 2030. In addition, the current promotion of electromobility based on the approach of reducing individual costs may not be fully convincing for the transformation of urban mobility, given the noticeable decline in demand-side interest in purchasing this category of vehicle. It is therefore necessary to look for other ways to demonstrate that reducing harmful substances in transport can be an economically beneficial process. The study, which aimed to estimate the costs of toxic emissions generated by all vehicle users travelling on the selected road section in a 25-year perspective, allowed the following conclusions to be drawn:

1. It was found that passenger cars dominated urban traffic, accounting for as much as 83.43% of all traffic participants. The smallest share was found among motorbike users, accounting for a mere 0.18%, and bus users, whose share was 1.94%.
2. Forecasts of annual traffic rates for all surveyed vehicle groups over the next 25 years indicate an upward trend. The highest growth rate was recorded for trucks with trailers and passenger cars. On the other hand, the lowest growth was recorded for bus-type vehicles.
3. The parameter of hourly traffic volume N1 will increase with the passing of the years. However, the coefficient on travel speed for the analysed groups of vehicles, despite the fact that the measurement methodology for volumes above 4000 does not provide for lower values than 30.6. However, the authors hypothesise that this will decrease.
4. The projections showed that for each vehicle category, there will be an increase in the cost of toxic emissions over the projected 25-year period. The largest increase will be

for trucks with trailers and passenger cars. In total, for all vehicle categories, emission costs will increase from €3,745,229 to €7,443,384.
5. Using the selected methods described in the 'Blue Book-Road Infrastructure', it is possible to identify the categories of vehicles for which it is necessary to take action as soon as possible, in the form of replacing the engines currently in use with low-emission ones.
6. The cost analyses presented can provide a basis for starting a discussion on the introduction of traffic restrictions or additional charges for this category of vehicles, which contribute most to the environmental costs of cities in line with the smart city concept.

The authors postulate that engaging in a discussion on the links between the smart city concept and the cost of toxic emissions could be a key argument in achieving the transport transformation policy goals implemented by all European Union cities, given that, according to experts, even increasingly stringent emission standards will not eliminate the use of conventionally fuelled vehicles in the coming years. Thus, in order to realise the demands for smart city and low-carbon concepts, in addition to the obvious support for measures to develop electromobility, the environmental aspects of the economic costs incurred should be pointed out, so that efforts to tackle urban air pollution are nevertheless widely supported by both cities and citizens themselves.

It should be emphasised that the research presented focused on estimating the costs of toxic emissions generated by all users of conventionally fuelled vehicles travelling on a selected section of urban road in the next 25 years. Against the background of academic considerations, the question of whether the presented research results can be directly implied into other urban destinations should be answered. In the opinion of the researchers, the answer is, in part, yes, with the proviso that the size and type of vehicle population and infrastructure and road conditions in another city will be similar to those analysed. Furthermore, in the authors' opinion, the results of the research presented can directly support the arguments for the creation of clean transport zones in other urban destinations. Given that the idea of such zones is identical to the smart city concept promoted by the European Commission.

The following article analyses selected parameters in the form of annual traffic growth rates on the analysed road section, road speeds, and analyses of annual air pollution costs for selected vehicle categories using the selected research method. Certainly, in the near future, much more extensive analyses will be needed, in particular covering parameters that affect fuel consumption and exhaust emissions and that relate, among other things, to the characteristics of the vehicle itself, the changing road infrastructure. In addition, further research on this topic should focus on aspects related to the analysis of changes in CO_2 emission standards for new passenger cars and new light commercial vehicles. The analysis of changes related to the increasing share of low-emission vehicles in the public transport sector, i.e., buses and taxis, should also be an important element of further research in this area.

In order to effectively address the problem of pollution caused by vehicles, it is necessary to implement a comprehensive modernisation programme for the transport sector. This includes the introduction of a system of subsidies and tax breaks for vehicle owners who choose to replace internal combustion engines with low-emission ones, such as hybrid, electric or fuel-cell-based engines, which can encourage a faster transition. It will be crucial to provide long-term support schemes to cover part of the purchase costs of new technologies. It is also important that electric and fuel-cell vehicles require dedicated infrastructure. Governments and private entities should invest in expanding charging stations for electric vehicles and a network of hydrogen refuelling stations, making these technologies more accessible to drivers. Another challenge is regulation and emission standards. The gradual introduction of more stringent emission standards for internal combustion vehicles will force manufacturers to adapt production to environmental requirements. Examples include CO_2 and NO_x emission limits, which will gradually be tightened, accelerating the transition to greener solutions. Promoting the benefits of low-emission vehicles and informing the public about the health and environmental impacts of transport can help raise awareness and acceptance of these changes. One of the main challenges that the transition will face is the high cost of electric vehicles and fuel cells. In particular, this technology, based mainly on the use of hydrogen, emits only water, eliminating the emission of harmful substances such as nitrogen oxides (NO_x), particulate matter (PM), and carbon dioxide (CO_2), which are typical of combustion vehicles [7–9]. However, with technological advances and effective support programmes, the prices can be gradually reduced and the availability of these technologies increased. The change in propulsion technology also requires a rethink of industrial and employment issues, particularly in the vehicle manufacturing and service sector. Worker retraining programmes and the development of new jobs in the area of low-carbon technologies will be necessary. Putting these measures in place will allow for a sustainable and effective reduction in environmental pollution caused by transport, while addressing potential challenges.

In conclusion, the presented research on the connection between the smart city concept and the costs of the emission of toxic components of exhaust gases, using the example of a selected population of motor vehicles in urban traffic, does not fully exhaust the essence of the issue. However, it represents an area for further research and analysis in this matter. Certainly, this topic represents a novel approach to understanding both the nature of the impact of the continued use of traditionally powered vehicles on the development plan of not only the electromobility itself, but also the role that the smart city concept will play in this process. Therefore, such analyses will be the subject of future work, in order to determine and identify the key factors for the realisation of the smart city concept and the transformation of modes of transport, not only in the studied geographical destination, but also in other Central and Eastern European countries.

Author Contributions: Conceptualisation, M.B., W.L. and M.Ś.-S.; methodology, M.B. and W.L.; software, M.B. and W.L.; validation, M.B. and W.L.; formal analysis, M.B. and W.L.; investigations, M.B. and W.L.; resources, M.B., W.L. and M.Ś.-S.; data curation, M.B. and W.L.; writing—preparation of the original draft, M.B., W.L. and M.Ś.-S.; writing—reviewing and editing, M.B., W.L. and M.Ś.-S.; visualisation, M.B.; supervision, W.L.; project administration, M.B. and W.L.; obtaining financing, M.B., W.L. and M.Ś.-S. All authors have read and agreed to the published version of the manuscript.

Funding: The research was financed as part of a project carried out within the Faculty of Economics of the West Pomeranian University of Technology in Szczecin under the name "Green Lab." Research and innovation.

Data Availability Statement: The data are included in the article.

Conflicts of Interest: The authors declare that there are no conflicts of interest.

Abbreviations

EU	European Union
ICT	Information and Communication Technology
NOx	Nitrogen Oxides
PM	Particulate Matter
VOC	Volatile Organic Compounds
ICE	Internal Combustion Engine
CNG	Compressed Natural Gas
NEDC	New Energy Demonstration City

References

1. Khomenko, S.; Cirach, M.; Pereira-Barboza, E.; Mueller, N.; Barrera-Gómez, J.; Rojas-Rueda, D.; de Hoogh, K.; Hoek, G.; Nieuwenhuijsen, M. Premature mortality due to air pollution in European cities: A health impact assessment. *Lancet Planet. Health* **2021**, *5*, e121–e134. [CrossRef] [PubMed]
2. Sicard, P.; Agathokleous, E.; De Marco, A.; Paoletti, E.; Calatayud, V. Urban population exposure to air pollution in Europe over the last decades. *Environ. Sci. Eur.* **2021**, *33*, 28. [CrossRef] [PubMed]

3. Vašalić, D.; Ivković, I.; Mladenović, D.; Sekulić, D.; Miličević, D.; Suljovrujić, E. Prediction of Fuel and Exhaust Emission Costs of Heavy-Duty Vehicles Intended for Gas Transportation. *Sustainability* **2024**, *16*, 5407. [CrossRef]
4. Andrych-Zalewska, M.; Chlopek, Z.; Merkisz, J.; Pielecha, J. Comparison of gasoline engine exhaust emissions of a passenger car through the WLTC and RDE Type Approval Tests. *Energies* **2022**, *15*, 8157. [CrossRef]
5. Fernandes, P.; Ferreira, E.; Macedo, E.; Coelho, M.C. Unraveling roundabout dynamics: Analysis of driving behavior, vehicle performance, and exhaust emissions. *Transp. Res. Part D Transp. Environ.* **2024**, *133*, 104308. [CrossRef]
6. Borucka, A.; Wiśniowski, P.; Mazurkiewicz, D.; Świderski, A. Laboratory measurements of vehicle exhaust emissions in conditions reproducing real traffic. *Measurement* **2021**, *174*, 108998. [CrossRef]
7. Jia, C.; He, H.; Zhou, J.; Li, K.; Li, J.; Wei, Z. A performance degradation prediction model for PEMFC based on bi-directional long short-term memory and multi-head self-attention mechanism. *Int. J. Hydrogen Energy* **2024**, *60*, 133–146. [CrossRef]
8. Jia, C.; Zhou, J.; He, H.; Li, J.; Wei, Z.; Li, K. Health-conscious deep reinforcement learning energy management for fuel cell buses integrating environmental and look-ahead road information. *Energy* **2024**, *290*, 130146. [CrossRef]
9. Jia, C.; Liu, W.; He, H.; Chau, K.T. Deep reinforcement learning-based energy management strategy for fuel cell buses integrating future road information and cabin comfort control. *Energy Convers. Manag.* **2024**, *321*, 119032. [CrossRef]
10. Mei, H.; Wang, L.; Wang, M.; Zhu, R.; Wang, Y.; Li, Y.; Bao, X. Characterization of exhaust CO, HC and NOx emissions from light-duty vehicles under real driving conditions. *Atmosphere* **2021**, *12*, 1125. [CrossRef]
11. Kumar, P.; Morawska, L.; Martani, C.; Biskos, G.; Neophytou, M.; Di Sabatino, S.; Bell, M.; Norford, L.; Britter, R. The rise of low-cost sensing for managing air pollution in cities. *Environ. Int.* **2015**, *75*, 199–205. [CrossRef] [PubMed]
12. Caragliu, A.; Del Bo, C.; Nijkamp, P. Smart Cities in Europe. *J. Urban Technol.* **2011**, *18*, 65–82. [CrossRef]
13. Manville, C.; Cochrane, G.; Cave, J.; Millard, J.; Pederson, J.K.; Thaarup, R.K.; Liebe, A.; Wissner, M.; Massink, R.; Kotterink, B. Maping Smart Cities in the EU, Study. In *Directorate General for Internal Policies, Policy Department A: Economic and Scientific Policy*; European Parliament: Brussels, Belgium, 2014.
14. Czupich, M.; Kola-Bezka, M.; Ignasiak-Szulc, A. Factors and barriers to implementing the smart city concept in Poland. *Stud. Ekon. Sci. J. Univ. Econ. Katow.* **2016**, *276*, 224.
15. Augustyn, A. Sustainable urban development in the light of smart city ideas. *Univ. Bialyst.* **2020**, *8*, 8–16.
16. Hollands, R.G. Will the real smart city please stand up? *City: Analysis of Urban Trend, Culture, Theory. Policy Action* **2008**, *12*, 304–316. [CrossRef]
17. Lewicki, W.; Stankiewicz, B.; Olejarz-Wahba, A.A. The role of intelligent transport systems in the development of the idea of smart city. In *Smart and Green Solutions for Transport Systems: 16th Scientific and Technical Conference" Transport Systems. Theory and Practice 2019" Selected Papers 16*; Springer International Publishing: Cham, Switzerland, 2020; pp. 26–36.
18. Stawasz, D.; Sikora-Fernandez, D.; Turała, M. The smart city concept as a determinant of decision-making related to city functioning and development. *Zesz. Nauk. Uniw. Szczecińskiego Stud. Inform.* **2012**, *721*, 97–109.
19. Jankowska, M. Smart city as a concept for sustainable urban development-the example of Vienna. *Stud. I Pr. WNEIZ US* **2015**, *42*, 173–182. [CrossRef]
20. Wach-Kloskowska, M.; Rześny-Cieplińska, J. Smart and sustainable transport development as an element of the smart city concept-Polish and European examples. *Stud. Miej.* **2018**, *30*, 99–108. [CrossRef]
21. Voronina, R. Introduction of the 'smart city' philosophy into the management of the transport system of Lviv. *City Reg. Transp.* **2016**, *4*, 26–32.
22. Batty, M.; Axhausen, K.W.; Giannotti, F.; Pozdnoukhov, A.; Bazzani, A.; Wachowicz, M.; Ouzounis, G.; Portugali, Y. Smart cities of the future. *Eur. Phys. J.* **2012**, *214*, 481–518. [CrossRef]
23. Letkiewicz, A.; Szulc, K. W kierunku green smart city-case study Trójmiasta. *Contemp. Econ. Electron. Sci. J.* **2022**, *13*, 59–73. [CrossRef]
24. Kiełbaska, P.; Wronkowski, D. Application of the Smart City concept in the Lodz area. *Wydział Ekon.-Socjol.* **2018**, 201–209.
25. Rudewicz, J. Industry and technology towards realising the smart city vision [smart city] Industry and technology. *Proc. Comm. Ind. Geogr. Pol. Geogr. Soc.* **2019**, *4*, 195–212.
26. Rosiński, J. Small and medium-sized enterprises in the smart city space: A business model for the organisation of the leisure industry. *Wars. Sch. Econ.* **2020**, 395–422.
27. Bodnar, E.; Budna, K.; Szpilko, D. Modern technologies used in municipal waste management in the smart city. *Akad. Zarządzania* **2023**, *7*, 113–142. [CrossRef]
28. Glebova, E.; Lewicki, W. Smart cities' digital transformation. In *Smart Cities and Tourism: Co-Creating Experiences, Challenges and Opportunities*; Goodfellow Publishers: Oxford, UK, 2022; p. 548.
29. Wolniak, R. Smart mobility as part of the smart city concept. *Manag. Qual.* **2023**, *5*, 208–222.
30. Nam, T.; Pardo, T.A. Conceptualizing Smart City with Dimensions of Technology, People, and Institutions. In Proceedings of the 12th Annual International Digital Government Research Conference: Digital Government Innovation in Challenging Times, College Park, MD, USA, 12–15 June 2011; pp. 283–286.
31. Glaeser, E. *Triumph of the City: How Our Greatest Invention Makes Us Richer, Smarter, Greener, Healthier, and Happier*; Penguin Press: London, UK, 2011.
32. Neirotti, P.; De Marco, A.; Cagliano, A.C. Current trends in Smart City initiatives: Some stylised facts. *Cities* **2014**, *38*, 25–36. [CrossRef]

33. Polese, F.; Botti, A.; Monda, A.; Grimaldi, M. Smart City as a Service System: A Framework to Improve Smart Service Management. *J. Serv. Sci. Manag.* **2019**, *12*, 89605. [CrossRef]
34. Chłopek, Z.; Polichnowski, T. Simulation studies of pollutant emissions from urban vehicle engines. *J. Kones* **2002**, *9*, 37–44.
35. Chłopek, Z.; Biedrzycki, J.; Lasocki, J. Uniqueness of pollutant emissions under conditions simulating passenger car traffic in cities. *TTS Technol.* **2015**, 272–276.
36. Chłopek, Z. Analysis of emissions from a traction internal combustion engine. *Mech. Mech. Eng.* **2005**, *9*, 43–68.
37. Kuczyński, S.; Liszka, K.; Łaciak, M.; Kyć, K.; Olinyk, A.; Szurlej, A. Impact of the use of alternative fuels in transport, with particular emphasis on CNG, on the reduction of air pollutant emissions. *Polityka Energetyczna–Energy Policy J.* **2016**, *19*, 91–104.
38. Kijewska, K. Importance of last-mile solutions for sustainable urban supply chains, Innovations in Production Management and Engineering. *Oficyna Wydawnicza Pol. Tow. Zarządzania Prod. Opole* **2018**, 736–746.
39. Chłopek, Z.; Pawlicki, M.; Sypowicz, R. Statistical analysis of pollutant emission intensities from an internal combustion engine under conditions simulating dynamic use. *Automot. Arch.* **2005**, 77–87.
40. Wojtal, R. Traffic-related air pollution in cities. *Urban Reg. Transp.* **2018**, *1*, 12–17.
41. Zych, M. Selected external costs of passenger transport in Polish voivodship cities. *Young Res. Achiev.* **2014**, *4*, 837–842.
42. Ziółkowski, R. Effectiveness of traffic calming measures in the context of environmental pollution. *J. Civ. Eng. Environ. Archit.* **2016**, *XX*, 573–580.
43. Filipowicz, J.; Filipowicz, P.; Zaprawa, K. Exhaust emissions from public transport buses. *Saf. Ecol.* **2017**, 52–55.
44. Bojanowska, M. Automotive pollution in the environment. *Buses Technol. Oper. Transp. Syst.* **2011**, *10*, 77–83.
45. Kwestarz, M.; Muszyńska, M. Electric mobility as an opportunity to reduce emissions in the road transport sector. *New Energy* **2017**, *4*, 68–70.
46. Zawieska, J.; Skotak, K. Sustainable transport policy in Polish cities in the context of air quality and emissions from the transport sector Part II. *Urban Reg. Transp.* **2015**, *12*, 11–17.
47. Kuropka, J. Natural gas as a green fuel as an opportunity to improve air pollution in cities. *State Soc.* **2004**, *2*, 305–312.
48. Król, E. Comparison of emissions of electric and combustion powered vehicles. *Drives Control.* **2017**, *7*, 140–143.
49. Grabański, S.; Igliński, H.; Szymczak, M. Low-carbon transport in urban e-commerce services, Economic Studies. *Zesz. Nauk.* **2019**, *381*, 76–85.
50. Hulicka, A. Green city-sustainable city. *Geogr. Work* **2015**, *141*, 73–85.
51. Bylinko, L. Concepts and tools for reducing urban traffic in an environmental context. *Mark. Mark.* **2017**, *7*, 65–72.
52. Skiba, M.; Bazan-Krzywoszańska, A.; Stępińska–Kaczmarczyk, A. Emission of pollutants in urban structure. A model of current environmental and energy efficiency challenges in adapting cities to climate change. *PUA Space Urban. Archit.* **2023**, *1*, 114–125. [CrossRef]
53. Puliafito, S.E. Emissions and air concentrations of pollutants for urban area sources. *Mecánica Comput.* **2005**, 1390–1407.
54. Armas, O.; Lapuerta, M.; Mata, C. Methodology for the analysis of pollutant emissions from a city bus. *Meas. Sci. Technol.* **2012**, *23*, 2247–2259. [CrossRef]
55. Guo, B.; Feng, Y.; Lin, J.; Wang, X. New energy demonstration city and urban pollutant emissions: An analysis based on a spatial difference-in-differences model. *Int. Rev. Econ. Financ.* **2024**, *91*, 287–298. [CrossRef]
56. López-Martínez, J.M.; Jiménez, F.; Páez-Ayuso, F.J.; Flores-Holgado, M.N.; Arenas, A.N.; Arenas-Ramirez, B.; Aparicio-Izquierdo, F. Modelling the fuel consumption and pollutant emissions of the urban bus fleet of the city of Madrid. *Transp. Res. Part D Transp. Environ.* **2017**, *52*, 112–127. [CrossRef]
57. Rosero, F.; Fonseca, N.; López, J.M.; Casanova, J. Real-world fuel efficiency and emissions from an urban diesel bus engine under transient operating conditions. *Appl. Energy* **2020**, *261*, 114442. [CrossRef]
58. Gómez, A.; Fernández-Yáñez, P.; Soriano, J.A.; Sánchez-Rodríguez, L.; Mata, C.; García-Contreras, R.; Armas, O.; Cárdenas, M.D. Comparison of real driving emissions from Euro VI buses with diesel and compressed natural gas fuels. *Fuel* **2021**, *289*, 119836. [CrossRef]
59. Keramydas, C.; Papadopoulos, G.; Ntziachristos, L.; Lo, T.-S.; Ng, K.-L.; Wong, H.-L.A.; Wong, C.K.-L. Real-world measurement of hybrid buses' fuel consumption and pollutant emissions in a metropolitan urban road network. *Energies* **2018**, *11*, 2569. [CrossRef]
60. Zhang, Y.; Lou, D.; Tan, P.; Hu, Z. Particulate emissions from urban bus fueled with biodiesel blend and their reducing characteristics using particulate after-treatment system. *Energy* **2018**, *155*, 77–86. [CrossRef]
61. Mataa, C.; Gómezb, A.; Armasb, O. The influence of ethanol-diesel blend on pollutant emissions from different bus fleets under acceleration transitions. *Fuel* **2018**, *209*, 322–327. [CrossRef]
62. Townsend, A. *Smart Cities: Big Data, Civic Hackers, and the Quest for a New Utopia*; WW Norton & Company: New York, NY, USA, 2013; pp. 1–320.
63. Knox, P.; Pinch, S. *Urban Social Geography: An Introduction*; Taylor & Francis: Abingdon, UK, 2010; pp. 1–392. [CrossRef]
64. Sheller, M.; Urry, J. Mobilizing the new mobilities paradigm. *Appl. Mobilities* **2016**, *1*, 10–25. [CrossRef]
65. Seo, J.; Park, S. Optimizing model parameters of artificial neural networks to predict vehicle emissions. *Atmos. Environ.* **2023**, *294*, 119508. [CrossRef]
66. Agarwal, A.K.; Mustafi, N.N. Real-world automotive emissions: Monitoring methodologies, and control measures. *Renew. Sustain. Energy Rev.* **2021**, *137*, 110624. [CrossRef]

67. Mądziel, M. Vehicle emission models and traffic simulators: A review. *Energies* **2023**, *16*, 3941. [CrossRef]
68. Colvile, R.N.; Hutchinson, E.J.; Mindell, J.S.; Warren, R.F. The transport sector as a source of air pollution. *Atmos. Environ.* **2001**, *35*, 1537–1565. [CrossRef]
69. Salman, M.Y.; Hasar, H. Review on environmental aspects in smart city concept: Water, waste, air pollution and transportation smart applications using IoT techniques. *Sustain. Cities Soc.* **2023**, *94*, 104567. [CrossRef]
70. Olejarz, A.A.; Kędzior-Laskowska, M. How Much Progress Have We Made towards Decarbonization? Policy Implications Based on the Demand for Electric Cars in Poland. *Energies* **2024**, *17*, 4138. [CrossRef]

Disclaimer/Publisher's Note: The statements, opinions and data contained in all publications are solely those of the individual author(s) and contributor(s) and not of MDPI and/or the editor(s). MDPI and/or the editor(s) disclaim responsibility for any injury to people or property resulting from any ideas, methods, instructions or products referred to in the content.

Article

Utilizing Artificial Neural Network Ensembles for Ship Design Optimization to Reduce Added Wave Resistance and CO_2 Emissions

Tomasz Cepowski

Faculty of Navigation, Maritime University of Szczecin, 1-2 Wały Chrobrego St., 70-500 Szczecin, Poland; t.cepowski@pm.szczecin.pl

Abstract: Increased maritime cargo transportation has necessitated stricter management of emissions from ships. The primary source of this pollution is fuel combustion, which is influenced by factors such as a ship's added wave resistance. Accurate estimation of this resistance during ship design is crucial for minimizing exhaust emissions. The challenge is that, at the preliminary parametric design stage, only limited geometric data about the ship is available, and the existing methods for estimating added wave resistance cannot be applied. This article presents the application of artificial neural network (ANN) ensembles for estimating added wave resistance based on dimensionless design parameters available at the preliminary design stage, such as the length-to-breadth ratio (L/B), breadth-to-draught ratio (B/T), length-to-draught ratio (L/T), block coefficient (CB), and the Froude number (Fn). Four different ANN ensembles are developed to predict this resistance using both complete sets of design characteristics (i.e., L/B, B/T, CB, and Fn) and incomplete sets, such as L/B, CB, and Fn; B/T, CB, and Fn; and L/T, CB, and Fn. This approach allows for the consideration of CO_2 emissions at the parametric design stage when only limited ship dimensions are known. An example in this article demonstrates that minor modifications to typical container ship designs can significantly reduce added wave resistance, resulting in a daily reduction of up to 2.55 tons of CO_2 emissions. This reduction is equivalent to the emissions produced by 778 cars per day, highlighting the environmental benefits of optimizing ship design.

Keywords: artificial neural networks (ANNs); CO_2 emissions; ship design; optimization; added wave resistance; hull shape parameters

1. Introduction

The 2021 United Nations Conference on Trade and Development (UNCTAD) report [1] noted that over 80% of global trade volume, particularly in goods, is facilitated via maritime routes. This heavy reliance on maritime transport, driven by increasing international trade, has led to a growing share of shipping emissions in total anthropogenic emissions [2]. Moreover, a subsequent UNCTAD report in 2023 [3] highlighted a significant rise in greenhouse gas emissions from the maritime sector, with a 20% increase observed over the past decade. The report also projects that global trade will continue to grow at an annual rate of over 2% between 2024 and 2028.

The primary source of pollution associated with maritime transport is the combustion of fuels, particularly fuel oils such as marine diesel and heavy fuel oils (HFOs) [4,5]. The shipping industry is undertaking various efforts to reduce emissions by introducing alternative energy sources, such as hybrid propulsion systems and low-emission fuels; however, the combustion of traditional fossil fuels still dominates in maritime operations, contributing to the majority of emissions [6,7]. To mitigate CO_2 emissions from shipping, the International Maritime Organization (IMO) introduced the Energy Efficiency Design Index (EEDI) [8]. The EEDI is a key regulatory tool that compels designers to control CO_2

emissions at the design and construction stages of ships. As a result, ship designers must optimize vessel designs to meet increasingly stringent emission standards [9].

Therefore, one of the primary challenges in designing maritime transport vessels is the effective management of environmental pollution emissions resulting from the combustion of ship fuels [10,11]. During the design phase, one of the most effective measures to reduce fuel consumption and CO_2 emissions is minimizing the hydrodynamic resistance of the hull [12]. This approach is highly efficient, as reducing hull resistance directly lowers engine power requirements, decreases the vessel's energy demand, and consequently reduces the consumption of any type of energy source, especially fossil fuels. To achieve this reduction, optimizing the hull's dimensions and shape is crucial. Accurate estimation of resistance based on the hull's shape is essential for this optimization. An additional factor contributing to total hull resistance is wave resistance, which results from the interaction between waves and the ship [13–15]. An IMO report [16] revealed that a ship only sails in wave conditions below 1 m in height for 3% of the time of its operation, while wave activity impacts the vessel for the remaining 97% of the time. This continuous wave interaction increases the ship's wave resistance and, consequently, raises CO_2 emissions.

Thus, one of the particularly challenging components for estimating total hull resistance is added resistance in waves. This resistance is associated with ship motion in high seas and can account for approximately 30–50% of the total ship resistance [17]. The increase in added wave resistance leads to elevated fuel consumption and augmented exhaust emissions. This phenomenon is influenced by factors including the dimensions and shape of the hull, as well as the Froude number.

The added wave resistance is determined using theoretical, semiempirical, or fully empirical methods. Theoretical methods include far- and near-field approaches, such as the momentum and energy method by Maruo [18,19], the direct pressure integration method by Boese [20], the radiated energy method by Gerritsma and Beukelman [21], and the developments of these methods by Salvesen et al. [22]. Fujii and Takahashi [23] proposed the National Maritime Research Institute of Japan (NMRI) formula for diffraction-dominated added resistance in short waves. Faltinsen et al. [24] proposed an asymptotic formula to calculate the added wave resistance in short waves. Jinkine and Ferdinande [25] proposed a radiation formula to calculate the added resistance in long waves.

Semiempirical methods have been developed based on the above methods and experimental studies. These methods include the STAWAVE-1 and STAWAVE-2 approaches developed by Boom et al. [26], based on the Maruo method, and the SHOPERA-NTUA-NTU-MARIC (SNNM) method developed by Liu and Papanikolaou [27], a combination and simplification of the Faltinsen and Jinkine–Ferdinande methods, all applied in ref. [28]. Lang and Mao [29] proposed a new semiempirical approach by combining the NMRI method for added resistance in short waves and the Jinkine and Ferdinande method for added wave resistance in long waves. This new method was found to be more flexible and to better fit model test results compared to earlier methods, based on experiments with 11 ships in regular head waves. Kim et al. [30] recently proposed a method that integrates the approaches of Lang and Mao with those of Liu and Papanikolaou [31], employing a combining function to seamlessly link the results to wave conditions and heading. Cepowski [32] proposed an alternative approach that employs an ensemble of artificial neural networks (ANNs), trained with data from model tests, to predict added wave resistance using ship principal dimensions parameters as inputs.

Semiempirical methods, as opposed to theoretical methods, are characterized by limitations concerning, among others, the Froude number, wave direction, hydromechanical parameters, and geometric parameters of the ship's hull. Among the geometric parameters, the main limitation of these methods is the ship length. According to ref. [28], the STAWAVE-2 method can be used for ships with lengths ranging from 50 to 400 m, while the SNNM method is applicable for ships with lengths ranging from 75 to 400 m. The Lang and Mao [29], and Cepowski [32] methods are suitable for ships with lengths ranging from 122 to 355 m, 90 to 355 m, and 105 to 360 m, respectively. Applying these methods to estimate the added wave resistance of ships with dimensions outside of the specified ranges may lead to unreliable results.

In ship design, the estimation of added wave resistance is a crucial aspect that largely depends on the primary dimensions of the hull. The ship design process consists of several stages and is an iterative process [13–15]. In the preliminary design phase, during the parametric stage, the basic dimensions of the ship, such as length (L), breadth (B), draught (T), and hull shape, are determined and described using the block coefficient (CB) of the underwater part [15]. In this phase, these dimensions are sequentially determined based on dimensionless ratios, such as L/B, B/T, and possibly L/T [15]. For instance, in the case of deadweight-type ships, the length, Froude number (Fn), and CB are first determined based on the ship's mass and speed. The B is then derived using the L/B or B/T ratio, followed by determining the T based on the B/T or L/T ratio [15].

In the subsequent stages of the design process, based on these dimensions, the geometric shape of the ship's hull is designed. Considering that the added wave resistance of a ship largely depends on these primary hull dimensions, modeling this resistance is most beneficial at the parametric design stage, particularly when selecting the proportions of the main dimensions. Therefore, design methods are needed that allow for the accurate estimation of added wave resistance based on individually selected ratios of the main dimensions, which can be applied across the full range of ship sizes without restricting the length of the ship. The above methods, such as those described in refs. [27,29,31,32], rely on ship dimensions that are not available to the designer during the preliminary parametric design phase, as they require more geometric parameters than can be determined at that stage. Specifically, there is a lack of methods for estimating added resistance that rely only on limited basic parameters, such as Fn, CB, and L/B ratio. This lack prevents accurate estimation of the increase in CO_2 emissions caused by wave interaction with the ship during the early parametric design phase, and thus, hinders effective hull shape optimization at this stage of design.

As mentioned above, the most effective method for reducing a ship's energy demand, and consequently minimizing fuel consumption and CO_2 emissions resulting from that combustion, is the minimization of hull resistance during the early design stage. Therefore, accurate methods for estimating added resistance using only basic design parameters such as Fn, CB, and L/B are required for optimizing the hull in terms of CO_2 emissions caused by increased wave resistance at the preliminary parametric design stage. The existing semi-empirical methods for predicting added wave resistance are not suitable for estimating the increase in CO_2 emissions at this design stage, as they require complete data, such as Fn, CB, L, B, T, and other parameters, which are not yet determined at this point. Furthermore, current semi-empirical methods utilize ship dimensions in full scale rather than model scale, which significantly limits their applicability to vessels of specific lengths.

Thus, the primary goal of this research was to address these issues by developing artificial neural networks (ANNs) capable of accurately estimating added wave resistance based solely on individually selected main dimension ratios, making them applicable across the full range of ship sizes. These estimates can be expressed as follows:

$$CAW = f_1(L/B, CB, Fn, \lambda/L), \qquad (1)$$

$$CAW = f_2(B/T, CB, Fn, \lambda/L), \qquad (2)$$

$$CAW = f_3(L/T, CB, Fn, \lambda/L), \qquad (3)$$

where f_1, f_2, and f_3 are mathematical functions, L/B is the ratio of the length to the model's breadth (which are perpendicular), B/T is the ratio of the breadth to the draft, L/T is the ratio of the ship's length to the draft (perpendicular), CB is the block coefficient, Fn is the Froude number, λ/L is the ratio of the wavelength to the ship length, and CAW, a non-dimensional added wave resistance coefficient, can be defined as follows:

$$CAW = \frac{R}{\zeta_a^2 \rho g \frac{B^2}{L}}.\qquad(4)$$

where R is the added wave resistance in a regular wave, ζ_a is the amplitude of a regular wave, B is the model breadth, L is the model length, ρ is the seawater density, and g is the gravity acceleration.

Additionally, it was assumed that the ANNs would be constructed based on the results of model tests using only dimensionless parameters of the ship model. This approach allows for the estimation of added wave resistance at the model scale, resulting in more universal estimates without ship size limitations.

The second aim of this research is to identify the non-dimensional shape parameter that has the greatest impact on the accuracy of added wave resistance estimation. To effectively identify these parameters, it was assumed that the study will also develop a neural network to predict the added resistance based on the following parameters:

$$CAW = f_4\,(L/B,\,B/T,\,CB,\,Fn,\,\lambda/L),\qquad(5)$$

where f_4 is the mathematical function.

Artificial neural networks (ANNs) are increasingly used in ship theory to predict various experimental measurement outcomes. The development and application of neural networks have enabled researchers to analyze and interpret large amounts of data in a more efficient and accurate manner. The use of neural networks in ship theory has significantly improved the accuracy of prediction models, leading to advancements in the design and operation of ships. The utilization of artificial neural networks has significantly improved the accuracy of predictive models, leading to advancements in various areas of ship design and operation. For instance, Yangjun and Yonghwan [33] applied an ANN to predict sloshing model test results for floating units larger than standard-sized LNG carriers. The ANN was developed using more than 540 terabytes of experimental data that were mined, organized, cleaned, and analyzed from various cargo holds, vessels, environmental conditions, operational conditions, and experimental conditions. Sun et al. [34] employed an ANN model to estimate the resistance of ships in ice-covered waters, utilizing appropriate ship and ice parameters. The developed ANN model offers a more accurate means for predicting ice resistance than conventional semiempirical formulas in the design of polar ships. Furthermore, Dyer et al. [35] utilized an ANN model to estimate the remaining service life of offshore oil and gas platforms. They incorporated a comprehensive dataset that represented both natural and engineered aspects of offshore systems. Their results demonstrated the ability of ANNs to generate highly accurate predictions, suggesting strategies for life extension, maintenance, and risk minimization for offshore platforms. Artificial neural networks are mathematical models based on the architecture of the human brain and the idea that the human brain is a collection of interconnected neurons. ANNs are used to solve various problems, including regression, where the goal is to predict a numerical value based on other variables. In the development of artificial neural networks, assessing their reliability remains a challenge due to the heuristic approach used in designing architecture and training methods.

2. Materials and Methods

2.1. Data

The study utilized 919 measurements from model tests of 20 transport ship models. Data were collected from the following publicly available literature sources [21,36–54]. The added wave resistance values are presented as dimensionless values for the added resistance coefficient (CAW) in relation to the wavelength (λ/L). The design characteristics of these models are presented in Table 1.

Table 1. Main particulars of the ship models.

Type of Ship	L/B	B/T	L/T	CB [-]	Fn [-]
Bulk carrier	5.7	2.7	15.4	0.829	0.1, 0.15
Bulk carrier	5.3	3.2	17.1	0.84	0.172
Containership	6.9	3.0	20.6	0.559	0.15, 0.2, 0.25, 0.3
Containership	6.8	2.8	19.3	0.568	0.15, 0.2, 0.25, 0.3
Containership	8.4	3.0	24.9	0.598	0.24
Containership	6.7	3.2	21.6	0.602	0.183
Containership	6.5	3.7	24.2	0.643	0.204
Containership	7.1	3.0	21.3	0.651	0.26
Containership	7.5	2.9	21.4	0.66	0.247
Containership	7.0	3.5	24.5	0.661	0.139
Cruise ship	6.9	4.5	30.6	0.654	0.223
Fast cargo ship	6.7	4.4	29.3	0.503	0.15, 0.2, 0.25, 0.3
Fast cargo ship	6.7	2.5	16.7	0.563	0.15, 0.2, 0.25, 0.3
Gas carrier	6.4	3.9	25.2	0.77	0.188
Product carrier	6.2	2.9	18.2	0.757	0.177
Product carrier	5.9	2.8	16.5	0.79	0.15
RoPax	5.1	4.2	21.4	0.549	0.087
S-Cb87	5.5	4.4	24.3	0.83	0.147
S-Cb87	5.5	2.8	15.4	0.87	0.142
Series 60	7.5	2.5	18.8	0.6	0.266, 0.283
Series 60	7.3	2.5	18.1	0.65	0.237, 0.254
Series 60	7.0	2.5	17.5	0.7	0.207, 0.222
Series 60	6.8	2.5	16.9	0.75	0.177, 0.195
Series 60	7.5	2.5	18.8	0.8	0.147, 0.165
SR221C	5.5	3.0	16.6	0.803	0.15
Tanker	5.4	3.2	17.6	0.835	0.156
VLCC	5.5	2.8	15.4	0.808	0.142
VLCC	5.5	4.4	24.4	0.81	0.147
VLCC	5.5	2.8	15.4	0.84	0.142
VLCC	5.5	4.4	24.4	0.84	0.147
VLCC	5.5	2.8	15.4	0.87	0.142
VLCC	5.5	4.4	24.4	0.87	0.147
Min	5.1	2.5	15.4	0.50	0.09
Max	8.4	4.5	30.6	0.87	0.27

2.2. Research Method

In this study, ensembles of ANNs were used instead of individual ANNs to increase the reliability and accuracy of predictions. Each ANN ensemble consisted of five elementary neural networks, and the collective response of each ANN ensemble was calculated as the arithmetic mean of the responses from all five networks in the ensemble. It was assumed that four ANN ensembles would be developed to estimate the added wave resistance coefficient, incorporating the same input variables as the independent variables presented in Equations (1)–(4).

The accuracy of the ANN estimates was calculated using the Pearson correlation coefficient (PCC), determination coefficient (R^2), root mean square error (RMSE), and relative root mean square error (RRMSE). These are expressed as follows:

$$\text{PCC} = \frac{\Sigma(y_e - \bar{y})(y_e - \overline{y_e})}{\sqrt{\Sigma(y - y)^2(y_e - \overline{y_e})^2}}, \tag{6}$$

$$R^2 = \frac{\Sigma(y_e - \bar{y})^2}{\Sigma(y - \bar{y})^2}, \tag{7}$$

$$\text{RMSE} = \sqrt{\frac{\Sigma(y - y_e)^2}{n}}, \tag{8}$$

$$\mathrm{RRMSE} = \frac{\mathrm{RMSE}}{\overline{y_e}} \times 100, \tag{9}$$

where y is the measured output value, y_e is the estimated output value, and n is the number of measurements.

Figure 1 presents a block diagram illustrating the process of creating an ANN ensemble. For each network ensemble, training, testing, and validation datasets were created through randomization, containing 70%, 15%, and 15% of the total data, respectively. The training set contained data used to train each network in the ANN ensemble, the validation set contained data used to evaluate the network during training, and the test set contained data to assess network performance after training. The values of each input variable and the output variable were normalized to a range of 0 to 1 using the min-max normalization technique.

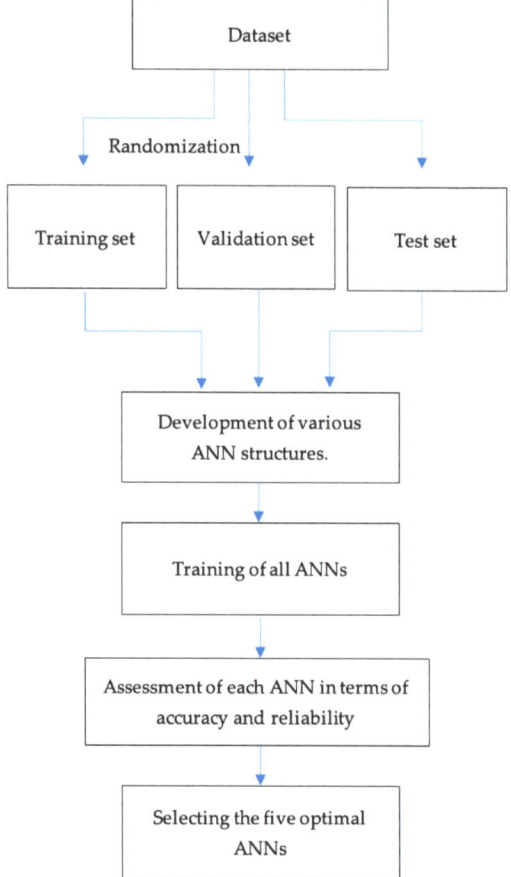

Figure 1. Block diagram for developing an ANN ensemble.

A large set of ANNs with a multilayer perceptron (MLP) architecture featuring one input layer, one hidden layer, and one output layer was then randomly generated. The number of neurons in the hidden layer and the activation functions for both the hidden and output layers were chosen randomly. Additionally, the weights and biases of the neurons were randomly initialized and adjusted during training to minimize the error function.

The Broyden–Fletcher–Goldfarb–Shanno (BFGS) algorithm [55–58] was used to train the networks. The mean absolute error (MAE) was used as the error function in this algorithm, which is defined as follows:

$$\text{MAE} = \frac{\sum_{i=1}^{n}\left|out_{(i)} - out_{p(i)}\right|}{n} \tag{10}$$

where out is the output value, out_p is the predicted output value, and n is the number of instances.

The early stopping technique was used to avoid overfitting during network training. According to this method, the network was trained repeatedly for a specified number of cycles as long as the test error decreased. When the test error began to increase, the training was stopped. After training the networks, five optimal ANNs were selected, characterized by the highest estimation accuracy in the test set.

Additionally, various strategies were employed in the optimization process of the neural networks to increase the accuracy and reliability of the predictions. Network architectures were randomly generated with different numbers of neurons in the hidden layer and various activation functions, allowing for the exploration of a wide range of configuration possibilities. The training process involved optimizing the weights and biases of the neurons using the BFGS algorithm, ensuring the minimization of the error function. To avoid overfitting, the early stopping technique was applied, which halted the training process when the error on the test set began to increase.

For each ANN ensemble, after the training process was completed, five optimal models were selected from the generated set of networks. These models were characterized by the highest accuracy on the test set and minimal differences in RMSE errors across the training, validation, and test sets. These networks formed the final version of the ANN ensembles used for predicting the added wave resistance coefficient (CAW).

3. Results

3.1. Artificial Neural Networks for Estimating Added Wave Resistance Based on Ship Design Parameters

As a result of the research, four ensembles of neural networks were created to approximate the added wave resistance coefficient based on the ship's design parameters, as shown below:

$$CAW \approx ANN1(L/B, B/T, CB, Fn, \lambda/L), \tag{11}$$

$$CAW \approx ANN2(L/B, CB, Fn, \lambda/L), \tag{12}$$

$$CAW \approx ANN3(B/T, CB, Fn, \lambda/L), \tag{13}$$

$$CAW \approx ANN4(L/T, CB, Fn, \lambda/L), \tag{14}$$

Due to the complex mathematical model of the calculations, the author developed a computer program called "WaveResist 2.0 (WIN/MAC)", in which the above four ensembles of artificial neural networks were implemented. The program was written in Pascal using the cross-platform IDE Lazarus [59]. Binary files for Windows and Mac environments, as well as the source code, are available in the open research data repository Mendeley Data [60]. Figure 2 shows the user interface of this program.

Table 2 presents the structures, RMSE errors, PCC coefficients, number of iterations, and activations of all the neural networks included in the above ensembles. All neural networks achieved good convergence with the number of neurons in the hidden layer ranging from 6 to 10, with an average of 9 neurons. The networks in the ANN1 and ANN4 groups achieved convergence the fastest, with an average of 208 and 204 iterations, respectively. The ANN2 and ANN3 ensembles required an average of 245 and 261 iterations, respectively. During these iterations, the early stopping technique was used to prevent overfitting. The average RMSE values in the test set for the networks included in the individual ensembles ANN1, ANN2, ANN3, and ANN4 were low and amounted to 0.91, 1.16, 1.03, and 0.97, respectively. The average Pearson correlation coefficients for the test set for these networks were 0.93, 0.89, 0.91, and 0.92, respectively. This indicates that these networks estimate the measured values highly effectively, with the highest estimation accuracy achieved by the networks in the ANN1 and ANN4 ensembles. The differences between the RMSE errors and PCC coefficients between the test, validation, and training sets for all 20 elementary networks did not exceed 10%, indicating the absence of overfitting and confirming the effectiveness of the early stopping technique.

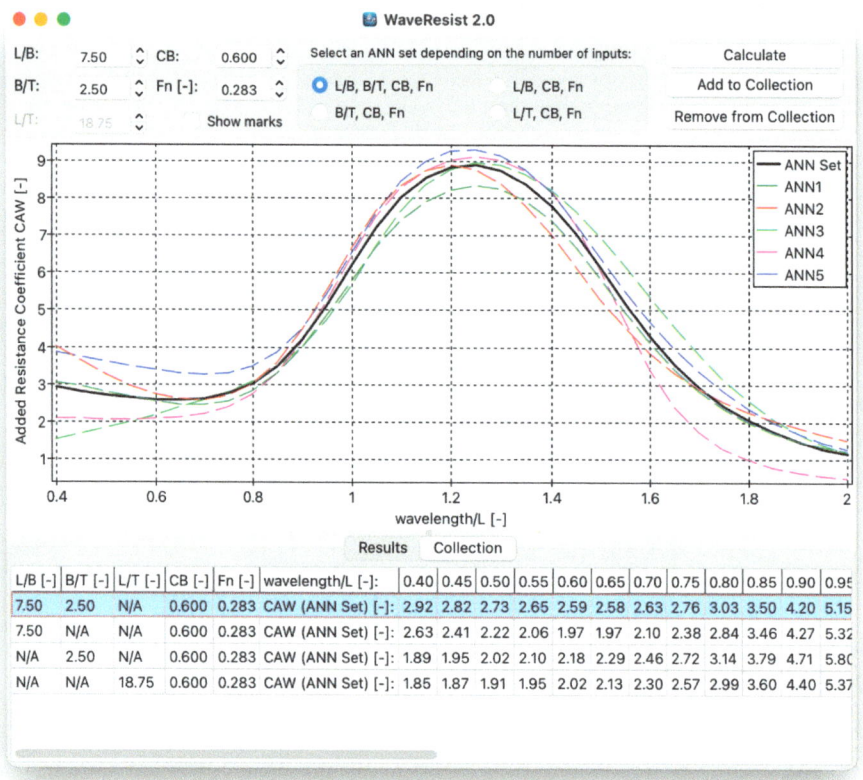

Figure 2. WaveResist computer program user interface.

Table 3 shows the statistical parameters of the ANN1, ANN2, ANN3, and ANN4 ensembles, consisting of 20 elementary neural networks described in Table 2.

The response of each network ensemble was the arithmetic mean of the responses of the five elementary neural networks that formed the ensemble.

The values of these parameters presented in Table 3 and calculated for the test set indicate that the ANN1 ensemble is characterized by the highest estimation quality, as confirmed by the lowest RMSE (0.75) and RRMSE (16.1%) values and the highest determination coefficients (R^2 = 0.91) and Pearson correlation coefficients (0.95). The ANN4 ensemble also exhibits high quality, with a low RMSE (0.89) and RRMSE (19.1%) and high determination coefficients ($R^2 = 0.89$) and Pearson correlation coefficients (0.94), making it the second-best ensemble. The ANN3 ensemble has worse performance compared to ANN1 and ANN4, but its results indicate good quality (RMSE = 0.92, RRMSE = 19.9%, $R^2 = 0.86$, and Pearson correlation = 0.93). The ANN2 ensemble shows the highest RMSE (1.11) and RRMSE (24%) values among all the analyzed ensembles, and its determination coefficient ($R^2 = 0.80$) and Pearson correlation coefficient (0.90) are the lowest. According to earlier studies [61,62], estimates can be classified based on the RRMSE criterion: excellent (below 10%), good (between 10% and 20%), acceptable (between 20% and 30%), and unacceptable (above 30%). According to this criterion, the estimates obtained using the ANN1, ANN3, and ANN4 ensembles can be classified as good. Although ANN2 has slightly lower performance compared to the other ensembles, it still falls within the category of acceptable estimates.

Table 2. Structures, RMSE and PCC values, activation functions, and iterations of the artificial neural networks included in the ANN ensembles for the training, test, and validation datasets.

ANN Ensemble	Inputs	Structure	RMSE [-]			PCC [-]			Iterations	Activation	
			Train	Test	Valid	Train	Test	Valid		Hidden Layer	Output Layer
ANN1	$L/B, B/T, CB, Fn, \lambda/L,$	5-9-1	0.96	0.94	0.91	0.93	0.93	0.94	213	Sigm	Sigm
		5-8-1	0.91	0.92	0.86	0.94	0.93	0.94	157	Tanh	Sigm
		5-9-1	0.84	0.89	0.90	0.95	0.94	0.94	266	Tanh	Sigm
		5-10-1	0.85	0.89	0.91	0.95	0.94	0.94	225	Tanh	Sigm
		5-10-1	0.90	0.88	0.93	0.94	0.94	0.94	181	Sigm	Sigm
		mean	*0.89*	*0.91*	*0.90*	*0.94*	*0.93*	*0.94*	*208*		
ANN2	$L/B, CB, Fn, \lambda/L,$	4-9-1	1.12	1.19	1.03	0.90	0.88	0.92	373	Sigm	Lin
		4-10-1	1.11	1.16	1.01	0.90	0.89	0.93	223	Sigm	Lin
		4-10-1	1.07	1.13	1.00	0.91	0.89	0.93	266	Sigm	Lin
		4-9-1	1.07	1.11	1.04	0.91	0.90	0.92	223	Tanh	Sigm
		4-6-1	1.15	1.20	1.09	0.90	0.88	0.91	140	Tanh	Sigm
		mean	*1.11*	*1.16*	*1.03*	*0.90*	*0.89*	*0.92*	*245*		
ANN3	$B/T, CB, Fn, \lambda/L,$	4-10-1	0.99	1.05	0.97	0.93	0.91	0.93	206	Sigm	Sigm
		4-8-1	1.03	1.06	1.02	0.92	0.91	0.92	233	Tanh	Sigm
		4-10-1	0.93	0.97	1.01	0.93	0.92	0.92	337	Sigm	Sigm
		4-8-1	0.99	1.00	1.02	0.92	0.92	0.92	290	Tanh	Tanh
		4-9-1	1.04	1.05	1.01	0.92	0.91	0.92	240	Tanh	Tanh
		mean	*0.99*	*1.03*	*1.01*	*0.92*	*0.91*	*0.92*	*261*		
ANN4	$L/T, CB, Fn, \lambda/L,$	4-10-1	0.85	0.84	0.82	0.95	0.94	0.95	277	Tanh	Sigm
		4-9-1	0.97	0.98	0.98	0.93	0.92	0.93	168	Sigm	Sigm
		4-8-1	1.03	1.03	1.00	0.92	0.91	0.92	122	Tanh	Sigm
		4-7-1	0.99	1.01	1.01	0.92	0.92	0.92	192	Sigm	Lin
		4-9-1	0.99	1.00	1.00	0.93	0.92	0.93	261	Tanh	Tanh
		mean	*0.96*	*0.97*	*0.96*	*0.93*	*0.92*	*0.93*	*204*		

The above data show that the average RMSE values of the networks included in the individual ensembles ANN1, ANN2, ANN3, and ANN4 are similar to the RMSE values of these ensembles. Table 3 shows that the differences between the errors and correlation coefficients of the predictions of all the neural network ensembles for the test, validation, and training data are small, indicating the absence of clear signs of overfitting and the effectiveness of the early stopping algorithm used during neural network training. The ensembles are characterized by high and consistent Pearson correlation coefficients (PCC) and determination coefficients (R^2), indicating good generalization. The ANN2 ensemble, although having the lowest PCC and R^2 values, also shows similar results for all datasets, with differences not exceeding 12%, suggesting the absence of overfitting and good generalization properties.

Table 3. PCC, R^2, RMSE, and RRMSE values for all the ANN ensembles across training, test, and validation datasets.

	ANN1	ANN2	ANN3	ANN4
PCC train [-]	0.96	0.92	0.94	0.94
PCC valid [-]	0.95	0.93	0.94	0.95
PCC test [-]	0.95	0.90	0.93	0.94
R^2 train [-]	0.92	0.84	0.88	0.89
R^2 valid [-]	0.90	0.87	0.89	0.90
R^2 test [-]	0.91	0.80	0.86	0.87
RMSE train [-]	0.72	1.05	0.89	0.87
RMSE valid [-]	0.81	0.97	0.88	0.85
RMSE test [-]	0.75	1.11	0.92	0.89
RRMSE train [%]	15.7	22.8	19.4	19.0
RRMSE valid [%]	18.0	21.6	19.6	18.9
RRMSE test [%]	16.1	24.0	19.9	19.1

The regression plots and residuals for the testing data for all neural network ensembles, which are given in Figure 3, show that the trend line between the calculated and predicted CAW values runs almost along the linear y = x (45°) curve. This indicates that, overall, all ensembles demonstrated high effectiveness. However, the slightly lower constant value (0.3902) and higher slope of the trend line for ANN1 (0.908) suggest that it is closest to the linear y = x (45°) curve compared to the trend lines for the other networks, where the constant is 0.9296, 0.5923, and 0.5888, and the slope is 0.8104, 0.8677, and 0.8756 for the ANN2, ANN3, and ANN4 ensembles, respectively. Among these remaining network ensembles, the trend line for ANN4 is the closest to the linear y = x (45°) curve. Figure 4 shows that the residuals were randomly distributed for all networks for the test data. The range of residuals for the ANN2 and ANN3 ensembles was wider than for the ANN1 and ANN4 ensembles. The smallest range of residuals was exhibited by the predictions made using the ANN1 ensemble. Figure 5 compares the CAW values predicted by the neural network ensembles with the measurements for various ship models using only test data. These graphs show that the ensembles effectively predict CAW for the test data, which were not used in training these networks, and exhibit appropriate trends.

In this study, a sensitivity analysis was performed to assess the impact of individual variables on the outputs of the ANN ensembles. This analysis involved examining how the network's error changes when the values of each input variable were replaced by their means, effectively removing any information provided by that variable. Once the modified data were introduced into the network, the final prediction error was calculated. The larger the increase in this error, the more sensitive the network is to the given variable, indicating its importance. A minor increase, or even a decrease in the error, may suggest that the variable contributes little to the model or might even degrade its performance. Based on the error increase, a ranking of variable importance was also established. Table 4 presents the multiple increase in RMSE of the ANN ensembles after the removal of each input variable. In this table, the numbers in parentheses indicate the ranking of the variable's importance. The results show that in all ensembles, all input variables are significant, as their removal increases the prediction error by a factor ranging from 3.31 to 14.49 times the estimation error with all variables present. For all ensembles, the variable λ/L is the most important, with its removal causing up to a 14-fold increase in error for the ANN1 ensemble. The variables Fn and CB are also of substantial importance, though their influence varies depending on the ensemble, being particularly significant for the ANN2 and ANN3 ensembles. The variables L/B, B/T, and L/T are of somewhat lower importance, but their impact on prediction performance is still notable, as they increase the prediction error by three, four, or even five times.

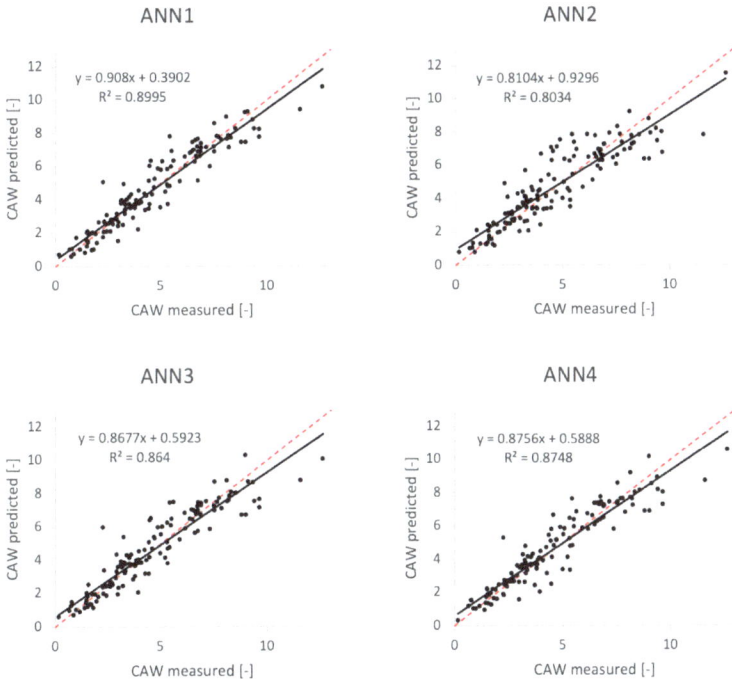

Figure 3. Regression plot for all the ANN ensembles for the test data, with the black solid line representing the regression between predicted and experimental CAW values, and the red dashed line representing the ideal y = x (45°) line.

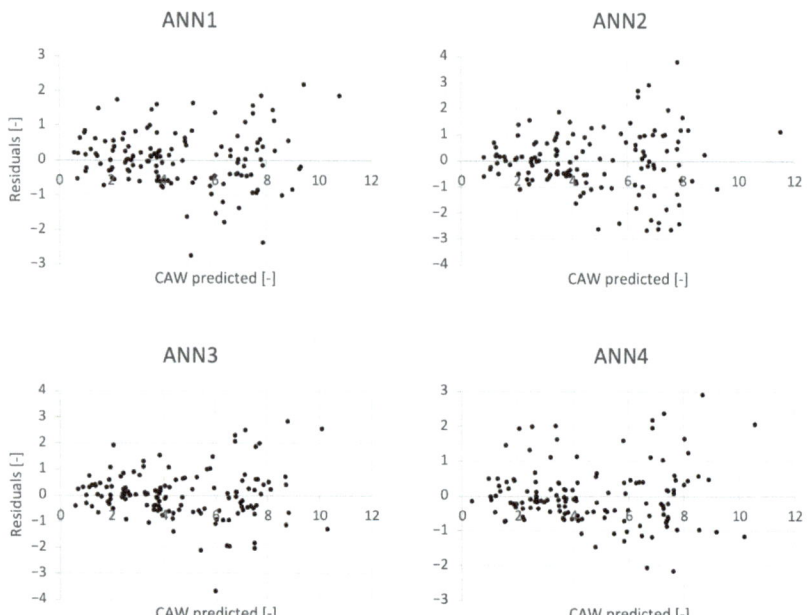

Figure 4. Residual plot for all the ANN ensembles for the test data.

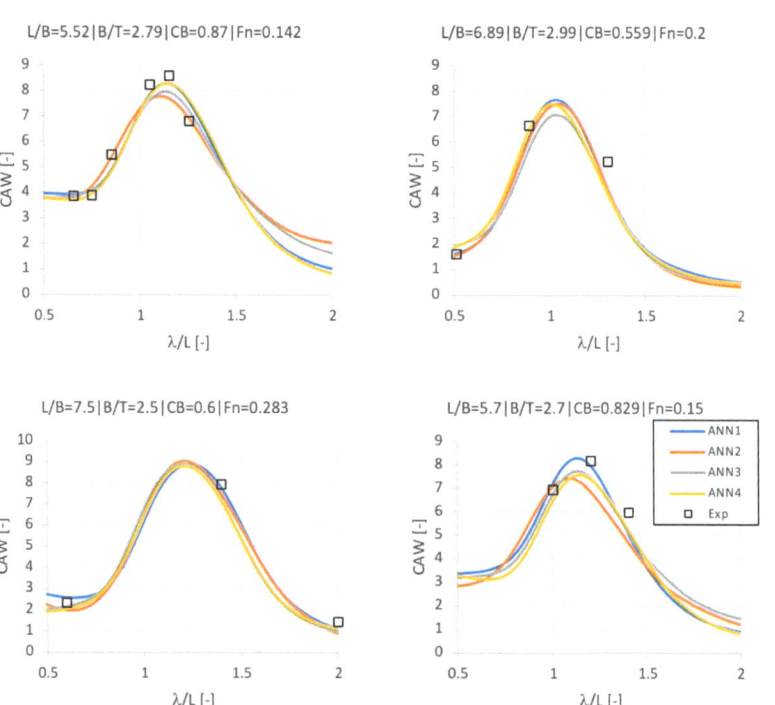

Figure 5. Comparison of the estimations using the ANN ensembles with measurements for test cases.

Table 4. Sensitivity analysis of the ANN ensembles.

	λ/L	Fn	CB	L/B	B/T	L/T
ANN1	14.49 (1)	4.88 (5)	7.19 (2)	5.08 (4)	5.14 (3)	n/a
ANN2	11.20 (1)	4.48 (2)	3.92 (3)	3.36 (4)	n/a	n/a
ANN3	11.68 (1)	5.17 (2)	4.00 (3)	3.31 (4)	n/a	n/a
ANN4	12.38 (1)	3.53 (3)	4.07 (2)	n/a	n/a	3.42 (4)

Note: 'n/a' denotes that the respective parameters were not included as input variables for the corresponding ANN ensembles.

3.2. Utilizing Artificial Neural Network Ensembles for Ship Design Optimization to Reduce Added Wave Resistance and CO_2 Emissions

In this section, the practical application of the ANN2 ensemble is examined in the context of solving a sample design task at the preliminary ship design stage. It is assumed that the design process is in the parametric phase of determining the ship's breadth, with only the following parameters known: ship length $L = 200$ m, Froude number $Fn = 0.26$, block coefficient $CB = 0.65$. The design constraint resulting from other design criteria concerns the L/B ratio, which is restricted to the range of 7.0 to 7.3. The objective of this task is to optimize the L/B with respect to minimizing CO_2 emissions and to determine the extent to which CO_2 emissions will be reduced as a result of this optimization.

For marine engines powered by heavy fuel, daily CO_2 emissions can be determined using the following relationship:

$$CO_2 = 3.206 \times FC \ [t/day]. \tag{15}$$

where FC represents the daily fuel consumption, which is calculated using the following:

$$FC = 0.001 \times FOC \times \Delta PP \times 24 \ [t/day], \tag{16}$$

where FOC is the fuel consumption rate, which in this design task is set as $FOC = 0.175$ kg/kWh/hour, and ΔPP is the increment of the propulsion power, which is calculated as follows:

$$\Delta PP = \frac{\Delta P_E}{\eta} \ [kW], \tag{17}$$

where η denotes the total propulsion efficiency, which in this design task is set as $\eta = 0.645$, ΔP_E is the increment of effective power, which is determined as follows:

$$\Delta P_E = R_{AW} \times v \ [kW], \tag{18}$$

where v is the velocity of the ship [m/s], R_{AW} is the mean added wave resistance in irregular waves, which is calculated as follows:

$$R_{AW} = 2\zeta_a^2 \rho g \frac{B^2}{L} \int_0^\infty CAW(\omega) S_{\zeta\zeta}(\omega) d\omega, \tag{19}$$

where R_{AW} represents the mean additional resistance, ζ_a is the wave amplitude, ω denotes the circular wave frequency, L is the ship length, B is the ship breadth, ρ is the seawater density, g represents the acceleration due to gravity, and $S_{\zeta\zeta}$ is the wave spectral function.

Figure 6 shows the estimated values of the CAW coefficient, calculated using the ANN2 ensemble. This figure indicates that variants with $L/B = 7.2$ and 7.3 are characterized by the lowest added wave resistance. The mean added resistance RAW was calculated for waves with a significant height $H_s = 1$ to 3 m and a period of $T_1 = 10$ s. The following wave spectral function was employed in accordance with ref. [63]:

$$S_{\zeta\zeta}(\omega) = A\omega^{-5} exp\left(B\omega^{-4}\right), \tag{20}$$

where A and B are coefficients calculated as follows:

$$A = \frac{173 H_s^2}{T_1^4}, \tag{21}$$

$$B = \frac{691}{T_1^4}. \tag{22}$$

where H_s is the significant wave height and T_1 is the average zero-up-crossing wave period.

Tables 5 and 6, as well as Figure 7, present the estimated values of added wave resistance and daily CO_2 emissions for all design variants, depending on the wave height. Figure 8 illustrates the CO_2 emissions values as a function of the L/B ratio, calculated for a ship sailing in waves with a height of $H_s = 3$ m and a characteristic wave period T1 = 10 s. An analysis of these tables and figures indicates that the variant with $L/B = 7.3$ is characterized by the lowest added wave resistance and fuel consumption, while variants with $L/B = 7.0$ exhibit the highest. The maximum differences in added wave resistance and daily CO_2 emissions between these design variants, for a wave height of 3 m, are 5.4 kN and 2.55 tons/day, respectively. However, the $L/B = 7.2$ variant also demonstrates low CO_2 emissions, only slightly higher by 0.02 tons/day compared to the $L/B = 7.3$ variant.

The analysis shows that the optimal design in terms of CO_2 emissions for a ship sailing in waves is the variant with an L/B ratio of 7.3, which corresponds to a ship breadth B equal to the following:

$$B = \frac{L}{L/B} = \frac{200\ m}{7.3} = 27.40\ m. \tag{23}$$

The difference in CO_2 emissions between the best and worst variants amounts to 2.55 tons/day. This difference indicates that even minor modifications to ship design parameters can have a significant impact on added wave resistance and the subsequent increase in CO_2 emissions.

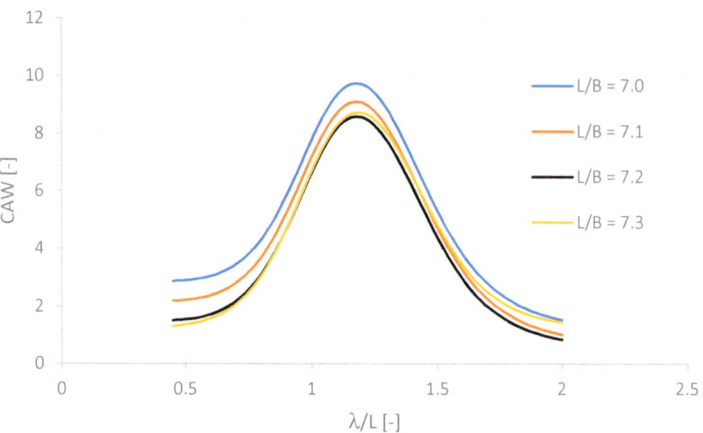

Figure 6. Estimation of the added wave resistance coefficient (CAW) using the ANN2 ensemble for design variants.

Table 5. Estimated added wave resistance values for all the design variants depending on the significant wave height (H_s) and characteristic wave period $T_1 = 10$ s using the ANN2 ensemble.

H_s [m]	L/B = 7.0	L/B = 7.1	L/B = 7.2	L/B = 7.3
1	2.40	2.07	1.80	1.80
2	9.60	8.28	7.22	7.20
3	21.61	18.64	16.24	16.20

Table 6. Estimated daily CO_2 emissions values for all the design variants depending on the significant wave height (H_s) and characteristic wave period $T_1 = 10$ s using the ANN2 ensemble.

H_s [m]	L/B = 7.0	L/B = 7.1	L/B = 7.2	L/B = 7.3
1	1.13	0.98	0.85	0.85
2	4.53	3.91	3.40	3.40
3	10.19	8.79	7.66	7.64

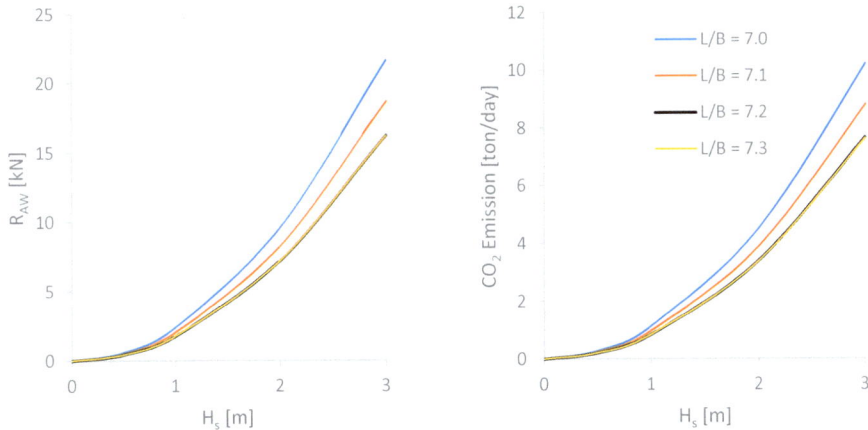

Figure 7. Estimated wave-added resistance and daily CO_2 emissions values for all the design variants depending on the significant wave height (H_s) and characteristic wave period $T_1 = 10$ s.

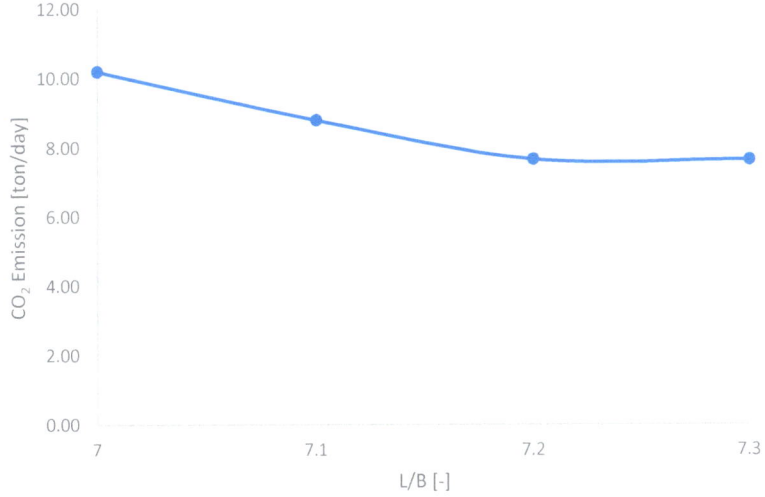

Figure 8. Estimated daily CO_2 emissions values for all design variants as a function of the L/B ratio, for the significant wave height $H_s = 3$ m and the characteristic wave period $T_1 = 10$ s.

4. Discussion

The design analysis presented in this article demonstrates that even minor modifications to a ship's design parameters can have a substantial impact on added wave resistance, which subsequently leads to increased CO_2 emissions. Given the large quantities of fuel consumed by ships, such small design modifications can yield significant environmental benefits, far greater than those achieved through changes in other modes of transportation, particularly when considering the mass of the cargo being transported. This study shows that in the case of a typical container ship, improper design modifications could result in a daily increase in CO_2 emissions of up to 2.55 tons, assuming the vessel operates in wave conditions with a significant wave height of 3 m. This amount of CO_2 emissions is equivalent to the emissions generated by 778 gasoline-powered cars, each driving 10,000 km annually at an average fuel consumption rate of 5.2 L per 100 km. Considering that approximately 105,000 cargo ships navigate global seas and oceans each year [3], even a small reduction in added wave resistance and consequently in CO_2 emissions could have a considerable positive impact on environmental protection.

The ANN1 ensemble, with input parameters L/B, B/T, CB, Fn, and λ/L, exhibits the lowest RMSE (0.75) and RRMSE (16.1%), the highest R^2 (0.91), and Pearson correlation coefficient (0.95). The combination of these parameters provides the best results, suggesting they have a strong influence on added wave resistance. The ANN2 ensemble, with input parameters L/B, CB, Fn, λ/L, has higher RMSE (1.11), RRMSE (24%), the lowest R^2 (0.80), and the lowest Pearson correlation (0.90). The absence of the B/T parameter worsens the results, indicating its significant impact on added wave resistance. The ANN3 ensemble, with input parameters B/T, CB, Fn, λ/L, has RMSE (0.92), RRMSE (19.9%), R^2 (0.86), and Pearson correlation (0.93). The absence of the L/B parameter worsens the results, but not as significantly as in the case of ANN2, suggesting that the L/B parameter also has a significant impact, but slightly less than B/T. The ANN4 ensemble, with input parameters L/T, CB, Fn, λ/L, shows low RMSE (0.89), RRMSE (19.1%), high R^2 (0.87), and Pearson correlation coefficient (0.94). This indicates that replacing the L/B and B/T parameters with L/T also yields good results comparable to the model containing all the considered parameters.

These results show that the set of parameters L/B, B/T, CB, Fn, and λ/L together allows for the most accurate estimation of added wave resistance, as evident in the results for the ANN1 ensemble. These findings confirm the studies by Maruo [18,19], Boom et al. [26], and Liu and Papanikolaou [27,31] that the B/T parameter appears to be particularly significant, as its absence in ANN2 significantly degrades the results. Replacing L/B and B/T with the L/T ratio in combination with CB, Fn, and λ/L also yields good results, suggesting its significance and potential applicability at the parametric design stage, where the ship's breadth and draught are known. Using the L/T parameter allows for the optimization of draught concerning the ship's added resistance at the design stage, where the length, the Froude number, and CB are known.

The average RMSE values for the test data for the individual networks within the ensembles were 0.91, 1.16, 1.03, and 0.97 for the ANN1, ANN2, ANN3, and ANN4 networks, respectively. Meanwhile, the RMSE values for the ensemble estimates were 0.75, 1.11, 0.92, and 0.89, respectively, indicating an improvement in prediction accuracy compared to the average RMSE values for individual networks in some cases. Additionally, the average Pearson correlation coefficients for the elementary networks for the test data were 0.93, 0.89, 0.91, and 0.92, respectively. In contrast, the Pearson correlation coefficients of the ensemble responses were 0.95, 0.90, 0.93, and 0.94, respectively, indicating an even higher agreement between predictions and measurements compared to the results obtained for individual networks. This suggests that using neural network ensembles instead of individual networks can lead to increased prediction accuracy and improved consistency with benchmark values. The average RMSE values for the network ensembles were generally lower than for individual networks, confirming that the ensemble approach can reduce prediction error. Additionally, the higher Pearson correlation coefficients for the network ensembles indicate a better model fit to the test data.

The developed neural network ensembles are based on fundamental dimensionless design characteristics and can be used at an early design stage, e.g., to optimize the ship's breadth or draught based on the length-to-breadth ratio, the Froude number, and the hull block coefficient concerning added wave resistance. An example design task demonstrates how to select the optimal ship breadth to minimize added resistance and CO_2 emissions at the preliminary design stage when only the ship's length, the Froude number, and the hull block coefficient are known. In such cases, with incomplete design data, the application of the STAWAVE-2 and SNNM methods is not possible, as these methods rely on all the main ship dimensions and have size limitations. Therefore, using the presented method in cases where not all the design characteristics are known can be helpful in more accurate estimates of resistance, fuel consumption, and emission indicators. This is important because added wave resistance depends, among other things, on the main hull dimensions, so modeling this resistance at the design stages (where these dimensions are iteratively determined) can bring the most benefit. Furthermore, in design work, there is often a need to verify the obtained solutions using other alternative methods. The methods proposed in this article can be useful to the designer for this purpose and may help reduce the likelihood of design errors.

5. Conclusions

The research demonstrated that neural networks trained based on model test results can accurately predict the coefficient of added wave resistance (CAW) using various dimensionless combinations of main dimension ratios, the Froude number (Fn), and the block coefficient (CB). The main innovation presented in this article is the development of accurate methods for predicting CAW based on sets of incomplete design characteristics, such as L/B, CB, Fn; B/T, CB, Fn; and L/T, CB, Fn.

The prediction accuracy of the ANN ensembles was assessed using RMSE, RRMSE, R^2, and the Pearson correlation coefficient. The ANN1 ensemble, which included all input parameters (L/B, B/T, CB, Fn, and λ/L), demonstrated the highest performance, with an RMSE of 0.75, RRMSE of 16.1%, R^2 of 0.91, and a Pearson correlation coefficient of 0.95. In contrast, the ANN2 ensemble, which excluded the B/T parameter, exhibited greater errors, with an RMSE of 1.11, RRMSE of 24%, and lower correlation values (R^2 of 0.80 and a Pearson correlation coefficient of 0.90), highlighting the critical importance of the B/T ratio. The ANN3 ensemble, which omitted the L/B parameter, achieved intermediate performance, with an RMSE of 0.92, RRMSE of 19.9%, R^2 of 0.86, and a Pearson correlation coefficient of 0.93, suggesting that the L/B parameter has a notable influence, though less than that of B/T. The ANN4 ensemble, where L/T replaced L/B and B/T, also produced favorable results, with an RMSE of 0.89, RRMSE of 19.1%, R^2 of 0.87, and a Pearson correlation coefficient of 0.94.

The observed prediction errors can be attributed to several factors. Although parameters such as L/B, CB, and Fn have a significant influence on the predictions, a primary limitation is the insufficient availability of measurement data for a more diverse range of ship types. The current models have been trained on a limited dataset, primarily consisting of standard hull forms. Expanding the dataset in future work to incorporate measurements from a wider variety of ship designs, particularly those featuring more innovative or unconventional hull forms, could facilitate the development of more accurate and reliable prediction models.

Since the artificial neural networks were trained using these dimensionless design characteristics of ship models, they can be applied to ships of any length, provided that the values of these ratios fall within the training set limits (i.e., L/B from 5.1 to 8.4, B/T from 2.5 to 4.5, CB from 0.5 to 0.87, and Fn from 0.09 to 0.27). This allows for estimating added wave resistance, and thus the ship's energy demand, fuel consumption, and CO_2 emissions, already at the early stages of ship design. As a result, it becomes possible to optimize the ship's design parameters to reduce CO_2 emissions at the preliminary parametric design stage. One limitation of the presented neural networks is that they were trained based on standard hull shapes, which restricts their application in predicting added resistance for new, atypical, or innovative hull shapes.

The design calculation example presented in this article proved that the proposed methods can be used for preliminary design analyses when only the ship length, the Froude number, and the block coefficient are known. This analysis demonstrated that even a minor modification to a typical-sized container ship design can significantly reduce additional wave-induced resistance, leading to a decrease in CO_2 emissions due to lower fuel consumption. This example illustrates that, within specified design constraints, selecting the most advantageous design variant can result in a reduction of up to 2.55 tons of CO_2 emissions per day compared to the least favorable variant. Therefore, the method can find practical application in design offices for parametric design and preliminary project analyses when all ship dimensions or their ratios are not known. This is particularly useful when the application of semiempirical methods, such as STAWAVE-2 and SNNM, is not possible because they require knowledge of all main dimensions. Based on the validation results, it can be concluded that all ensembles of artificial neural networks are effective methods for estimating CAW, but the final choice of the network set may depend on the number of input variables available at a given design stage.

The research confirmed that ratios such as B/T and L/B have a significant impact on added wave resistance. In particular, the B/T parameter proved to be crucial; its absence significantly worsened the results. Replacing L/B or B/T with the L/T ratio improves the accuracy of CAW estimation. Although the research was conducted on a limited number of measurements, future studies on a larger number of ships may change this view. For now, it appears that the key main dimension ratio affecting added resistance is B/T.

This article utilized four ensembles of elementary artificial neural networks to predict added wave resistance. The response obtained from each ensemble was the average of the responses from the individual networks, which increased the estimation accuracy and provided a more reliable assessment. This study found that predictions from neural network ensembles were characterized by higher Pearson correlation coefficients, and, in some cases, lower root mean square errors compared to predictions from individual networks. This confirms that integrating results from multiple neural networks leads to more reliable and accurate predictions, which is crucial in practical machine learning applications.

Funding: This research received no external funding.

Data Availability Statement: During the study, the WaveResist software was developed. This tool is available in the Mendeley database: Cepowski, Tomasz (2024), 'WaveResist 2.0 (WIN/MAC): An Artificial Neural Network Tool for Ship Wave Resistance Coefficient Calculation', Mendeley Data, V2, https://doi.org/10.17632/nggvmdpbrt.2, https://data.mendeley.com/datasets/nggvmdpbrt/2 (accessed on 10 June 2024).

Conflicts of Interest: The author declares no conflicts of interest.

Abbreviations

ANN	Artificial neural network
ANN1	Neural network ensemble with inputs L/B, B/T, CB, Fn, and λ/L
ANN2	Neural network ensemble with inputs L/B, CB, Fn, and λ/L
ANN3	Neural network ensemble with inputs B/T, CB, Fn, and λ/L
ANN4	Neural network ensemble with inputs L/T, CB, Fn, and λ/L
BFGS	Broyden–Fletcher–Goldfarb–Shanno algorithm
B	Breadth of the ship
B/T	Breadth-to-draught ratio
CAW	Non-dimensional added wave resistance coefficient
CB	Block coefficient
EEDI	Energy Efficiency Design Index
Fn	Froude number
H_s	Significant wave height
IMO	International Maritime Organization
ITTC	International Towing Tank Conference
L	Length of the ship
L/B	Length-to-breadth ratio
L/T	Length-to-draught ratio
MAE	Mean absolute error
MLP	Multilayer perceptron
NMRI	National Maritime Research Institute
PCC	Pearson correlation coefficient
RMSE	Root mean square error
RRMSE	Relative root mean square error
R^2	Coefficient of determination
SNNM	SHOPERA-NTUA-NTU-MARIC method
STAWAVE-2	State-of-the-art wave model—2nd version
T	Draught of the ship
UNCTAD	United Nations Conference on Trade and Development
VLCC	Very large crude carrier
λ	Wavelength of a regular wave
λ/L	Wavelength-to-ship length ratio

References

1. UNCTAD Review of Maritime Transport (UNCTAD/RMT/2021). United Nations Conference on Trade and Development. 2021. Available online: https://unctad.org/publication/review-maritime-transport-2021 (accessed on 1 February 2024).
2. Fu, X.; Chen, D.; Wang, X.; Li, Y.; Lang, J.; Zhou, Y.; Guo, X. The impacts of ship emissions on ozone in eastern China. *Sci. Total Environ.* **2023**, *903*, 166252. [CrossRef] [PubMed]
3. UNCTAD Review of Maritime Transport (UNCTAD/RMT/2023). United Nations Conference on Trade and Development. 2023. Available online: https://unctad.org/publication/review-maritime-transport-2023 (accessed on 1 February 2024).
4. Moreno-Gutiérrez, J.; Pájaro-Velázquez, E.; Amado-Sánchez, Y.; Rodríguez-Moreno, R.; Calderay-Cayetano, F.; Durán-Grados, V. Comparative analysis between different methods for calculating on-board ship's emissions and energy consumption based on operational data. *Sci. Total Environ.* **2019**, *650*, 575–584. [CrossRef]
5. Yang, N.; Deng, X.; Liu, B.; Li, L.; Li, Y.; Li, P.; Tang, M.; Wu, L. Combustion Performance and Emission Characteristics of Marine Engine Burning with Different Biodiesel. *Energies* **2022**, *15*, 5177. [CrossRef]
6. Bei, Z.; Wang, J.; Li, Y.; Wang, H.; Li, M.; Qian, F.; Xu, W. Challenges and Solutions of Ship Power System Electrification. *Energies* **2024**, *17*, 3311. [CrossRef]
7. Skoko, I.; Stanivuk, T.; Franic, B.; Bozic, D. Comparative Analysis of CO_2 Emissions, Fuel Consumption, and Fuel Costs of Diesel and Hybrid Dredger Ship Engines. *J. Mar. Sci. Eng.* **2024**, *12*, 999. [CrossRef]
8. IMO. Resolution Res MEPC.364(79) Guidelines on the Method of Calculation of the Attained Energy Efficiency Design Index (EEDI) for New Ships. 2022. Available online: https://wwwcdn.imo.org/localresources/en/KnowledgeCentre/IndexofIMOResolutions/MEPCDocuments/MEPC.364(79).pdf (accessed on 16 May 2024).
9. Cepowski, T.; Kacprzak, P. Reducing CO_2 Emissions through the Strategic Optimization of a Bulk Carrier Fleet for Loading and Transporting Polymetallic Nodules from the Clarion-Clipperton Zone. *Energies* **2024**, *17*, 3383. [CrossRef]
10. Barberi, S.; Sambito, M.; Neduzha, L.; Severino, A. Pollutant Emissions in Ports: A Comprehensive Review. *Infrastructures* **2021**, *6*, 114. [CrossRef]
11. Lu, B.; Ming, X.; Lu, H.; Chen, D.; Duan, H. Challenges of decarbonizing global maritime container shipping toward net-zero emissions. *NPJ Ocean Sustain.* **2023**, *2*, 11. [CrossRef]

12. Cepowski, T. Determination of regression formulas for main tanker dimensions at the preliminary design stage. *Ships Offshore Struct.* **2018**, *14*, 320–330. [CrossRef]
13. Watson, D.G.M. *Practical Ship Design*; Elsevier Science: Amsterdam, The Netherlands, 1998; Volume 1.
14. Rawson, K.J.; Tupper, E.C. *Basic Ship Theory: Ship Dynamics and Design*; Butterworth-Heinemann: Oxford, UK, 2001.
15. Papanikolaou, A. *Ship Design—Methodologies of Preliminary Design*; Springer: Berlin/Heidelberg, Germany, 2014. [CrossRef]
16. International Maritime Organization (IMO). *Explanatory Notes to the Interim Guidelines on the Second Generation Intact Stability Criteria*; MSC.1/Circ.1652; IMO: London, UK, 2023.
17. Arribas, F.P. Some methods to obtain the added resistance of a ship advancing in waves. *Ocean Eng.* **2007**, *34*, 946–955. [CrossRef]
18. Maruo, H. The drift of a body floating on waves. *J. Ship Res.* **1960**, *4*, 1–10.
19. Maruo, H. *Resistance in Waves*; 60th Anniversary Series; The Society of Naval Architects of Japan: Minato-ku, Tokyo, 1963.
20. Boese, P. Eine einfache Methode zur Berechnung der Widerstandserhöhung eines Schiffes im Seegang. *J. Schiffstechnik* **1970**. [CrossRef]
21. Gerritsma, J.; Beukelman, W. Analysis of the Resistance Increase in Waves of a Fast Cargo Ship 12. *Int. Shipbuild. Prog.* **1972**, *19*, 285–293. [CrossRef]
22. Salvesen, N.; Tuck, E.O.; Faltinsen, O. Ship motions and sea loads. *Trans. Soc. Nav. Arch. Mar. Eng.* **1970**. Available online: https://trid.trb.org/View/495 (accessed on 10 May 2024).
23. Fujii, H.; Takahashi, T. Experimental study on the resistance increase of a ship in regular oblique waves. In Proceedings of the 14th International Towing Tank Conference, Ottawa, ON, Canada, September 1975; pp. 351–360.
24. Faltinsen, O.M.; Minsaas, K.J.; Liapis, N.; Skjørdal, S.O. Prediction of resistance and propulsion of a ship in a seaway. In Proceedings of the 13th Symposium on Naval Hydrodynamics, Tokyo, Japan, 6–10 October 1980.
25. Jinkine, V.; Ferdinande, V. A method for predicting the added resistance of fast cargo ships in head waves. *Int. Shipbuild. Prog.* **1974**, *21*, 149–167. [CrossRef]
26. Boom H van den Huisman, H.; Mennen, F. New Guidelines for Speed/Power Trials, SWZ/Maritime, Jan/Feb. 2013. Available online: https://www.researchgate.net/profile/Henk-Van-Den-Boom/publication/336530190_New_Guidelines_for_SpeedPower_Trials_Level_playing_field_established_for_IMO_EEDI/links/5dad58ee299bf111d4bf69a7/New-Guidelines-for-Speed-Power-Trials-Level-playing-field-established-for-IMO-EEDI.pdf (accessed on 15 April 2024).
27. Liu, S.; Papanikolaou, A. Regression analysis of experimental data for added resistance in waves of arbitrary heading and development of a semi-empirical formula. *Ocean. Eng.* **2020**, *206*, 107357. [CrossRef]
28. ITTC. *Recommended Procedures and Guidelines—Preparation, Conduct and Analysis of Speed/Power Trials*; ITTC: Zürich, Switzerland, 2021.
29. Lang, X.; Mao, W. A practical speed loss prediction model at arbitrary wave heading for ship voyage optimization. *J. Mar. Sci. Appl.* **2021**, *20*, 410–425. [CrossRef]
30. KKim, Y.-R.; Esmailian, E.; Steen, S. A meta-model for added resistance in waves. *Ocean Eng.* **2022**, *266 Pt 2*, 112749. [CrossRef]
31. Liu, S.; Papanikolaou, A. Improvement of the prediction of the added resistance in waves of ships with extreme main dimensional ratios through numerical experiments. *Ocean Eng.* **2023**, *273*, 113963. [CrossRef]
32. Cepowski, T. The use of a set of artificial neural networks to predict added resistance in head waves at the parametric ship design stage. *Ocean Eng.* **2023**, *281*, 114744. [CrossRef]
33. Ahn, Y.; Kim, Y. Data mining in sloshing experiment database and application of neural network for extreme load prediction. *Mar. Struct.* **2021**, *80*, 103074. [CrossRef]
34. Sun, Q.; Zhang, M.; Zhou, L.; Garme, K.; Burman, M. A machine learning-based method for prediction of ship performance in ice: Part I. ice resistance. *Mar. Struct.* **2022**, *83*, 103181. [CrossRef]
35. Dyer, A.S.; Zaengle, D.; Nelson, J.R.; Duran, R.; Wenzlick, M.; Wingo, P.C.; Bauer, J.R.; Rose, K.; Romeo, L. Applied machine learning model comparison: Predicting offshore platform integrity with gradient boosting algorithms and neural networks. *Mar. Struct.* **2022**, *83*, 103152. [CrossRef]
36. Strøm-Tejsen, J.; Yeh, H.Y.H.; Moran, D.D. Added resistance in waves. *Trans. Soc. Nav. Archit. Mar. Eng.* **1973**, *81*, 109–143.
37. Nakamura, S.; Hosoda, R.; Naito, S.; Nema, K. Propulsive performance of a container ship in waves (2nd report). *J. Kansai Soc. Nav. Archit.* **1975**, *1*, 45–55.
38. Nakamura, S. Added Resistance and Propulsive Performance of Ships in Waves. In *International Seminar on Wave Resistance, Japan*; The Society of Naval Architects of Japan: Minato-ku, Tokyo, 1976.
39. Journee, J.M.J. *Motions, Resistance and Propulsion of a Ship in Longitudinal Regular Waves*; Ship Hydromechanics Laboratory, Delft University of Technology: Delft, The Netherlands, 1976.
40. O'Dea, J.F.; Kim, Y.H. Added Resistance Power of a Frigate in Regular Waves. In *Ship Performance Department Report*; David Taylor-NSRDC/SPD-0964-02; David W. Taylor Naval Ship Research and Development Center: Bethesda, MD, USA, 1981; pp. 1–28.
41. Kadomatsu, K. Study on the Required Minimum Output of Main Propulsion Engine Considering Maneuverability in Rough Sea. Ph.D. Thesis, Ship Design Lab., Yokohama National University, Yokohama, Japan, 1988.
42. Kashiwagi, M.; Sugimoto, K.; Ueda, T.; Yamazaki, K.; Arihama, K.; Kimura, K.; Yamashita, R.; Itoh, A.; Mizokami, S. An analysis system for propulsive performance in waves. *J. Kansai. Soc. Nav. Arch. Jpn.* **2004**, *241*, 67–82.

43. Tsujimoto, M.; Shibata, K.; Kuroda, M.; Takagi, K. A Practical Correction Method for Added Resistance in Waves. *J. Jpn. Soc. Nav. Arch. Ocean Eng.* **2008**, *8*, 177–184. [CrossRef]
44. Guo, B.; Steen, S. Added resistance of a VLCC in short waves. In Proceedings of the International Conference on Offshore Mechanics and Arctic Engineering, OMAE, Vancouver, BC, Canada, 22–27 June 2010.
45. Bunnik, T.; Van, D.; Kapsenberg, G.; Shin, Y.; Huijsmans, R.; Deng, G.B.; Delhommeau, G.; Kashiwagi, M.; Beck, B. A Comparative study on state-of-the-art prediction tools for seakeeping. In Proceedings of the 28th ONR Symposium on Naval Hydrodynamics, Pasadena, CA, USA, 12–17 September 2010; Volume 1, pp. 242–255.
46. Ichinose, Y.; Tsujimoto, M.; Shiraishi, K.; Sogihara, N. Decrease of ship speed in actual seas of a bulk carrier in full load and ballast conditions. *J. Jpn. Soc. Nav. Arch. Ocean Eng.* **2012**, *15*, 37–45. [CrossRef]
47. Sadat-Hosseini, H.; Wu, P.; Carrica, P.; Kim, H.; Toda, Y.; Stern, F. CFD verification and validation of added resistance and motions of KVLCC2 with fixed and free surge in short and long head waves. *Ocean. Eng.* **2013**, *59*, 240–273. [CrossRef]
48. Söding, H.; Shigunov, V.; Schellin, T.E.; El Moctar, O. A Rankine Panel Method for Added Resistance of Ships in Waves. *J. Offshore Mech. Arct. Eng.* **2014**, *136*, 031601. [CrossRef]
49. Simonsen, C.D.; Otzen, J.F.; Nielsen, C.; Stern, F. CFD prediction of added resistance of the KCS in regular head and oblique waves. In Proceedings of the 30th Symposium on Naval Hydrodynamics, Hobart, Australia, 2–7 November 2014.
50. Ley, J.; Sigmund, S.; El Moctar, O. Numerical Prediction of the Added Resistance of Ships in Waves. In Proceedings of the A.S.M.E. International Conference on Offshore Mechanics and Arctic Engineering, San Francisco, CA, USA, 8–13 June 2014; Volume 2: CFD and VIV. V002T08A069. [CrossRef]
51. Oh, S.; Yang, J.; Park, S.-H. Computational and experimental studies on added resistance of AFRAMAX-class tankers in head seas. *J. Soc. Nav. Arch. Korea* **2015**, *52*, 471–477. [CrossRef]
52. Sprenger, F.; Maron, A.; Delefortrie, G.; Hochbaum, A.; Fathi, D. *Mid-Term Review of Tank Test Results. SHOPERA Deliverable D3.2*; National Technical University of Athens: Athens, Greece, 2015.
53. Yasukawa, H.; Matsumoto, A.; Ikezoe, S. Wave height effect on added resistance of full hull ships in waves. *J. Jpn. Soc. Nav. Arch. Ocean Eng.* **2016**, *23*, 45–54.
54. Lee, J.-H.; Kim, Y.; Kim, B.-S.; Gerhardt, F. Comparative study on analysis methods for added resistance of four ships in head and oblique waves. *Ocean Eng.* **2021**, *236*, 10955. [CrossRef]
55. Broyden, C.G. The convergence of a class of double-rank minimization algorithms 1. General considerations. *IMA J. Appl. Math.* **1970**, *6*, 76–90. [CrossRef]
56. Fletcher, R. A new approach to variable metric algorithms. *Comput. J.* **1970**, *13*, 317–322. [CrossRef]
57. Shanno, D.F. Conditioning of quasi-newton methods for function minimization. *Math. Comput.* **1970**, *24*, 647–656. [CrossRef]
58. Goldfarb, D. A family of variable-metric methods derived by variational means. *Math. Comput.* **1970**, *24*, 23–26. [CrossRef]
59. Lazarus. The Professional Free Pascal RAD IDE (2.2.0). 2023. Available online: https://www.lazarus-ide.org (accessed on 23 January 2023).
60. Cepowski, T. WaveResist 2.0 (WIN/MAC): An Artificial Neural Network Tool for Ship Wave Resistance Coefficient Calculation. Mendeley Data, V2. 2024. Available online: https://data.mendeley.com/datasets/nggvmdpbrt/2 (accessed on 10 June 2024).
61. Citakoglu, H.; Coşkun, Ö. Comparison of hybrid machine learning methods for the prediction of short-term meteorological droughts of Sakarya Meteorological Station in Turkey. *Environ. Sci. Pollut. Res.* **2022**, *29*, 75487–75511. [CrossRef]
62. Demir, V.; Citakoglu, H. Forecasting of solar radiation using different machine learning approaches. *Neural Comput. Appl.* **2023**, *35*, 887–906. [CrossRef]
63. ITTC. *Recommended Procedures and Guidelines, Seakeeping Experiments*; ITTC: Zürich, Switzerland, 2014.

Disclaimer/Publisher's Note: The statements, opinions and data contained in all publications are solely those of the individual author(s) and contributor(s) and not of MDPI and/or the editor(s). MDPI and/or the editor(s) disclaim responsibility for any injury to people or property resulting from any ideas, methods, instructions or products referred to in the content.

Article

Towards Simpler Approaches for Assessing Fuel Efficiency and CO_2 Emissions of Vehicle Engines in Real Traffic Conditions Using On-Board Diagnostic Data

Fredy Rosero, Carlos Xavier Rosero * and Carlos Segovia

Faculty of Engineering in Applied Sciences, Universidad Técnica del Norte, Ibarra 100102, Ecuador; farosero@utn.edu.ec (F.R.); cmsegovia@utn.edu.ec (C.S.)
* Correspondence: cxrosero@utn.edu.ec

Abstract: Discrepancies between laboratory vehicle performance and real-world traffic conditions have been reported in numerous studies. In response, emission and fuel regulatory frameworks started incorporating real-world traffic evaluations and vehicle monitoring using portable emissions measurement systems (PEMS) and on-board diagnostic (OBD) data. However, in regions with technical and economic constraints, such as Latin America, the use of PEMS is often limited, highlighting the need for low-cost methodologies to assess vehicle performance. OBD interfaces provide extensive vehicle and engine operational data in this context, offering a valuable alternative for analyzing vehicle performance in real-world conditions. This study proposes a straightforward methodology for assessing vehicle fuel efficiency and carbon dioxide (CO_2) emissions under real-world traffic conditions using OBD data. An experimental campaign was conducted with three gasoline-powered passenger vehicles representative of the Ecuadorian fleet, operating as urban taxis in Ibarra, Ecuador. This methodology employs an OBD interface paired with a mobile phone data logging application to capture vehicle kinematics, engine parameters, and fuel consumption. These data were used to develop engine maps and assess vehicle performance using the vehicle-specific power (VSP) approach based on the energy required for vehicle propulsion. Additionally, VSP analysis combined with OBD data facilitated the development of an energy-emission model to characterize fuel consumption and CO_2 emissions for the tested vehicles. The results demonstrate that OBD systems effectively monitor vehicle performance in real-world conditions, offering crucial insights for improving urban transportation sustainability. Consequently, OBD data serve as a critical resource for research supporting decarbonization efforts in Latin America.

Keywords: engine mapping; fuel consumption; CO_2 emissions; urban taxis

Citation: Rosero, F.; Rosero, C.X.; Segovia, C. Towards Simpler Approaches for Assessing Fuel Efficiency and CO_2 Emissions of Vehicle Engines in Real Traffic Conditions Using On-Board Diagnostic Data. *Energies* **2024**, *17*, 4814. https://doi.org/10.3390/en17194814

Academic Editors: Kazimierz Lejda, Artur Jaworski and Maksymilian Mądziel

Received: 7 September 2024
Revised: 21 September 2024
Accepted: 24 September 2024
Published: 26 September 2024

Copyright: © 2024 by the authors. Licensee MDPI, Basel, Switzerland. This article is an open access article distributed under the terms and conditions of the Creative Commons Attribution (CC BY) license (https://creativecommons.org/licenses/by/4.0/).

1. Introduction

Urban transportation significantly contributes to greenhouse gas (GHG) emissions and air pollution in cities worldwide, presenting critical environmental and public health challenges. Global energy consumption for road transport is expected to rise by 28% between 2022 and 2050, signaling a worsening scenario in pollution [1]. In Latin America, the energy–mix is predominantly fossil fuel-based, accounting for approximately 65% of the region's total energy matrix. Road transport is the primary GHG emitter in the region, consuming nearly 37% of the total energy supply [2]. Addressing emissions from road transport is thus an urgent task that requires considering each region's unique political, social, technical, and economic context.

In each region, vehicle performance and engine efficiency are influenced by various factors, which can be categorized into four main groups: (i) vehicle characteristics (exogenous factors like fuel type, engine condition, and engine load level; and endogenous factors like engine speed, powertrain configuration, and air/fuel mixture) [3], (ii) driver behavior (e.g., driving style, gear-shifting pattern) [4], (iii) traffic conditions (e.g., average speed, number of stops) [5], and (iv) road conditions (endogenous factors like road type, road grade, intersection signal configuration; and exogenous factors like altitude, ambient temperature, humidity) [6,7]. In Latin America, unique regional factors also play a significant role. For instance, cities in Andean countries such as Ecuador, Colombia, and Peru are often situated in mountainous regions at altitudes exceeding 2000 m above sea level. Additionally, the vehicle technology in these markets tends to be less advanced than in Europe or the United States. These regional differences significantly impact vehicle emissions and fuel efficiency, highlighting the need for tailored solutions.

Implementing emission standards and fuel regulations has driven advancements in testing procedures and equipment over recent decades. Initially, vehicles were tested under controlled laboratory conditions using chassis dynamometers and constant volume sampler (CVS) gas analyzers to measure diluted emissions, especially for light-duty vehicles (LDVs). For heavy-duty vehicles (HDVs), engine testing was conducted on engine benches with CVS analyzers [8]. However, discrepancies between laboratory vehicle performances and real-world traffic conditions were reported in several studies. As a result, starting in 2015, regulatory frameworks began including real-world traffic evaluations and monitoring using portable emissions measurement systems (PEMS) and on-board diagnostic (OBD) data [9,10]. Laboratory testing also evolved to incorporate more demanding driving cycles, such as the Worldwide Harmonized Light Vehicles Test Cycle (WLTC) [11,12]. Additionally, regulatory authorities now propose vehicle assessments before and after commercialization, including "In-Service Conformity" (ISC) tests [13]. Despite these advances, regulatory changes are not progressing simultaneously across all regions. Only countries like Chile and Brazil have a clear roadmap for adopting Euro IV-equivalent regulations in Latin America. In contrast, countries like Ecuador and Costa Rica still rely on Euro II and III standards [14]. Given the limited access to PEMS equipment, OBD data provide a valuable alternative for analyzing vehicle performance under real-world conditions in the region.

Researchers utilize various methodologies to monitor and assess vehicle emissions, including emission modeling, vehicle simulators, the vehicle-specific power (VSP) approach—which accounts for the vehicle's tractive energy—engine map development, and real-world testing using PEMS [15]. Each method has unique advantages and applications. For example, advanced mathematical models were developed to estimate fuel consumption in flexible fuel vehicle engines by considering fundamental fuel properties [16]. The VSP approach was used to assess heavy vehicles' driving cycle distributions and emissions, offering insights into their environmental impact [17]. The PEMS-based experimental data were utilized to analyze real-world traffic emissions for taxis powered by gasoline and biofuel in various Chinese cities, highlighting the variability in urban driving conditions [18]. Machine learning techniques, such as Artificial Neural Networks (ANN), were also employed to model a six-cylinder marine diesel engine by mapping its thermal performance characteristics [19]. Furthermore, comprehensive procedures for creating engine maps using dynamometer test data were established, allowing a detailed vehicle performance simulation [20].

In Latin America, evaluating vehicle performance under real-world traffic conditions is often limited by technical and economic constraints [21]. In countries like Ecuador, vehicle homologation processes, where state agencies must verify a vehicle's performance before commercialization, are often conducted solely through documentation, bypassing actual laboratory or road tests. Moreover, many countries in the region have yet to establish emission factors in terms of mass (g/km), and outdated legislation continues to assess vehicle performance based on emission concentrations (ppm, %V). The poor fuel quality further hinders governments from implementing stricter emission standards [22]. Monitoring and controlling in-use vehicles through ISC tests are also largely absent. Given these challenges, low-cost methodologies for evaluating vehicle performance are highly recommended for Latin America. Using cost-effective OBD interfaces, which provide access to extensive vehicle information like engine performance data, exhaustive after-treatment system status, and various operational parameters, are valuable tools for researchers and policymakers in the region. OBD systems, already regulated in regions like Europe and the United States [12,23], are reliable for assessing and monitoring vehicles operating under real-world traffic conditions. Therefore, OBD data serve as a crucial resource for studies supporting the decarbonization efforts in Latin America.

This research makes a significant contribution by addressing the gaps in the current literature by developing a straightforward and low-cost methodology for assessing vehicle fuel efficiency and carbon dioxide (CO_2) emissions under real-world traffic conditions using OBD data. It provides practical insights to improve the sustainability of urban transport, especially in regions with technical and economic limitations like Latin America. The methodology encompasses three key aspects: (i) generating engine maps using OBD data collected during real-world driving; (ii) comparing vehicle performance through engine operation patterns and the VSP approach; (iii) and developing and evaluating a VSP-based emissions model for characterizing fuel consumption and CO_2 emissions in Latin American traffic. An experimental campaign involved three gasoline passenger vehicles representative of the Ecuadorian fleet, tested during urban taxi operations in Ibarra, Ecuador. The data on vehicle kinematics, engine parameters, and fuel consumption were collected using an OBD-II interface connected to a mobile phone application, with navigation data referenced via a global positioning system (GPS) logger.

The remainder of this paper is organized as follows. Section 2 discusses the roles and functions of PEMS and OBDs in emission monitoring. Section 3 details the proposed methodology, including the experimental setup and procedures. Section 4 discusses the results and their implications. Finally, Section 5 concludes the paper and suggests avenues for future research.

2. PEMS and OBD

PEMS and OBD systems are two crucial vehicle emission monitoring and control technologies. These systems provide essential data that enhance the understanding of vehicle performance under real-world conditions, facilitating more effective regulatory compliance and emission reduction strategies.

PEMS are advanced devices designed to measure and record on-board vehicle emissions on a second-by-second basis in real time under real-world operating conditions [24]. As detailed in Table 1, PEMS consist of several key components, including emission sensors, a data acquisition system, a GPS, an on-board computer, and a power supply. The emission sensors detect and measure the concentrations of gaseous pollutants such as CO_2, nitrous oxides (NO_x), carbon monoxide (CO), and total hydrocarbon (THC). The data acquisition system collects and stores this information for subsequent analysis. The GPS unit records the vehicle's location, allowing for a contextual emission data analysis. The on-board computer integrates and processes the collected data, generating comprehensive reports on vehicle emission performance. The power supply ensures the continuous operation of the PEMS components.

Conversely, OBD systems are crucial for diagnosing and monitoring vehicle performance and emission control systems [25]. Table 2 summarizes the key components and functions of OBD systems. The core of the OBD system is the engine control module (ECM), which monitors engine performance and exhaust gas treatment processes. A network of sensors throughout the vehicle collects engine and exhaust condition data, which the ECM uses to monitor and regulate these systems. The diagnostic interface allows the connection of external diagnostic tools, enabling the retrieval and analysis of stored performance data and fault data from the ECM. Additionally, the malfunction indicator light (MIL) alerts drivers to detected faults, while the fault code log stores specific diagnostic codes for identified problems, enabling the accurate identification and resolution of technical issues.

Overall, PEMS and OBD systems offer comprehensive tools for monitoring and controlling vehicle emissions. These systems are designed to address all aspects of emissions and control, providing an all-encompassing solution. PEMS are particularly effective for capturing emission data during real-world driving conditions, while OBD systems deliver detailed diagnostics and continuous performance monitoring. Integrating these technologies is essential for ensuring that vehicles meet emission standards and developing strategies to reduce overall vehicle emissions. However, in Latin America, where access to PEMS equipment is limited, using OBD systems is a viable and cost-effective alternative for assessing fuel efficiency and CO_2 emissions under real-world operating conditions.

Table 1. Operation of PEMS.

Component	Description	Function
Emission sensors	Vibration-proof gas sensors that measure pollutant concentrations using Heated Non-Dispersive Hot Infrared (HNDIR), Hot-Flame Ionization (HFID), and Heated Chemiluminescence (HCLD) analyzers.	Monitor real-time levels of NO_x, CO_2, CO, THC, and other pollutants.
Exhaust flowmeter	A tail-pipe attachment with a Pitot tube that measures the flow of exhaust gases.	Measures exhaust gas flow to convert emission concentrations into mass emissions.
Data acquisition system	Unit that collects and stores data from the emission sensors.	Stores data for subsequent analysis and comparison with regulatory standards.
GPS	Global Positioning System that records the vehicle's location.	Associates emission data with precise geographical locations for contextual analysis.
On-board computer	Processor that integrates and analyzes data collected by sensors and the GPS.	Performs calculations and generates reports on emission performance under real-world conditions.
Power supply	Battery that powers all PEMS components.	Provides the necessary energy for continuous system operation.

Table 2. Operation of OBD systems.

Component	Description	Function
Engine control module (ECM)	Central unit managing engine performance and exhaust treatment systems.	Monitors and controls the operation of engine components and emission systems.
Vehicle sensors	Includes oxygen, temperature, pressure, and airflow sensors.	Collects data on the condition and performance of various engine and exhaust components.
Diagnostic interface	Port for connecting external diagnostic tools.	Allows extraction and analysis of performance and fault data stored in the ECM.
Malfunction indicator light (MIL)	Dashboard light that alerts the driver to detected faults.	Informs the driver about issues requiring attention or maintenance.
Fault code register	System storing specific codes for each type of detected fault.	Facilitates identification and resolution of technical problems through specific diagnostic codes.

3. Proposed Methodology

The proposed methodology, shown in Figure 1, encompasses several crucial stages for assessing vehicle fuel efficiency and CO_2 emissions under real-world traffic conditions. This methodology is tested through an experiment that follows these outlined steps:

1. Vehicles representative of the local fleet are chosen for analysis. The vehicles' real-world traffic operation is recorded while they function as urban transportation (city taxi service) rather than being tested on a specific route.
2. A GPS logger and an OBD interface connected to a mobile phone application are installed to collect real-time data on vehicle and engine operating parameters, enabling the recording of critical data required to analyze fuel consumption and CO_2 emissions.
3. The instrumented vehicles are driven on various routes based on demand from taxi users, primarily in urban areas, to collect data representative of real-world driving scenarios.

4. The collected data from the GPS logger and OBD interface undergo synchronization and smoothing processes to ensure the accuracy and reliability for subsequent analysis.
5. Engine maps are generated to visualize the relationship between engine load, speed, and performance parameters (e.g., fuel consumption and CO_2 rates). This involves creating grid maps for engine load and speed ranges and plotting two-dimensional contour maps based on the collected data.
6. Metrics such as the relative frequency of engine speed and load, VSP values, fuel efficiency, and CO_2 emission factors are calculated using the collected data and developed engine maps.

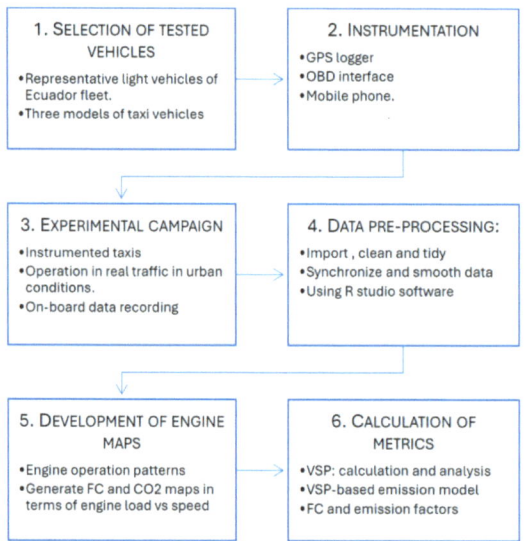

Figure 1. The methodology proposed in this study for the vehicle emissions analysis process.

Although this study did not directly validate OBD-derived fuel consumption estimates against actual vehicle fuel consumption using PEMS or flow meters, previous research demonstrated the reliability of OBD data. Studies showed that errors associated with OBD fuel consumption estimates for spark ignition and diesel vehicles typically remain below 4% [25,26]. Moreover, the use of OBD data has been widely regulated by international legislation [12,23], reinforcing its credibility as a tool for vehicle performance assessment. Despite the absence of direct validation in this study, the methodology remains a cost-effective and scientifically robust approach, especially valuable for regions with limited access to more advanced technologies like PEMS. This makes the OBD-based approach particularly applicable for real-world emissions analysis in resource-constrained areas, such as Latin America, where cost-effective yet accurate methodologies are essential.

3.1. Selection of Tested Vehicles

Three gasoline-powered passenger vehicles—Chevrolet Aveo Activo, Chevrolet Sail, and Hyundai Accent—were selected to serve as taxis representing the Ecuadorian fleet. These vehicles were chosen for their suitability for fuel consumption analysis through the OBD II system and because they align with the typical passenger capacity, weight, engine torque, power, and gearbox configurations standard in Ecuador's automotive segment. In Ecuador, the transport sector consumes 49% of the total energy, primarily from diesel and gasoline, and grows annually by up to 25% [27]. This sector was responsible for approximately 50% of the country's GHG emissions in 2021. The taxi drivers participate voluntarily, providing a reliable basis for analysis. Detailed specifications of the tested vehicles are presented in Table 3.

Table 3. Technical specifications of the tested vehicles.

Component	Vehicle 1 (V1)	Vehicle 2 (V2)	Vehicle 3 (V3)
Fuel type	Gasoline	Gasoline	Gasoline
Model name	Chevrolet Aveo Activo	Chevrolet Sail	Hyundai Accent
Model year	2008	2018	2011
Gross vehicle weight (t)	1.535	1470	1.560
Weight (t)	1.125	1012	1.203
Length/Width/Height (m)	4.04/1.67/1.495	4.29/1.69/1.505	4.3/1.7/1.45
Max passengers	5	5	5
Axle configuration	4 × 2	4 × 2	4 × 2
Gearbox	Manual (5)	Manual (5)	Manual (5)
Number of engine cylinders	4 in-line	4 in-line	4 in-line
Engine total displacement (cm^3)	1598	1398	1598
Engine maximum power (kW@rev/min)	76@5800	76.06@6000	90@6300
Engine peak torque (N.m@rev/min)	145@3600	129.74@4200	155@4200
Fuel injection type	Indirect injection	Indirect injection	Indirect injection
Compression ratio	9.5:1	10.2:1	9.5:1

3.2. Instrumentation

Instruments are required to measure three types of information: vehicle kinematics (e.g., position and speed), vehicle and engine operating parameters (e.g., engine load and speed values), and instantaneous fuel consumption rate data. A GPS logger device, the *GL-770*, was installed in the test vehicles to record latitude, longitude, altitude, and time at 1 Hz. This device was non-intrusive, allowing the collection of vehicle kinematic information without altering the ordinary course of operation of tested vehicles. Additionally, an OBD interface device, the *ELM Electronics 327*, was connected to a mobile phone application named *Torque Pro* to record vehicle fuel consumption data and engine operating parameters. Overall, the ELM 327 connects to the vehicle's ECU via the OBD2 diagnostic port, reading its operating parameters in real time and sending them via Bluetooth to the *Torque Pro* application.

The ELM Electronics 327 interface is a diagnostic tool designed for vehicles equipped with OBD II and CAN systems, commonly found in vehicles manufactured after 1996 in the United States, Europe, and Asia, with a 16-pin diagnostic connector. This interface can read diagnostic trouble codes (DTCs), clear them, and retrieve real-time sensor data from the engine through the vehicle's ECM [28]. Additionally, the *ELM 327* could turn off the check engine light by clearing DTCs and providing information on the sensors installed in the vehicle. It offers a real-time display of sensor data such as intake manifold pressure, engine speed, vehicle speed, fuel system status, air flow rate, oxygen sensors, and the fuel consumption rate. This interface is compatible with many OBD II-compliant vehicles [25,26]. This device typically uses wireless communication methods like Bluetooth or Wi-Fi, making it highly versatile for monitoring vehicle performance under real-world driving conditions.

The mobile application *Torque Pro* is a versatile diagnostic tool for real-time vehicle performance monitoring through an OBD-II interface, such as *ELM327*. It enables users to access and log data such as engine speed, fuel consumption, and kinematic vehicle parameters. Additionally, Torque Pro allows users to read and interpret DTCs generated by a vehicle's systems when issues arise. With access to detailed sensor data and parameter identification codes (PIDs), the mobile application offers advanced diagnostic and monitoring capabilities, making it ideal for assessing vehicle performance under real-world operating conditions.

Figure 2 illustrates the schematic of the instrument installation used to evaluate the performance of the taxis. The installation of the ELM 327 interface with the Torque Pro application was completed through the following seven steps: (i) the *ELM 327* interface was connected to the OBDII port of the tested vehicle to extract the data from the ECM; (ii) the Torque Pro application was installed on a mobile device, which served as the primary data recording platform; (iii) a Bluetooth pairing was established between the *ELM 327* interface and the mobile device to enable data transmission; (iv) *Torque Pro* was launched, and the OBDII adapter connection type was configured to Bluetooth, ensuring seamless communication with the interface; (v) the connection was verified to ensure the successful data reception from the vehicle; (vi) specific PIDs were selected within *Torque Pro* for recording, focusing on critical variables such as vehicle kinematics (e.g., longitude, latitude, altitude, distance, GPS speed, and wheel speed sensor) and engine operation parameters (e.g., engine load, engine speed, throttle position, intake manifold pressure, intake air temperature, and engine coolant temperature); and (vii) *Torque Pro* was configured to record data at a frequency of 1 Hz. After completing these configurations, the OBD interface and mobile application were available to start data recording for the experimental campaign. Notably, the GPS logger *GL-770* operated independently, capturing positional data separately from the OBD system, enabling a comprehensive data collection to assess vehicle emissions and fuel efficiency under real-world driving conditions.

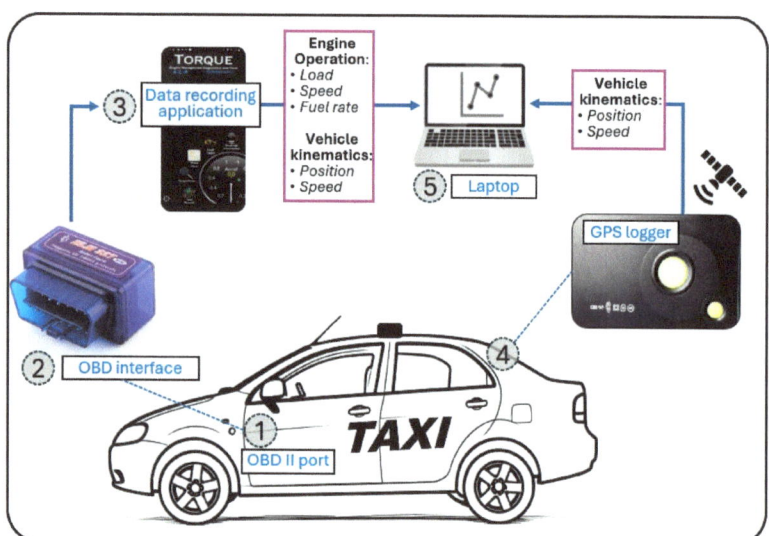

Figure 2. Schematic of the instrumentation.

3.3. Experimental Campaign

The experimental campaign was conducted in Ibarra, Ecuador, a city in the Imbabura province with approximately 200,000 residents. Due to a significant annual growth rate of around 8%, the city faces challenges with traffic management and mobility, which includes approximately 90% of urban conditions (0–60 km/h). The remaining 10% includes suburban areas (60–90 km/h) and highways (>90 km/h). This approach was designed to collect data representative of real-world driving situations.

The campaign began with the meticulous installation and assembly of monitoring equipment, following in-depth discussions with members of a taxi company who voluntarily participated in the project. The data collection process was carefully planned, with two weeks allocated for this purpose. The data was collected in February, covering peak and off-peak hours, using three taxi models running on commercially available gasoline. Before each session, the taxis were warmed up for at least 30 min to ensure optimal operating conditions. A single driver operated each taxi throughout this study to minimize variability due to driving style. Consistent weather conditions during the tests helped mitigate the influence of environmental factors. Additionally, the taxis' air conditioning systems were activated during all tests. Table 4. provides an overview of the operating conditions for the tested vehicles during the experimental campaign.

Table 4. Overview of the operating conditions for tested vehicles in the experimental campaign.

Tested Vehicle	Average Trip Speed (km/h)	Average Moving Speed (km/h)	Maximum Trip Speed (km/h)	Average Positive Accel. (m/s²)	Total Idle Period (%)	Total Accel. Period (%)	Total Decile. Period (%)	Total Cruise Period (%)
Vehicle 1	13.98	22.01	83.32	0.458	37.77	28.17	23.92	10.06
Vehicle 2	15.31	22.09	82.50	0.450	31.56	30.44	25.81	12.10
Vehicle 3	16.40	22.53	67.12	0.460	27.96	31.45	27.56	12.97
All vehicles	15.23	22.21	77.64	0.457	32.43	30.02	25.76	11.71

3.4. Data Pre-Processing

Data pre-processing used R Studio software version 2023.12.1 to address the differences in initial recording times between the GPS and OBD devices used in the experimental campaign. The pre-processing involved synchronizing the signals from these devices based on vehicle speed, which was recorded second-by-second. The speed profiles were then smoothed using a moving window filter, ensuring the accuracy and reliability of the results presented in this study.

3.5. Development of Engine Maps

Developing an engine map requires three key variables. Commonly, torque and engine speed serve as the axes for the engine maps. At the same time, a third variable (e.g., brake thermal efficiency (BTE), fuel consumption, and emission rates) defines the map's specific focus. The methodology for engine map development involved two primary stages: (i) creating grid engine maps based on ranges of engine load and speed with averaged data values and (ii) generating two-dimensional contour maps from the grid data. These processes were carried out using *R Studio*.

The construction of grid engine maps was based on the methodology outlined in [29]. First, grids were established according to predefined engine load and speed intervals, after which the remaining operational data (the third variable) were assigned to their respective grids. The data within each grid were then averaged to produce a single representative value for each speed–load combination. Outliers were identified and removed by analyzing each data group's relative frequency and standard deviation. Once filtered, the averaged data were discretized for color visualization on the engine maps. Then, to create the contour engine maps, R Studio with the ggplot2 R package was used [30,31]. The data from the grid maps served as the input for generating these contours. The third variable's intervals were clearly defined to improve visualization, and a divergent color scheme was employed. This approach resulted in detailed contour maps for various engine parameters, including BTE, brake-specific fuel consumption (BSFC), brake-specific CO_2 emissions, fuel consumption, and CO_2 emission rates.

3.6. Calculation of Metrics

3.6.1. Estimation of CO_2 Emission Rates

The CO_2 emission rate (ER_{CO_2}) for each tested vehicle (i) was calculated by

$$ER_{CO_2 i} = FC_i \cdot \frac{44}{12 + \left(\frac{H}{C}\right)}, \tag{1}$$

which is derived from the stoichiometric combustion of gasoline (C_8H_{18}). Given that the fuel used is gasoline, the assumed values were 18 for hydrogen (H) and 8 for carbon (C).

The instantaneous fuel consumption rate $(FC_{i,t})$ was expressed in (g/s), serving as the basis for determining the CO_2 emission rate.

3.6.2. Fuel Consumption and Emission Factors

To estimate the performance of the tested vehicles, their distance-specific fuel consumption and CO_2 emissions factors were calculated by

$$\overline{FC}_i^d = \frac{\sum_{t=1}^{T_i}(FC_{i,t})}{\sum_{t=1}^{T_i}(d_t)}, \qquad (2)$$

$$\overline{EF}_{CO_{2i}}^d = \frac{\sum_{t=1}^{T_i}\left(ER_{CO_{2i,t}}\right)}{\sum_{t=1}^{T_i}(d_t)}, \qquad (3)$$

where $\left(\overline{FC}_i^d\right)$ is the estimated distance-specific FC for each tested vehicle (i); (d_t) is the distance travelled; and (T_i) is the total time of travel. In Equation (3), $\left(\overline{EF}_{CO_{2i}}^d\right)$ is the estimated distance-specific CO_2 emission factor by vehicle (i), while $\left(ER_{CO_{2i,t}}\right)$ is the instantaneous CO_2 emission rate expressed in (g/s).

3.6.3. VSP Calculation

VSP is a crucial metric representing the power required by a vehicle's engine. It has been widely used to analyze the performance of LDVs [32] and HDVs [33]. This metric is defined as the vehicle's traction power per unit mass, expressed in kilowatts per ton (kW/t). VSP effectively reflected the relationship between traction power and fuel efficiency, as well as emissions, providing a valuable tool for developing some emission models for analyzing vehicle performance. The VSP was calculated as follows,

$$VSP = v[gC_R\cos\theta + g\sin\theta + a\cdot(1+\varepsilon_i)] + 0.5\rho\frac{C_D A}{m}(v+v_m)^2 v, \qquad (4)$$

where v is the vehicle speed (m/s), g represents the gravitational acceleration (9.81, m/s^2), m is the total mass (kg) of the vehicle, C_R is the rolling resistance coefficient (0.0150, dimensionless), θ is the road grade (°), a is the vehicle acceleration (m/s^2), ε_i is the mass factor for the rotational masses, (0.1, dimensionless), ρ is air density (0.995, kg/m^3), C_D is the air drag coefficient (0.32, dimensionless) [34], A is the frontal area of the tested vehicle (m^2) and v_m is the wind speed (0, m/s). The VSP bins were defined by combining the VSP and vehicle speed intervals, resulting in 28 modes, as shown in Table 5.

Table 5. Vehicle specific power (VSP) operating mode bins.

Description		Vehicle Speed (km/h)		
VSP Range (kW/t)		$v < 1.6$	$1.6 \leq v < 40$	$40 \leq v \leq 80$ *
VSP < −4	Bin 0	Bin 1	Bin 101	Bin 201
−4 ≤ VSP < −2	deceleration or braking	idling	Bin 102	Bin 202
−2 ≤ VSP < 1			Bin 103	Bin 203
−1 ≤ VSP < 0			Bin 104	Bin 204
0 ≤ VSP < 1			Bin 105	Bin 205
1 ≤ VSP < 2			Bin 106	Bin 206
2 ≤ VSP < 4			Bin 107	Bin 207
4 ≤ VSP < 6			Bin 108	Bin 208
6 ≤ VSP < 8			Bin 109	Bin 209
8 ≤ VSP < 10			Bin 110	Bin 210
10 ≤ VSP < 12			Bin 111	Bin 211
12 ≤ VSP < 14			Bin 112	Bin 212

* Experimental trips did not include vehicle speeds greater than 80 km/h.

4. Results and Discussion

To address the research questions of this study, the results are organized into five parts: (a) engine operations patterns, (b) engine maps for FC and CO_2 emissions, (c) comparative analysis of tested vehicles based on VSP, (d) VSP-based model performance, and (e) an overview of FC and emission factors derived from OBD data.

4.1. Engine Operation Patterns

Figure 3 shows typical engine operating patterns of tested vehicles (Aveo, Sail, and Accent) regarding engine load versus engine speed. Each data point represents a second-by-second operating record for each car. In the case of Vehicle 1 (Aveo), depicted in orange, there is a clear positive correlation between engine speed and engine load, indicating that as engine speed increases, engine load also tends to increase. Most of the data for this vehicle are concentrated in engine load ranges below 25% and speed ranges under 2500 rev/min.

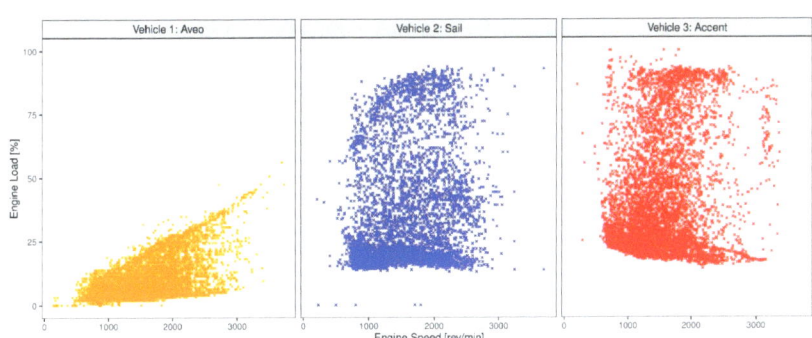

Figure 3. Typical engine operation patterns by load and speed for tested vehicles.

In contrast, Vehicle 2 (Sail), depicted in blue, shows a broader distribution of data points across engine loads and speeds. This vehicle primarily operates in the 1000–2000 rpm range with engine loads between 12 and 30%. Finally, Vehicle 3 (Accent), shown in red, exhibits a similar trend to Vehicle 2 but with a slightly more dispersed pattern, suggesting more varied operating conditions.

Vehicle 1 (Aveo) exhibits a distinct engine load profile compared to the other two vehicles. Notably, when Vehicle 1 operates under idle conditions, its engine load is near zero, which contrasts sharply with Vehicles 2 and 3, which maintain an engine load of around 20% at idle. This variation at idle conditions highlights the relative nature of engine load, which can differ depending on the car manufacturer. According to the work in [25], assuming that engine load represents engine torque requires verifying the linearity between these two parameters using a chassis dynamometer. This verification is essential for accurately interpreting the engine load data. Furthermore, identifying the torque value under idle conditions is necessary to adjust the offset between engine load and torque parameters. Hence, the engine maps in this study are expressed in terms of engine load instead of torque.

4.2. Engine Maps Based on OBD Data

Figure 4 presents engine maps for the three tested vehicle engines, illustrating how the fuel consumption rate (measured in g/s, represented by a color gradient from yellow to blue) varies across different operational ranges. The three engine maps demonstrate that fuel consumption is proportional to engine load demand, reaching up to 2 g/s.

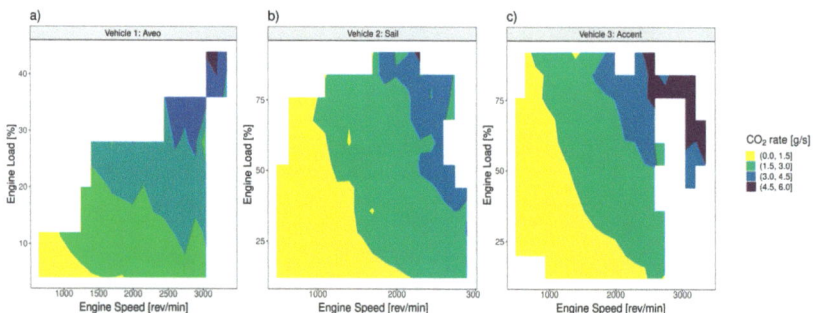

Figure 4. Engine maps illustrating the CO2 emission rates for the Aveo (**a**), Sail (**b**) and Accent (**c**) vehicles.

For Vehicle 1, the engine primarily exhibits fuel consumption rates between 0.4 and 1.2 g/s in medium engine load and speed regions. During idling conditions common in urban driving, fuel consumption decreases to around 0.4 g/s (yellow). Vehicles 2 and 3 operate over more loads and speeds than Vehicle 1. Consequently, the engines from Vehicles 2 and 3 show a broader operational pattern, with significant zones in the medium fuel rate range (green and blue, 0.4–1.2 g/s). Additionally, these engines exhibit fuel rates near 0.4 g/s under idling conditions.

Overall, the engine maps indicate higher engine loads and speeds in average fuel consumption rates. Maximum fuel consumption rates can be up to five times higher than the minimum rates observed under idling conditions. The fuel consumption rates in this study align with those reported for light vehicles in China [35] over a decade ago. However, compared to more recent European studies [15,36], the observed fuel consumption rates are higher. This difference may be due to the lower engine technology of the vehicles evaluated. Furthermore, many modern vehicles in Latin America are often equipped with older technologies, contributing to increased fuel consumption rates.

4.3. Comparative Analysis Based on Vehicle-Specific Power

Figure 5 illustrates the correlation between instantaneous VSP values and the fuel consumption rate for the tested vehicles. In this graph, each point represents a vehicle operation record captured second by second. As previously mentioned, VSP is an energy-based variable that accounts for the vehicle's tractive power and mass, enabling an absolute comparison of vehicle performance.

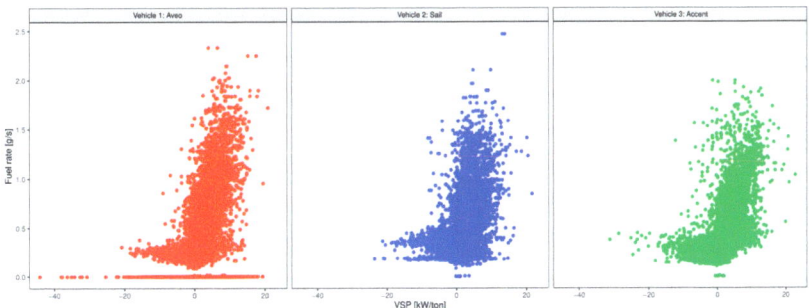

Figure 5. Correlation between fuel rate and VSP for tested vehicles.

Overall, Figure 5 shows a general increasing trend for all three vehicles, indicating that as VSP rises, the fuel consumption rate increases. Vehicle 1 exhibits a slightly broader distribution of data points compared to Vehicles 2 and 3, with a notably higher number of points near fuel rates close to 0 g/s. These records indicate fuel cut-off conditions and may correspond to more efficient engine control module (ECM) management during downhill or deceleration events.

The observed increasing trends based on VSP are consistent with findings from previous studies on light-duty vehicles [35,37] and buses [6,33]. This further validates using VSP as a critical indicator for analyzing, comparing, and characterizing vehicle fuel consumption patterns. Consequently, VSP was highly useful in developing emission models, such as the Motor Vehicle Emission Simulator (MOVES) developed by the United States Environmental Protection Agency (EPA) [38] and COPERT (Computer Programme to Calculate Emissions from Road Transport) in Europe [39]. However, the application of VSP in Latin America was limited to a few previous studies conducted in Mexico [7] and Colombia [5].

Figure 6 presents the VSP frequency distribution for tested vehicles, highlighting distinct operational differences. Vehicle 3 (green) shows the highest peak frequency, followed by Vehicle 2 (blue) and Vehicle 1 (red), suggesting that Vehicle 3 spends more time idling than the others. As VSP values shift into the negative range, indicating deceleration or downhill slopes, the frequency distribution gradually decreases for all vehicles. Vehicle 2 (blue) exhibits a slightly higher frequency in these harmful bins, implying that it operates during more extended periods while decelerating or descending compared to Vehicles 1 and 3.

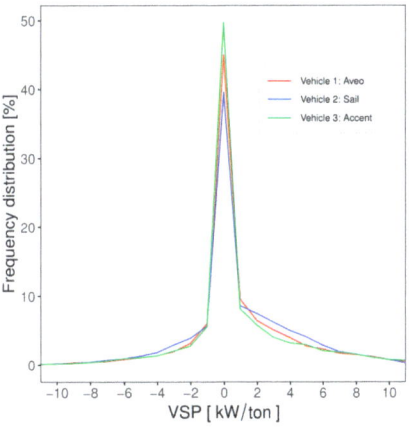

Figure 6. VSP frequency distribution for tested vehicles.

In the positive VSP range, which corresponds to acceleration or an increased power demand, the frequency distribution sharply declines for all vehicles. While Vehicles 1 (red) and 2 (blue) display similar trends, Vehicle 3 (green) exhibits a slightly higher frequency, indicating that it spends more time accelerating or under higher power demands. These trends underscore the operational differences, with Vehicle 3 more frequently idling and accelerating, while Vehicle 2 has a greater tendency for deceleration.

All three vehicles show frequency distributions more significant than 38% for VSP values near zero, primarily reflecting idling conditions. This finding aligns with their primary use as taxis in urban environments and is consistent with previous studies reporting similar VSP values for taxis under idling conditions [37,40,41].

Figure 7a,b provide a comparative analysis of fuel consumption rate and CO_2 emission rate, respectively, as functions of VSP bins for the tested vehicles. The VSP mode bins were defined by combining VSP and vehicle speed, as detailed in the Methodology section. Distinct patterns emerge when examining the trends across different speed intervals.

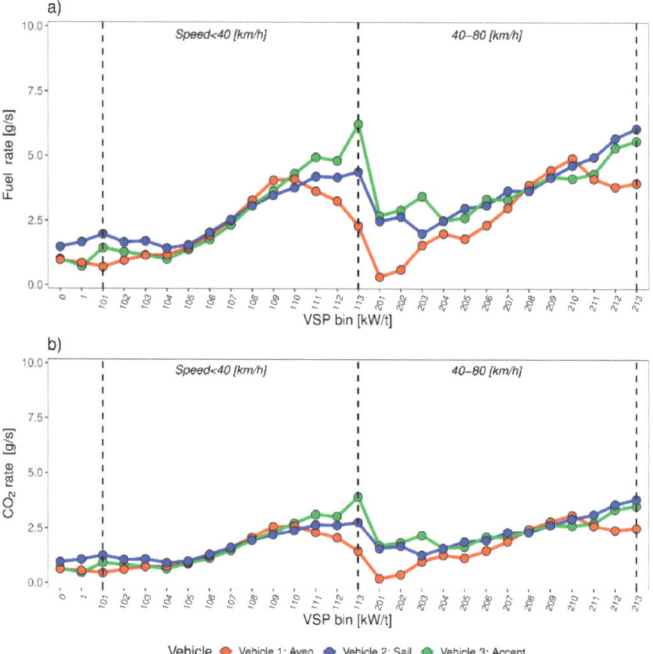

Figure 7. Fuel emission rate (**a**) and CO_2 emission rate (**b**) by VSP bin for tested vehicles.

Under idling and deceleration conditions (bins 0 and 1), Vehicle 3 consistently exhibited higher fuel consumption and CO_2 emissions than the other vehicles, indicating lower engine efficiency even under minimal load conditions. For speeds below 40 km/h, both Vehicle 2 and Vehicle 3 displayed pronounced peaks in fuel consumption and CO_2 emissions at VSP bin 113, corresponding to a high-load operating condition. In contrast, Vehicle 1 maintained significantly lower rates throughout this speed range.

Within the 40–80 km/h range, Vehicle 2 showed increased fuel consumption and CO_2 emissions as VSP increased, while Vehicle 3 demonstrated a more moderate response. Vehicle 1 continued to exhibit the most efficient performance. Vehicle 1 consistently displayed lower fuel consumption across all VSP bins, particularly at lower speeds. Conversely, Vehicle 3 demonstrated a higher sensitivity to VSP variations.

These findings underscore the critical influence of vehicle design and engine efficiency on fuel consumption and emissions under diverse driving conditions. Moreover, the data presented in Figure 7a,b provide essential insights for developing an accurate energy-based emissions model. Notably, this study's fuel consumption and CO_2 emissions rates are higher than those reported in previous studies based on VSP bins [3,40]. These discrepancies may be attributed to the older engine technologies prevalent in vehicles marketed in Latin America, as discussed previously.

4.4. Performance of the VSP-Based Emission Model

Figure 8a,b present an example of a travel section for K iteration, comparing actual data (red) and VSP-simulated data (blue) for fuel consumption (Figure 8a) and CO_2 emissions (Figure 8b) of Vehicle 3. The continuous lines represent second-by-second instantaneous values, while the dashed lines compare the cumulative values over time. In both figures, the actual and simulated data follow similar trends, indicating reasonable accuracy of the simulation model in capturing the dynamic behavior of fuel consumption and CO_2 emissions. However, some discrepancies are noticeable.

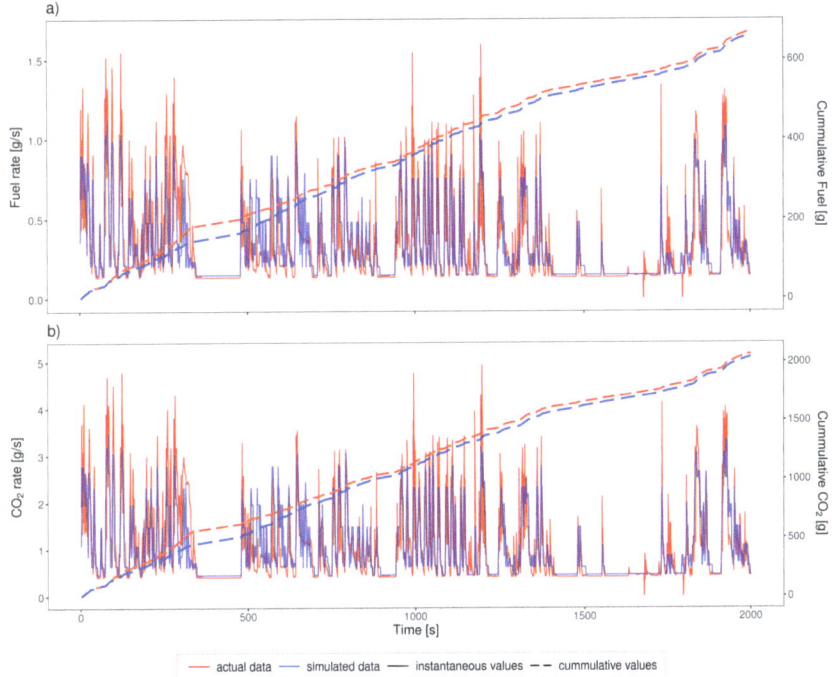

Figure 8. Comparison of measured and simulated data using the VSP model for a typical trip of the Hyundai Accent vehicle: (**a**) Fuel consumption and (**b**) CO_2 emissions.

The actual data show more significant variability and higher peaks, suggesting that the real driving conditions introduce more fluctuations than those captured by the simulation. For cumulative values, fuel consumption and CO_2 emissions exhibit closely matching trends between actual and simulated data, although the simulated values tend to underestimate the actual cumulative totals slightly. This underestimation implies that, although the simulation model effectively captures overall trends, further refinement is necessary to better account for the higher variability observed under real-world conditions. Figure 8a,b demonstrate that VSP model simulations provide a valuable approximation of vehicle fuel consumption and emissions.

Table 6 presents a comparative analysis of the validation results for a VSP-based model applied to the three tested vehicles. The VSP model is validated using the k-fold-cross-validation approach, focusing on three metrics: relative error, Pearson correlation, and Root Mean Squared Error (RMSE) for fuel consumption and CO_2 emissions. The model demonstrates a robust predictive accuracy across all three vehicles, with relatively low relative errors and moderate to high correlation coefficients. When comparing actual and predicted cumulative values for fuel consumption and CO_2 emissions, the relative errors are below 3%. These results align with those reported in other VSP-based model simulations for light-duty vehicles [37,42], trucks [43], and buses [6], which typically report errors in the range of 4–8%.

The proposed VSP-based model exhibits an acceptable correlation for second-by-second data, with Pearson correlation coefficients ranging from 60% to 80% and reasonable RMSE values. A notable advantage of this VSP-based model, developed using OBD data, is its capability to predict second-by-second fuel consumption and CO_2 emissions at a microscopic level, provided that speed profiles, driving cycles, or VSP distributions are available. This approach is particularly relevant for Latin America, where the use of some emission models, such as the 'Passenger Car and Heavy Duty Emission Model' (PHEM) and the 'Motor Vehicle Emissions Simulator' (MOVES), developed in different spatial domains like Europe or North America, may result in significant discrepancies in estimated emissions for Latin American cities due to differing driving patterns, emission data, and congestion levels [5]. Consequently, the VSP-based model offers a valuable and contextually appropriate tool for predicting emissions in Latin America, making it a crucial asset for regional environmental assessment and policy development.

Table 6. Overview of the validation model from iterations for FC and CO_2 emissions for tested vehicles.

Average Measures	V1: Aveo		V2: Sail		V3: Accent	
	Fuel	CO_2	Fuel	CO_2	Fuel	CO_2
Relative error (%) *	2.27	2.26	2.29	2.28	0.76	0.76
Correlation	0.77	0.77	0.61	0.60	0.84	0.84
RMSE (g/s)	0.15	0.47	0.20	0.62	0.13	0.42

* Calculated from cumulative values.

4.5. Overview of Vehicle Performance Based on OBD Data

Table 7 provides a comprehensive overview of the engine performance characteristics for tested vehicles, focusing on metrics such as VSP, fuel consumption and CO_2 emission rates, and both distance-based and energy-based emission factors. Several vital trends emerge when comparing the three vehicles. Vehicle 2 exhibits the highest VSP, indicating a greater power output than its weight. However, this increased power is accompanied by slightly higher fuel consumption and CO_2 emissions than Vehicles 1 and 3. Conversely, Vehicle 3 demonstrates the lowest VSP and fuel consumption and CO_2 emissions when considering energy-based emission factors, suggesting a more fuel-efficient design.

Table 7. Overall engine performance of fuel efficiency, energy emissions, emission rates, and distance emissions by vehicle.

Tested Vehicle	Average Value (kW/ton)	Average Emission Rates (g/s)		Distance-Emission Factors (g/km)		Energy-Emission Factors (g/kWh) *	
	VSP	FC	CO_2	FC	CO_2	FC	CO_2
Vehicle 1	0.897	0.344	1.06	80.75	249.35	699.31	2159.2
Vehicle 2	0.937	0.384	1.18	84.20	260.00	757.71	2339.6
Vehicle 3	0.812	0.318	0.98	82.01	253.25	600.11	1852.9
All vehicles	0.917	0.348	1.07	82.32	254.20	685.71	2117.2

* is relative to traction energy.

Regarding distance-based emission factors, Vehicle 1 shows the lowest fuel consumption and CO_2 emissions, while Vehicle 3 presents the highest values. Notably, despite Vehicle 2's higher VSP, its distance-based and energy-based emission factors are not proportionally elevated, indicating that a higher power output does not necessarily translate to higher emissions per unit of distance or energy.

The distance-based CO_2 emission factors identified (249–260 g CO_2/km), which predominantly reflect full urban operating conditions of LDVs in Latin America, are consistent with results reported for gasoline taxis in China [44] with values around 230 g CO_2/km. However, the CO_2 emission factors observed in this study are significantly higher than those reported in previous studies [3,40] for light-duty gasoline vehicles, including urban, suburban, and motorway sections, with values ranging from 130 to 190 g CO_2/km. This discrepancy may be attributed to differences in vehicle technology, driving patterns, and the specific urban conditions of this study.

5. Conclusions and Future Work

This study demonstrates the feasibility of using low-cost OBD data to develop detailed engine maps and analyze urban taxis' fuel efficiency and CO_2 emissions under real-world driving conditions. The proposed methodology, which combines a VSP analysis with OBD-collected data, enables a detailed assessment of vehicles without the need for more expensive equipment like PEMS.

The results confirm that the OBD-based approach can effectively monitor vehicle performance in real-world urban traffic conditions, providing valuable insights for improving the sustainability of urban transport. Furthermore, the methodology is advantageous in regions like Latin America, where access to advanced equipment such as PEMS is limited. However, while the method captures general trends, cumulative fuel consumption and CO_2 emissions are slightly underestimated. This suggests that the model could be refined to better account for the fluctuations observed in real-world conditions.

The implications of these findings are significant for urban transportation policies, especially in resource-constrained regions such as Latin America. The demonstrated feasibility of using low-cost OBD data provides a practical solution that can support policymakers in several key areas: (i) improving the accuracy of local emissions inventories, (ii) establishing more region-specific emission control regulations based on local geography and vehicle fleet characteristics, (iii) encouraging the renewal of fuel-inefficient vehicle fleets, and (iv) more effectively targeting fuel subsidies.

In the future, it is recommended that this methodology be extended to other vehicle categories and geographic areas to validate its broader applicability. Additionally, integrating advanced techniques such as machine learning could significantly improve the model's predictive accuracy. A direct comparison with other methodologies, such as those based on PEMS, would also be valuable in further validating the precision and utility of this approach in a vehicle emission assessment. These potential avenues for future improvements make the future of this research attractive.

While this study introduces an innovative, low-cost method for assessing vehicle performance using OBD data and mobile technology, it has limitations. A fundamental limitation is the inability to directly validate fuel consumption estimates from the OBD interface with actual vehicle fuel consumption. Future research should address this by comparing OBD estimates with precise fuel measurements obtained through flowmeters, ensuring a more accurate assessment of vehicle performance and emissions in real-world conditions.

Author Contributions: This paper was a collaborative effort among all authors. All authors participated in the analysis, discussed the results, and wrote the paper. All authors have read and agreed to the published version of the manuscript.

Funding: This research was funded by the Universidad Técnica del Norte, within the project 0000001223.

Data Availability Statement: The original contributions presented in the study are included in the article, further inquiries can be directed to the corresponding author.

Conflicts of Interest: The authors declare no conflicts of interest.

References

1. Frey, H.C. Trends in Onroad Transportation Energy and Emissions. *J. Air Waste Manag. Assoc.* **2018**, *68*, 514–563. [CrossRef] [PubMed]
2. Blanco Bonilla, A.; Ejecutivo, S.; Castillo, T.; García, F.; Mosquera, L.; Rivadeneira, T.; Segura, K.; Yujato, M.; Gain, K.; Guerra, L.; et al. *Panorama Energético de Amaérica Latina El El Caribe 2022*; Olade: Quito, Ecuador, 2022.
3. Yuan, W.; Frey, H.C.; Wei, T.; Rastogi, N.; VanderGriend, S.; Miller, D.; Mattison, L. Comparison of Real-World Vehicle Fuel Use and Tailpipe Emissions for Gasoline-Ethanol Fuel Blends. *Fuel* **2019**, *249*, 352–364. [CrossRef]
4. Gallus, J.; Kirchner, U.; Vogt, R.; Benter, T. Impact of Driving Style and Road Grade on Gaseous Exhaust Emissions of Passenger Vehicles Measured by a Portable Emission Measurement System (PEMS). *Transp. Res. Part D Transp. Environ.* **2017**, *52*, 215–226. [CrossRef]
5. Rodríguez, R.A.; Virguez, E.A.; Rodríguez, P.A.; Behrentz, E. Influence of Driving Patterns on Vehicle Emissions: A Case Study for Latin American Cities. *Transp. Res. Part D Transp. Environ.* **2016**, *43*, 192–206. [CrossRef]
6. Rosero, F.; Fonseca, N.; López, J.-M.; Casanova, J. Effects of Passenger Load, Road Grade, and Congestion Level on Real-World Fuel Consumption and Emissions from Compressed Natural Gas and Diesel Urban Buses. *Appl. Energy* **2021**, *282*, 116195. [CrossRef]
7. Giraldo, M.; Huertas, J.I. Real Emissions, Driving Patterns and Fuel Consumption of in-Use Diesel Buses Operating at High Altitude. *Transp. Res. Part D Transp. Environ.* **2019**, *77*, 21–36. [CrossRef]
8. EPA. *Greenhouse Gas Emissions Standards and Fuel Efficiency Standards for Medium- and Heavy-Duty Engines and Vehicles*; EPA: Washington, DC, USA, 2011; Volume 76.
9. EPA; NHTSA; DOT. *Greenhouse Gas Emissions and Fuel Efficiency Standards for Medium- Heavy-Duty Engines and Vehicles—Phase 2*; EPA: Washington, DC, USA, 2016; Volume 81, pp. 73478–74274.
10. EU Commission Regulation (EU) No 582/2011 of 25 May 2011 Implementing and Amending Regulation (EC) No 595/2009 of the European Parliament and of the Council with Respect to Emissions from Heavy Duty Vehicles (Euro VI) and Amending Annexes I and III to Direct. Available online: https://eur-lex.europa.eu/eli/reg/2011/582/oj (accessed on 14 May 2024).
11. Luján, J.M.; Bermúdez, V.; Dolz, V.; Monsalve-Serrano, J. An Assessment of the Real-World Driving Gaseous Emissions from a Euro 6 Light-Duty Diesel Vehicle Using a Portable Emissions Measurement System (PEMS). *Atmos. Environ.* **2018**, *174*, 112–121. [CrossRef]
12. European Commission and Council of the European Union Commission Regulation (EU) 2016/427 Amending Regulation (EC) No 692/2008 as Regards Emissions from Light Passenger and Commercial Vehicles (Euro 6) (Text with EEA Relevance). *Off. J. Eur. Union* **2016**, *82*, 1–98.
13. EU. *Commission Regultaion (EU) 2019/318 of 19 February 2019 Amending Regulation (EU) 2017/2400 and Directive 2007/46/EC of the European Parliament and of the Council as Regards the Determination of the CO2 Emissions and Fuel Consumption of Heavy-Duty Vehicle*; European Union: Maastricht, The Netherlands, 2019; Volume 2011.
14. Acevedo, H.; Delgado, O.; Pineda, L.; Pettigrew, S. Hoja De Ruta Para Descarbonizar El Transporte De Carga En América Latina Entre 2025 Y 2050. *Atmos. Environ.* **2023**, *188*, 1–30.
15. Mera, Z.; Varella, R.; Baptista, P.; Duarte, G.; Rosero, F. Including Engine Data for Energy and Pollutants Assessment into the Vehicle Specific Power Methodology. *Appl. Energy* **2022**, *311*, 118690. [CrossRef]
16. Kroyan, Y.; Wojcieszyk, M.; Kaario, O.; Larmi, M. Modelling the End-Use Performance of Alternative Fuel Properties in Flex-Fuel Vehicles. *Energy Convers. Manag.* **2022**, *269*, 116080. [CrossRef]
17. Zhang, M.; Cheng, W.; Shen, Y. Designing Heavy-Duty Vehicles' Four-Parameter Driving Cycles to Best Represent Engine Distribution Consistency. *IEEE Access* **2020**, *8*, 212079–212093. [CrossRef]
18. He, L.; Hu, J.; Yang, L.; Li, Z.; Zheng, X.; Xie, S.; Zu, L.; Chen, J.; Li, Y.; Wu, Y. Science of the Total Environment Real-World Gaseous Emissions of High-Mileage Taxi Fl Eets in China. *Sci. Total Environ.* **2019**, *659*, 267–274. [CrossRef] [PubMed]
19. Castresana, J.; Gabiña, G.; Martin, L.; Basterretxea, A.; Uriondo, Z. Marine Diesel Engine ANN Modelling with Multiple Output for Complete Engine Performance Map. *Fuel* **2022**, *319*, 123873. [CrossRef]
20. Dekraker, P.; Barba, D.; Moskalik, A.; Butters, K. Constructing Engine Maps for Full Vehicle Simulation Modeling. *SAE Technol. Pap. Ser.* **2018**, *1*, 4. [CrossRef]
21. Acevedo, H.; Delgado, O. Mecanismos de Bajo Costo Para La Verificación y Control de Emisiones Vehiculares En Colombia. 2023. Available online: https://theicct.org/wp-content/uploads/2023/05/heavy-vehicles-Colombia-costs-may23.pdf (accessed on 10 March 2024).
22. Sierra, J.C. Estimating Road Transport Fuel Consumption in Ecuador. *Energy Policy* **2016**, *92*, 359–368. [CrossRef]
23. EPA. *EPA Finalizes Regulations Requiring Onboard Diagnostic Systems on 2010 and Later Heavy-Duty Engines Used in Highway Applications Over 14,000 Pounds; Revisions to Onboard Diagnostic Requirements for Diesel Highway Heavy-Duty Applications Under 14,000 Pound*; EPA: Washington, DC, USA, 2008.

24. Horiba. On Board Emission Measurement System OBS-2200 Instruction Manual. 2006. Available online: https://www.horiba.com/int/automotive/products/detail/action/show/Product/obs-one-gs-unit-28/ (accessed on 22 June 2024).
25. Alessandrini, A.; Filippi, F.; Ortenzi, F. Consumption Calculation of Vehicles Using OBD Data. *2012 Int. Emiss. Invent. Conf. "Emission Invent.—Meet. Challenges Posed by Emerg. Glob. Natl. Reg. Local Air Qual. Issues"* **2012**, 1–21. Available online: https://static.horiba.com/fileadmin/Horiba/Products/Automotive/Emission_Measurement_Systems/OBS-ONE/OBS-ONE_Brochure_English.pdf (accessed on 25 March 2024).
26. Posada, F.; Bandivadekar, A. Global Overview of On-Board Diagnostic (OBD) Systems for Heavy-Duty Vehicles [White Paper]; 2015. Available online: https://theicct.org/wp-content/uploads/2021/06/ICCT_Overview_OBD-HDVs_20150209.pdf (accessed on 3 March 2024).
27. Instituto de Investigación Geológico y Energético de Ecuador. *Balance Energético Nacional Del Ecuador 2021*; Instituto de Investigación Geológico y Energético de Ecuador: Quito, Ecuador, 2022.
28. European Commission. Commission Regulation (EU) 2017/1151 of 1 June 2017 Supplementing Regulation (EC) No 715/2007 of the European Parliament and of the Council on Type-Approval of Motor Vehicles with Respect to Emissions from Light Passenger and Commercial Vehicles (Euro 5 A). *Off. J. Eur. Union* **2017**, 1–643. Available online: https://eur-lex.europa.eu/legal-content/EN/TXT/PDF/?uri=CELEX:32007R0715 (accessed on 15 January 2024).
29. Rosero, F.; Fonseca, N.; López, J.-M.J.-M.; Casanova, J. Real-World Fuel Efficiency and Emissions from an Urban Diesel Bus Engine under Transient Operating Conditions. *Appl. Energy* **2020**, *261*, 114442. [CrossRef]
30. RStudio. RStudio Desktop. Available online: https://s3.amazonaws.com/rstudio-ide-build/electron/windows/RStudio-2023.09.0-463.exe (accessed on 26 January 2024).
31. Wickham, H. *Ggplot2: Elegant Graphics for Data Analysis*; Tidiverse, Ed.; Springer: New York, NY, USA, 2016; ISBN 978-3-319-24277-4.
32. Jiménez-Palacios, J.L. Understanding and Quantifying Motor Vehicle Emissions with Vehicle Specific Power and TILDAS Remote Sensing. Ph.D. Thesis, Massachusetts Institute of Technology (MIT), Cambridge, MA, USA, 1999.
33. Zhai, H.; Frey, H.C.; Rouphail, N.M. A Vehicle-Specific Power Approach to Speed- and Facility-Specific Emissions Estimates for Diesel Transit Buses. *Environ. Sci. Technol.* **2008**, *42*, 7985–7991. [CrossRef]
34. Alves, J.; Baptista, P.C.; Gonçalves, G.A.; Duarte, G.O. Indirect Methodologies to Estimate Energy Use in Vehicles: Application to Battery Electric Vehicles. *Energy Convers. Manag.* **2016**, *124*, 116–129. [CrossRef]
35. Song, G.; Yu, L. Estimation of Fuel Efficiency of Road Traffic by Characterization of Vehicle-Specific Power and Speed Based on Floating Car Data. *Transp. Res. Rec. J. Transp. Res. Board* **2009**, *2139*, 11–20. [CrossRef]
36. Bishop, J.D.K.; Stettler, M.E.J.; Molden, N.; Boies, A.M. Engine Maps of Fuel Use and Emissions from Transient Driving Cycles. *Appl. Energy* **2016**, *183*, 202–217. [CrossRef]
37. He, W.; Duan, L.; Zhang, Z.; Zhao, X.; Cheng, Y. Analysis of the Characteristics of Real-World Emission Factors and VSP Distributions—A Case Study in Beijing. *Sustainability* **2022**, *14*, 11512. [CrossRef]
38. EPA. *Exhaust Emission Rates for Heavy—Duty On—Road Vehicles in MOVES2014 Exhaust Emission Rates for Heavy—Duty On—Road Vehicles in MOVES2014*; EPA: Washington, DC, USA, 2015.
39. Li, F.; Zhuang, J.; Cheng, X.; Li, M.; Wang, J.; Yan, Z. Investigation and Prediction of Heavy-Duty Diesel Passenger Bus Emissions in Hainan Using a COPERT Model. *Atmosphere* **2019**, *10*, 106. [CrossRef]
40. Yao, Z.; Cao, X.; Shen, X.; Zhang, Y.; Wang, X.; He, K. On-Road Emission Characteristics of CNG-Fueled Bi-Fuel Taxis. *Atmos. Environ.* **2014**, *94*, 198–204. [CrossRef]
41. Ghaffarpasand, O.; Talaie, M.R.; Ahmadikia, H.; TalaieKhozani, A.; Shalamzari, M.D.; Majidi, S. On-Road Performance and Emission Characteristics of CNG-Gasoline Bi-Fuel Taxis/Private Cars at the Roadside Environment. *Atmos. Pollut. Res.* **2020**, *11*, 1743–1753. [CrossRef]
42. Zhai, Z.; Song, G.; Yu, L. How Much Vehicle Activity Data Is Needed to Develop Robust Vehicle Specific Power Distributions for Emission Estimates? A Case Study in Beijing. *Transp. Res. Part D Transp. Environ.* **2018**, *65*, 540–550. [CrossRef]
43. Zhang, S.; Yu, L.; Song, G. Emissions Characteristics for Heavy-Duty Diesel Trucks Under Different Loads Based on Vehicle-Specific Power. *Transp. Res. Rec. J. Transp. Res. Board* **2017**, *2627*, 77–85. [CrossRef]
44. Hu, J.; Wu, Y.; Wang, Z.; Li, Z.; Zhou, Y.; Wang, H.; Bao, X.; Hao, J. Real-World Fuel Efficiency and Exhaust Emissions of Light-Duty Diesel Vehicles and Their Correlation with Road Conditions. *J. Environ. Sci.* **2012**, *24*, 865–874. [CrossRef]

Disclaimer/Publisher's Note: The statements, opinions and data contained in all publications are solely those of the individual author(s) and contributor(s) and not of MDPI and/or the editor(s). MDPI and/or the editor(s) disclaim responsibility for any injury to people or property resulting from any ideas, methods, instructions or products referred to in the content.

Article

Analysis of Non-Road Mobile Machinery Homologation Standards in Relation to Actual Exhaust Emissions

Natalia Szymlet, Michalina Kamińska, Andrzej Ziółkowski * and Jakub Sobczak

Faculty of Civil Engineering and Transport, Institute of Combustion Engines and Powertrains,
Poznan University of Technology, Piotrowo 3, PL-60965 Poznan, Poland; natalia.szymlet@put.poznan.pl (N.S.);
michalina.kaminska@put.poznan.pl (M.K.); jakub.sobczak@student.put.poznan.pl (J.S.)
* Correspondence: andrzej.j.ziolkowski@put.poznan.pl

Abstract: This article presents issues related to the current approval procedures in the group of off-road vehicles. Our research aimed to demonstrate significant differences between actual railway vehicle operation and stationary homologation tests regarding exhaust emissions. The research cycle consisted of analyzing emissions of toxic compounds from exhaust systems under real operating conditions, supplemented by a temporal share analysis based on the denormalized NRTC test upon which the tested object was homologated. Based on the conducted analyses, a significant difference was found between the actual operation of the tested railway vehicle and the stationary homologation test. By interpreting emission intensities within the parameter ranges of the propulsion unit's operation, key areas with a significant impact on the vehicle's overall emissions were identified. Based on the obtained results, a critical opinion is expressed regarding current homologation standards for the off-road vehicle group and the necessity for further empirical research in the area of actual operation of the tested vehicle group.

Keywords: non-road mobile machinery; rail bus; exhaust emission; NRTC cycle; real operating conditions

Citation: Szymlet, N.; Kamińska, M.; Ziółkowski, A.; Sobczak, J. Analysis of Non-Road Mobile Machinery Homologation Standards in Relation to Actual Exhaust Emissions. *Energies* **2024**, *17*, 3624. https://doi.org/10.3390/en17153624

Academic Editors: Maksymilian Mądziel, Kazimierz Lejda and Artur Jaworski

Received: 19 June 2024
Revised: 17 July 2024
Accepted: 18 July 2024
Published: 24 July 2024

Copyright: © 2024 by the authors. Licensee MDPI, Basel, Switzerland. This article is an open access article distributed under the terms and conditions of the Creative Commons Attribution (CC BY) license (https://creativecommons.org/licenses/by/4.0/).

1. Introduction

Tests on emissions of harmful exhaust gases are carried out in laboratory conditions as dynamometric tests. However, such measurements are only intended to reflect the actual operating conditions of vehicles, and the results obtained are most often used as input data in the process of modeling the impact of road or track traffic on air quality. Measurements of this type are divided into two groups. The first group includes those measurements performed on an engine dynamometer, during which the engine installed directly on the test stand, equipped only with the equipment necessary for proper operation, is analyzed. This type of measurement is performed on heavy vehicles. The second group includes vehicle measurements carried out on a chassis dynamometer equipped with special rollers simulating road conditions. These tests are performed for light vehicles. The advantage of laboratory tests is primarily the ability to carry out many measurement cycles at relatively low unit costs. However, the data obtained are only approximate because the measurements do not include the impact of road conditions, the driver's driving style, or the ambient air quality. Therefore, most test cycles performed on dynamometers do not reflect the actual operating conditions of vehicles in real conditions, which is confirmed by the literature [1–3].

Therefore, for several years, scientists have been developing a number of alternative methods that allow us to obtain more reliable results. The first step was to develop dynamic test cycles simulating specific driving conditions [4,5]. In a further stage, tests were carried out in real operating conditions. Currently, approval regulations require measurements of harmful exhaust gas compounds for passenger cars and heavy vehicles in road conditions in specially developed RDE tests using exhaust gas analyzers from the PEMS group. However, for rail vehicles, there are no specific regulations regarding emission tests of harmful exhaust gases, which is why they are carried out only in laboratory conditions. However, the latest regulations for non-healthy vehicles contain a provision informing us about the need for tests in real conditions, but no test procedure has been established and no limit values for individual harmful exhaust gas compounds have been provided.

The tests performed in real operating conditions are aimed at verifying the ecological indicators of vehicles in a wide range of operations of their drive systems. Measurements of this type are performed primarily to determine the impact of the used drive systems and vehicle movement parameters on the emission of harmful compounds and to indicate the differences between approval procedures and their actual operation. The development and minimization of measurement equipment for pollution testing allows for increasingly more accurate analyses in real operating conditions. Moreover, the application possibilities of this type of device are constantly increasing. Thanks to this, it is possible to take into account the specificity of the traffic of each type of vehicle (road, track, off-road).

2. Railway Passenger Rolling Stock in Poland

Air pollution is currently one of the biggest problems facing humanity, and continuous economic development contributes to the emergence of new sources of pollution. According to a World Health Organization (WHO) report published in 2023 [6], approximately 6.7 million people died prematurely in 2019 due to diseases of the respiratory and circulatory systems and cancers caused by air pollution.

One of the possible ways to counteract excessive exhaust emissions is to reduce the share of passenger vehicles in favor of the development of public transport. According to EEA data [7], in 2019, passenger vehicles were responsible for 60.6% of total CO_2 emissions in the European Union. In order to reduce the share of cars, city authorities are increasingly introducing restrictions that prevent or limit the use of combustion cars in city centers. Instead, public transport solutions are proposed, thanks to which citizens can move around cities. Due to rising housing prices and the continuous expansion of urban agglomerations, young people decide to live in nearby, smaller towns, which most often involves the need to travel by passenger car to their workplaces. A favorable solution in such a case is rail transport. Thanks to the use of rail transport, it is possible to connect the city center with nearby urban centers, which allows for quick access compared with travel by passenger vehicle. Such railway connections are most often implemented using electrified or diesel traction units when the railway line does not have an overhead contact line.

A diesel multiple unit (DMU) is a type of train that consists of at least two units that constitute one coherent whole during operation, and its main source of drive is the internal combustion engine. DMUs are used to transport passengers on main railway lines in order to deliver passengers from smaller towns to larger trans-shipment points, and they serve connections on regional and agglomeration railway lines, which are most often not electrified. Currently, multi-faceted research is being carried out on this type of vehicle and engine [8–18]. DMUs are also called rail buses; however, this concept also includes single-motor wagons, which are one separate unit equipped with a drive source. Diesel ignition engines are most often used in DMUs due to the nature of the operation of rail vehicles, which require high torque. In order to reduce the emission of toxic compounds from exhaust system pollutants, a design solution is usually used in which a selective reduction system is installed in the exhaust system for catalytic reduction (Selective Catalytic Reduction, SCR), which allows NO_x to be reduced by 90% to nitrogen (N_2) and water (H_2O) [19,20]. To carry out this process, AdBlue (trade name) is used, which consists of 32.5% urea and 67.5% distilled water. AdBlue is injected into the exhaust manifolds of diesel engines to remove impurities and neutralize NO_x [21,22].

According to data collected by the Office of Rail Transport, in 2021, 246 diesel multiple units were in operation throughout Poland [23]. A summary of changes in the number of DMUs in operation over the last 19 years is presented in Figure 1. Although efforts are being made to reduce exhaust emissions and reduce the number of combustion vehicles in order to replace them with electric versions, the number of combustion rail buses is constantly increasing. In just one year (2020–2021), the number of DMUs in operation increased by 37 units.

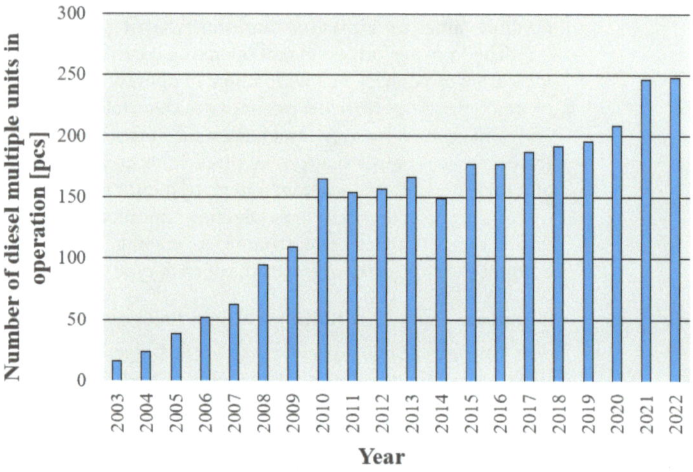

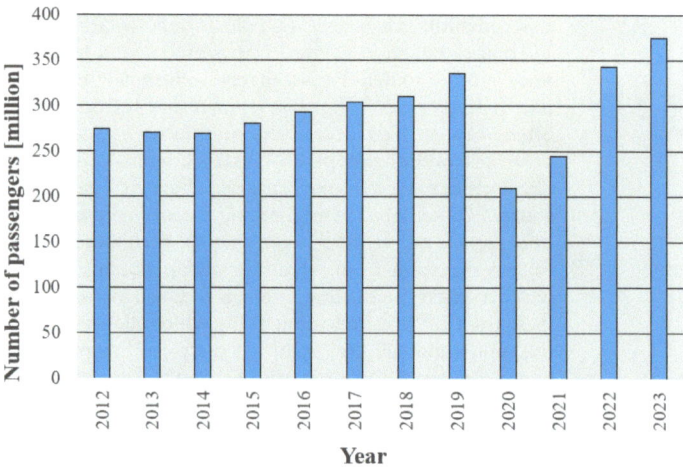

Figure 1. Number of diesel multiple units operated in Poland in 2003–2021 [23].

This is caused by an increase in the number of passengers, as people are more and more willing to choose to travel by train, which is confirmed by the data presented in Figure 2. It should be remembered that the decrease in the number of passengers in 2020–2021 was caused by restrictions related to movement caused by the COVID-19 pandemic (Coronavirus Disease). After the end of the pandemic in 2022, the highest number of passengers transported in 10 years was recorded, and the data from 2023 are even more optimistic, because in the period from January to July 2023 the number of passengers amounted to almost 213 million, while last year in this period it was only 188.60 million passengers, which significantly shows greater interest in rail transport. Over time, new regional connections are also being created, which increases the number of vehicles that need to serve passengers on railway lines.

Figure 2. Number of passengers transported by rail in Poland in 2012–2022 [24].

3. Legal Regulations Regarding Testing and Approval of Non-Road Vehicles

The European Union pursues a policy aimed at reducing the negative impact of means of transport on the natural environment while ensuring the safety and comfort of their users. For this reason, regulations and directives are introduced that define guidelines for achieving the goals set by the European Union. Taking into account the regulations governing the requirements for engines used in diesel multiple units regarding permissible emissions of toxic compounds, two regulations can be distinguished. The first one is included in the report of the European Rail Research Institute (ERRI), formerly the Office de Recherches et d'Essais—Railway Research and Testing Office (ORE), while the second one is included in the Union Internationale des Chemins de fer—International Union of Railways (UIC) cards [25,26]. However, they do not have international legal status, and it is only an external obligation of all railways associated with the UIC. These regulations are still used to assess the engines used in diesel multiple units [27], and the permissible values are presented in Table 1.

Table 1. Exhaust emission limits according to UIC 623-2 and UIC 624 [25,26].

Date of Introduction	Emission (g/kWh)			
	CO	NO$_x$	HC	PM
Until 31 December 2002 UIC I	3	12	0.8	1.6
From 1 January 2003 UIC II	2.5	6	0.6	0.25
$p \leq 560$ kW		9.5 (n > 1000 rpm)		
$p > 560$ kW	3	9.9 (n $\leq$ 1000 rpm)	0.8	0.25
From 1 January 2003				
$p \leq 560$ kW	2	4.6	0.5	0.15
$p > 560$ kW	2	6	0.5	0.2

The current EU regulations imposing requirements on emissions of toxic substances from exhaust gases are established in directives that have legal force. These directives specify in detail the standards regarding permissible emissions of harmful components from exhaust systems. These regulations apply both to vehicles traveling on roads and to railway rolling stock, which is equipped with various types of internal combustion engines, which is important in the case of the use of car engines in railway vehicles with internal combustion drives, e.g., in internal combustion multiple units. The most important directives relating to DMUs include the following:

- Directive 97/68/EC of 16 December 1997 (as amended), which contains provisions on emission standards and procedures for the approval of engines installed in non-road machines [19];
- Directive 2004/26EC of 21 April 2004 (as amended), which amends Directive 97/68/EC in order to bring members of the community closer to reducing emissions from combustion engines [27];
- Regulation 2016/1628 of the European Parliament and of the EU Council of 14 September 2016, which repeals Directive 97/68/EC, introducing the Stage V emission standard [28].

Table 2 shows the emission limits of toxic exhaust gases from exhaust systems that apply to off-road vehicles, including rail vehicles. On 16 December 1997, the first EU directive was introduced regulating the permissible emission values of toxic exhaust gases for off-road vehicles. The regulations were implemented in two stages. In 1999, the Stage I standard was introduced, and in 2002, Stage II. Directive 2004/26/EC introduced further emission standards divided into Stage IIIA and Stage IIIB. The Stage IIIB standard introduced a stringent hydrocarbon emission standard, lowering the permissible value to 0.19 g/kWh compared with the previous Stage II standard, where the value was 1 g/kWh. In addition, in Stage IIIB, a particulate emission limit of 0.025 g/kWh was introduced, which was to force the use of particulate filters in vehicles. The Stage IV standard introduced in 2014 significantly reduced the permissible value of nitrogen oxide emissions to 0.4 g/kWh, which contributed to the increased use of SCR systems in rail vehicles. The latest emission standard for off-road vehicles, Stage V, was intended to contribute to increased control over particulate matter and limited the permissible PM emission to 0.015 g/kWh [29]. The permissible values of toxic compound emissions are determined using laboratory tests developed to measure emissions from engines for NRMM. The main test cycle for NRMM vehicles is the ISO 8178 test cycle, known as the Non-Road Steady Cycle (NRSC). This engine dynamometer test cycle includes several sequences of steady states with different weighting factors. The NRSC cycle for the Stage IIIA standard was the only test defining the allowable emission limits. Starting with the Stage IIIB standard, the requirement to conduct the Non-Road Transient Cycle (NRTC) test was introduced, and both test cycles (NRSC and NRTC) are now conducted in parallel during engine testing for NRMM vehicles.

Table 2. Emission limits of toxic exhaust gases for non-road combustion engines with a power of 130 kW $\geq p \leq$ 560 kW [29].

Date of Introduction		Emission (g/kWh)			
		CO	HC	NO_x	PM
Stage I	1999	5.0	1.3	9.2	0.54
Stage II	2002	3.5	1.0	6.0	0.2
Stage IIIA	2006	3.5		4.0	0.2
Stage IIIB	2011	3.5	0.19	2.0	0.025
Stage IV	2014	3.5	0.19	0.4	0.025
Stage V	2019	3.5	0.19	0.4	0.015

The currently applicable worldwide approval test for non-road vehicles is the Non-Road Transient Cycle (NRTC) test. The test cycle is mandatory for engines used in NRMM vehicles approved for Stage IIIB and newer [30–32]. The main assumption of this test is to test the engine in such a way as to approximate its actual operating conditions. Two cycles are performed during the test (Figure 3). The first is a cold start cycle, which involves performing the test after conditioning the engine at laboratory temperature (20–30 °C). The second cycle is a hot start performed 20 min after warm conditioning beginning immediately after the first cycle is completed. To summarize, the test sequence includes a cold start cycle following natural or forced engine cooling, followed by warm conditioning and a hot start cycle. As a result of this study, we obtained the total emission of harmful compounds in exhaust gases. The work produced by the engine over the entire cycle is determined by integrating the power with respect to the cycle time, using torque and rotational speed feedback signals from the dynamometric brake. The concentration of gas components for the entire cycle can be determined in two ways:

1. The concentration in the raw exhaust gas can be determined by integrating the exhaust gas analyzer signal in accordance with the procedure described in the Directive [33];
2. In the undiluted exhaust gas in a Constant Volume Sample (CVS) system, diluting the total flow by integration or collecting a sample into bags in accordance with the procedure described in the Directive [33].

To determine the concentration of particulate matter, an exhaust gas sample is taken on an appropriate filter using the total or partial flow dilution method. The flow rate of the diluted or raw exhaust gas over the entire cycle shall be determined depending on the method used. To determine the mass of each tested pollutant per kilowatt hour, the mass emission value refers to the operation of the tested facility. Emissions expressed in g/kWh are measured during each cycle, both cold start and hot start. Total weighted emissions are determined by applying a weight of 10% for a cold start and 90% for a hot cycle. Total emissions of pollutants must not exceed the limit values [27,33,34].

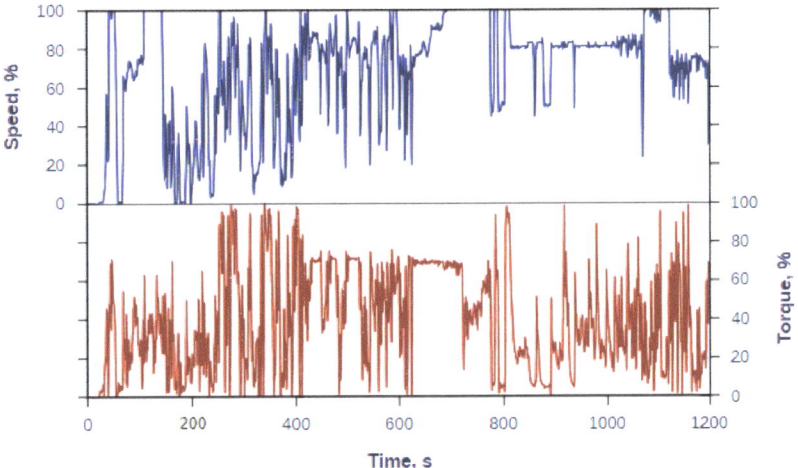

Figure 3. Normalized speed and torque over the NRTC cycle [29].

To run an NRTC on a specific engine, engine torque and crankshaft speed reference levels must be determined in advance. It is therefore necessary to carry out the destandardization procedure in accordance with the procedure described in Commission Directive 2010/26/EU [19]. The engine crankshaft speed is determined based on the formula:

$$n = \%n_{nor}(n_{ref} - n_{idle})/100 + n_{idle} \qquad (1)$$

where:

n_{nor} is the normalized rotational speed,
n_{idle} is the idle speed, and
n_{ref} is the reference speed calculated according to the formula:

$$n_{ref} = n_{50} + 0.95(n_{70} + n_{50}) \qquad (2)$$

where:

n50 is the minimum speed at which the engine reaches 50% of its rated power, and n_{70} is the maximum speed at which the engine reaches 70% of the rated power. Then, the torque is denormalized and determined based on formula:

$$Mo = (\%Mo_{nor} \times Mo_{max})/100 \qquad (3)$$

where:

Mo_{nor} is the normalized torque, and
Mo_{max} is the maximum torque.

4. Research Methodology

4.1. Research Object

A diesel multiple unit intended for operation on agglomeration and regional railway lines was used to test the emission of toxic exhaust gases and the operation of an internal combustion engine in real operating conditions [35–40]. The examined facility was selected in such a way that it could be compared whether the emission standards set by the European Union are realistically reflected in the actual operation of rail vehicles in the form of an improvement in the amount of harmful toxic compounds emitted into the atmosphere in exhaust gases. Therefore, a rail bus operating in Poland, meeting the Stage IIIB emission standard, was subjected to testing. The vehicle is equipped with two CI engines with a power of 390 kW and a maximum torque of 2300 Nm.

4.2. Measuring Equipment

Homologation tests of vehicles from the Non-Road Mobile Machinery (NNRM) group, including rail vehicles, are carried out exclusively on special engine dynamometers. The main reason for the approval of the engines of off-road vehicles is the difficulties in carrying out tests in real operating conditions, especially due to the dimensions of the vehicles [36]. Many scientists in their research works [41–47] indicate that standardized measurements on test benches are unable to reflect the impact of actual operating conditions on the emission results of harmful exhaust gases, as is the case with RDE tests. However, continuous technological development creates new opportunities to perform measurements during the daily operation of NNRM vehicles. One of them is the Axion R/S+ mobile measuring device, which was used during our research.

The equipment from the PEMS group (Axion R/S+, Figure 4), developed by the American company Global MRV, has passed a rigorous assessment of the Environmental Technology Verification (ETV) program conducted by the United States Environmental Protection Agency (USEPA). It is characterized by a weight of approximately 18 kg and compact dimensions, which allow us to conduct research in hard-to-reach places, including in non-road vehicles. The analyzers used in the equipment allow for the measurement of emissions of harmful compounds from exhaust systems, such as CO, CO_2, HC, NO_x, PM, and O_2.

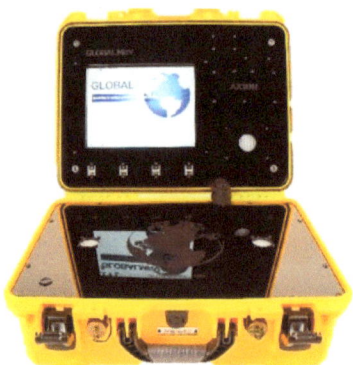

Figure 4. Research equipment—Axion R/S+ [48].

A Nondipersive Infrared Sensor (NDIR) analyzer is used to measure CO, CO_2, and HC concentrations. The operation of this analyzer is based on the spectrometric method, in which the measurement photometer records the total absorption of radiation in a relatively small wavelength band specific to a specific gas compound. Measurements of particulate matter are made by a method based on laser scattering (Laser Scatter). NO_x and O_2 are checked using electrochemical analyzers, where an electrical signal is obtained after the electrochemical transformation of the measured substances. The equipment is also equipped with a Global Positioning System (GPS). The data measurement process takes place at a frequency of 1 Hz. The obtained test measurements are processed and corrected, and, on their basis, road and unit emissions of measured pollutants can be calculated.

In order to record vehicle operating parameters (rotational speed, torque, pressures, and temperatures) during the tests, the TEXA Navigator TXTs TRUCK (Treviso, Italy) diagnostic device was used. The measurement module is connected to the socket of the on-board diagnostic (OBD) system and to a computer with specially installed software IDC5, where data from the tested vehicle are recorded. Figure 5 shows the connection of research equipment to the research facility.

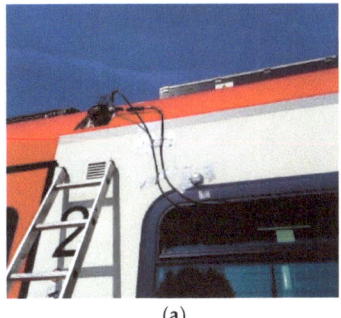  

(a) (b) (c)

Figure 5. (**a**) Connecting the probe to the exhaust gas system. (**b**) Connecting the probe to the equipment. (**c**) Axion R/S+ measuring equipment.

However, PEMS equipment has certain limitations. This primarily applies to the fact that the analyzers during testing activities must be installed on the test vehicle and the measurement probes must be placed directly in the exhaust system. Installing the equipment is usually very time-consuming, and it is not always possible to use exhaust gas flow meters, which make testing more efficient. For this reason, flow characteristics should be determined using information about the pressure in the intake manifold, the temperature behind the turbocharger, and the crankshaft rotational speed. Additionally, the device used for testing is characterized by a limited measurement range, as shown in Table 3. Another problem may be the relative measurement accuracy of ±2–4%, which is problematic mainly in the case of particulate matter, which has small values.

Table 3. Technical data on the Axion R/S+ device [48].

Exhaust Component	Measurement Range	Measurement Accuracy	Distribution	Type of Measurement	Measurement Time (s)
CO_2	0–16%	±0.3% absolute ±4% relative	0.01 vol.%	NDIR	<3.5
CO	0–10%	±0.02% absolute ±3% relative	0.001 vol.%	NDIR	<3.5
HC	0–4000 ppm	±8 ppm absolute ±3% relative	1 ppm	NDIR	<3.5
NO *	0–4000 ppm	±25 ppm absolute ±3% relative	1 ppm	E-chem	<5
PM	0–300 mg/m^3	±2%	0.01 mg/m^3	Laser Scatter	2

* The NOx emissions value was estimated based on the assumption that the amount of nitrogen dioxide (NO_2) in exhaust gases is approximately 5% of the measured NO value for gasoline and approximately 10% for diesel oil.

4.3. Research Route

Tests in real vehicle operating conditions took place on a route used for daily passenger transport by a railway operator from Greater Poland (Figure 6). The journey was carried out in the form of a standard driving style, i.e., one in which the driver tries to reach the maximum traveling speed as quickly as possible. The tests were conducted in clear weather conditions, which allowed for the elimination of external factors that could affect the increase in the emission of toxic compounds during the tests. Such factors mainly include precipitation, which causes the tracks to become wet, potentially leading to slippage between the rail vehicle and the infrastructure. In such cases, the train operator must brake and accelerate the vehicle cautiously, preventing the tests from being conducted in a standard driving style. The air temperature during the tests was approximately 20 °C (readings from meteorological stations), which was favorable for conducting the tests, as the temperature did not significantly impact the train's operation, unlike in low temperatures during winter or high temperatures in summer, where additional systems such as air conditioning are activated for passenger comfort and safety.

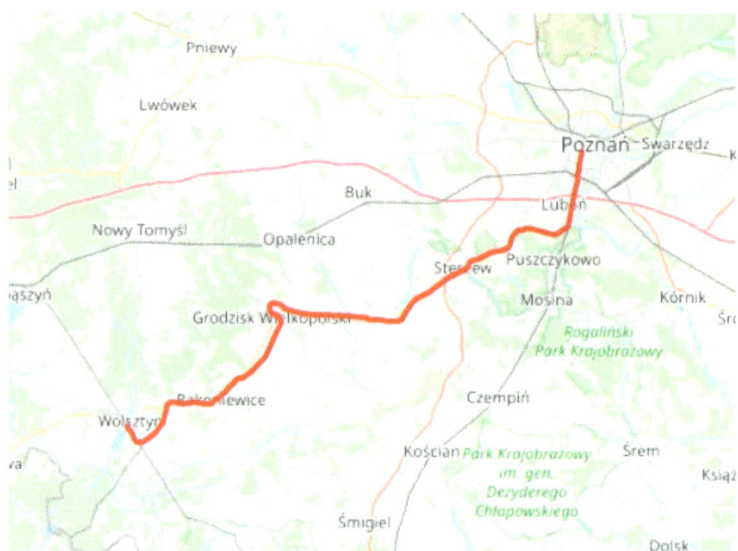

Figure 6. View of the research route.

The length of the route is 79.63 km and there are 19 stops. The total stop time during the measurements was 655 s, of which the longest single stop was 149 s. The railway line from Wolsztyn to Luboń near Poznań is a single-track line and the traffic on it is a shuttle, i.e., one train runs from Wolsztyn towards Poznań, and the second runs from Poznań to Wolsztyn. The result of this solution is the forced stoppage of one of the trains at the station in order to let the other train pass. In the case of the described research facility at work, the train of the opposite direction was waiting at the station, which allowed for the elimination of an excessive waiting time, which often reaches about 10 min on this route. Figure 7 shows the train's driving profile. The largest height difference was 48 m.

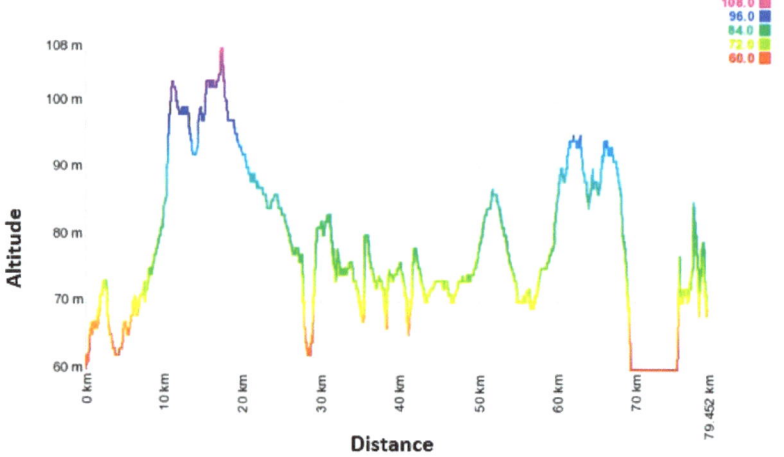

Figure 7. Driving profile.

The measurement run was carried out during the actual passenger service on the above-described test route. This allowed for maintaining realistic operating parameters of the rail vehicle, including stop times at stations during the journey according to the railway operator's schedule, as well as the traffic characteristics on the given route. By conducting measurements during the actual rail service, realistic values of toxic compound emissions were obtained, which may differ slightly from those of an empty train set. During the measurements, the tested vehicle carried approximately 98 passengers.

5. Research Results

5.1. Introduction to Analysis

Density characteristics of their operation time have been used for many years for vehicle drive systems. This mathematical tool allows you to obtain information about phenomena occurring in specific speed and acceleration ranges while taking into account the variability in these parameters for the indicated range.

By determining the time share density characteristics, it is possible to determine the time shares of operation during the tests carried out, as well as to determine the emissions of harmful compounds in the torque and rotational speed ranges.

In the analysis of traffic parameters, certain assumptions were made regarding the determination of the characteristics of operating time shares for the tested vehicles. One-sided closed intervals were used in the time share characteristics, which allowed for greater readability and accuracy of the analyses performed. In order to properly analyze this object, the value of M = 10 Nm was assumed for the read negative torque values.

Exhaust emission standards for all groups of off-road vehicles are expressed as a unit of mass related to the work performed by the drive system. As shown in the works [41–47], for tests in real operating conditions this may result in underestimation of the obtained indicators by taking into account the internal combustion engine's own starts. This means that the calculations take into account the indicated work, and it should be effective work. The reason for this is reading the torque value from the on-board diagnostic system, which is calculated based on the fuel injection time and pressure.

Non-road vehicles, due to their various purposes, cannot be assessed ecologically in terms of the distance traveled. This conclusion applies to the entire group of these machines in general. However, certain subgroups of these machines, e.g., passenger rail vehicles, can be expressed in such coefficients. This makes it easier to compare the environmental friendliness of this means of transport to the group of passenger cars, motorcycles, or suburban and intercity buses.

5.2. Analysis of Operating Parameters in the Aspect of Applicable Approval Tests

Based on the external characteristics of the tested engine, its characteristic reference speeds were determined, which are n_{idle} = 600 rpm, n_{50} = 960 rpm, and n_{70} = 2600 rpm. Then, the reference speed of the tested engine was determined, which was nref = 2518 rpm.

After denormalization, the NRTC cycle was obtained for the tested engine, which is shown in Figure 8.

In order to compare the operating conditions of the combustion engine in real operating conditions to the conditions of the NRTC homologation test, the shares of the drive unit's operating time in the speed and torque ranges were determined. The tested combustion engine of a rail vehicle in the NRTC test most often operated at a crankshaft speed in the range of 2000 rpm to 2200 rpm in the full load range, where the total share of operating time was 32.71% (Figure 9). The time share of idling during the test was 3.88%. The total time share of the test in the actual operating area of the internal combustion engine was only 30.29%.

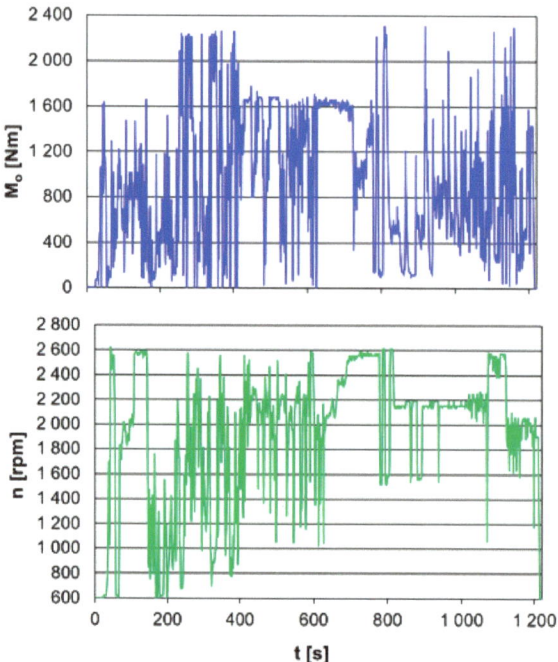

Figure 8. NRTC cycle for the test facility.

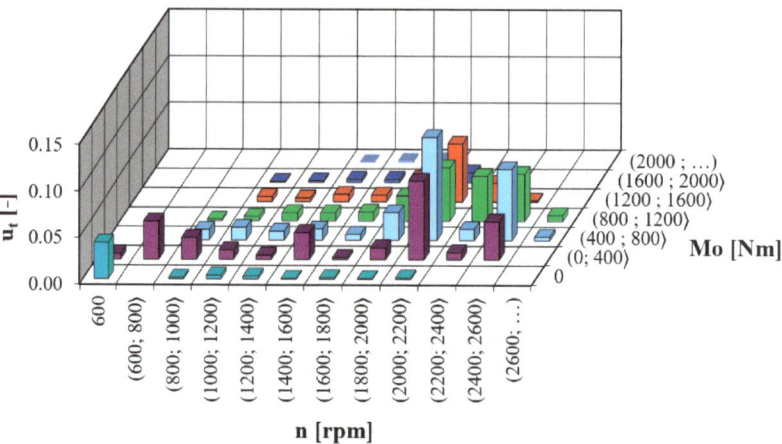

Figure 9. Shares of operating time in the ranges of crankshaft rotational speed and torque in the NRTC test.

The profile of the instantaneous power generated by the internal combustion engine indicates that the test object frequently reached the engine's rated power during the test, which influenced the obtained total work result of 50.36 kWh (Figure 10). Significant changes in engine speed and torque resulting from the test cycle characteristics contributed to dynamic changes in the power generated by the engine. High power values were achieved through the engine operating at high speeds and variable torque.

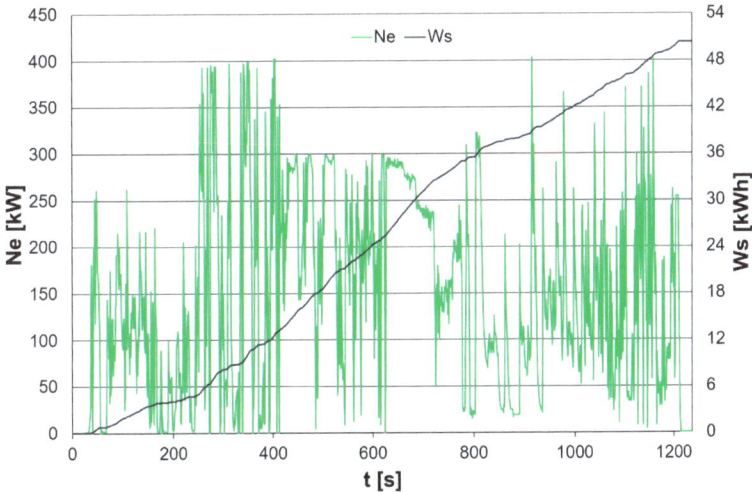

Figure 10. Power curve along with the total work of the combustion engine in the NRTC cycle.

Similarly, Figure 11 shows the characteristics of the share of operating time for individual operating intervals of the combustion engine during travel in real operating conditions. In this case of the research object, the engine operating parameters were concentrated in three ranges. In the range of torque (… Nm, 400 Nm⟩ and rotational speed ⟨800 rpm, 1000 rpm⟩ and ⟨1000 rpm, 1200 rpm⟩, 29.96% and 13.75% of the total operating time were recorded, respectively. When the vehicle was accelerating, the engine worked mainly at the rotational speed in the range of ⟨1600 rpm, 1800 rpm⟩ and ⟨2000 Nm, … Nm⟩, and its time share was 25.14%.

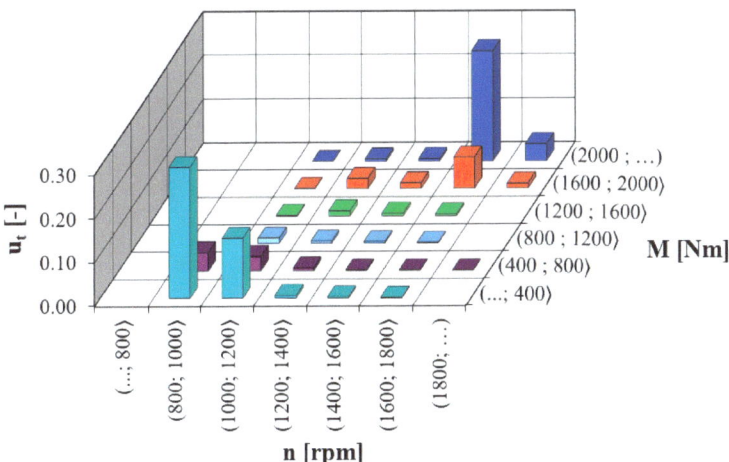

Figure 11. Share of operating time in the rotational speed and torque ranges for the research object.

To complement the analysis, a comparison was made of the difference in the share of engine operating time in the NRTC cycle (subtractor) to actual operation (subtrahend). In Figure 12, the red line shows the range of engine operating parameters during its actual operation in a rail vehicle. During the RDE tests, the combustion engine operated in the rotational speed range from 800 rpm to 2000 rpm, which is a much narrower range of its possible operation. The main reason why the engine in a rail vehicle does not work to its maximum extent is the limitations resulting from the construction of the entire drive unit. The combustion engine must be synchronized with the main gear, which then transmits torque to the reversing gear via the cardan shaft. Therefore, in order for all elements of the drive unit of a rail vehicle to cooperate, the combustion engine must operate in a smaller range of rotational speeds.

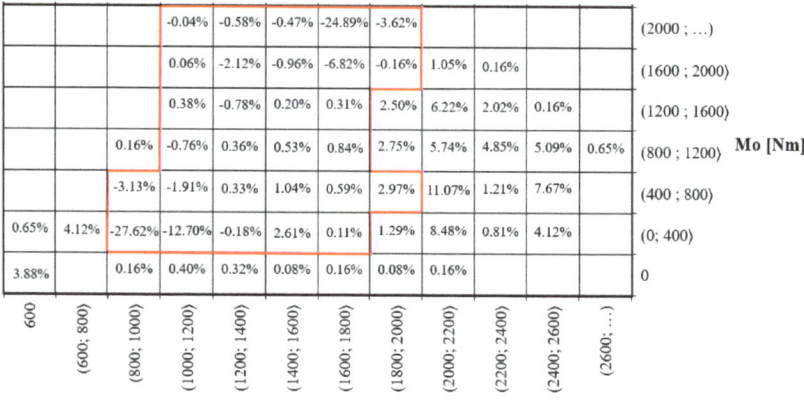

Figure 12. Difference in the share of the internal combustion engine operating time of the research facility from the RDE tests and the NRTC test cycle.

The largest difference of −27.62% was observed in the range of rotational speed (800 rpm, 1000 rpm) and torque (0 Nm, 400 Nm). In this range, the combustion engine operated during stops, which in the characteristics of railway traffic on agglomeration lines is a significant part of the entire rail vehicle journey. An equally large difference (−24.89%) was obtained in the range of rotational speed (1600 rpm, 1800 rpm) and torque (2000 Nm, … Nm), where the diesel multiple unit accelerated to the maximum permitted speed on a given section of the route after a stop. The difference in the share of engine operating time during the NRTC cycle above the rotational speed range obtained during tests in real operating conditions was 59.45%.

Significant differences between the NRTC test and the actual operating conditions of the rail vehicle arise from the nature of the test object's operation. The NRTC test cycle aims to reflect real driving conditions through dynamic changes in torque and engine speed. Consequently, a larger portion of the test is conducted at high engine speeds and variable torque. However, the actual conditions of the tested vehicle differ significantly from those presented in the homologation test. The tested object primarily operated within specific ranges of high engine speed and torque, resulting from the driver's driving style and the vehicle's design. The NRTC test cycle includes a small time proportion representing engine operation at low speeds and idle, whereas the tested vehicle in actual operating conditions frequently had to stop at railway stations, leading to a high percentage of idle operation.

5.3. Analysis of Emissions of Toxic Compounds in Real Operating Conditions from a Combustion Railbus

Section 4.2 of this article demonstrates significant differences in the time distribution between the NRTC test and actual operating conditions. It was shown that, compared with the actual operation of the vehicle, the NRTC test only accounts for 30.29% of the total time distribution. Such significant differences undermine the reliability of the laboratory test, which cannot accurately reflect the actual conditions of the tested vehicle, resulting in a lack of reliability in determining permissible values of toxic compound emissions. For this purpose, an analysis of the emission intensity of harmful compounds was conducted within the ranges of actual operating parameters of the tested object's propulsion unit.

The emission values of harmful compounds were related to the real operating conditions of the rail vehicle engine. The analysis of the emission intensity of the tested toxic compounds in the ranges of torque and crankshaft rotational speed makes it possible to learn about the influence of engine operating parameters on the content of harmful compounds emitted into the atmosphere from the exhaust systems.

Analyzing the characteristics of the CO_2 emission intensity, a close relationship between the development of the value of this relationship and the engine operating parameters is noticeable (Figure 13). The area in which significant values of CO_2 emission intensity occurred is primarily the range of high torques (1600 Nm, . . . Nm) in the entire range of crankshaft rotational speeds. The maximum value (54.6 g/s) was recorded in the field described by the intervals (1800 rpm, . . . rpm) and (1600 Nm, 2000 Nm⟩ for the crankshaft rotational speed and vehicle load, respectively. The average value of this compound in the entire driving test was 27.78 g/s.

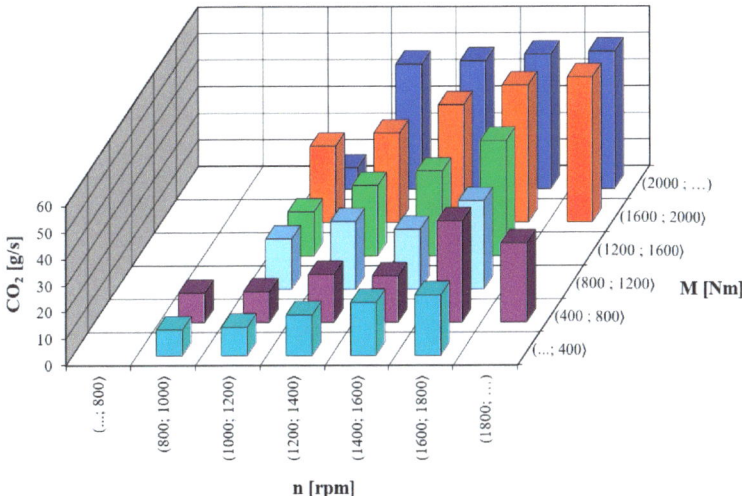

Figure 13. CO_2 emission intensity in the torque and rotational speed ranges of the research object.

A similar trend was obtained for the CO and HC emission rates (Figures 14 and 15), which is undoubtedly related to the amount of fuel injected. Its increase causes a temporary global and local oxygen deficiency, which in turn leads to the values of these compounds being relatively higher. The maximum value of CO, which amounted to 249.00 mg/s, was recorded during the highest engine load in the range (2000 Nm, . . . Nm) and the rotational speed described in the range (1800 rpm, . . . rpm). The average value for this compound in the entire test along the measurement route was 135.10 mg/s. In the case of the HC emission intensity characteristics, the highest value (53.20 mg/s) was recorded in the same engine operating area. The average HC emission intensity in the entire test was 28.51 mg/s.

In the case of the NO_x emission intensity (Figure 16), much lower values of this compound can be observed at lower rotational speeds in the range (1000 rpm, 1200 rpm⟩, where the average intensity was 135.5 mg/s. One of the factors influencing the EGR exhaust gas recirculation system was responsible for this distribution of flows. This system operates when the appropriate conditions are met, i.e., when the engine is operating in the medium speed range. Then, some of the exhaust gas enters the intake channel and less oxygen enters the cylinder for the combustion process. The SCR system also contributed to reducing nitrogen oxide emissions, as it does not cope well with high engine loads.

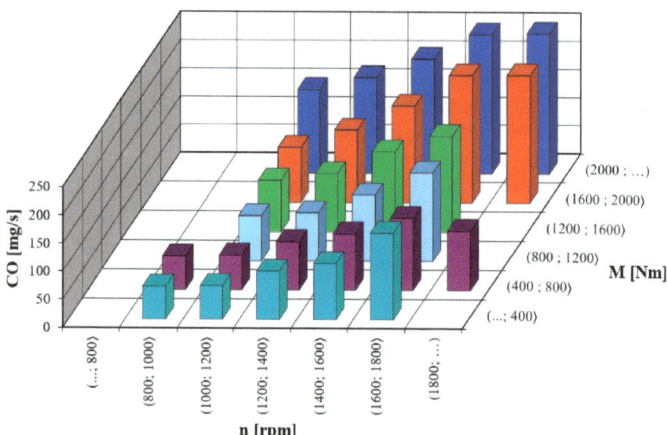

Figure 14. CO emission intensity in the torque and rotational speed ranges of the research object.

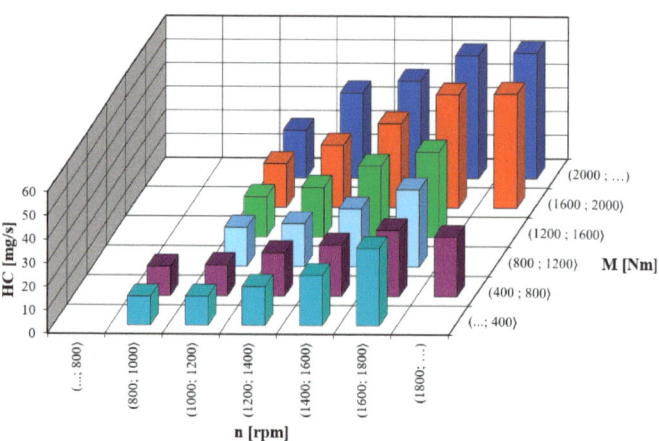

Figure 15. HC emission intensity in the torque and rotational speed ranges of the research object.

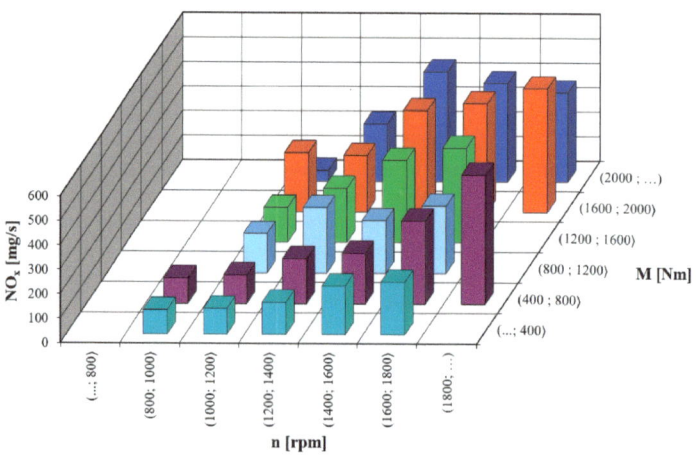

Figure 16. NO_x emission intensity in the torque and rotational speed ranges of the research object.

NO₂ is one of the two main pollutants in cities. Short-term exposure to this exhaust gas component in high concentrations leads to irritation of the respiratory tract and may cause inflammation, which is most often manifested by coughing, mucus production, and shortness of breath. It has also been shown that there is a relationship between NO_2 contained in the air and abnormal lung development and respiratory infections in children, as well as the impact on proper functioning in adulthood. Additionally, research is being conducted [49] on the adverse health effects of this compound, which ultimately lead to a reduction in life expectancy. However, it has not yet been confirmed whether these effects are caused by NO_2 itself or by other pollutants emitted at the same time during exhaust emissions [49].

Significant values of PM emission intensity (Figure 17) were observed primarily in the range of maximum rotational speed (1800 rpm, ... rpm) and high torque (1600 Nm, 2000 Nm⟩ and (2000 Nm, ... Nm). The reason for the increased emission of particulate matter there is a DPF filter which, in order to burn them, must have appropriate conditions, i.e., the exhaust gas temperature must be high enough to burn the remaining carbon. It is very difficult to achieve high exhaust gas temperatures in a low speed range, as a result of which a large proportion of the particulate matter is deposited on the filter until a higher temperature occurs, which the test object reached in the above-mentioned crankshaft speed and torque range.

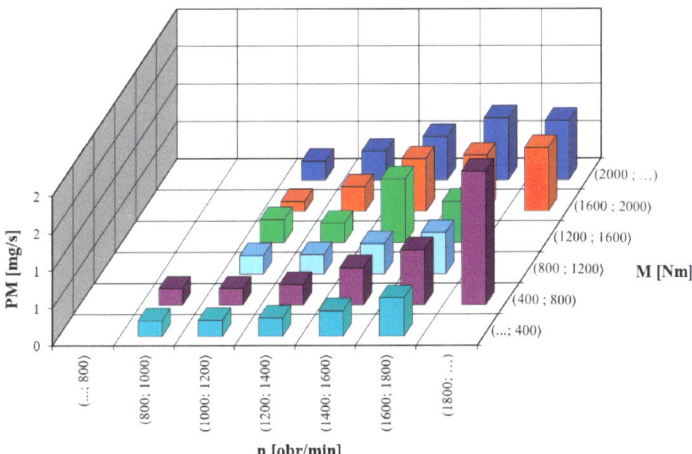

Figure 17. PM emission intensity in the torque and rotational speed ranges of the research object.

The obtained results on specific emissions of CO, HC, and NO_x during tests in real operating conditions can be related to the approval requirements for combustion engines specified in standards, in this case to the Stage III B standard. For comparison, the emission factor kj must be determined [50]. The emission factor is defined by the quotient:

$$k_j = e_{real,j}/e_{limit,j}, \quad (4)$$

where:

j is a toxic compound for which an emission factor has been determined,

$e_{real,j}$ are specific emissions determined from tests in real operating conditions (g/kWh), and

$e_{limit,j}$ are permissible specific emissions in accordance with standards (g/kWh).

The presented results show that the most problematic emissions for a railbus in real operating conditions are the emissions of HC and NO_x (Figure 18). The k_{HC} coefficient was 1.51, which indicates that the permissible value for this compound has been exceeded. The DOC reactor installed in the vehicle had a positive effect on CO oxidation ($k_{CO} < 1$), but at the cost of exceeding the permissible value of hydrocarbons. In the case of the k_{NOx} coefficient, its value was 1.18, which also indicates that the permissible value established in the standards was exceeded. The k_{PM} factor, also when the k_{CO} is significantly below the permissible value, can be regarded as a positive impact of the use of additional exhaust gas treatment systems, such as a particulate filter, for newer generations of vehicles (Stage IIIB standard).

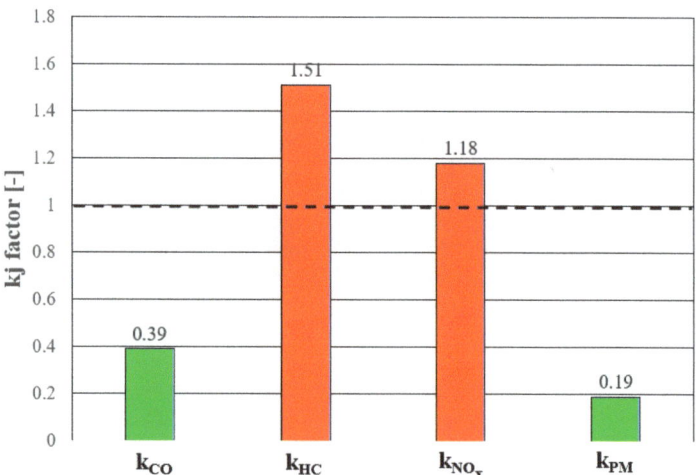

Figure 18. Specific emission factors k_j.

6. Conclusions

From our results, we can draw the following conclusions:

- The specific nature of the operation of rail vehicles, their large population, and the power of the drive units used in them play a significant role in the impact on the natural environment;
- The characteristics of the share of the research facility's operating time in actual operation showed that the largest share in the traffic of these vehicles is its standstill, the value of which was 30.00% compared with the value of 2.34% for the NRTC approval test. The difference is therefore 27.7%, so steps are necessary to modify the applicable laboratory test for vehicles of this category, with particular emphasis on the engine operating characteristics during typical passenger transport (numerous stops at railway stops);
- The analysis of the determined differences between the values of the operating time share for real conditions and the NRTC approval test showed that the largest values of individual intervals were 27.62% (idle speed) and 24.08% for the average crankshaft rotational speed and the largest interval of torque values. Significant discrepancies prove that the applicable approval tests are unrepresentative and that they need to be modified for the group of machines in question or the testing legislation in real operating conditions;
- The measurements of rail vehicles were carried out in real operating conditions, which is an original research methodology for this group of vehicles. Current approvals for off-road vehicles are performed only for engines in special laboratory conditions, which makes it impossible to obtain reliable information on the impact of various external factors (Figure 19). This is confirmed by the results obtained. A comparative analysis of road emission factors determined on the basis of tests in real operating conditions with the permissible values of homologation tests showed significant exceedances of HC and NO_x limits;

- The presented scheme for carrying out measurements in real operating conditions should be continued in order to obtain the actual emissions of a vehicle that can operate in various conditions. Thanks to field tests, it is possible to measure and relate the obtained results of emissions of harmful compounds to atmospheric conditions, vehicle traffic parameters, the number of passengers, etc.

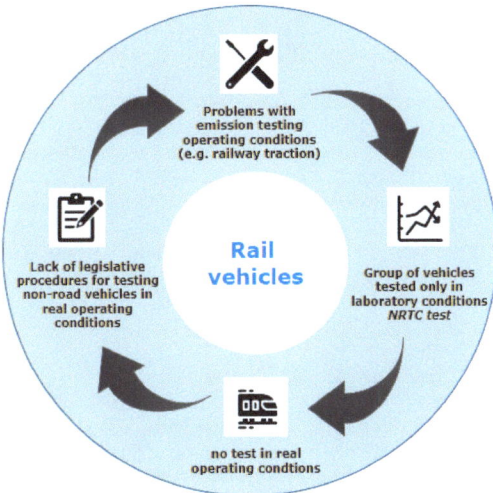

Figure 19. Article summary and cause-and-effect diagram of homologation regulations for rail vehicles.

Author Contributions: Conceptualization, J.S., N.S. and M.K.; methodology, J.S. and N.S.; formal analysis, J.S.; investigation, J.S.; data curation, J.S.; writing—original draft preparation, J.S. and N.S.; writing—review and editing, N.S.; visualization, J.S. and N.S.; supervision, A.Z. and N.S.; project administration A.Z.; funding acquisition, A.Z. All authors have read and agreed to the published version of the manuscript.

Funding: The study presented in this article was performed within statutory research (Contract No. 0415/SBAD/0351).

Data Availability Statement: The original contributions presented in the study are included in the article. Further inquiries can be directed to the corresponding authors.

Conflicts of Interest: The authors declare no conflicts of interest.

References

1. Andrews, G.E.; Li, H.; Wylie, J.A.; Zhu, G.; Bell, M.; Tate, J. *Influence of Ambient Temperature on Cold-Start Emissions for a Euro 1 SI Car Using In-Vehicle Emissions Measurement in an Urban Traffic Jam Test Cycle*; Society of Automotive Engineers: Warrendale, PA, USA, 2005.
2. Li, H.; Andrews, G.E.; Zhu, G.; Daham, B.; Bell, M.; Tate, J.; Ropkins, K. Impact of ambient temperatures on exhaust thermal characteristics during cold start for real world si car urban driving tests (No. 2005-01-3896). *SAE Tech. Pap.* **2005**, *1*, 3896.
3. Samuel, S.; Morrey, D.; Fowkes, M.; Taylor, D.H.C.; Austin, L.; Felstead, T.; Latham, S. The most significant vehicle operating parameter for real-world emission levels. *SAE Trans.* **2004**, *113*, 407–425.
4. Andr'e, M. The ARTEMIS European driving cycles for measuring car pollutant emissions. *Sci. Total Environ.* **2004**, *334–335*, 73–84. [CrossRef] [PubMed]
5. Stahel, W.; Pritscher, L.; De Haan, P. *Neues EMPA-Standardmessprogramm, Sonderuntersuchung der "Real-World"-Fahrzyklen*; BUWAL: Bern, Switzerland, 2000.
6. World Health Organization. *World Health Statistics 2023: Monitoring Health for the SDGs, Sustainable Development Goals*; World Health Organization: Geneva, Switzerland, 2023.
7. European Environment Agency. *Greenhouse Gas Emissions from Transport in Europe, 2022*; European Environment Agency: Copenhagen, Denmark, 2023.

8. Latha, H.; Prakash, K.; Veerangouda, M.; Maski, D.; Ramappa, K. A Review on Scr System for Nox Reduction in Diesel Engine. *Int. J. Curr. Microbiol. Appl. Sci.* **2019**, *8*, 1553–1559. [CrossRef]
9. Bajerlein, M.; Karpiuk, W.; Smolec, R. Use of Gas Desorption Effect in Injection Systems of Diesel Engines. *Energies* **2021**, *14*, 244. [CrossRef]
10. Bor, M.; Borowczyk, T.; Idzior, M.; Karpiuk, W.; Smolec, R. Analysis of Hypocycloid Drive Application in a High-Pressure Fuel Pump. *EDP Sci.* **2017**, *118*, 00020. [CrossRef]
11. Lijewski, P.; Merkisz, J.; Fuc, P.; Kozak, M.; Rymaniak, L. Air Pollution by the Exhaust Emissions from Construction Machinery under Actual Operating Conditions. *Appl. Mech. Mater.* **2013**, *390*, 313–319. [CrossRef]
12. Merkisz, J.; Idzior, M.; Lijewski, P.; Fuc, P.; Karpiuk, W. The Analysis of the Quality of Fuel Spraying in Relation to Selected Rapeseed Oil Fuels for the Common Rail System. In Proceedings of the Ninth Asia-Pacific International Symposium on Combustion and Energy Utilization, Beijing, China, 2–6 November 2008; pp. 352–356.
13. Merkisz, J.; Mizera, J.; Bajerlein, M.; Rymaniak, L.; Maj, P. The Influence of Laser Treatment and the Application of Reduced Pressure Force Piston Rings on the Engine Exhaust Emissions under the Conditions of Engine Lubrication with Different Engine Oils. *Appl. Mech. Mater.* **2014**, *518*, 102–107. [CrossRef]
14. Rymaniak, L.; Daszkiewicz, P.; Merkisz, J.; Bolzhelarskyi, Y.V. Method of Determining the Locomotive Engine Specific Fuel Consumption Based on Its Operating Conditions. *AIP Publ.* **2019**, *2078*, 1.
15. Sawczuk, W.; Jüngst, M.; Ulbrich, D.; Kowalczyk, J. Modeling the Depth of Surface Cracks in Brake Disc. *Materials* **2021**, *14*, 3890. [CrossRef]
16. Sawczuk, W.; Merkisz-Guranowska, A.; Rilo Cañás, A.-M. Assessment of Disc Brake Vibration in Rail Vehicle Operation on the Basis of Brake Stand. *Eksploatacja i Niezawodność* **2021**, *23*, 2. [CrossRef]
17. Sawczuk, W.; Merkisz-Guranowska, A.; Ulbrich, D.; Kowalczyk, J.; Cañás, A.-M.R. Investigation and Modelling of the Weight Wear of Friction Pads of a Railway Disc Brake. *Materials* **2022**, *15*, 6312. [CrossRef]
18. Tomaszewski, S.; Tomaszewski, F.; Stawecki, W.; Urbański, P.; Far, M.; Bolzhelarskyi, Y. Optimization of the Design of Relay Valves for Rail Vehicles Braking Systems in the Context of Train Traffic Safety. *EDP Sci.* **2019**, *294*, 03013. [CrossRef]
19. European Union. EC Directive 97/68/EC of the European Parliament and of the Council. *Off. J. Eur. Union* **1997**, *L 353*, 2–286.
20. Jaworski, A.; Mateichyk, V.; Kuszewski, H.; Mądziel, M.; Woś, P.; Babiarz, B.; Śmieszek, M.; Porada, S. Towards Cleaner Cities: An Analysis of the Impact of Bus Fleet Decomposition on PM and NO_X Emissions Reduction in Sustainable Public Transport. *Energies* **2023**, *16*, 6956. [CrossRef]
21. Balogh, R.; Ionel, I. Experimental Approach Concerning the Selective Catalytic Reduction of NOx for Diesel Engines of Romanian Railway Transport. *Optoelectron. Adv. Mater. Rapid Commun.* **2011**, *5*, 1123–1129.
22. Picus, C.M.; Mihai, I.; Suciu, C. Experimental Investigations upon Ultrasound Influence on Calefaction of AdBlue in Selective Catalytic Reduction Systems (SCR). *Micromachines* **2023**, *14*, 1488. [CrossRef]
23. www.utk.gov.pl. Available online: https://dane.utk.gov.pl/sts/przewozy-pasazerskie/dane-eksploatacyjne/21369,Przewozy-pasazerskie.html (accessed on 18 June 2024).
24. www.utk.gov.pl. Available online: https://dane.utk.gov.pl/sts/przewozy-pasazerskie/tabor-pasazerski/16737,Tabor-kolejowy-przewoznikow-pasazerskich.html (accessed on 16 June 2024).
25. International Union of Railways UIC 624 Railway Code. *Research on Exhaust Gas Emissions of Traction Combustion Engines*; International Union of Railways: Paris, France, 2006.
26. International Union of Railways UIC 623-2 Railway Code. *Homologation Tests of Combustion Engines of Propulsion Vehicles*; International Union of Railways: Paris, France, 2005.
27. European Union. EC Directive 2004/26/EC of the European Parliament and of the Council of 21 April 2004 Amending Directive 97/68. *Off. J. Eur. Union* **2004**, *L 225*, 3–107.
28. European Union. EP Committee Regulation 2016/1628 of the European Parliament and of the EU Council of 14 September 2016, Which Repeals Directive 97/68/EC, Introducing the Stage V Emission Standard 2021. Available online: https://eur-lex.europa.eu/eli/reg/2016/1628/oj (accessed on 16 June 2024).
29. *DieselNet Emission Standards, European Union: Nonroad Engines*; European Union: Brussels, Belgium. Available online: https://dieselnet.com/standards/#eu (accessed on 16 June 2024).
30. Merkisz, J.; Siedlecki, M. Specific Emissions Analysis for a Combustion Engine in Dynamometer Operation in Relation to the Thermal State of the Exhaust Gas Aftertreatment Systems in a Modified NRSC Test. *EDP Sci.* **2017**, *118*, 00027. [CrossRef]
31. Siedlecki, M.; Lijewski, P.; Weymann, S. Analysis of Tractor Particulate Emissions in a Modified NRSC Test after Implementing a Particulate Filter in the Exhaust System. *EDP Sci.* **2017**, *118*, 00028. [CrossRef]
32. Siedlecki, M.; Szymlet, N.; Fuć, P.; Kurc, B. Analysis of the Possibilities of Reduction of Exhaust Emissions from a Farm Tractor by Retrofitting Exhaust Aftertreatment. *Energies* **2022**, *15*, 7963. [CrossRef]
33. European Union. COMMISSION DIRECTIVE 2010/26/EU of 31 March 2010 Amending Directive 97/68/EC of the European Parliament and of the Council on the Approximation of the Laws of the Member States Relating to Measures against the Emission of Gaseous and Particulate Pollutants from Internal Combustion Engines to be Installed in Non-Road Mobile Machinery; European Union: Brussels, Belgium, 2010.

34. European Union. *COMMISSION DIRECTIVE 2012/46/EU of 6 December 2012 Amending Directive 97/68/EC of the European Parliament and of the Council on the Approximation of the Laws of the Member States Relating to Measures against the Emission of Gaseous and Particulate Pollutants from Internal Combustion Engines to be Installed in Non-Road Mobile Machinery*; European Union: Brussels, Belgium, 2012.
35. Bajerlein, M.; Rymaniak, L. The Reduction of Fuel Consumption on the Example of Ecological Hybrid Buses. *Appl. Mech. Mater.* **2014**, *518*, 96–101. [CrossRef]
36. Lijewski, P.; Merkisz, J.; Ziółkowski, A.; Rymaniak, Ł. Analiza Emisji Gazów Wylotowych Wyznaczonej w Teście NTE z Ciężkiego Pojazdu Samochodowego Na Podstawie Badań Przeprowadzonych w Rzeczywistych Warunkach Jego Eksploatacji. *Combust. Engines* **2013**, *52*, 696–700.
37. Merkisz, J.; Pielecha, J. *Nanoparticle Emissions from Combustion Engines*; Springer: New York, NY, USA, 2015; ISBN 3-319-15928-3.
38. Merkisz, J.; Pielecha, J.; Pielecha, I. Road Test Emissions Using On-Board Measuring Method for Light Duty Diesel Vehicles. *Jordan J. Mech. Ind. Eng.* **2011**, *5*, 89–96.
39. Rymaniak, L. Comparison of the Combustion Engine Operating Parameters and the Ecological Indicators of an Urban Bus in Dynamic Type Approval Tests and in Actual Operating Conditions. *EDP Sci.* **2017**, *118*, 00009. [CrossRef]
40. Szymlet, N.; Lijewski, P.; Sokolnicka, B.; Siedlecki, M.; Domowicz, A. Analysis of Research Method, Results and Regulations Regarding the Exhaust Emissions from Two-Wheeled Vehicles under Actual Operating Conditions. *J. Ecol. Eng.* **2020**, *21*, 1. [CrossRef]
41. Rosero, F.; Fonseca, N.; López, J.-M.; Casanova, J. Real-World Fuel Efficiency and Emissions from an Urban Diesel Bus Engine under Transient Operating Conditions. *Appl. Energy* **2020**, *261*, 114442. [CrossRef]
42. Rymaniak, L.; Ziolkowski, A.; Gallas, D. Particle Number and Particulate Mass Emissions of Heavy Duty Vehicles in Real Operating Conditions. *EDP Sci.* **2017**, *118*, 00025. [CrossRef]
43. Bajerlein, M.; Rymaniak, L.; Swiatek, P.; Ziolkowski, A.; Daszkiewicz, P.; Dobrzynski, M. Modification of a Hybrid City Bus Powertrain in the Aspect of Lower Fuel Consumption and Exhaust Emissions. *Appl. Mech. Mater.* **2014**, *518*, 108–113. [CrossRef]
44. Markowski, J.; Pielecha, J.; Jasiński, R.; Kniaziewicz, T.; Wirkowski, P. Development of Alternative Ship Propulsion in Terms of Exhaust Emissions. *EDP Sci.* **2016**, *10*, 00140. [CrossRef]
45. Mądziel, M. Instantaneous CO_2 Emission Modelling for a Euro 6 Start-Stop Vehicle Based on Portable Emission Measurement System Data and Artificial Intelligence Methods. *Environ. Sci. Pollut. Res.* **2024**, *31*, 6944–6959. [CrossRef]
46. Mądziel, M. Modelling CO_2 Emissions from Vehicles Fuelled with Compressed Natural Gas Based on On-Road and Chassis Dynamometer Tests. *Energies* **2024**, *17*, 1850. [CrossRef]
47. Rymaniak, Ł.; Kamińska, M.; Szymlet, N.; Grzeszczyk, R. Analysis of Harmful Exhaust Gas Concentrations in Cloud behind a Vehicle with a Spark Ignition Engine. *Energies* **2021**, *14*, 1769. [CrossRef]
48. GlobalMRV Axion R/S+ PEMS. Available online: https://www.globalmrv.com/products/axion-rs-2-pems/ (accessed on 12 July 2024).
49. www.gov.uk. Available online: https://www.gov.uk/government/statistics/air-quality-statistics/ntrogen-dioxide (accessed on 12 July 2024).
50. Rymaniak, Ł. Analiza Wpływu Rodzaju Układu Napędowego i Parametrów Ruchu Autobusów Miejskich Na Ekologiczne Wskaźniki Pracy. Ph.D. Thesis, Politechnika Poznańska, Poznań, Poland, 2016; 137p.

Disclaimer/Publisher's Note: The statements, opinions and data contained in all publications are solely those of the individual author(s) and contributor(s) and not of MDPI and/or the editor(s). MDPI and/or the editor(s) disclaim responsibility for any injury to people or property resulting from any ideas, methods, instructions or products referred to in the content.

Article

Reducing CO$_2$ Emissions through the Strategic Optimization of a Bulk Carrier Fleet for Loading and Transporting Polymetallic Nodules from the Clarion-Clipperton Zone

Tomasz Cepowski *[ID] and Paweł Kacprzak [ID]

Faculty of Navigation, Maritime University of Szczecin, 1-2 Wały Chrobrego St., 70-500 Szczecin, Poland
* Correspondence: t.cepowski@pm.szczecin.pl

Abstract: As global maritime cargo transportation intensifies, managing CO$_2$ emissions from ships becomes increasingly crucial. This article explores optimizing bulk carrier fleets for transporting polymetallic nodules (PMNs) from the Clarion-Clipperton Zone (CCZ) to reduce CO$_2$ emissions. Our analysis shows that larger bulk carriers, despite greater drifting forces from environmental conditions, emit less CO$_2$ over the entire transport mission, including loading and transit. Deploying large ships in global maritime trade could significantly reduce CO$_2$ emissions. This study also introduces a novel artificial neural network (ANN) model to estimate drifting forces during loading operations and proposes a new method for estimating CO$_2$ emissions, considering environmental conditions and ship seakeeping properties. These findings highlight the importance of fleet size optimization and effective operational planning in achieving environmental sustainability in maritime transport.

Keywords: maritime carbon management; polymetallic nodules transportation; fleet optimization; artificial neural networks; CO$_2$ emissions

Citation: Cepowski, T.; Kacprzak, P. Reducing CO$_2$ Emissions through the Strategic Optimization of a Bulk Carrier Fleet for Loading and Transporting Polymetallic Nodules from the Clarion-Clipperton Zone. *Energies* **2024**, *17*, 3383. https://doi.org/10.3390/en17143383

Academic Editors: Kazimierz Lejda, Artur Jaworski and Maksymilian Mądziel

Received: 13 June 2024
Revised: 5 July 2024
Accepted: 8 July 2024
Published: 10 July 2024

Copyright: © 2024 by the authors. Licensee MDPI, Basel, Switzerland. This article is an open access article distributed under the terms and conditions of the Creative Commons Attribution (CC BY) license (https:// creativecommons.org/licenses/by/ 4.0/).

1. Introduction

The 2021 United Nations Conference on Trade and Development report [1] highlights maritime transport as the cornerstone of international trade and the global economic framework. The report indicates that over 80% of global trade volume, particularly in goods, occurs via sea routes. This significant reliance on maritime transportation, in the context of expanding international trade and coupled with numerous pollution control measures targeting other sources, has led to an increasingly substantial contribution of ship emissions to total anthropogenic emissions [2]. Nunes et al. [3] observed that the maritime sector is likely to continue its significant growth alongside global trade. However, the full extent of its impact on the environment, societal aspects, and human health remains unclear. Moreno-Gutiérrez et al. [4] identified the combustion of fuel in ship engines as the primary source of ship-emitted pollutants. Hoang et al. [5] argue that the pursuit of intelligent strategies through the utilization of renewable energy sources, clean fuels, smart grids, and measures for efficient energy use is beneficial for achieving the main IMO objectives, particularly reductions in future CO$_2$ emissions. Fan et al. [6], in their analysis of ship energy management, justified that managing energy savings and emission reductions is a systemic issue that should be comprehensively considered from multiple perspectives, such as ship design, operational management, and performance assessment. The technical modernization of the ship can also contribute to reducing CO$_2$ emissions. Barone et al. [7] conducted simulations on a ship equipped with five electric chillers, two auxiliary boilers, two reverse osmosis devices, and two multi-stage flash desalination devices. The simulation results show remarkable energy savings obtained through the proposed optimization approach, corresponding to a fuel consumption reduction of 2.5 kt/y (−1.6 M$/y) and to avoided CO$_2$ emissions of 8.0 kt/y. The application of solar and wind energy devices affects the reduction in CO$_2$ emissions. Nyanya et al. [8], in their analysis using a bulk carrier as an example, demonstrated that considering the optimal sail angle and optimizing the available deck area through the installation of solar and wind systems allowed for the maximum utilization of renewable energy, which contributed to a 36% reduction in carbon dioxide emissions when compared to the same ship without innovative technologies. The conclusions showed that if the ship's speed were reduced to 56% of its original speed, the ship could sail exclusively on renewable energy collected on board. Ytreberg et al. [9] reported that maritime shipping contributes to several environmental challenges, including deteriorations in air quality and human well-being, leading to numerous quantifiable and unquantifiable harms. Their findings demonstrate that transporting goods solely in the Baltic Sea incurs costs due to air quality deterioration, which exceed those associated with climate change impacts.

It is worth noting that oceans and seas serve not only as crucial transport routes for global trade, but also as valuable sources of natural raw materials upon which many industries depend. Recent studies suggest that the deep-sea floor at depths of 4000–6000 m represents a significant source of polymetallic nodules (PMNs), comprising minerals such as manganese (Mn), copper (Cu), cobalt (Co), nickel (Ni), and some rare earth elements [10]. These nodules are considered a promising source of metallic raw materials, with diverse mining systems currently being developed for their extraction [11]. Cunningham [12] concluded that the extraction of nodules from the ocean floor could provide a more stable and decentralized alternative to the current cobalt supplies. The increased cobalt supply from nodule extraction may offset the harmful impacts on the marine environment. Studies by Gollner et al. and Bonifácio et al. [13,14] noted that the extraction process could have adverse environmental effects, impacting biodiversity in the area. Nodule extraction disrupts the composition of seabed sediments, which may lead to long-term negative consequences considering the slow pace of natural habitat recovery. The CC zone features 300 morphotypes, including numerous groups of invertebrates such as corals, crustaceans, and fish [15]. Stratmann [16] argues that conducting extraction operations could lead to an 18% loss of organisms (taxa). B. Li et al. [17] suggests the application of ore separation technology on extraction ships and transporting the separated sediments to land for comprehensive utilization on a separate ship. However, Katona et al. [18] argues that the long-term environmental impact of nodule extraction on marine ecosystems might be lesser when compared to terrestrial mining.

According to recent findings [19], it is anticipated that nodules extracted from the ocean floor will be directly loaded onto high-capacity extraction vessels and transferred to transport ships that will deliver the cargo to destination ports.

During the transfer of nodules from the extraction unit to the transport ship on the ocean surface, the marine environment, i.e., wind and waves, interacts with the transport ship. This can lead to excessive ship swaying, bending, or a loss of stability, as indicated by Kacprzak [20]. In such events, it becomes necessary to suspend the transfer, potentially disrupting the extraction of the nodules from the ocean floor and the operation of the extraction system. Therefore, the transport ship should have the appropriate maritime characteristics to ensure high-efficiency loading and the ability to operate in challenging weather conditions for as long as possible over the mining field, thus minimizing the risk of downtime. An inappropriate selection of transport ships might disrupt the entire extraction system.

According to Kacprzak [21], the transport of PMNs to the destination port can be achieved by both small and large bulk carriers. Research by this author shows that loading onto smaller bulk carriers would lead to shorter loading times and good longitudinal strength, which are significant advantages. However, the disadvantage of these units is their smaller transport capacity, increased ship movement, and poorer stability. On the other hand, large bulk carriers have the advantages of high transport capacity, minimal ship movement, and good stability, with the downside being longer loading times and reduced longitudinal strength.

In addition to the technical limitations with regard to safe loading, a key aspect is the environmental pollution caused by the transporting ship during loading and transport operations. The total CO_2 emissions into the marine environment by the transporting ship depends on the amount of fuel being burned [22]. This amount mainly depends on the power of the dynamic positioning (DP) system which stabilizes the bulk carrier at the loading site, as well as environmental conditions. The required power of the DP system depends on many factors, including the technical parameters of the ship and environmental conditions. Among the technical properties, the size of the ship is most important; according to Jurdziński [23], larger bulk carriers are characterized by a larger windage area, exposed to wind action, and a larger wetted surface, exposed to wave action, which may result in greater drift force. An increase in drift force leads to an increase in the energy needed to maintain the ship position, and thus an increase in pollutant emissions.

In turn, the amount of fuel burned during transport depends on the technical–operational parameters and environmental conditions. The most important technical–operational parameters include ship speed, capacity, the hydro-mechanical properties of the hull, engine efficiency, propulsion system and propeller, type and quality of fuel, draught, trim and hull, and the propeller condition from the fouling process, as mentioned in [24–26]. Environmental conditions affecting fuel consumption include waves, wind, ocean currents, air, and water temperature, as noted in [27]. The reloading and transport of PMNs can be carried out by a fleet of small, medium, and large bulk carriers. Research by Kacprzak [21] has shown that transporting a specified amount of PMNs from the Clarion-Clipperton Zone (CCZ) to the destination requires the consideration of the environmental impact on the technical properties of ships, including stability, motions, and longitudinal strength. Therefore, the bulk carrier fleet should consist of appropriately selected units of various sizes. The key concern is what impact the use of the entire fleet to transport the specified amount of PMNs to different destinations, on routes of varying distances, will have on emissions to the marine environment. Will the use of a fleet of small, medium, or large bulk carriers ensure lower exhaust emissions? How does the distance to the destination port from the CCZ affect this choice? Therefore, the main goal of this study is to answer these questions, i.e., to determine which fleet of bulk carriers (small, medium, large) will be most efficient in terms of environmental CO_2 emissions during the loading and transport of PMNs from the CCZ to various ports around the world.

Studies [28,29] have shown that a large ship can transport a larger amount of cargo at once, burning relatively less fuel during transport to the destination, which translates into less environmental pollution. Therefore, to assess the energy efficiency of the ship, many indicators have been developed, such as

- the energy efficiency operational indicator (EEOI), which is calculated as the ratio of CO_2 emissions to the amount of cargo carried and the distance traveled [30],
- the energy efficiency design index (EEDI), which is expressed as the amount of carbon dioxide CO_2 emitted into the environment based on the estimated fuel consumption in relation to ship capacity and speed [31],
- the annual efficiency ratio (AER), which was created to measure the energy efficiency of a ship based on its annual activity, taking into account the total fuel consumption and distance traveled [32],
- the deadweight tonnage (DWT) efficiency, which determines the efficiency of a ship by the ratio of its cargo capacity (carrying capacity) to fuel consumption,
- the carbon intensity indicator (CII), which measures CO_2 emissions per unit of transport and distance [33].

Chen et al. [34] argue that the ship's EEOI calculated at full load is naturally lower than the equivalent at partial load. The energy efficiency of the fleet shows a slight increase (at least by 1%) due to the implementation of the Ship Energy Efficiency Management Plan. Lee [35] evaluated the possibility of achieving the 40% CO_2 emission reduction target by 2030 through the implementation of the EEDI and the EEXI for three types of ships as follows: bulk carriers, container ships, and tankers. The results show that achieving the 40%

CO_2 emission reduction target by 2030 is possible for ships regulated by the EEDI, while the calculated CO_2 emissions for ships regulated by the EEXI range from 17.4% to 24.6%. This difference arises from the fact that these calculations are based on the maximum speed limit of the EPL's engine output, which exceeds the actual operational speed.

These indicators can be used to assess CO_2 emissions during the transport of nodules. However, they do not consider the energy supplied to the system stabilizing ship movement during reloading operations. During this operation, the transport ship is stationary or moves at a low speed along with the extraction ship, and the DP system stabilizes any ship movement caused by the drift force resulting from wave and wind action. There is no appropriate indicator which assesses the energy efficiency of this process. Therefore, the second goal of the research is to develop a method for estimating CO_2 emissions that considers the impact of environmental conditions on ship loading operations at sea with a cargo of nodules.

2. Materials and Methods

2.1. Data

The subject of the study was a fleet of bulk carriers comprising small, medium, and large ships. Representatives of the three groups of ships were selected for the study. The parameters of these ships are presented in Table 1.

Table 1. Technical parameters of the studied bulk carriers, where L—length between perpendiculars, B—breadth, T—draught, CB—block coefficient, D—displacement, DWT—deadweight, and A_W—lateral plane of the vessel above the plane of draught.

Main Particulars	Ship Size		
	A	B	C
L [m]	103.9	185	217
B [m]	18.2	24.4	32.26
T [m]	7.057	11.01	14.02
L/B [-]	5.71	7.58	6.73
B/T [-]	2.58	2.22	2.30
L/T [-]	14.72	16.80	15.48
CB [-]	0.80	0.82	0.85
D [t]	11,036	41,900	85,700
DWT [t]	7600	32,000	73,600
A_W [m^2]	746	1711	1977

To compare the efficiency of the different ship types, assumptions regarding the mining system and the locations of the destination ports to which the mined PMNs could be transported were made. It was assumed that the loading of PMNs would take place in the CCZ and would be carried out by the mining ship Hidden Gem. The technical characteristics of this ship are presented in Table 2.

Table 2. Parameters of the mining ship Hidden Gem.

L [m]	228
B [m]	42
T [m]	12
CB [-]	0.82
D [t]	96,504
DWT [t]	61,042

Assumptions regarding the mining system were adopted in accordance with [36] and are presented below:
- the number of planned days for transshipment operations—250 days,
- the assumed annual transport of PMNs—2,000,000 tons,
- the mining capacity of the mining ship—333 tons/h,
- the cargo carrying capacity of the mining ship Hidden Gem—20,000–25,000 t.

It was assumed that the transportation of PMNs from the CCZ would be carried out to the destination ports along the routes described in Figure 1.

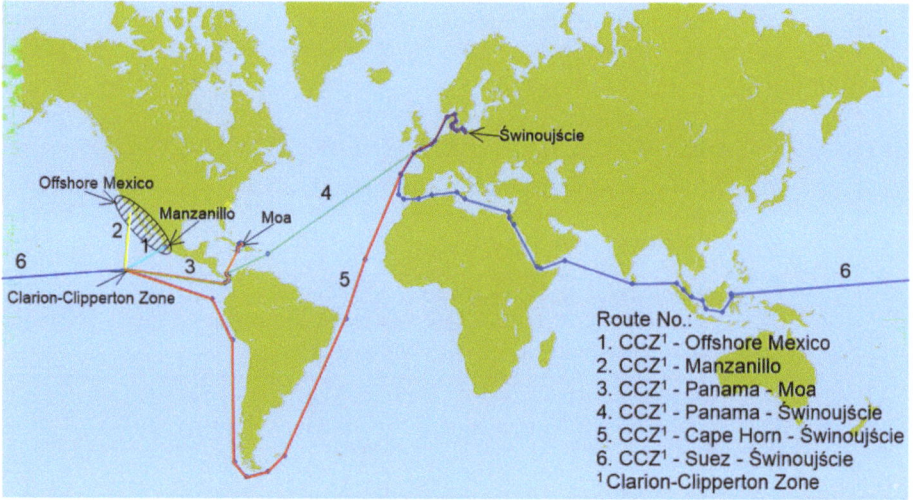

Figure 1. Location of destination ports.

The distribution of wave parameters in the CCZ was developed based on data from [37] and is presented in Table 3. These distributions relate to the occurrence frequency of these parameters throughout the year.

Table 3. Annual frequency of wave parameter occurrence p_{wave} in the CCZ. Data adapted from [37], where Hs—significant wave height and T_z—average zero-up-crossing wave period.

Hs (m)	T_z (s)										
	<4	4 to 5	5 to 6	6 to 7	7 to 8	8 to 9	9 to 10	10 to 11	11 to 12	12 to 13	>13
7 to 8	0	0	0	0	0	0.0001	0.0001	0.0001	0.0001	0.0001	0
6 to 7	0	0	0	0	0	0.0002	0.0005	0.0007	0.0006	0.0004	0.0002	0.0001
5 to 6	0	0	0	0	0.0003	0.0014	0.0027	0.0031	0.0023	0.0013	0.0006	0.0002
4 to 5	0	0	0	0.0003	0.0025	0.0081	0.0129	0.012	0.0077	0.0037	0.0014	0.0005
3 to 4	0	0.0001	0.0023	0.015	0.037	0.0455	0.0339	0.0176	0.007	0.0023	0.0006
2 to 3	0	0.0008	0.0143	0.0623	0.1054	0.0928	0.0514	0.0204	0.0064	0.0017	0.0004
1 to 2	0.0001	0.0057	0.0474	0.11	0.1092	0.0605	0.0223	0.0062	0.0014	0.0003	0
0 to 1	0.0005	0.0064	0.0191	0.0185	0.0082	0.0021	0.0004	0.0001	0	0	0

The distribution of wind parameters in the CCZ was adopted in accordance with data from [38] and is presented in Table 4.

To estimate the drift force (F_{wy}), results from ship model tests, the dimensions of which are presented in Table 5, were utilized. These results are based on the studies [39–44] and include the nondimensional drift force coefficient (C_{wy}), calculated using the following equation:

$$C_{wy} = \frac{F_{wy}}{\rho \cdot g \cdot \varsigma_a^2 \cdot L} \quad (1)$$

where

C_{wy}—nondimensional drift force coefficient,
F_{wy}—mean drift force in a regular wave,
ρ—seawater density,
g—gravity acceleration,
ς_a—amplitude of a regular wave,
L—length of the ship.

All data from these model tests are publicly available, and the references for each dataset are provided in Table 5. Figure 2 shows the values of the drift force coefficient (C_{wy}) in regular waves for these ship models at zero speed as a function of the ratio of the wavelength (λ) to the length of the ship (L). Based on the nondimensional coefficient C_{wy}, calculated using Equation (1) for the ship model scale, it is possible to determine the drift force F_{wy} for a full-scale ship of length L.

Table 4. Distribution of wind speed. Data adapted from [38], where Hs—the significant wave height, v_W—the wind velocity, and p_{wind}—the annual frequency of wind occurrence.

Hs (m)	v_W (m/s)	p_{wind} [-]
1	7.3	0.240
2	10.0	0.270
3	12.1	0.170
4	15.4	0.115
5	19.3	0.091
6	22.1	0.037
7	24.9	0.037
8	27.0	0.020

Table 5. Design characteristics of ship model subjected to model test on mean drift force, where L—length between perpendiculars, B—breadth, T—draught, and CB—block coefficient.

Ship	L/B (-)	B/T (-)	L/T (-)	CB (-)	References
VLCCb	5.52	3.01	16.62	0.81	[43,44]
KVLCC2	5.52	2.79	15.4	0.81	[40,41]
Bulk carrier	5.7	2.7	15.39	0.83	[39]
Scb84	5.52	2.79	15.4	0.84	[42]

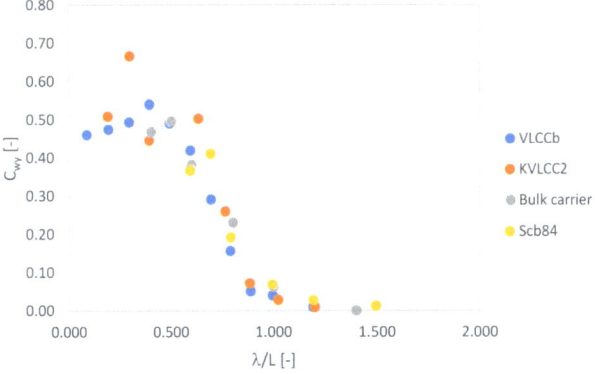

Figure 2. Drift force coefficients of four ships in regular beam waves at zero speed.

2.2. General Research Framework

To achieve the research objective, the estimation of CO_2 emissions during the loading and transportation of PMNs to various long-range destinations was identified as a key step. This estimation was conducted for each type of bulk carrier fleet, designated as A, B, and C, as described in Table 1, depending on the location of the destination port to which the PMNs were to be delivered.

Accordingly, the total CO_2 emission is made up of emissions during loading and transportation as follows:

$$CO_2 = CO_{2load} + CO_{2transp}, \qquad (2)$$

where

CO_2—total CO_2 emissions by bulk carrier fleets A/B/C,

CO_{2load}—CO_2 emissions by bulk carrier fleets A/B/C during loading,

$CO_{2transp}$—CO_2 emissions during transportation to the destination port using the bulk carrier fleets A/B/C.

In this study, it was assumed that loading from the mining ship could be executed concurrently with the extraction of PMNs and continuously without breaks for nodule reloading.

2.2.1. Estimation of CO_2 Emissions during Polymetallic Nodules Loading in the Clarion-Clipperton Zone

Figure 3 illustrates a general block diagram for the estimation of CO_2 emissions while nodule loading in the CCZ. The calculation of CO_2 emissions during the loading process is based on the fuel consumption of the transporting bulk carrier while it loads PMNs from the extraction ship as follows:

$$CO_2 = EF \cdot FC \; [t] \qquad (3)$$

where

FC—fuel consumption,

EF—emission factor, whose value for a marine engine can be assumed as approximately 3.206 kg CO_2/kg for marine diesel oil (MDO) or approximately 3.114 kg CO_2/kg for heavy fuel oil (HFO).

The fuel consumption during this operation stems from the necessity to generate thrust via the propeller of the DP system to balance out environmental forces, thereby maintaining the ship position during loading under the specified environmental conditions. This fuel consumption (FC) was calculated based on the general relation as follows:

$$FC = P \cdot t \cdot FOC \qquad (4)$$

where

P—power supplied to the DP system [kW],

t—time under specified environmental conditions [s],

FOC—fuel consumption rate.

Equation (4) finds extensive application in the analysis and estimation of fuel consumption, both under operational conditions and in the design of ship propulsion systems. A crucial component of this Equation is the fuel consumption rate (FOC), whose values can be derived from the nominal fuel consumption of the marine engine or from real-time fuel flow measurements under operational conditions. The typical FOC values range from 0.170 to 0.185 kg/kWh for marine diesel oil (MDO) and marine gas oil (MGO), and from 0.195 to 0.230 kg/kWh for heavy fuel oil (HFO), depending on the engine type and the operational conditions.

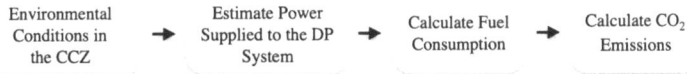

Figure 3. General research scheme for the calculation of CO_2 emissions during loading.

The required power supplied to the propellers of the DP system can be calculated based on the following formula [45]:

$$P = Tp/Cp \tag{5}$$

where

P—propeller power [kW],

Tp—thrust force of the DP system propeller required to overcome environmental forces [kN],

Cp—power-to-thrust efficiency ratio for the given propeller; for tunnel thrusters, Cp = 0.12–0.16.

The calculation of propeller thrust force generally considers the wind drift force, wave drift force, and current drift force. Given that the CCZ, located in the Pacific between Hawaii and Mexico, is known for its slow-moving ocean currents with a maximum speed of approximately 0.3 m/s [37] and given the negligible speed of the ship, the influence of ocean currents on the ship was omitted in this study. Therefore, the thrust force (Tp) was calculated as follows:

$$Tp = F_w + F_{wy} \tag{6}$$

where

F_w—force from wind pressure,

F_{wy}—force from wave drift.

The magnitude of forces from wind and waves depends on their direction relative to the ship. Given the lack of detailed information about wave and wind directions in the CCZ, this study assumes that waves and wind impact the ship unfavorably, i.e., perpendicular to the beam. The force from wind pressure is calculated as follows:

$$F_w = \frac{1}{2} \cdot v_w^2 \cdot \rho \cdot A_w \tag{7}$$

where

v_w—wind speed according to Table 4,

ρ—air density,

A_w—the lateral plane of the vessel above the plane of draught according to Table 1.

The wave drift force was calculated based on the following relation:

$$F_{wy} = L \cdot \rho \cdot g \cdot \int S_{\zeta\zeta}(\omega) \cdot C_{wy}(\omega) d\omega \tag{8}$$

where

F_{wy}—mean wave drift force from irregular waves [kN],

L—ship length [m],

ρ—seawater density,

g—acceleration due to gravity,

$S_{\zeta\zeta}$—wave spectral function,

C_{wy}—mean nondimensional sway force in regular wave.

The wave spectrum function was implemented according to the ITTC guidelines [46] as follows:

$$S_{\zeta\zeta}(\omega) = A\omega^{-5} \exp\left(-B\omega^{-4}\right) \tag{9}$$

where

A, B—coefficients calculated as follows:

$$A = \frac{173 \, H_S^2}{T_z^4} \qquad (10)$$

$$B = \frac{691}{T_z^4} \qquad (11)$$

H_S—significant wave height,
T_z—zero-up-crossing wave period,
ω—circular frequency.

To calculate the drift force, numerical methods, such as the far-field and near-field pressure methods, as well as experimental studies, are used. Far-field methods, initiated by Maruo [47], further developed by Newman, Salvesen, Gerritsma, and Beukelman [48–50] and most recently by Kashiwagi [51], are based on considerations of the energy of diffracted (reflected and transmitted) and radiated waves, momentum flux at infinity, and work conducted by the body in the near field, leading to a constant force acting through the total rate of momentum change. Alternatively, near-field methods, initiated by Boese [52], calculate second-order constant forces/moments through the direct integration of hydrodynamic, steady second-order pressure acting on the wetted surface of the body. More accurate numerical implementations were later introduced by Faltinsen et al. [53], and then by Papanikolaou and Nowacki [54], which led to an increase in precision. These methods have evolved and been validated based on experimental studies, mainly in terms of the longitudinal component, referred to as additional wave resistance. However, the validation of these methods for a transverse component is limited, mainly due to the scarcity of experimental research. Recently, Liu S and Papanikolaou [55] approximated the mean sway force with an empirical formula, only utilizing main ship particulars and wave parameters.

2.2.2. Application of Artificial Neural Networks to Estimate Mean Nondimensional Sway Force in Regular Waves

The application of artificial neural networks (ANNs) to estimate the nondimensional drift force coefficient in regular waves (C_{wy}) for bulk carriers A, B, and C was assessed using the results of model studies described in Table 5, which are presented in Figure 2. An ANN was used for the estimation process. An additional objective of this study was to examine the effectiveness of ANN use to predict drift force based on the basic dimensions of the ship.

In the field of maritime engineering, the application of ANNs has been increasingly acknowledged for its critical role in enhancing the predictability of the experimental measurements pertinent to ship theory. The development and implementation of ANNs have enabled scientists and engineers to process and interpret extensive datasets efficiently and accurately, significantly improving the accuracy of predictive models in ship design and operational strategies. The integration of ANNs in naval architecture has notably enhanced the accuracy of prognostic models, facilitating substantial advancements in various aspects of ship construction and functionality.

For example, research by Cepowski [56,57] utilized Multilayer Perceptron (MLP) structured neural networks to estimate the additional wave resistance of ships based on fundamental design parameters. This approach resulted in remarkably accurate estimations when compared to outcomes from model experiments. Similarly, Yangjun and Yonghwan [58] applied an ANN to predict sloshing effects in model tests of floating structures larger than conventional LNG carriers. The development of this ANN was based on a thorough analysis of over 540 terabytes of experimental data, covering a wide range of variables, such as cargo hold dimensions, vessel types, and environmental and operational conditions. Furthermore, Dyer et al. [59] employed an ANN to estimate the remaining service life of offshore oil and gas platforms, incorporating a comprehensive dataset representing both natural and engineered aspects of offshore systems. Their findings highlighted the capacity of ANNs to provide highly accurate predictions, indicating life extension, maintenance, and risk minimization strategies for offshore platforms.

In this study, the development of an ANN to estimate the C_{wy} coefficient utilized ship data from Table 5 and model study results presented in Figure 2. Following [56,57], a Multilayer Perceptron (MLP)-structured ANN was adopted.

Figure 4 illustrates the procedure for developing the ANN. The first step involved normalizing the dataset to the [0, 1] range to accelerate learning, avoid numerical issues such as overflow or underflow, prevent feature domination, facilitate weight initialization, and enhance the performance of algorithms based on gradients or distances. Following this procedure, a set of alternative ANNs was developed, differing in the number of neurons in the hidden layer and activation functions. In each ANN, to detect overfitting, the hold-out method with an additional validation set was applied, consisting of the following steps:

1. randomly splitting the data into three parts: training set (50%), testing set (25%), and validation set (25%),
2. training the model on the training set,
3. applying early stopping to prevent overfitting by monitoring the error on the testing set,
4. testing the model on the testing set after each epoch and calculating performance metrics,
5. validating the model on the validation set after training is complete,
6. calculating final performance metrics and comparing them with the results from the testing set to assess the model's generalization capability,
7. analyzing the results from the validation set to evaluate the model's ability to generalize to new data.

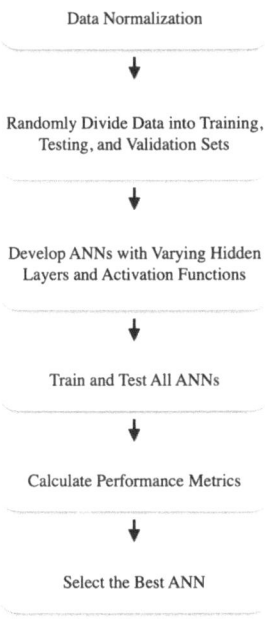

Figure 4. Research scheme for developing an ANN.

In the hold-out procedure, the training and testing process involved early stopping to prevent overfitting by halting the training process if errors increased. The early stopping technique consisted of the following steps:

1. input data: feeding the network with training dataset inputs,
2. prediction: calculating the network's predictions (outputs),
3. error calculation: computing the difference between the predictions and the actual output values using an error function,
4. repetition: repeating steps 1 and 2 for all input–output pairs in the training dataset,
5. weight adjustment: using a learning algorithm to adjust the weights of the neurons to minimize the prediction error,
6. testing: feeding the network with all cases from the testing dataset, obtaining predictions, and comparing them with the actual output values from the dataset to calculate the network error,
7. error comparison: comparing the current network error with the error from the previous cycle. If the error has decreased, the training continues; otherwise, the process is stopped.

The mean absolute error (MAE) was used as the error function in this algorithm, which is defined as follows:

$$\text{MAE} = \frac{\sum_{i=1}^{n}\left|out_{(i)} - out_{p(i)}\right|}{n} \quad (12)$$

where out is the output value, out_p is the predicted output value, and n is the number of instances.

The efficacy of the proposed ANN was assessed using the following established performance metrics: Pearson correlation coefficient (PCC), determination coefficient (R^2), RMSE, and relative RMSE (RRMSE), as delineated in [60,61]. Values of RMSE and RRMSE approaching zero, along with R^2 and PCC values nearing one, indicated a robust alignment between forecast outcomes and empirical data. According to a previous study [62,63], THE estimation accuracy was divided into THE FOLLOWING four categories based on the RRMSE criterion: excellent (below 10%), good (10% to 20%), acceptable (20% to 30%), and unacceptable (above 30%). The performance metrics were calculated as follows:

$$\text{PCC} = \frac{\sum(y_e - \bar{y})(y_e - \bar{y_e})}{\sqrt{\sum(y - y)^2(y_e - \bar{y_e})^2}} \quad (13)$$

- Determination coefficient (R^2):

$$R^2 = \frac{\sum(y_e - \bar{y})^2}{\sum(y - \bar{y})^2} \quad (14)$$

- Root mean square error (RMSE):

$$\text{RMSE} = \sqrt{\frac{\sum(y - y_e)^2}{n}} \quad (15)$$

- Relative root mean square error (RRMSE):

$$\text{RRMSE} = \frac{\text{RMSE}}{\bar{y_e}} \times 100 \quad (16)$$

In the final stage, the best ANN was selected based on two criteria as follows:
- highest accuracy of prediction relative to measured data,
- smallest difference between the performance metrics calculated for the training, testing, and validation sets.

2.2.3. Application of Artificial Neural Networks to Estimate Mean Nondimensional Sway Force in Regular Waves

The general Equation (3) presented above for calculating CO_2 emissions is based on Equation (4) for estimating fuel consumption. A key element in calculating fuel consumption is the power supplied to the DP system over time under specified environmental conditions, such as wind and waves. To comprehensively include the impact of wave and wind interaction on fuel consumption, this article expands Equation (4) and proposes a new approach to calculating fuel consumption by considering the impact of wind and waves as follows:

$$FC = FOC \cdot t \cdot \left(\sum_{i=1}^{8} \sum_{j=1}^{11} p_{wave(H_{S(i)}, T_{z(j)})} \cdot P_{wave(H_{S(i)}, T_{z(j)})} + \sum_{i=1}^{8} p_{wind(i)} \cdot P_{wind(i)} \right) \quad (17)$$

where

P_{wave}—power needed to counteract the drift force caused by the wave height Hs and period T_z [kW],

p_{wave}—annual frequency of the wave height (Hs) and period (T_z) in the CCZ according to Table 3 [-],

P_{wind}—power needed to counteract the drift force caused by wind [kW],

p_{wind}—annual frequency of wind occurrence at a given speed according to Table 4 [-].

The above formula takes fuel consumption calculations into account in the case of the wave interaction time at a given height Hs and period T_z and the time of wind interaction at a given speed. These times are calculated based on the wave occurrence frequency and wind parameters presented in Tables 3 and 4.

Equation (17) can be used to calculate fuel consumption and considers the full range of wave and wind parameters, as well as taking a limited range of wave and wind conditions into account, e.g., for waves up to 3 m and corresponding wind speeds.

2.2.4. CO_2 Emissions during Transport

The estimation of CO_2 emissions during maritime transport is based on the energy efficiency design index (EEDI), which, according to the IMO guidelines [31], is defined as the ratio of the CO_2 emitted by a ship per nautical mile to its transport capacity and speed. The calculations considered the threshold values of the EEDI that must be met by each bulk carrier.

Figure 5 illustrates the block diagram for estimating CO_2 emissions during the transport of PMNs to the destination port. The CO_2 emission was calculated using the following formula:

$$CO_{2transport} = EEDI_c \cdot DWT \cdot d_T \quad (18)$$

where

DWT—deadweight tonnage of the bulk carrier from fleet A/B/C [t],

d_T—total distance [NM],

$EEDI_c$—the required energy efficiency design index for bulk carriers A/B/C according to the International Maritime Organization [31] guidelines, calculated as follows:

$$EEDI_c = \frac{1-X}{100} \cdot 961.79 \cdot DWT^{-0.477} \quad (19)$$

where

$EEDI_c$—the required energy efficiency design index, expressed as the number of grams of CO_2 emissions per nautical mile per ton of cargo,

X—is the reduction factor, whose value for phase 3, according to [31], which starts from 2025, is 30.

The $EEDI_c$ index, calculated using Equation (19), serves as the reference line for determining the required emission reductions within the framework of the energy efficiency design index (EEDI). In calculating the $EEDI_c$, various technical and operational factors were considered, including deadweight tonnage (DWT), ship type, technical specifications, and operational speed. Weather conditions were also indirectly accounted for through the inclusion of the sea margin and historical data on the actual fuel consumption and CO_2 emissions of ships, which formed the basis for developing the $EEDI_c$. Figure 6 shows the values of $EEDI_c$ for different types of ships depending on the DWT, assuming a reduction factor X, applicable from 2025.

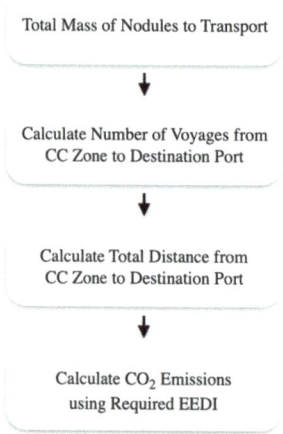

Figure 5. General research scheme for calculating CO_2 emissions during transport.

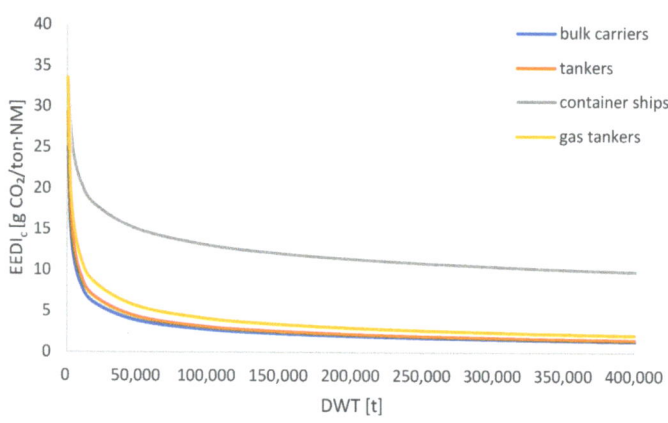

Figure 6. Values of EEDIc for different types of ships as a function of DWT, assuming a reduction factor X, applicable from 2025.

The total distance d_T was calculated based on the formula:

$$d_T = n_{load} \cdot d \tag{20}$$

where
  d—distance from the CCZ to the destination port,
  n_{load}—number of transfers from the extraction ship to the transporting bulk carrier.

The number of PMN transfers from the extraction ship to the transporting bulk carrier was calculated based on the total assumed mass of nodules to be transferred and the deadweight tonnage of carriers from fleets A, B, and C as follows:

$$n_{load} = M/DWT \tag{21}$$

where
M—total mass of PMNs to be transported, M = 2,000,000 t.

3. Results
3.1. Prediction of a Nondimensional Sway Force Coefficient Using an ANN

A nondimensional sway force coefficient, C_{wy}, was predicted using a Multilayer Perceptron (MLP) neural network structure. The model utilized the following ratios and coefficients that impact the sway force coefficient, C_{wy}:

- L/B (length-to-breadth ratio),
- B/T (breadth-to-draught ratio),
- L/T (length-to-draught ratio, as a combination of L/B and B/T),
- CB (block coefficient),
- λ/L (wavelength-to-ship length ratio).

These features were chosen based on their physical significance and impact on ship hydrodynamics, as supported by the literature [39–44]. This selection ensured that the model utilized the most informative and relevant variables to accurately predict the C_{wy} coefficient.

The training dataset was sourced from publicly available ship model tests, as reported in the literature [39–44]. The design characteristics of the ship models used in these tests are presented in Table 5. The dataset consisted of 32 cases. To evaluate the model's performance and prevent overfitting, early stopping was employed during the network's training. Additionally, the dataset was randomly divided into the following three parts using the hold-out method: training (sixteen cases), testing (eight cases), and validation (eight cases). This approach allowed for the detection and prevention of overfitting by providing separate datasets for model training, validation, and final performance evaluation.

To optimize the learning process, the data were normalized to a range from 0 to 1. This normalization aimed to accelerate learning, avoid numerical issues such as overflow or underflow, prevent feature domination, facilitate weight initialization, and improve the performance of gradient-based algorithms.

During the training of the ANN, additional hyperparameters were used as follows: a learning rate of 0.1, a batch size of 32, and a maximum of 100 epochs. The BFGS optimization algorithm [64–67] was employed, which has proven to be highly effective for training ANNs to predict additional wave resistance [57]. The most accurate neural network identified through this research contained five neurons in the input layer, six neurons in the hidden layer, and one neuron in the output layer. All neurons were activated using a logistic function as follows:

$$\sigma(x) = \frac{1}{1+e^x} \tag{22}$$

where $\sigma(x)$ is the value of the activation function for the input x.

Graphically, the general structure of this ANN is illustrated in Figure 7. Normalization coefficients and weights are provided in Appendix A.

Additionally, during training, early stopping was utilized to prevent overfitting by halting the training process if errors increased for the training and testing data. This process was stopped after 54 epochs, with the stopping condition being a change in error of 1×10^{-5}.

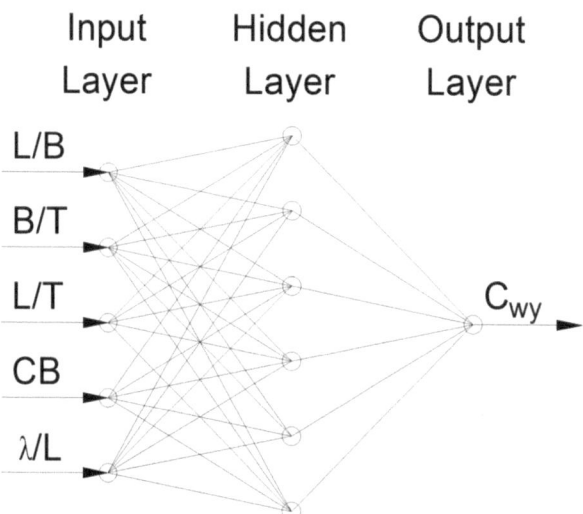

Figure 7. The general structure of an ANN.

To further increase the reliability of the results, four different performance evaluation metrics were applied as follows: Pearson correlation coefficient (PCC), determination coefficient (R^2), root mean square error (RMSE), and relative root mean square error (RRMSE). Table 6 presents these model evaluation metrics for the ANN. This table shows that the PCC values are close to 1, indicating a strong linear correlation between the predicted and actual values for all datasets. High PCC values suggest that the model predictions are well correlated with the actual data. For the training set, the PCC value is 0.98, for the validation set, it is 0.99, and for the test set, it is 0.97. The R^2 values are also high, indicating that the model explains the variability of the data well. High R^2 values mean that the model fits the data well for the training, validation, and test sets, where the R^2 values are 0.95, 0.99, and 0.95, respectively. Low RMSE values indicate small root mean squared errors, meaning that the model predictions are close to the actual values. The RMSE for all datasets is 0.05. RRMSE expressed as a percentage also shows relatively small errors relative to the range of data, with values of 19.58% for the training set, 16.36% for the validation set, and 14.44% for the test set. Lower RRMSE values for the validation and test sets suggest good overall model performance and no signs of overfitting. Despite the small number of measurements, with only 33 data points, the network exhibits good metrics. Based on these parameters, it can be concluded that the neural network model performs very well. High PCC and R^2 values, along with low RMSE and RRMSE values, indicate accurate and stable predictions for the training, validation, and test sets. The relatively low RMSE indicates that the average deviation of the model predictions from the actual values is small. According to a previous study [48,49], the estimation accuracy can be divided into four categories based on the RRMSE criterion as follows: excellent (below 10%), good (10% to 20%), acceptable (20% to 30%), and unacceptable (above 30%). Thus, the RRMSE values indicate good precision within the model. The model appears to be well-fitted without signs of overfitting. These model evaluation metrics for the test and validation sets demonstrate the sufficient generalization capability of the ANN for new, unseen data, which is crucial for the practical application of the network.

Figure 8 presents a residuals plot against the predicted values, indicating that the largest portion of data yields estimation errors ranging from −0.15 to 0.1.

Figure 9 shows scatter plots which indicate that the trend line between the measured and predicted C_{wy} values follows the linear y = x line (45°) relatively closely, with slopes of 0.959, 1.0345, and 1.096, and low constants of 0.0097, 0.0003, and 0.0068 for the training, test, and validation data, respectively. This means that predictions using the ANN are characterized by a fairly accurate prediction ability.

Table 6. Model evaluation metrics for the ANN.

PCC train [-]	0.98
PCC valid [-]	0.99
PCC test [-]	0.97
R^2 train [-]	0.95
R^2 valid [-]	0.99
R^2 test [-]	0.95
RMSE train [-]	0.05
RMSE valid [-]	0.05
RMSE test [-]	0.05
RRMSE train [%]	19.58
RRMSE valid [%]	16.36
RRMSE test [%]	14.44

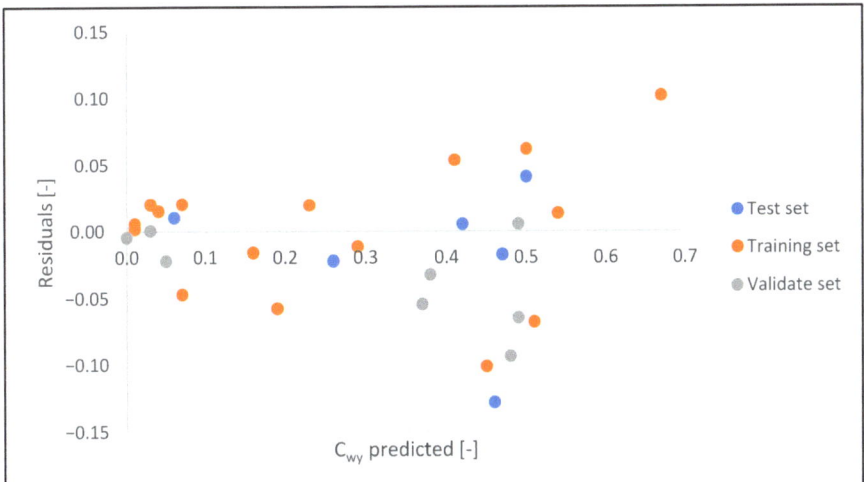

Figure 8. Residuals vs. predicted values.

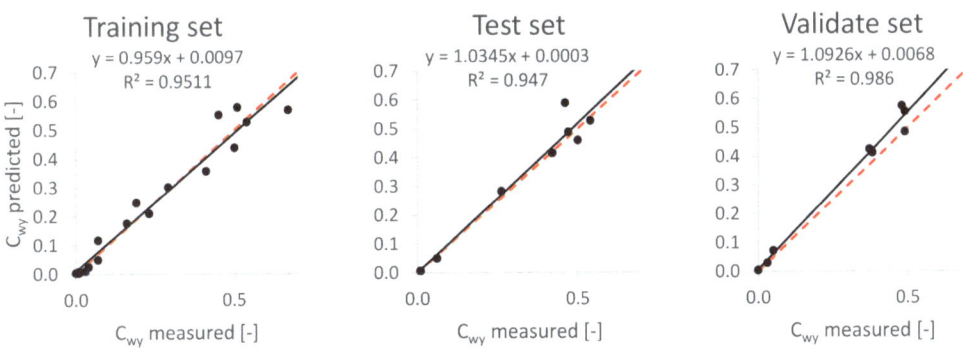

Figure 9. Scatter plots of predicted vs. measured C_{wy} values.

Figure 10 compares predictions made by the ANN with results from a model test. These plots show that the predictions obtained with the ANN are consistent with the test results and exhibit corresponding trends.

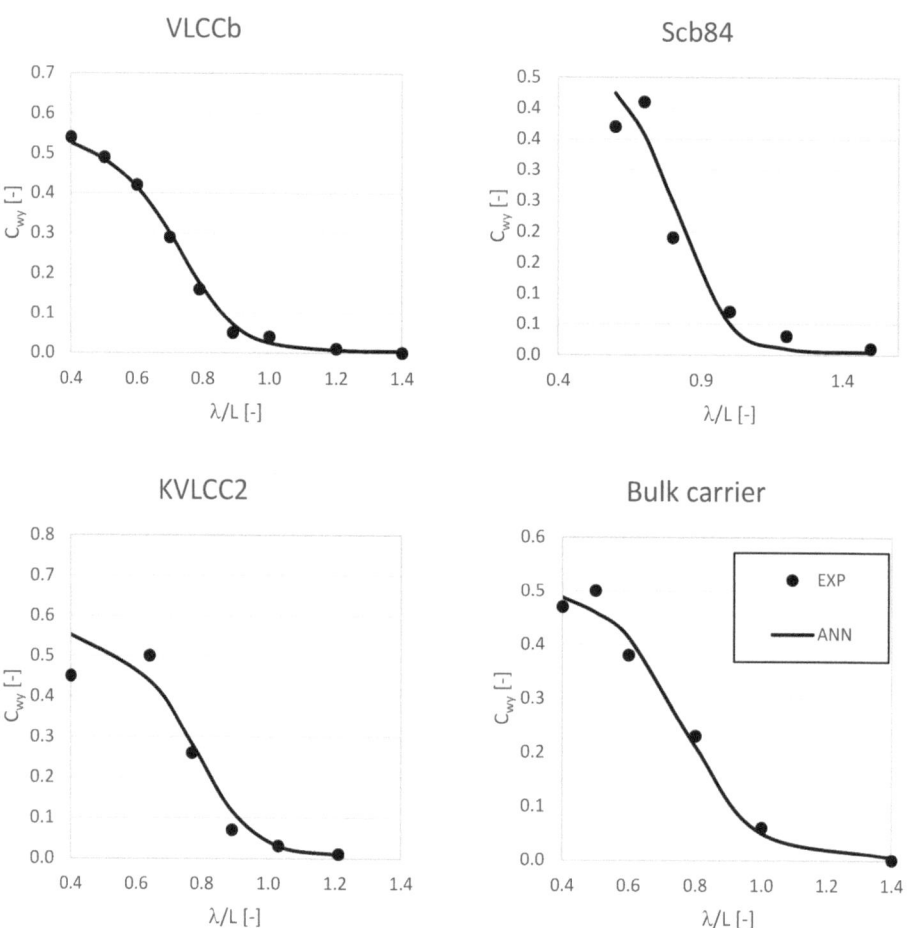

Figure 10. Comparison of ANN predictions with model test results.

Figure 11 presents predictions for fleets A, B, and C in relation to the data used for training the neural network. This Figure indicates that these predictions exhibit trends similar to those observed in the model test.

3.2. CO_2 Emissions during Nodule Loading in the CCZ

Using the developed ANN and Equations (8)–(11), wave-induced drift force values for bulk carriers A, B, and C were estimated by considering the significant wave height (H_s) and the average zero-up-crossing wave period (T_z) in the CCZ, as shown in Table 3. The results are presented in Tables A4–A6 in Appendix B. By applying Equation (7) and considering the wind speed distribution shown in Table 4, along with the characteristics of bulk carriers A, B, and C, the wind-induced drift forces were estimated. These are depicted in Figure 12.

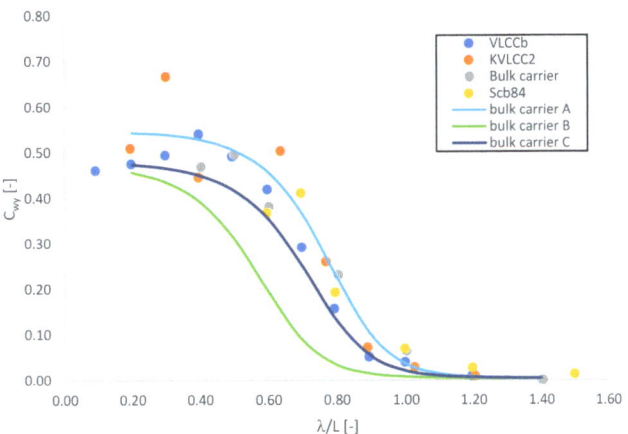

Figure 11. Comparison of ANN predictions for bulk carriers types A, B, and C with model test results.

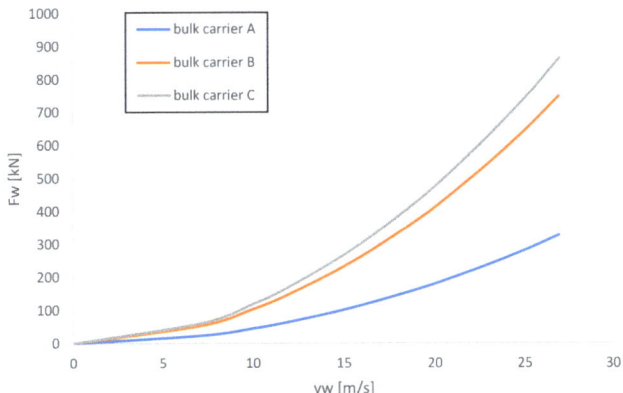

Figure 12. Drift force values from wind for bulk carriers A, B, and C which correspond to wind velocity (v_W).

When considering the most frequently occurring wave heights in the CCZ, which are 4 m according to Table 3, it is suggested that the largest drift forces, as shown in the figures and tables, are as follows:

- 775 kN, 1287 kN, and 1688 kN from waves,
- 106 kN, 244 kN, and 282 kN from wind,

for bulk carriers A, B, and C, respectively.

The thrust force (T_p) and the power (P) supplied to the propellers of the DP system were calculated using Equations (5) and (6). A power-to-thrust efficiency ratio (C_p) of 0.14 was assumed for the power calculations. Then, using Equations (3) and (17), the fuel consumption and the resulting CO_2 emissions of bulk carriers A, B, and C were calculated for 250 days a year during concrete loading operations in the CCZ.

Detailed values of fuel consumption resulting from the operation of the DP system to counteract the drift force caused by waves and wind on bulk carriers A, B, and C over 250 days a year during concrete loading in the CCZ can be found in Tables A7–A10 of

Appendix B. The values of fuel consumption, which depend solely on Hs, are presented collectively in Figure 13. This Figure illustrates that fuel consumption increases with wave height up to 3 m, reaching a maximum of 600 t, 1250 t, and 2100 t for bulk carriers A, B, and C, respectively. Beyond this level, fuel consumption decreases despite the increase in drift forces. This reduction in fuel consumption occurs because the amount of time the vessel spends in these conditions decreases as the wave height increases, given the lower occurrence of such large waves.

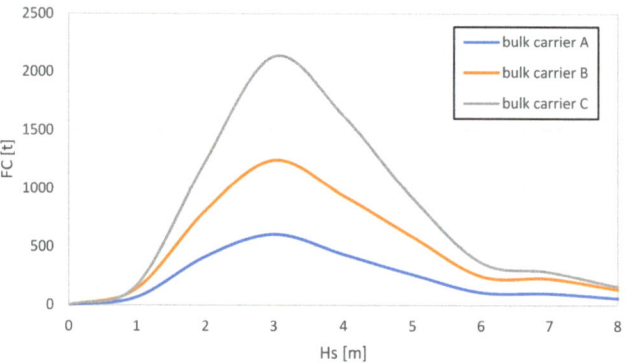

Figure 13. The values of fuel consumption (FC) [t] resulting from the operation of the DP system to counteract the drift force caused by wave and wind action on bulk carriers A, B, and C, corresponding to significant wave height (Hs).

Meanwhile, Figure 14 shows the cumulative CO_2 emissions from fuel combustion on ships due to wind and wave action as a function of wave height. For wave heights of up to 4 m, the cumulative CO_2 emissions are calculated to be 4900 t, 10,000 t, and 16,500 t for bulk carriers A, B, and C, respectively. These numbers demonstrate that significantly lower amounts of CO_2 will be emitted when using type A bulk carriers for the transshipment of extracted concrete. Utilizing type B and C bulk carriers for this mission will result in an increase in CO_2 emissions by 100% and 240%, respectively.

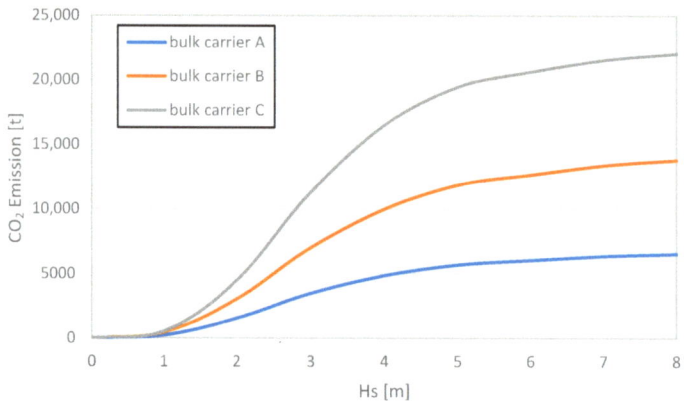

Figure 14. Cumulative CO_2 emissions due to combined wave and wind action in relation to significant wave height (Hs).

3.3. CO_2 Emissions during Transport

The number of shipments was calculated using Equation (21), based on the predetermined total mass of PMNs to be transported and the cargo capacities of bulk carriers A,

B, and C. The results indicated that there were 264 shipments for bulk carrier A, 63 for B, and 27 for C. Subsequently, utilizing the established routes shown in Figure 1, the distances that bulk carriers A, B, and C need to traverse to reach their designated destination ports were calculated based on Equations (20) and (21). The varying distances, according to the planned routes, are displayed in Table 7, which shows that type A bulk carriers are required to cover the longest distance, while type C has the shortest distance to traverse. These discrepancies increase with route length.

Table 7. Total distance to be covered by A, B, and C bulk carriers.

Route Number According to Figure 1	Route Length from the CC Zone to the Destination Port [NM]	Total Distance [NM]		
		A	B	C
1	2044	1,077,083	255,475	108,321
2	2400	1,264,800	300,000	127,200
3	6971	3,673,506	871,325	369,442
4	16,696	8,798,581	2,086,950	884,867
5	22,512	11,863,666	2,813,963	1,193,120
6	26,291	13,855,462	3,286,400	1,393,434

In accordance with a report by the International Maritime Organization [31], the threshold values for the energy efficiency design index (EEDI) were calculated using Equation (19), resulting in 9.48 for bulk carrier A, 4.78 for B, and 3.21 for C. This analysis shows that the largest bulk carriers have the lowest CO_2 emissions. Comparatively, type A and B bulk carriers emit nearly 200% and 50% more CO_2 than those of type C, respectively.

Based on these data, the total CO_2 emissions of bulk carriers A, B, and C, which transport the predetermined mass of PMNs to ports via designated routes, were calculated. The results, graphically presented in Figure 15, indicate that type C bulk carriers will emit the least amount of CO_2. Transporting the same amount of extracted PMNs to the farthest destination port using type C bulk carriers will result in CO_2 emissions of 998,768 tones, compared to 502,449 tones for type B and as much as 329,338 tones for type A. Therefore, transporting PMNs from the CC zone to the unloading port using type A and B bulk carriers will cause an increase in CO_2 emissions by 203% and 53%, respectively, when compared to using type C bulk carriers.

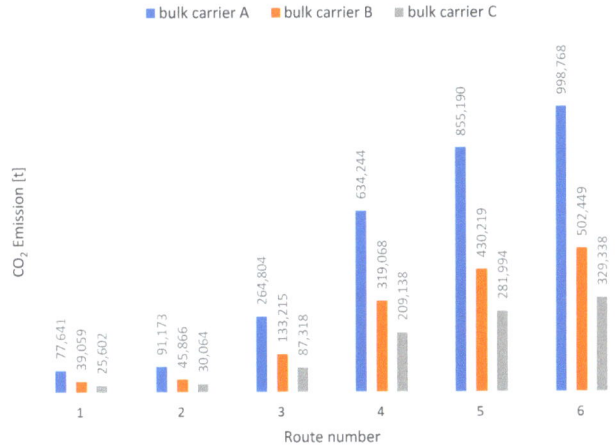

Figure 15. Total CO_2 emissions of A, B, and C bulk carriers during transportation.

3.4. Assessment of CO_2 Emissions from Bulk Carriers during Nodule Loading and Maritime Transport

Based on the CO_2 emissions of A, B, and C bulk carriers during the loading of PMNs in the CCZ and during transportation, the total CO_2 emissions for the entire mission were calculated. The calculations were conducted under the assumption that the loading in the CCZ is carried out with wave heights up to 3 m and wind speeds up to 12 m/s. The results are presented in Figure 16. On route number 1, which is the shortest, type A, B, and C bulk carriers emit 81,111, 46,058, and 36,902 tons of CO_2, respectively. On route number 6, which is the longest, A, B, and C bulk carriers emit 1,002,238, 509,449, and 340,638 tons of CO_2 together, respectively. Hence, it follows that, regardless of route length, the smallest bulk carriers emit the largest amount of CO_2, and the largest bulk carriers emit the least. Therefore, the amount of CO_2 emitted decreases with an increase in bulk carrier size. Moreover, the further the distance to the unloading port of PMNs, the greater the differences in CO_2 emissions in favor of large bulk carriers. On the shortest route, type A and B bulk carriers emit 120% and 25% more CO_2 than type C bulk carriers, respectively. On the longest route, type A and B bulk carriers emit 194% and 50% more CO_2 than type C bulk carriers, respectively.

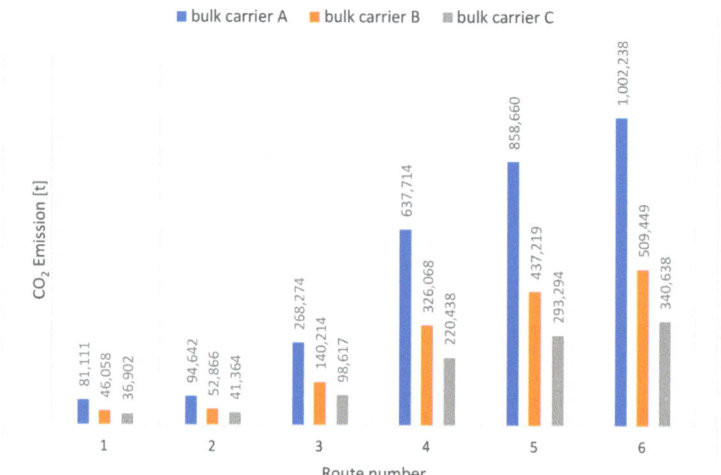

Figure 16. Total CO_2 emissions of A, B, and C bulk carriers during loading and transportation.

On the shortest route, the differences in CO_2 emissions between small and large bulk carriers are relatively minimal. Figure 17 shows the cumulative CO_2 emissions during the extraction and transportation of nodules to the nearest destination port, depending on the wave height (and corresponding wind speed) prevailing during the loading of nodules in the CC zone.

For the smallest bulk carriers, the impact of wind and waves on CO_2 emissions is minimal. This is due to the small size of these ships and the need for them to cover longer distances compared to larger ships, which significantly increases CO_2 emissions during the voyage. With wave heights up to 1 m and wind speeds up to 7 m/s, these ships emit 77,850 thousand tons of CO_2, and with wave heights of 5 m, they emit 83,300 thousand tons of CO_2, indicating an increase of only about 7%.

Environmental conditions have a greater impact on medium and large bulk carriers, as the influence of waves and wind on these ships is more significant, increasing CO_2 emissions, although the relative emissions during transport are lower. Loading nodules in wave heights up to 5 m and wind speeds up to 19 m/s results in annual CO_2 emissions of 51,000 and 45,000 tons for bulk carriers type B and C, respectively. In contrast, loading nodules in wave heights up to 1 m and wind speeds up to 7 m/s results in annual CO_2 emissions of 39,500 and 26,100 tons for bulk carriers type B and C, respectively. When compared to emissions at 5 m wave heights, this represents an increase of approximately 29% for bulk carrier type B and as much as 72% for bulk carrier type C. For wave heights up to 2 m and wind speeds up to 10 m/s, the annual emissions are 42,000 and 30,000 tons of CO_2 for bulk carriers type B and C, respectively, which corresponds to an increase of approximately 21% for bulk carrier type B and 50% for bulk carrier type C when compared to emissions at 5 m wave heights.

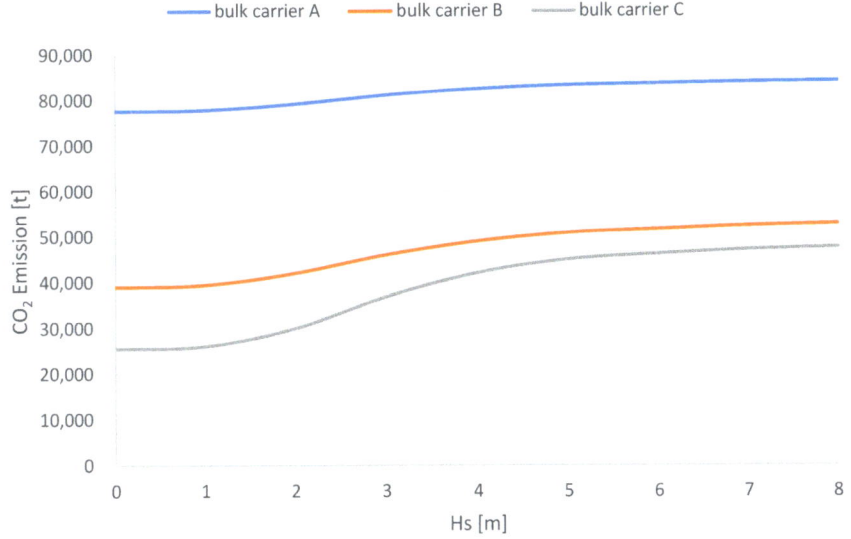

Figure 17. Cumulative CO_2 emissions of A, B, and C bulk carriers during loading and transportation on route no. 1, as influenced by environmental conditions represented by significant wave height (Hs).

Loading all types of ships in worse weather conditions, i.e., wave heights greater than 5 m and wind speeds above 19 m/s, does not significantly increase CO_2 emissions, as such conditions are relatively rare and the duration of the trips in these conditions is short.

Assuming that loading will be conducted in wave conditions up to 1–2 m due to technical issues related to ensuring safe loading, as indicated by [20,21], on the shortest route and on the largest bulk carriers, a significant impact of waves and wind on CO_2 emissions can be expected. However, on longer routes, this impact is not substantial.

4. Discussion

The analysis presented in the article indicates that the smallest type A bulk carriers, although characterized by a lower drift force, generate higher CO_2 emissions when compared to larger vessels. This is associated with their lower transport efficiency and the greater number of trips required to deliver nodules to the destination port, resulting in higher fuel consumption and, consequently, higher CO_2 emissions. In contrast, the largest type C bulk carriers, despite the greater drift force caused by marine environmental effects during loading, emit less CO_2 overall, thanks to their higher transport efficiency. These results suggest that selecting larger vessels for the loading and transporting of nodule cargo can positively contribute to reducing the environmental impact of the mission by reducing carbon dioxide emissions.

These studies were conducted with the assumption that all types of ships can conduct loading operations efficiently, even in challenging weather conditions. However, research conducted by Kacprzak [20] has shown that the loading of PMNs on small bulk carriers can only be carried out under limited weather conditions. This means that the use of a fleet of small bulk carriers for the loading and transport of PMNs may lead to an extension of the time required to complete these operations due to the lower number of days in a year that are suitable for loading. These findings suggest that using large bulk carriers for these tasks may be a more efficient solution than using small ones.

Figures 15–17 show that the overall CO_2 emission related to the transport mission primarily depends on the emissions generated during sea transport. This clearly demonstrates that, although smaller ships may emit less CO_2 during the loading operations themselves, their lower transport efficiency, and the need to make a significantly larger number of voyages and thus cover a longer distance, means that when considering the entire venture, their environmental impact is negatively greater than that of large bulk carriers. Combined with the fact that small ships burn relatively more fuel per unit of transported cargo than large ones, this leads to the conclusion that the use of a fleet of large ships could be key from the point of view of ecological efficiency. This discovery may have significant implications for ecological policy in maritime transport. It shows that efforts to limit CO_2 emissions in maritime transport should not only include the search for new technologies and fuels, but also the optimization of fleet size and operation planning. Using large bulk carriers for the transportation of all types of cargo could significantly reduce global CO_2 emissions from maritime transport, presenting both a challenge and an opportunity for the industry.

Scaling the research findings to encompass general maritime transport clearly shows the substantial potential for reducing CO_2 emissions. This is particularly relevant given the feasibility of the annual transportation of up to 11 billion tons of cargo via sea, as reported in UNCTAD [68]. For example, Figure 18 depicts the hypothetical CO_2 emissions generated by ships carrying a volume of cargo along a typical route spanning an average of 5000 nautical miles in relation to the ship deadweight capacity. The illustration reveals that utilizing only a fleet of the smallest bulk carriers, each with a deadweight of 1000 tons, would culminate in an annual CO_2 emission totaling 1,960,843,563 tons. Conversely, only employing a fleet composed of the largest bulk carriers, each boasting a deadweight of 320,000 tons, would significantly reduce the annual CO_2 emission to merely 125,166,010 tons. Therefore, if we assume hypothetically that all the global cargo transported annually was only carried by the smallest bulk carriers, this will increase the emissions by 1,835,677,553 tons of CO_2 annually. This roughly equates to the annual burning of 1.142 billion passenger cars with an average consumption of 6 L of diesel per 100 km, assuming that they each travel 10,000 km annually. This example highlights the need for further research and analysis to better understand and utilize the potential for the efficient use of large bulk carriers in ecological maritime transport. The analysis presented in the article potentially represents a significant advancement in understanding the complexities of CO_2 emissions in maritime transport. It emphasizes that the ecological efficiency of this sector depends not only on the employed technologies, which primarily involve alternative fuels [69–71], but also on operational strategies and planning. Optimizing the size of ships could bring significant benefits to the environment, which should be considered in global efforts towards sustainable development and carbon footprint reduction. However, verifying this approach requires conducting more detailed research in this area, which may be explored in subsequent studies by the authors.

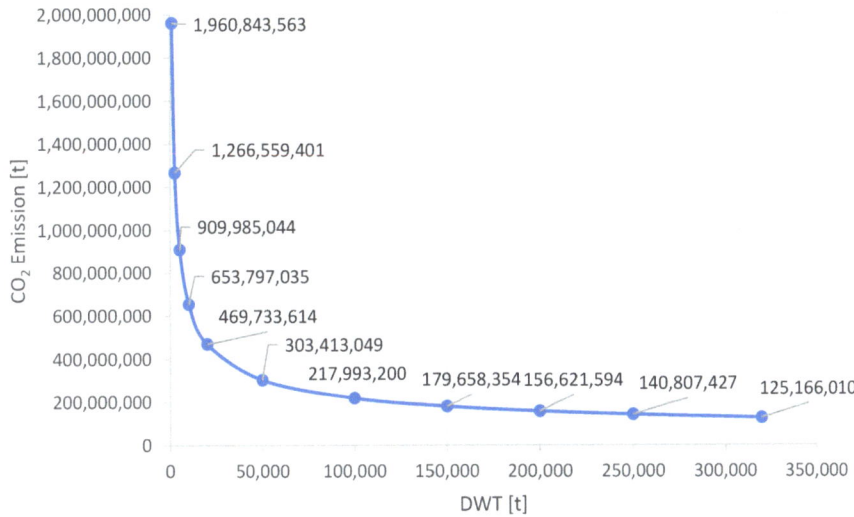

Figure 18. Comparative analysis of annual CO_2 emissions: smallest vs. largest bulk carriers on a 5000-nautical mile route transporting 11 billion tons of cargo.

5. Conclusions

The study confirmed the set objectives, providing significant insights into the impact of ship size on CO_2 emissions. It introduces new methods for assessing energy efficiency and the use of ANNs to estimate drifting force. The key innovations of the article include the following:

- the impact of bulk carrier size on CO_2 emissions, indicating the possibility of reducing the carbon footprint through the optimal selection of transport units,
- a new method to assess energy efficiency that takes environmental conditions into account, which can significantly contribute to the optimization of fuel consumption and the reduction of CO_2 emissions,
- the application of ANNs to estimate drifting force, offering new perspectives for ship design and operation, and highlighting the potential of artificial intelligence in the analysis of maritime operations.

The study demonstrated that large bulk carriers, despite the greater drift force acting on them, emit less CO_2 than smaller bulk carriers, both when considering the loading at sea of PMNs and their transport to the destination port. This indicates that preferring larger ships in global maritime trade can reduce environmental impacts due to their higher transport efficiency and lower CO_2 emissions.

The proposed method to estimate fuel consumption and CO_2 emissions, considering environmental conditions with the DP system, could lead to a new energy efficiency indicator. This indicator would aid in assessing and optimizing maritime operations from an environmental perspective, but also requires further research.

Using neural networks (ANNs) to estimate lateral drifting force in regular waves has proven effective, demonstrating its potential in maritime analysis. The developed ANN accurately predicts drifting force based on ship design parameters. However, practical use is limited by the narrow dimension ratio ranges of the training models. Expanding the training dataset with more diverse ship shapes could improve the method's universality, with further research required.

Author Contributions: All authors contributed equally to this editorial. All authors have read and agreed to the published version of the manuscript.

Funding: This research received no external funding.

Data Availability Statement: All data supporting the reported results are included in the attachments of this article. No new data were created during this study.

Conflicts of Interest: The authors declare no conflicts of interest.

Abbreviations

ANN	Artificial neural network
A_W	Lateral plane of the vessel above the plane of draught
B	Breadth of the ship
BFGS	Broyden–Fletcher–Goldfarb–Shanno algorithm
CB	Block coefficient
CCZ	Clarion-Clipperton Zone
CII	Carbon intensity indicator
CO_2	Carbon dioxide
D	Displacement
DP	Dynamic positioning
DWT	Deadweight tonnage
EEDI	Energy efficiency design index
EEOI	Energy efficiency operational indicator
H_s	Significant wave height
IMO	International Maritime Organization
L	Length between perpendiculars of the ship
MSE	Mean squared error
PCC	Pearson correlation coefficient
PMNs	Polymetallic nodules
RMSE	Root mean squared error
RRMSE	Relative root mean square error
SCG	Scaled conjugate gradient
SGD	Stochastic gradient descent
T	Draught of the ship
T_z	Average zero-up-crossing wave period
λ	Wavelength of a regular wave

Appendix A

Table A1. Normalization coefficients for the inputs and output of the ANN.

	Shift	Scale
L/B.	−30.6666666666666	5.55555555555554
B/T	−8.70967741935485	3.22580645161291
L/T	−12.5121951219512	0.81300813008130
CB	−27.0000000000001	33.3333333333334
1/L	−1.53846153846154	0.76923076923076
Output	0	1.49253731343284

Table A2. Neuron weights and biases in the hidden layer.

	Neuron in the Hidden Layer					
Input	1	2	3	4	5	6
L/B	0.301285	0.300698	0.193848	0.427797	0.160716	−0.070965
B/T	0.770441	0.370726	0.749137	0.094164	0.785890	−0.726824
L/T	0.794974	0.526635	1.035377	0.867195	1.019568	0.842477
CB	2.544873	1.573534	2.629856	2.722212	2.506663	−0.238512
1/L	3.111188	2.391626	3.641622	5.705001	4.086734	9.063809
bias	0.558491	0.400501	0.554879	−1.215970	0.480272	−4.862085

Table A3. Neuron weights and biases in the output layer.

	Output
Neuron 1	0.703693
Neuron 2	−0.501220
Neuron 3	1.014894
Neuron 4	−1.718017
Neuron 5	0.080677
Neuron 6	−6.121465
bias	1.356850

Appendix B

Table A4. Estimated drift force values from beam seas for bulk carrier type A using ANN.

Hs (m)	T_z (s)										
	<4	4 to 5	5 to 6	6 to 7	7 to 8	8 to 9	9 to 10	10 to 11	11 to 12	12 to 13	>13
7 to 8	2566	3303	2787	1960	1301	861	581	401	284	206	153
6 to 7	1965	2529	2134	1501	996	659	445	307	217	158	117
5 to 6	1443	1858	1568	1103	732	485	327	226	160	116	86
4 to 5	1002	1290	1089	766	508	336	227	157	111	80	60
3 to 4	642	826	697	490	325	215	145	100	71	52	38
2 to 3	361	464	392	276	183	121	82	56	40	29	21
1 to 2	160	206	174	123	81	54	36	25	18	13	10
0 to 1	40	52	44	31	20	13	9	6	4	3	2

Table A5. Estimated drift force values from beam seas for bulk carrier type B using ANN.

Hs (m)	T_z (s)										
	<4	4 to 5	5 to 6	6 to 7	7 to 8	8 to 9	9 to 10	10 to 11	11 to 12	12 to 13	>13
7 to 8	1232	3291	4043	3565	2717	1959	1393	996	722	532	398
6 to 7	943	2519	3095	2729	2080	1500	1067	763	553	407	305
5 to 6	693	1851	2274	2005	1528	1102	784	560	406	299	224
4 to 5	481	1285	1579	1393	1061	765	544	389	282	208	156
3 to 4	308	823	1011	891	679	490	348	249	181	133	100
2 to 3	173	463	568	501	382	275	196	140	102	75	56
1 to 2	77	206	253	223	170	122	87	62	45	33	25
0 to 1	19	51	63	56	42	31	22	16	11	8	6

Table A6. Estimated drift force values from beam seas for bulk carrier type C using ANN.

Hs (m)	T_z (s)										
	<4	4 to 5	5 to 6	6 to 7	7 to 8	8 to 9	9 to 10	10 to 11	11 to 12	12 to 13	>13
7 to 8	888	3352	5256	5689	5055	4040	3072	2293	1709	1283	974
6 to 7	680	2566	4024	4355	3871	3093	2352	1756	1309	982	746
5 to 6	499	1885	2957	3200	2844	2273	1728	1290	961	722	548
4 to 5	347	1309	2053	2222	1975	1578	1200	896	668	501	380
3 to 4	222	838	1314	1422	1264	1010	768	573	427	321	243
2 to 3	125	471	739	800	711	568	432	322	240	180	137
1 to 2	55	209	329	356	316	253	192	143	107	80	61
0 to 1	14	52	82	89	79	63	48	36	27	20	15

Table A7. Values of fuel consumption FC [t] resulting from the operation of the DP system to counteract the drift force caused by wave action on bulk carrier A.

Hs (m)	T$_z$ (s)										
	<4	4 to 5	5 to 6	6 to 7	7 to 8	8 to 9	9 to 10	10 to 11	11 to 12	12 to 13	>13
7 to 8	0	0	0	0	0	1	0	0	0	0	0
6 to 7	0	0	0	0	0	2	3	3	2	1	0	0
5 to 6	0	0	0	0	3	9	11	9	4	2	1	0
4 to 5	0	0	0	3	16	35	37	23	10	4	1	0
3 to 4	0	1	14	63	103	84	42	15	4	1	0
2 to 3	0	3	48	147	165	96	36	10	2	0	0
1 to 2	0	10	71	116	76	28	7	1	0	0	0
0 to 1	0	3	7	5	1	0	0	0	0	0	0

Table A8. Values of fuel consumption FC [t] resulting from the operation of the DP system to counteract the drift force caused by wave action on bulk carrier B.

Hs (m)	T$_z$ (s)										
	<4	4 to 5	5 to 6	6 to 7	7 to 8	8 to 9	9 to 10	10 to 11	11 to 12	12 to 13	>13
7 to 8	0	0	0	0	0	2	1	1	1	0	0
6 to 7	0	0	0	0	4	6	6	4	2	1	0
5 to 6	0	0	0	5	18	26	21	11	5	2	0
4 to 5	0	0	4	30	74	85	56	26	9	2	1
3 to 4	0	1	20	115	215	191	101	38	11	3	1
2 to 3	0	3	70	268	345	219	86	25	6	1	0
1 to 2	0	10	103	210	159	63	17	3	1	0	0
0 to 1	0	3	10	9	3	1	0	0	0	0	0

Table A9. Values of fuel consumption FC [t] resulting from the operation of the DP system to counteract the drift force caused by wave action on bulk carrier C.

Hs (m)	T$_z$ (s)										
	<4	4 to 5	5 to 6	6 to 7	7 to 8	8 to 9	9 to 10	10 to 11	11 to 12	12 to 13	>13
7 to 8	0	0	0	0	0	3	3	2	1	1	0
6 to 7	0	0	0	0	7	13	14	9	4	2	1
5 to 6	0	0	0	8	34	53	46	25	11	4	1
4 to 5	0	0	5	48	137	175	123	59	21	6	2
3 to 4	0	1	26	183	401	394	223	86	26	6	1
2 to 3	0	3	91	427	642	452	190	56	13	3	0
1 to 2	0	10	133	335	296	131	37	8	1	0	0
0 to 1	0	3	13	14	6	1	0	0	0	0	0

Table A10. Values of fuel consumption (FC) [t] resulting from the operation of the DP system to counteract the drift force caused by wind on bulk carriers A, B, and C.

Hs (m)	v$_W$ (m/s)	FC (t)		
		Bulk Carrier A	Bulk Carrier B	Bulk Carrier C
7 to 8	0	56	128	148
6 to 7	0	88	202	234
5 to 6	0	69	159	184
4 to 5	0	130	297	344
3 to 4	0	105	241	278
2 to 3	0	95	218	252
1 to 2	0	104	239	276
0 to 1	0	49	112	130

References

1. UNCTAD. Review of Maritime Transport (UNCTAD/RMT/2021). United Nations Conference on Trade and Development. 2021. Available online: https://unctad.org/publication/review-maritime-transport-2021 (accessed on 1 February 2024).
2. Fu, X.; Lang, J.; Wang, X.; Li, Y.; Lang, J.; Zhou, Y.; Guo, X. The impacts of ship emissions on ozone in eastern China. *Sci. Total Environ.* **2023**, *903*, 166252. [CrossRef] [PubMed]
3. Nunes, R.A.O.; Alvim-Ferraz, M.C.M.; Martins, F.G.; Sousa, S.I.V. Environmental and social valuation of shipping emissions on four ports of Portugal. *J. Environ. Manag.* **2019**, *235*, 62–69. [CrossRef] [PubMed]
4. Moreno-Gutiérrez, J.; Pájaro-Velázquez, E.; Amado-Sánchez, Y.; Rodríguez-Moreno, R.; Calderay-Cayetano, F.; Durán-Grados, V. Comparative analysis between different methods for calculating on-board ship's emissions and energy consumption based on operational data. *Sci. Total Environ.* **2019**, *650*, 575–584. [CrossRef] [PubMed]
5. Hoang, A.T.; Foley, A.M.; Nižetić, S.; Huang, Z.; Ong, H.C.; Ölçer, A.I.; Pham, V.V.; Nguyen, X.P. Energy-related approach for reduction of CO_2 emissions: A critical strategy on the port-to-ship pathway. *J. Clean. Prod.* **2022**, *355*, 131772. [CrossRef]
6. Fan, A.; Li, Y.; Liu, H.; Yang, L.; Tian, Z.; Li, Y.; Vladimir, N. Development trend and hotspot analysis of ship energy management. *J. Clean. Prod.* **2023**, *389*, 135899. [CrossRef]
7. Barone, G.; Buonomano, A.; Del Papa, G.; Maka, R.; Palombo, A. How to achieve energy efficiency and sustainability of large ships: A new tool to optimize the operation of on-board diesel generators. *Energy* **2023**, *282*, 128288. [CrossRef]
8. Nyanya, M.N.; Vu, H.B.; Schönborn, A.; Ölçer, A.I. Wind and solar assisted ship propulsion optimisation and its application to a bulk carrier. *Sustain. Energy Technol. Assess.* **2021**, *47*, 101397. [CrossRef]
9. Ytreberg, E.; Astrom, S.; Fridell, E. Valuating environmental impacts from ship emissions—The marine perspective. *J. Environ. Manag.* **2021**, *2021*, 111958. [CrossRef] [PubMed]
10. Amudha, K.; Bhattacharya, S.; Sharma, R.; Gopkumar, K.; Kumar, D.; Ramadass, G. Influence of flow area zone and vertical lift motion of polymetallic nodules in hydraulic collecting. *Ocean. Eng.* **2024**, *294*, 116745. [CrossRef]
11. Sha, F.; Xi, M.; Chen, X.; Liu, X.; Niu, H.; Zuo, Y. A recent review on multi-physics coupling between deep-sea mining equipment and marine sediment. *Ocean. Eng.* **2023**, *276*, 114229. [CrossRef]
12. Cunningham, A. Assessing the feasibility of deep-seabed mining of polymetallic nodules in the Area of seabed and ocean floor beyond the limits of national jurisdiction, as a method of alleviating supply-side issues for cobalt to US markets. *Miner. Econ.* **2024**, *37*, 207–226. [CrossRef]
13. Gollner, S.; Haeckel, M.; Janssen, F.; Lefaible, N.; Molari, M.; Papadopoulou, S.; Reichart, G.J.; Trabucho Alexandre, J.; Vink, A.; Vanreusel, A. Restoration experiments in polymetallic nodule areas. *Integr. Environ. Assess. Manag.* **2022**, *18*, 682–696. [CrossRef] [PubMed]
14. Bonifácio, P.; Kaiser, S.; Washburn, T.W.; Smith, C.R.; Vink, A.; Arbizu, P.M. Biodiversity of the Clarion-Clipperton Fracture Zone: A worm perspective. *Mar. Biodivers.* **2024**, *54*, 5. [CrossRef]
15. Uhlenkott, K.; Meyn, K.; Vink, A.; Martínez Arbizu, P. A review of megafauna diversity and abundance in an exploration area for polymetallic nodules in the eastern part of the Clarion Clipperton Fracture Zone (North East Pacific), and implications for potential future deep-sea mining in this area. *Mar. Biodivers.* **2023**, *53*, 22. [CrossRef]
16. Stratmann, T. Role of polymetallic-nodule dependent fauna on carbon cycling in the eastern Clarion-Clipperton Fracture Zone (Pacific). *Front. Mar. Sci.* **2023**, *10*, 1151442. [CrossRef]
17. Li, B.; Jia, Y.; Fan, Z.; Li, K.; Shi, X. Impact of the Mining Process on the Near-Seabed Environment of a Polymetallic Nodule Area: A Field Simulation Experiment in a Western Pacific Area. *Sensors* **2023**, *23*, 8110. [CrossRef] [PubMed]
18. Katona, S.; Paulikas, D.; Stone, G.S. Ethical opportunities in deep-sea collection of polymetallic nodules from the Clarion-Clipperton Zone. *Integr. Environ. Assess. Manag.* **2022**, *18*, 634–654. [CrossRef]
19. Shobayo, P.; van Hassel, E.; Vanelslander, T. Logistical Assessment of Deep-Sea Polymetallic Nodules Transport from an Offshore to an Onshore Location Using a Multiobjective Optimization Approach. *Sustainability* **2023**, *15*, 11317. [CrossRef]
20. Kacprzak, P. Assessment of cargo handling operation efficiency in the CCZ for standard bulk carriers in the view of significant amplitudes of roll as a limiting criterion. *Sci. J. Marit. Univ. Szczec.* **2023**, *74*, 55–64. [CrossRef]
21. Kacprzak, P. An analysis of shear forces, bending moments and roll motion during a nodule loading simulation for a ship at sea in the Clarion—Clipperton Zone. *Sci. J. Marit. Univ. Szczec.* **2021**, *65*, 9–20. [CrossRef]
22. Marques, C.H.; Belchiora, C.R.P.; Capracea, J.-D. Optimising the engine-propeller matching for a liquefied natural gas carrier T under rough weather. *Appl. Energy* **2018**, *232*, 187–196. [CrossRef]
23. Jurdziński, M. Processes of a freely drifting vessel. *TransNav* **2020**, *14*, 687–693. [CrossRef]
24. Bialystocki, N.; Konovessis, D. On the estimation of ship's fuel consumption and speed curve: A statistical approach. *J. Ocean. Eng. Sci.* **2016**, *1*, 157–166. [CrossRef]
25. Bal Beşikçi, E.; Arslan, O.; Turan, O.; Ölçer, A. An ANN based decision support system for energy efficient ship operations. *Comput. Oper. Res.* **2016**, *66*, 393–401. [CrossRef]
26. Petersen, J.P.; Jacobsen, D.J.; Winther, O. Statistical modelling for ship propulsion efficiency. *J. Mar. Sci. Technol.* **2011**, *17*, 30–39. [CrossRef]
27. Cepowski, T.; Drozd, A. Measurement-based relationships between container ship operating parameters and fuel consumption. *Appl. Energy* **2023**, *347*, 121315. [CrossRef]

28. Panagakos, G.; de Pessôa, T.S.; Dessypris, N.; Barfod, M.B.; Psaraftis, H.N. Monitoring the carbon footprint of dry bulk shipping in the EU: An early assessment of the MRV regulation. *Sustainability* **2019**, *11*, 5133. [CrossRef]
29. Kanberoğlu, B.; Kökkülünk, G. Assessment of CO_2 emissions for a bulk carrier fleet. *J. Clean. Prod.* **2021**, *283*, 124590. [CrossRef]
30. IMO. *Resolution MEPC.1/Circ.684 Guidelines for Voluntary Use of The Ship Energy Efficiency Operational Indicator (EEOI)*; IMO: London, UK, 2009.
31. IMO. *Resolution Res MEPC.364(79) Guidelines on the Method of Calculation of the Attained Energy Efficiency Design Index (EEDI) for New Ships*; IMO: London, UK, 2022.
32. IMO. *Resolution MEPC.346(78) Guidelines for the Development of a Ship Energy Efficiency. Management Plan (SEEMP)*; IMO: London, UK, 2022.
33. IMO. *Resolution MEPC.336(76) Guidelines on Operational Carbon Intensity Indicators and the Calculation Methods (CII GUIDELINES, G1)*; IMO: London, UK, 2021.
34. Chen, X.; Lv, S.; Shang, W.; Wu, H.; Xian, J.; Song, C. Ship energy consumption analysis and carbon emission exploitation via spatial-temporal maritime data. *Appl. Energy* **2024**, *360*, 122886. [CrossRef]
35. Lee, S.-S. Analysis of the effects of EEDI and EEXI implementation on CO2 emissions reduction in ships. *Ocean. Eng.* **2024**, *295*, 116877. [CrossRef]
36. Van Laar, G. Sustainable Transport of Polymetallic Nodules [TU Delft]. 2021. Available online: https://repository.tudelft.nl/islandora/object/uuid:ac42edba-a4ca-49ec-bb99-1c1b0b203f20 (accessed on 3 February 2024).
37. Lipton, I.T.; Nimmo, M.J. NI 43-101 Technical Report TOML Clarion Clipperton Zone Project, Pacific Ocean (Issue July). 2016. Available online: https://int.nyt.com/data/documenttools/2021-03-metals-company-technical-report-on-toml-mining-zone-plan/2d5350243bade994/full.pdf (accessed on 8 February 2024).
38. Dudziak, J. *Teoria Okrętu (II)*; Fundacja Promocji Przemysłu Okrętowego i Gospodarki Morskiej: Gdańsk, Poland, 2008; ISBN 978-83-60584-09-5.
39. Kadomatsu, K.; Inoue, Y.; Takarada, N. On the Required Minimum Output of Main Propulsion Engine for Large Fat Ship with Considering Manoeuverability in Rough Seas. *J. Soc. Nav. Arch. Jpn.* **1990**, *1990*, 171–182. [CrossRef] [PubMed]
40. Sprenger, F.; Maron, A.; Delefortrie, G.; Van Zwijnsvoorde, T.; Cura-Hochbaum, A.; Lengwinat, A.; Papanikolaou, A. Experimental Studies on Seakeeping and Manoeuvrability in Adverse Conditions. *J. Ship Res.* **2017**, *61*, 131–152. [CrossRef]
41. Shigunov, V.; El Moctar, O.; Papanikolaou, A.; Pottho, R.; Liu, S. International benchmark study on numerical simulation methods for prediction of manoeuvrability of ships in waves. *Ocean Eng.* **2018**, *165*, 365–385. [CrossRef]
42. Yasukawa, H.; Hirata, N.; Matsumoto, A.; Kuroiwa, R.; Mizokami, S. Evaluations of wave-induced steady forces and turning motion of a full hull ship in waves. *J. Mar. Sci. Technol.* **2018**, *24*, 1–15. [CrossRef]
43. Ueno, M.; Nimura, T.; Miyazaki, H.; Nonaka, K. Steady Wave Forces and Moment Acting on Ships in Manoeuvring Motion in Short Waves. *J. Soc. Nav. Arch. Jpn.* **2000**, *2000*, 163–172. [CrossRef] [PubMed]
44. Ueno, M.; Nimura, T.; Miyazaki, H.; Nonaka, K.; Haraguchi, T. Model Experiment on Steady Wave Forces and Moment Acting on a Ship a Rest. *J. Kansai Soc. Nav. Archit.* **2001**, *235*, 69–77.
45. Bortnowska, M. The Determination of Power Output of the Tracking Control System of a Mining Ship Zeszyty Naukowe NR 10 Akademii Morskiej W Szczecinie. *Sci. J. Marit. Univ. Szczec.* **2006**, *10*, 97–106.
46. ITTC. ITTC—Recommended Procedures and Guidelines-Seakeeping Experiments. In Proceedings of the 27th ITTC Seakeeping Committee, Copenhagen, Denmark, 31 August–5 September 2014; pp. 1–22.
47. Maruo, H. The drift of a body floating on waves. *J. Ship Res.* **1960**, *4*, 1–10.
48. Newman, J.N. The drift force and moment on ships in waves. *J. Ship Res.* **1967**, *11*, 51–60. [CrossRef]
49. Salvesen, N. Second-Order Steady State Forces and Moments on Surface Ships in Oblique Regular Waves. In International Symposium on Dynamics of Marine Vehicles and Structures in Waves. *Mech. Eng.* **1974**, *22*, 225–241.
50. Gerritsma, J.; Beukelman, W. Analysis of the resistance increase in waves of a fast cargo ship 12. *Int. Shipbuild. Prog.* **1972**, *19*, 285–293. [CrossRef]
51. Kashiwagi, M. Added Resistance, Wave-Induced Steady Sway Force and Yaw Moment on an Advancing Ship. *Ship Technol. Res. (Schistech.)* **1992**, *39*, 3–16.
52. Boese, P. Eine einfache Methode zur Berechnung der Widerstandserhöhung eines Schiffes im Seegang. *Ship Technol. Res.* **1970**, *258*, 17.
53. Faltinsen, O.M.; Minsaas, K.J.; Liapis, N.; Skjordal, S.O. Prediction of Resistance and Propulsion of a Ship in a Seaway. In Proceedings of the 13th Symposium on Naval Hydrodynamics, Tokyo, Japan, 6–10 October 1980; pp. 505–529.
54. Papanikolaou, A.; Nowacki, Z. Second-Order Theory of Oscillating Cylinders in a Regular Steep Wave. In Proceedings of the l3th ONR Symp., Tokyo, Japan, 6–10 October 1980; pp. 303–333.
55. Liu, S.; Papanikolaou, A. Prediction of the Side Drift Force of Full Ships Advancing in Waves at Low Speeds. *J. Mar. Sci. Eng.* **2020**, *8*, 377. [CrossRef]
56. Cepowski, T. The prediction of ship added resistance at the preliminary design stage by the use of an artificial neural network. *Ocean. Eng.* **2020**, *195*, 106657. [CrossRef]
57. Cepowski, T. The use of a set of artificial neural networks to predict added resistance in head waves at the parametric ship design stage. *Ocean. Eng.* **2023**, *281*, 114744. [CrossRef]

58. Ahn, Y.; Kim, Y. Data mining in sloshing experiment database and application of neural network for extreme load prediction. *Mar. Struct.* **2021**, *80*, 103074. [CrossRef]
59. Dyer, A.S.; Zaengle, D.; Nelson, J.R.; Duran, R.; Wenzlick, M.; Wingo, P.C.; Bauer, J.R.; Rose, K.; Romeo, L. Applied machine learning model comparison: Predicting offshore platform integrity with gradient boosting algorithms and neural networks. *Mar. Struct.* **2022**, *83*, 103152. [CrossRef]
60. Citakoglu, H.; Coşkun, Ö. Comparison of hybrid machine learning methods for the prediction of short-term meteorological droughts of Sakarya Meteorological Station in Turkey. *Environ. Sci. Pollut. Res.* **2022**, *29*, 75487–75511. [CrossRef]
61. Demir, V.; Citakoglu, H. Forecasting of solar radiation using different machine learning approaches. *Neural Comput. Applic.* **2023**, *35*, 887–906. [CrossRef]
62. Premalatha, M.; Naveen, C. Analysis of different combi- nations of meteorological parameters in predicting the horizontal global solar radiation with ann approach: A case study. *Renew. Sustain. Energy Rev.* **2018**, *91*, 248–258. [CrossRef]
63. Uncuoglu, E.; Citakoglu, H.; Latifoglu, L.; Bayram, S.; Laman, M.; Ilkentapar, M.; Oner, A.A. Comparison of neural network, Gaussian regression, support vector machine, long short-term memory, multi-gene genetic programming, and M5 Trees methods for solving civil enginee *Energies* ring problems. *Appl. Soft Comput.* **2022**, *129*, 109623. [CrossRef]
64. Broyden, C.G. The convergence of a class of double-rank minimization algorithms 1. General considerations. *IMA J. Appl. Math.* **1970**, *6*, 76–90. [CrossRef]
65. Fletcher, R. A new approach to variable metric algorithms. *Comput. J.* **1970**, *13*, 317–322. [CrossRef]
66. Goldfarb, D. A family of variable-metric methods derived by variational means. *Math. Comput.* **1970**, *24*, 23–26. [CrossRef]
67. Shanno, D.F. Conditioning of quasi-newton methods for function minimization. *Math. Comput.* **1970**, *24*, 647–656. [CrossRef]
68. UNCTAD. Review of Maritime Transport (UNCTAD/RMT/2023). United Nations Conference on Trade and Development. 2023. Available online: https://unctad.org/publication/review-maritime-transport-2023 (accessed on 2 March 2024).
69. Obydenkova, S.V.; Defauw, L.V.E.; Kouris, P.D.; Smeulders, D.M.J.; Boot, M.D.; van der Meer, Y. Environmental and Economic Assessment of a Novel Solvolysis-Based Biorefinery Producing Lignin-Derived Marine Biofuel and Cellulosic Ethanol. *Energies* **2022**, *15*, 5007. [CrossRef]
70. Bortuzzo, V.; Bertagna, S.; Bucci, V. Mitigation of CO_2 Emissions from Commercial Ships: Evaluation of the Technology Readiness Level of Carbon Capture Systems. *Energies* **2023**, *16*, 3646. [CrossRef]
71. Richter, S.; Braun-Unkhoff, M.; Hasselwander, S.; Haas, S. Evaluation of the Applicability of Synthetic Fuels and Their Life Cycle Analyses. *Energies* **2024**, *17*, 981. [CrossRef]

Disclaimer/Publisher's Note: The statements, opinions and data contained in all publications are solely those of the individual author(s) and contributor(s) and not of MDPI and/or the editor(s). MDPI and/or the editor(s) disclaim responsibility for any injury to people or property resulting from any ideas, methods, instructions or products referred to in the content.

Article

Visualisation Testing of the Vertex Angle of the Spray Formed by Injected Diesel–Ethanol Fuel Blends

Artur Krzemiński *[ID] and Adam Ustrzycki *[ID]

Faculty of Mechanical Engineering and Aeronautics, Rzeszow University of Technology, 35-959 Rzeszów, Poland
* Correspondence: artkrzem@prz.edu.pl (A.K.); austrzyc@prz.edu.pl (A.U.); Tel.: +48-177-432-550 (A.K.); +48-178-651-531 (A.U.)

Abstract: The internal combustion engine continues to be the main source of power in various modes of transport and industrial machines. This is due to its numerous advantages, such as easy adaptability, high efficiency, reliability and low fuel consumption. Despite these beneficial qualities of internal combustion engines, growing concerns are related to their negative environmental impacts. As a result, environmental protection has become a major factor determining advancements in the automotive industry in recent years, with the search for alternative fuels being one of the priorities in research and development activities. Among these, fuels of plant origin, mainly alcohols, are attracting a lot of attention due to their high oxygen content (around 35%). These fuels differ from diesel oil, for instance, in kinematic viscosity and density, which can affect the formation of the fuel spray and, consequently, the proper functioning of the compression–ignition engine, as well as the performance and purity of the exhaust gases emitted into the environment. The process of spray formation in direct injection compression–ignition engines is extremely complicated and requires detailed analysis of the fast-changing variables. This explains the need for using complicated research equipment enabling visualisation tests and making it possible to gain a more accurate understanding of the processes that take place. The present article aims to present the methodology for alternative fuel visualisation tests. To achieve this purpose, sprays formed by diesel–ethanol blends were recorded. A visualisation chamber and a high-speed camera were used for this purpose. The acquired video provided the material for the analysis of the changes in the vertex angle of the spray formed by the fuel blends. The test was carried out under reproducible conditions in line with the test methodology. The shape of the fuel spray is impacted by an increase in the proportional content of ethanol in the diesel and dodecanol blend. Based on the present findings, it is possible to note that the values of the vertex angle in the spray produced by the diesel–ethanol blend with the addition of dodecanol are most similar to those produced by diesel oil at an injection pressure of 100 MPa. The proposed methodology enables an analysis of the injection process based on the spray macrostructure parameters, and it can be applied in the testing of alternative fuels.

Keywords: visualisation testing; fuel spray vertex angle; diesel; ethanol; dodecanol; alternative fuels

Citation: Krzemiński, A.; Ustrzycki, A. Visualisation Testing of the Vertex Angle of the Spray Formed by Injected Diesel–Ethanol Fuel Blends. *Energies* **2024**, *17*, 3012. https://doi.org/10.3390/en17123012

Academic Editor: Adam Smoliński

Received: 24 May 2024
Revised: 12 June 2024
Accepted: 14 June 2024
Published: 18 June 2024

Copyright: © 2024 by the authors. Licensee MDPI, Basel, Switzerland. This article is an open access article distributed under the terms and conditions of the Creative Commons Attribution (CC BY) license (https://creativecommons.org/licenses/by/4.0/).

1. Introduction

Compression–ignition engines are widely used in motor vehicles and agricultural machinery [1]. This results from their easy adaptability, relatively high efficiency and reliability [2,3].

Such engines, however, are associated with certain drawbacks, e.g., emissions of pollutants such as nitrogen oxides (NOx), particulate matter (PM), carbon monoxide (CO) and other harmful substances. The emissions of NOx produced by diesel engines constitute 70% of the total emissions from combustion engines [4]. For this reason, increasingly stringent rules and regulations are being introduced worldwide with regard to the emissions of toxic compounds in exhaust gases [5]. The search for clean and renewable fuels is of key importance for mitigating the energy crisis and protecting the environment [6].

In an attempt to cut down the consumption of fossil fuels and reduce emissions of CO_2 and other pollutants generated by internal combustion engines, a shift from internal combustion engines to electric drive systems has been proposed. However, this solution presents drawbacks adversely affecting advancements in this area; these include long charging times, a lack of charging infrastructure, short range compared to internal combustion engines and the high cost of these vehicles [7]. As a result, compression–ignition engines are still widely used in transport, construction machinery and agriculture [8]. Therefore, it is necessary to carry out research in order to reduce the pollution generated by the combustion of fossil fuels. One solution for reducing toxic emissions lies in the advancement of biofuels that can be produced from biodegradable domestic, industrial and agricultural waste. The main objective of research and development operations related to this issue is to find fuels that are environmentally friendly and present an economic advantage over petroleum-based fuels. In the short term, the goal is to develop fuel blends based on diesel oil with increasing proportional contents of bio-components so that in the medium and long term, it will be possible to produce clean fuels without the use of petroleum products. According to the International Energy Agency, bioenergy production is expected to reach 100 EJ by 2050 [9].

Alternative fuels attracting a lot of attention include alcohols (ethanol, methanol and butanol) [10–12]. This is associated with a number of advantages presented by these liquids, such as high oxygen content and low viscosity. They are produced using biomass fermentation [13,14]. when used in blends with diesel oil, ethanol presents better characteristics than methanol. It blends more effectively into diesel oil and has a higher cetane number and superior lower calorific value [15]. Moreover, compared to methanol–diesel blends, the combustion of ethanol–diesel blends produces exhaust gas with a lower number of toxic compounds [16,17].

Ethanol can, for example, be produced from lignocellulose acquired from food industry waste, forestry waste (wood and waste from paper industry and logging waste), secondary waste (waste from food production and municipal solid waste) and agricultural waste [18]. On the other hand, ethanol has a low cetane number and lubricity compared to diesel oil, so it is beneficial to use ethanol in a mixture with diesel [17,19,20]. Blends of this type can be used in compression–ignition engines designed to run on typical diesel fuel [21].

Ethanol–diesel blends differ from diesel oil in physical properties such as density, viscosity and surface tension [22]. Density affects the mass of fuel injected and, consequently, fuel consumption [23]. Viscosity, in turn, impacts the range of the fuel spray [24], whereas a change in surface tension affects the size of droplets in the fuel spray [25]. Experimental studies have shown that an increase in the proportional volume of ethanol in the blend corresponds to a decrease in the cetane number, kinematic viscosity and calorific value [26,27].

Many studies assessing the effects of ethanol–diesel blends on exhaust emissions have reported reduced CO, NOx and PM [28,29]. The use of these blends also affects engine performance, leading to an increase in specific fuel consumption, a slight decrease in engine power and an increase in exhaust gas temperature [30,31].

The fuel atomisation process plays an important role in the effective operation of internal combustion engines. The quality and precision of the fuel spray largely determine the course of the combustion process. The atomisation process affects economic efficiency as well as exhaust toxicity and engine noise [32,33].

The process of fuel injection may be assessed by taking into account the indicators of fuel spray geometry. These indicators are most commonly measured during injection into the ambient air or into a chamber filled with air or nitrogen, where the back-pressure conditions reflect the true conditions in the combustion chamber [34].

Fuel spray can be recorded using machine vision systems. These are various testing methods designed to record an image of an object or a process emitting internal light or reflecting light that originates from an independent source on a photosensitive or magnetic medium in an analogue or digital format. The recorded object can be assessed in terms of its qualitative and quantitative characteristics as well as geometric and physical parameters (e.g., shape, changes in shape, brightness, and colour). Machine vision systems are an important tool enabling the analysis of liquid atomisation and, in particular, the spray formation and decomposition processes. These techniques include direct photography, where the object is illuminated directly or from behind, as well as filming with high-speed cameras. In the latter case, the image is recorded with a photo camera with fast shutter speed or using video technology. Illumination with diffuse light is usually applied. Depending on the operating speed of the camera, it is possible to record, with a sufficient resolution, phenomena such as the formation of a fuel spray, the effect of droplet disintegration, etc. Another type used for this purpose is ultra-high-speed photography, reaching 1.25 million frames per second.

Another type of direct photography is microphotography. It can be used to capture a very small area and to identify extremely fine details. It makes it possible to observe a flame inside the combustion chamber and changes in fuel spray. Another photography technique applies backlighting. In this case, a diffused light source is placed behind the object, as a result of which the background is white and the photographed objects are black. It has been used to obtain images of fuel spray and clouds of soot [35]. The specific varieties of backlight photography include silhouette microphotography and shadow photography. The measuring system in silhouette photography consists of lighting and recording units. The former is used to provide a high level of brightness and a short illumination time. Flash lamps, stroboscopic lamps or flash generators are used as a source of light. The recording system is a photo camera in which a microscope system is installed instead of a photo lens. The dimensions of the objects observed depend on the resolving power of the microscope. In this method, the following parameters are important: exposure time, distance of the camera from the fuel spray, sharpness of the droplet image and the depth of field of the microscope. Possible ways to avoid contamination include using protection with air flow or an aperture that opens together with the lighting unit [36].

Shadow photography involves examining the distribution of the refractive index of isotropic media. This method makes it possible to record the shadow created against a bright background by an object illuminated from behind. Light focused on the object is used for illumination. The shadow method can be used to examine the density distribution and to visualise the distribution of fuel vapour; it is sensitive to the medium concentration gradient. The trailing method, on the other hand, is additionally sensitive to the density and temperature gradients in the medium. It can be used to observe the structure of the injected fuel spray, the concentration of the air–fuel mixture, droplet diameters and the concentration of the non-vapourised fuel spray [37].

The high-speed photography method is characterised by short exposure times, achieved either by reducing the duration of the light flash or by using a camera for high-speed shooting in continuous light. High-speed imaging cameras have a variety of designs, depending on the required shooting frequency. Continuous high-speed recording of a stream of droplets enables the internal structure of the spray to be examined in time and space. In this way, it is possible to acquire dynamic characteristics comprising processes such as the development of the liquid spray pattern at the outlet of the atomiser, the breakdown of the fuel into droplets, the formation of inhomogeneities in the structure of the spray pattern and the separation of smaller droplets in large vortex structures, etc.

The purpose of this study is to measure the vertex angle of the fuel spray in a tray injection system, using direct high-speed photography. The vertex angle is one of the parameters defining fuel spray. It determines the degree of fuel distribution in the combustion chamber and depends on a number of factors such as the fuel discharge velocity, the back pressure of the medium, the dimensions of the outlet aperture and the physical properties of the fuel. So far, no uniform analytic formula has been adopted to determine this type of angle. Therefore, to enable measurement of the vertex angle, it is necessary to conduct experimental research. Due to the non-stationary nature and a lack of local homogeneity of the injection process, machine vision systems are particularly well-suited for recording spray macrostructure parameters [37]. In view of this, measurement of the fuel spray vertex angle was performed using an optical method, which involves filming the fuel spray in a constant-volume chamber with a high-speed camera. As an advantage, this is a non-contact recording method, as a result of which the injection process is not disturbed.

The main purpose of this paper is to present a methodology and a workstation for visualisation tests enabling assessments of alternative fuels. The study also aims to determine the fuel spray vertex angle relative to the proportional volume of ethanol in the blends investigated. This issue has been selected for this research following a review of numerous studies of ethanol–diesel blends, which mainly focused on their physical properties and toxic emissions in the exhaust gas. Unfortunately, there are few studies evaluating the effect of ethanol on the geometric parameters of the spray, which is key to establishing the optimum operating conditions for engines running on this type of fuel.

2. Methodology of the Research

The study examined the atomisation of various fuel blends injected using a five-hole spray nozzle. The visualisation tests were performed at the workstation shown in Figure 1, comprising the following components:

- Testing bench (basic unit of the station).
- Hydraulic feed system.
- Common Rail injection system.
- Injection visualisation chamber.
- High-speed camera to film the injection process.

The basic components of the system responsible for executing the injection include a low-pressure circuit, a high-pressure circuit and a controller.

A low-pressure fuel circuit could not be used because the test bench manufacturer permits only the use of the dedicated test oil. Therefore, the tests were performed using a separate hydraulic system, which enables the precise measurement of the injected fuel output through the injector and from the overflow. Apart from measuring the fuel output, the flowmeters applied also enable measurement of the basic parameters of the injected fuel, such as temperature and density. Furthermore, the system provides cooling to maintain a constant temperature of the fuel subjected to testing and to ensure stable conditions for the discharge of fuel by the high-pressure pump. For this, it is necessary to maintain the defined temperature in the tank, for instance, by means of a cooler for the fuel returning from the overflow of the high-pressure pump.

The test bench manual pressure valves were used to ensure adequate feed and overflow pressure of the high-pressure pump. Manometers with appropriate measuring ranges were used to control the defined pressure.

The high-pressure system mounted on the test bench consists of a high-pressure pump, a fuel tank, an injector installed in the visualisation chamber and high-pressure piping. An electric fuel pump transfers fuel from the tank to the high-pressure pump. The feed pressure of the high-pressure pump is regulated by an adjustable valve and measured by a manometer, similar to the pressure at the pump overflow, which is also set using an adjustable valve and controlled by a manometer. The fuel from the overflow passes through the cooler and into the tank. The flow of water through the cooler is regulated by a solenoid valve based on information about the temperature of the returning fuel.

The high-pressure pump forces the fuel into the tank, where the pressure is measured; the pressure regulators in the pump and in the busbar are controlled based on this information. The fuel is forced from the tank to the injector, which then discharges it into the visualisation chamber. In the tests, an electromagnetic injector marked 0445 110 083 is used and the diameter of the nozzle holes is 0.15 mm. The injected fuel flow rate and overflow volume are measured using flowmeters. The water flow in the coolers is regulated by means of solenoid valves monitored by the test bench control unit.

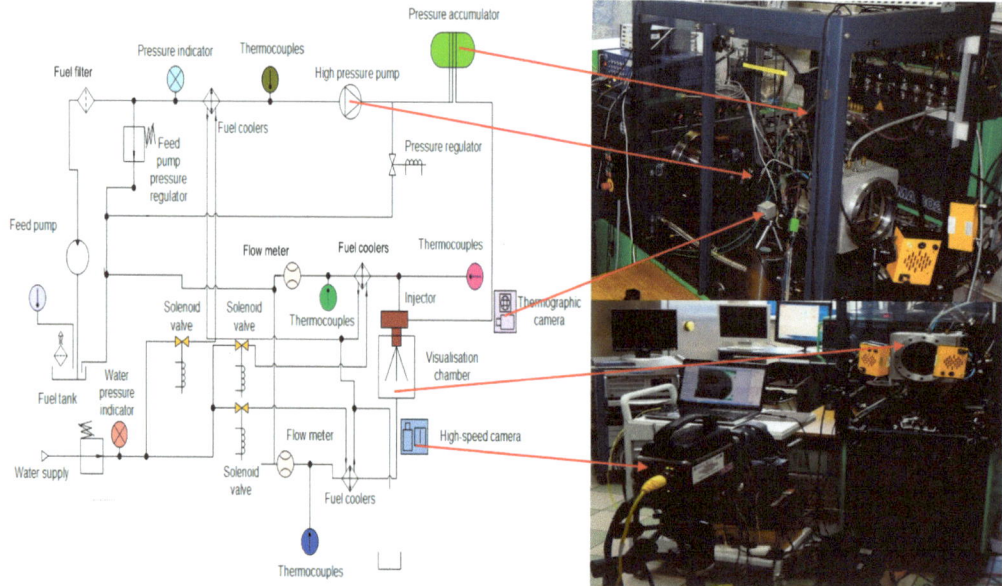

Figure 1. Schematic representation of the workstation for fuel spray visualisation testing.

The injector is monitored by a microprocessor controller, making it possible to change the following control signal parameters:

- Duration of the entire signal controlling the aperture of the injector.
- Duration of the sustained signal.
- Frequency of the sustained signal.
- Fullness of the sustained signal.

The controller enables up to three fuel doses to be injected during one revolution of the crankshaft, and it also makes it possible to apply continuous and single injections. The input signals of the injection control system, i.e., the position and speed signals of the high-pressure pump, are emitted by an encoder mounted on the shaft of the motor driving the pump. Based on these data and information provided by the user via the computer, the controller produces and sends the appropriate signal driving the solenoid injector.

The parameters of the injector control signal can be changed via the dialogue box provided by the software. A thermographic camera was used to monitor the injector temperature. The temperature of the injector was measured in the area marked with a white rectangle (Figure 2).

The basic components enabling optical testing include the visualisation chamber and the recording system. The visualisation chamber (Figure 3) consists of the body and glass panes, making it possible to adequately illuminate the spray. On the inner wall of the chamber, visible through the front glass pane, there is a disc with a graduation in the form of coaxial circles aligned with the injector; its purpose is to help quickly estimate the range of the fuel spray. The smallest circle has a diameter of 20 mm, and each subsequent circle is 20 mm larger in diameter.

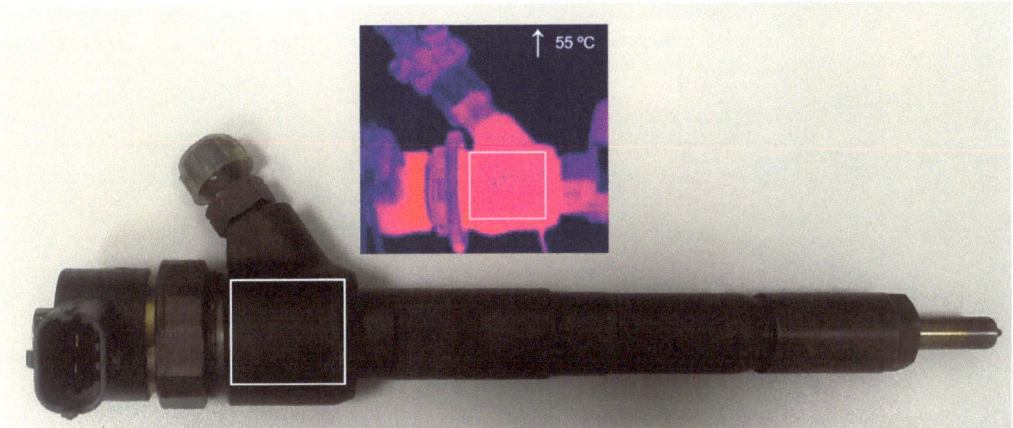

Figure 2. Image of injector frame taken with thermographic camera during the test.

By using high-strength glass inside the chamber, it is also possible to apply back-pressure of the medium (up to approximately 3 MPa), into which the fuel is injected, to better replicate the conditions within the engine chamber.

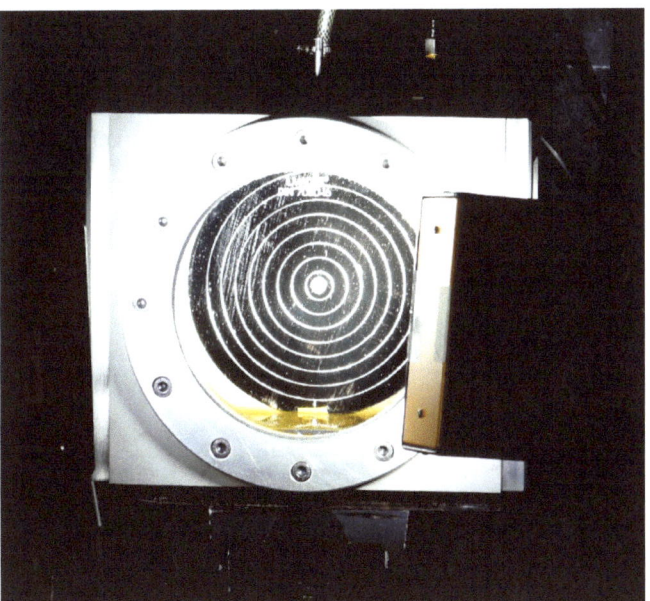

Figure 3. Illuminated visualisation chamber.

A high-speed camera was used to record the image of the fuel spray injected into the chamber containing air with atmospheric pressure. In order to perform tests that require high-speed filming, it is necessary to apply adequate illumination; for this purpose, two LED illuminators, 100 W each, were used. It is also important to select a lens that makes it possible to achieve adequate-quality recordings of the injected fuel spray. Hence, a full-frame, wide-angle lens with a fixed focal length of 20 mm and a maximum diameter of aperture of f/1.4 was used for this purpose.

Visualisation tests were performed to measure the vertex angle of the spray for seven samples of flammable liquids with different proportional volume contents of ethanol in diesel fuel blends (Table 1).

Table 1. List of the fuel blends tested.

No	Type of Fuel	Percentage [% v/v]		
		Dodecanol	Ethanol	Diesel Oil
1	ON	0	0	100
2	ON–ET5	5	5	90
3	ON–ET10	5	10	85
4	ON–ET15	5	15	80
5	ON–ET20	5	20	75
6	ON–ET25	5	25	70
7	ON–ET30	5	30	65

Each mixture with an ethanol content in the range of 5% to 30% v/v was supplemented with 5% dodecanol in order to obtain homogenous blends [14]. The samples were prepared from diesel oil, whose selected parameters are shown in Table 2, as well as denatured and dehydrated ethanol, whose main properties are shown in Table 3. Selected parameters of dodecanol, applied to improve the mixture's stability, are shown in Table 4.

Table 2. Selected properties of diesel oil [38].

Name of Parameter	Unit	Value
Cetane number	–	51.7
Kinematic viscosity (40 °C)	mm^2/s	2.51
Density (15 °C)	g/cm^3	0.833
Water content	%(m/m)	0.008
Sulphur content	mg/kg	5.9

Table 3. Selected ethanol parameters [39].

Name of Parameter	Unit	Value
Density	g/cm^3	0.789
Alcohol content (20 °C)	%	99.9
Kinematic viscosity (40 °C)	mm^2/s	1.13
Sulphur content	mg/kg	5.9

Table 4. Selected dodecanol parameters [40].

Name of Parameter	Unit	Value
Density (25 °C)	g/cm^3	0.843
Autoignition temperature	°C	275
Flash point	°C	134.8

The specified blends were subjected to visualisation tests designed to measure the vertex angle of the fuel spray. During the tests, atmospheric pressure was maintained in the constant-volume chamber. The program of the tests performed at the workstation was realised by the following sequence of activities:

- Cleaning the hydraulic system with compressed air and filling the system with the investigated fuel.
- Setting the appropriate injection pressure and time.
- Performing a procedure of heating the fuel in the system to a temperature of 40 °C ± 2 °C and the injector to a temperature of 55 °C ± 2 °C (the required temperatures are obtained through operation of the system under the specified test conditions).

- Performing a single injection into the visualisation chamber and recording it with the high-speed camera.
- Repeating the measurements a specified number of times to obtain results for statistical processing.
- Repeating the measurements for all the specified values of pressure in the system.
- After each experiment, the chamber was rinsed to prevent the formation of background fog from the previously sprayed fuel.
- After the tests had been completed, removing the fuel and flushing the system.

The recorded fuel sprays were examined by observation of individual frames to measure the vertex angle of the fuel spray. The measurement was performed for the frame preceding the end of the injection, defined as the spray breaking away from the atomiser. The vertex angle of the spray was measured between the two straight lines tracing the spray (Figure 4).

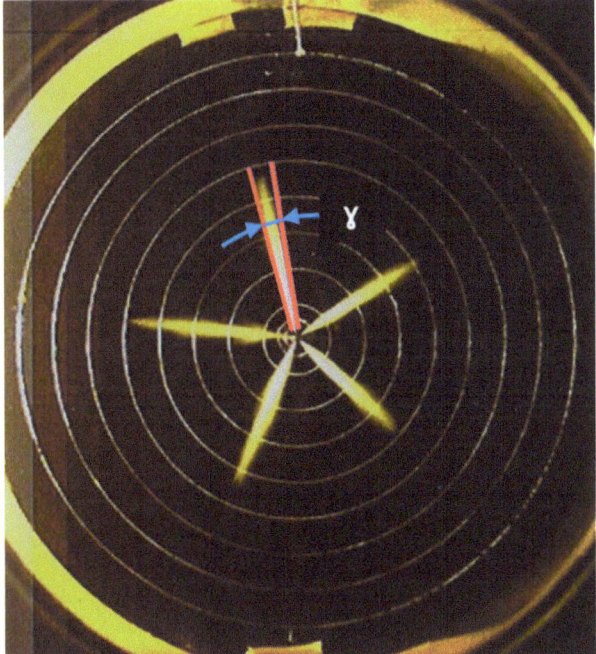

Figure 4. The method for measuring the vertex angle of the first spray: γ—the vertex angle measured for the first spray.

The basic parameters of the fuel blends, defined based on the authors' own measurements, are shown in Table 5. The related methodology has been reported in other publications [14,41].

As a result of a review of the related literature, which suggests that at least 5 to 10 measurements [42] should be performed to reduce uncertainty, the authors ultimately decided to carry out 7 trials.

Statistical analysis was applied to determine the measurement uncertainty. An expanded uncertainty was also calculated [43], with the expansion coefficients determined from Student's t-tables for a confidence interval of $\alpha = 0.05$.

Table 5. Selected parameters of the fuel blends [14,41].

Type of Fuel	Density (15 °C) [g/cm^3]	Kinematic Viscosity (40 °C) [mm^2/s]
ON	0.83041	2.5807
ON–ET5	0.82834	2.4294
ON–ET10	0.82660	2.3085
ON–ET15	0.82472	2.2143
ON–ET20	0.82299	2.1278
ON–ET25	0.82127	2.0662
ON–ET30	0.81999	1.9908

3. Results and Analysis

To assess all the blend samples prepared, the evolution of the spray produced by fuel injected into the visualisation chamber was filmed; based on that, it was possible to analyse the parameters of the sprays. The image was recorded with a speed of 24,096 fps and a resolution of 192 × 192. The tests were performed for three different values of injection pressure (75, 100 and 125 MPa). The injection time was selected in such a way as to ensure the fuel spray would not have an excessively large range and would not bounce off the walls of the visualisation chamber. The injection parameters applied during the tests are shown in Table 6.

Table 6. Parameters defined for fuel spray visualisation tests.

Speed of the High-Pressure Pump 1000 rpm			
Injection Time [μs]	Injection Pressure [MPa]	Fuel Tank Temperature [°C]	Injector Temperature [°C]
500	75	40 ± 2	55 ± 2
500	100	40 ± 2	55 ± 2
500	125	40 ± 2	55 ± 2

The film with the recorded fuel spray can be replayed as a video sequence, making it possible to examine the continuity of the fuel injection. Another method that can be used to make assessments involves viewing the individual frames in order to evaluate the geometric parameters of the spray (the spray vertex angle and the range of the spray tip). The vertex angle of the spray was measured at the maximum range of the spray tip. Selected images recorded during the fuel spray visualisation for some fuel blends are shown in Figures 5–11.

Figure 5. The recorded maximum range of spray for diesel oil at an injection pressure of 75, 100 and 125 MPa.

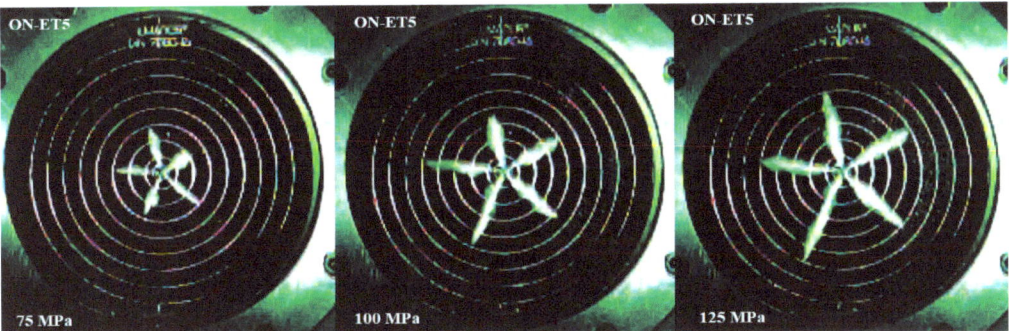

Figure 6. The recorded maximum range of spray for diesel blend with 5% v/v content of ethanol at an injection pressure of 75, 100 and 125 MPa.

Figure 7. The recorded maximum range of spray for diesel blend with 10% v/v content of ethanol at an injection pressure of 75, 100 and 125 MPa.

Figure 8. The recorded maximum range of spray for diesel blend with 15% v/v content of ethanol at an injection pressure of 75, 100 and 125 MPa.

Figure 9. The recorded maximum range of spray for diesel blend with 20% v/v content of ethanol at an injection pressure of 75, 100 and 125 MPa.

Figure 10. The recorded maximum range of spray for diesel blend with 25% v/v content of ethanol at an injection pressure of 75, 100 and 125 MPa.

Figure 11. The recorded maximum range of spray for diesel blend with 30% v/v content of ethanol at an injection pressure of 75, 100 and 125 MPa.

The mean values of the fuel spray vertex angle are presented in Figure 12. The findings show that in the case of injection pressure of 75 MPa, the lowest value of the spray vertex angle, i.e., 5.8°, was identified in ON–ET30, and the largest spray vertex angle of 12.5° was shown in ON–ET10. At an injection pressure of 100 MPa, the largest angle of 8.8° was identified in the case of ON–ET5, and the smallest of approximately 5.4° was found for ON–ET30. In the case of an injection pressure of 125 MPa, the findings show the smallest value of the vertex angle of approximately 3.6° produced by ON–ET30, whereas the largest value of 10.5° was identified for ON–ET5. It is possible to observe that in the tests with an injection pressure of 75 MPa, the spray vertex angle increases with the higher proportional volume of ethanol, up to a content of approximately 10%. After that, the spray vertex angle at this injection pressure decreases. In the case of an injection pressure of 100 MPa, the spray vertex angle decreases when the ethanol content in the blend is higher. At an injection pressure of 125 MPa, the spray vertex angle increases in the sample with a 5% v/v content of ethanol and then decreases with the higher contents. In the case of the blends with an ethanol content higher than 15% v/v, the spray is more concentrated, as reflected by the lower value of the spray vertex angle. At an injection pressure of 75 MPa, the highest value of the spray vertex angle amounting to 12.5° was identified in the case of the blend with 10% v/v content of ethanol; compared to diesel oil, this is an increase of approximately 50.7%. The lowest value of the angle, i.e., 5.8°, was measured in the sample with a 30% v/v content of ethanol, which is a nearly 30.2% decrease relative to diesel oil. In the tests with an injection pressure of 100 MPa, the smallest spray vertex angle of 5.41° was identified for the ON–ET30 fuel sample; this value was nearly 30.7% lower than the vertex angle shown for diesel oil. In the case of the ON–ET5 sample, the angle was 8.8°, which was nearly 13% higher than the ON sample. On the other hand, at an injection pressure of 125 MPa, there was a 69% increase in the spray vertex angle in the case of the sample with a 5% v/v content of ethanol in which the value of the vertex angle amounted to 10.5°. The smallest angle of 3.6° was identified in the case of the blend with a 30% v/v content of ethanol; this value was 73% lower than diesel oil.

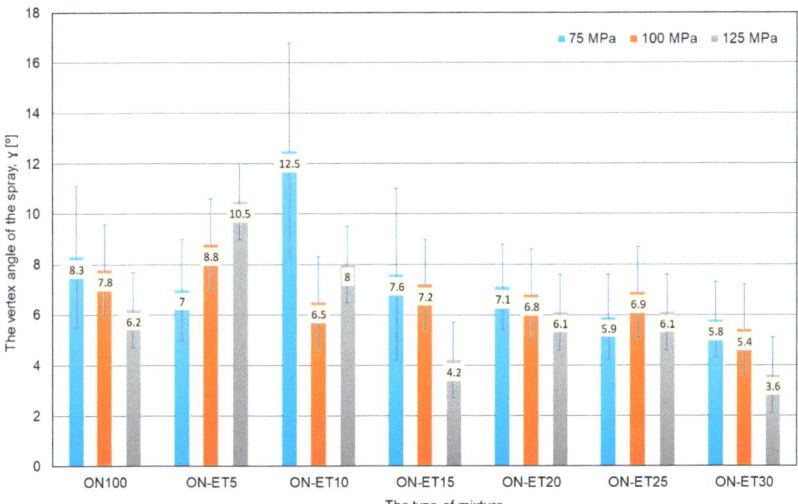

Figure 12. The compared mean values of the vertex angle of spray produced by flammable liquids investigated at different values of pressure in the injector system.

4. Conclusions

The characteristics of the fuel injection process in compression–ignition engines significantly impact the combustion process and all the resulting engine parameters, i.e., both the operational (engine torque, engine power, and specific fuel consumption) and environmental properties. These parameters are impacted by the macrostructure of the fuel spray, which depends on a number of factors, including the physical and chemical properties of the fuel (e.g., density and viscosity), the parameters of the injection process (e.g., injection pressure and the spray structure) and the design and precision engineering of the injector components.

The spray formation in direct injection compression–ignition engines is an extremely complicated process with a number of fast-changing variables that must be analysed in detail. Because of this, it is necessary to use advanced research equipment enabling visualisation tests and providing more accurate insight into the processes that take place. The proposed methodology enables such analysis and can be used to assess the macrostructure parameters of spray produced by alternative fuels, such as blends of diesel oil and alcohol.

Analysis of the findings based on the visualisation tests shows that:

- At an injection pressure of 75 MPa, only the ON–ET10 sample was shown to produce a spray vertex angle that was larger than commercial diesel oil (by approximately 50.7%), whereas the smallest vertex angle was identified in the case of ON–ET30 (the value was approximately 30% lower than the baseline diesel oil).
- At an injection pressure of 100 MPa, only the sample of the ON–ET5 blend was found to have a larger spray vertex angle than the diesel oil sample (by approximately 13%).
- The smallest spray vertex angle, at an injection pressure of 100 MPa, was identified in the case of ON–ET30; the value was approximately 31% lower than diesel oil.
- At an injection pressure of 125 MPa, a larger vertex angle than that produced by commercial diesel oil was identified in the case of the blends with 5% and 10% contents of ethanol; for the remaining blends, the value of this parameter was lower.
- At an injection pressure of 125 MPa, the largest spray vertex angle, compared to that produced by the baseline diesel oil, was identified in the case of the ON–ET5 blend (the value was approximately 69% higher).
- The smallest spray vertex angle at an injection pressure of 125 MPa was identified in the case of the ON–ET30 blend; the value was approximately 42% lower than the baseline fuel.

The present findings suggest that an increase in the proportional content of ethanol in blends with diesel oil and dodecanol may affect the shape of the fuel spray. It has also been shown that the values of the spray vertex angle identified in the case of ethanol–diesel blends with the addition of dodecanol are most similar to those observed in the case of diesel oil in the tests performed at an injection pressure of 100 MPa. Therefore, these blends can be used in compression–ignition engines since, in a wide range of operating conditions, the pressure values most commonly applied in typical injection systems roughly correspond to this value.

Author Contributions: Conceptualization, A.K. and A.U.; methodology, A.K. and A.U.; software, A.K. and A.U.; validation, A.U.; formal analysis, A.K. and A.U. investigation, A.U.; resources, A.K. and A.U.; data curation, A.K. and A.U.; writing—original draft preparation, A.K.; writing—review and editing, A.K. and A.U.; visualization, A.K.; supervision, A.U. All authors have read and agreed to the published version of the manuscript.

Funding: This research received no external funding. The APC was funded by Rzeszow University of Technology.

Institutional Review Board Statement: Not applicable.

Informed Consent Statement: Not applicable.

Data Availability Statement: Data are contained within the article.

Conflicts of Interest: The authors declare no conflicts of interest.

Abbreviations

CO	Carbon monoxide
CO_2	Carbon dioxide
HC	Hydrocarbons
NO_x	Nitrogen oxides
PM	Particulate matter

References

1. Zhang, Z.; Li, J.; Tian, J.; Dong, R.; Zou, Z.; Gao, S.; Tan, D. Performance, combustion and emission characteristics investigations on a diesel engine fueled with diesel/ethanol/n-butanol blends. *Energy* **2022**, *249*, 123733. [CrossRef]
2. Li, X.; Liu, Q.; Ma, Y.; Wu, G.; Yang, Z.; Fu, Q. Simulation Study on the Combustion and Emissions of a Diesel Engine with Different Oxygenated Blended Fuels. *Sustainability* **2024**, *16*, 631. [CrossRef]
3. Huang, Z.; Huang, J.; Luo, J.; Hu, D.; Yin, Z. Performance enhancement and emission reduction of a diesel engine fueled with different biodiesel-diesel blending fuel based on the multi-parameter optimization theory. *Fuel* **2022**, *314*, 122753. [CrossRef]
4. Zhang, Z.; Ye, J.; Tan, D.; Feng, Z.; Luo, J.; Tan, Y.; Huang, Y. The effects of Fe2O3 based DOC and SCR catalyst on the combustion and emission characteristics of a diesel engine fueled with biodiesel. *Fuel* **2021**, *290*, 120039. [CrossRef]
5. Valeika, G.; Matijošius, J.; Orynycz, O.; Rimkus, A.; Kilikevičius, A.; Tucki, K. Compression Ignition Internal Combustion Engine's Energy Parameter Research Using Variable (HVO) Biodiesel and Biobutanol Fuel Blends. *Energies* **2024**, *17*, 262. [CrossRef]
6. Channappagoudra, M.; Ramesh, K.; Manavendra, G. Comparative study of standard engine and modified engine with different piston bowl geometries operated with B20 fuel blend. *Renew. Energy* **2019**, *133*, 216–232. [CrossRef]
7. Di Iorio, S.; Catapano, F.; Magno, A.; Sementa, P.; Vaglieco, B.M. The Potential of Ethanol/Methanol Blends as Renewable Fuels for DI SI Engines. *Energies* **2023**, *16*, 2791. [CrossRef]
8. Sitnik, L.J.; Sroka, Z.J.; Andrych-Zalewska, M. The Impact on Emissions When an Engine Is Run on Fuel with a High Heavy Alcohol Content. *Energies* **2021**, *14*, 41. [CrossRef]
9. Zheng, F.; Cho, H.M. The Effect of Different Mixing Proportions and Different Operating Conditions of Biodiesel Blended Fuel on Emissions and Performance of Compression Ignition Engines. *Energies* **2024**, *17*, 344. [CrossRef]
10. EL-Seesy, A.I.; Waly, M.S.; He, Z.; El-Batsh, H.M.; Nasser, A.; El-Zoheiry, R.M. Enhancement of the combustion and stability aspects of diesel-methanol-hydrous methanol blends utilizing n-octanol, diethyl ether, and nanoparticle additives. *J. Clean. Prod.* **2022**, *371*, 133673. [CrossRef]
11. Gad, M.S.; El-Shafay, A.S.; Abu Hashish, H.M. Assessment of diesel engine performance, emissions and combustion characteristics burning biodiesel blends from jatropha seeds. *Process. Saf. Environ. Prot.* **2021**, *147*, 518–526. [CrossRef]
12. López Núñez, A.R.; Rumbo Morales, J.Y.; Salas Villalobos, A.U.; De La Cruz-Soto, J.; Ortiz Torres, G.; Rodríguez Cerda, J.C.; Calixto-Rodriguez, M.; Brizuela Mendoza, J.A.; Aguilar Molina, Y.; Zatarain Durán, O.A.; et al. Optimization and Recovery of a Pressure Swing Adsorption Process for the Purification and Production of Bioethanol. *Fermentation* **2022**, *8*, 293. [CrossRef]
13. Nour, M.; Kosaka, H.; Bady, M.; Sato, S.; Abdel-Rahman, A.K. Combustion and emission characteristics of DI diesel engine fuelled by ethanol injected into the exhaust manifold. *Fuel Process. Technol.* **2017**, *164*, 33–50. [CrossRef]
14. Krzemiński, A.; Ustrzycki, A. Effect of Ethanol Added to Diesel Fuel on the Range of Fuel Spray. *Energies* **2023**, *16*, 1768. [CrossRef]
15. Yilmaz, N.; Sanchez, T.M. Analysis of operating a diesel engine on biodiesel-ethanol and biodiesel-methanol blends. *Energy* **2012**, *46*, 126–129. [CrossRef]
16. Hansen, A.C.; Zhang, Q.; Lyne, P.W.L. Ethanol–diesel fuel blends—A review. *Bioresour. Technol.* **2005**, *96*, 277–285. [CrossRef] [PubMed]
17. Kuszewski, H.; Jaworski, A.; Ustrzycki, A. Lubricity of ethanol–diesel blends—Study with the HFRR method. *Fuel* **2017**, *208*, 491–498. [CrossRef]
18. Li, D.; Zhen, H.; Lu, X.; Zhang, W.; Yang, J. Physico-chemical properties of ethanol–diesel blend fuel and its effect on performance and emissions of diesel engines. *Renew. Energy* **2005**, *30*, 967–976. [CrossRef]
19. Laza, T.; Bereczky, A. Basic fuel properties of rapeseed oil-higher alcohols blends. *Fuel* **2011**, *90*, 803–810. [CrossRef]
20. Torres-Jimenez, E.; Svoljšak Jerman, M.; Gregorc, A.; Lisec, I.; Pilar Dorado, M.; Kegl, B. Physical and chemical properties of ethanol–diesel fuel blends. *Fuel* **2011**, *90*, 795–802. [CrossRef]
21. Pratas, M.J.; Freitas, S.V.D.; Oliveira, M.B.; Monteiro, S.C.; Lima, Á.S.; Coutinho, J.A.P. Biodiesel Density: Experimental Measurements and Prediction Models. *Energy Fuels* **2011**, *25*, 2333–2340. [CrossRef]
22. Kuszewski, H. Experimental investigation of the effect of ambient gas temperature on the autoignition properties of ethanol–diesel fuel blends. *Fuel* **2018**, *214*, 26–38. [CrossRef]
23. Anand, K.; Ranjan, A.; Mehta, P.S. Estimating the Viscosity of Vegetable Oil and Biodiesel Fuels. *Energy Fuels* **2010**, *24*, 664–672. [CrossRef]
24. Song, Y.; Li, R.; Qian, Y.; Zhang, L.; Wang, Z.; Liu, S.; Pei, Y.; An, Y.; Yue, H.; Meng, Y. Study on spray characteristics of biodiesel alternative fuels for in-cylinder environment of diesel engine. *J. Energy Inst.* **2024**, *113*, 101507. [CrossRef]
25. Lapuerta, M.; Rodríguez-Fernández, J.; Fernández-Rodríguez, D.; Patiño-Camino, R. Modeling viscosity of butanol and ethanol blends with diesel and biodiesel fuels. *Fuel* **2017**, *199*, 332–338. [CrossRef]
26. Murcak, A.; Haşimoğlu, C.; Cevik, İ.; Kahraman, H. Effect of injection timing to performance of a diesel engine fuelled with different diesel–ethanol mixtures. *Fuel* **2015**, *153*, 569–577. [CrossRef]
27. Barabás, I. Predicting the temperature dependent density of biodiesel–diesel–bioethanol blends. *Fuel* **2013**, *109*, 563–574. [CrossRef]
28. Masera, K.; Hossain, A.K.; Davies, P.A.; Doudin, K. Investigation of 2-butoxyethanol as biodiesel additive on fuel property and combustion characteristics of two neat biodiesels. *Renew. Energy* **2021**, *164*, 285–297. [CrossRef]
29. Kwanchareon, P.; Luengnaruemitchai, A.; Jai, S. In Solubility of a diesel–biodiesel–ethanol blend, its fuel properties, and its emission characteristics from diesel engine. *Fuel* **2007**, *86*, 1053–1061. [CrossRef]

30. Guarieiro, L.L.N.; de Almeida Guerreiro, E.T.; dos Santos Amparo, K.K.; Manera, V.B.; Regis, A.C.D.; Santos, A.G.; Ferreira, V.P.; Leao, D.J.; Torres, E.A.; de Andrade, J.B. Assessment of the use of oxygenated fuels on emissions and performance of a diesel engine. *Microchem. J.* **2014**, *117*, 94–99. [CrossRef]
31. Tutak, W.; Jamrozik, A.; Pyrc, M.; Sobiepański, M. A comparative study of co-combustion process of diesel-ethanol and biodiesel ethanol blends in the direct injection diesel engine. *Appl. Therm. Eng.* **2017**, *117*, 155–163. [CrossRef]
32. Lapuerta, M.; Armas, O.; Herreros, J.M. Emissions from a diesel–bioethanol blend in an automotive diesel engine. *Fuel* **2008**, *87*, 25–31. [CrossRef]
33. Pielecha, I.; Czajka, J.; Wisłocki, K.; Borowski, P. Research-based assessment of the influence of hydrocarbon fuel atomization on the for-mation of self-ignition spots and the course of pre-flame processes. *Combust. Engines* **2014**, *157*, 22–35. [CrossRef]
34. Hentschel, W. Optical diagnostics for combustion process development of direct-injection gasoline engines. *Proc. Combust. Inst.* **2000**, *28*, 1119–1135. [CrossRef]
35. Kamimoto, T.; Ahn, S.K.; Chang, Y.J.; Kobayashi, H.; Matsuoka, S. Measurement of Droplet Diameter and Fuel Concentration in a Non-Evaporating Diesel Spray by Means of an Image Analysis of Shadow Photographs. *SAE* **1984**, *93*, 372–380. [CrossRef]
36. Wisłocki, K. *Studium Wykorzystania Badań Optycznych do Analizy Procesów Wtrysku i Spalania w Silnikach o Zapłonie Samoczynnym*; Wydawnictwo Politechniki Poznańskiej: Poznań, Poland, 2004; Volume 387.
37. Orzechowski, Z.; Prywer, J. *Wytwarzanie i Zastosowanie Rozpylonej Cieczy*; Wydawnictwa Naukowo—Techniczne: Warsaw, Poland, 2008; ISBN 978-83-204-3416-3.
38. Available online: https://www.orlen.pl/pl/dla-biznesu/produkty/paliwa/oleje-napedowe/ekodiesel-ultra-klasa-2 (accessed on 13 June 2024).
39. *Safety Data Sheet: Ethyl Alcohol Completely Contaminated*; Alpinus: Solec Kujawski, Poland, 2016.
40. Available online: https://www.carlroth.com/pl/pl/aliphatic-alcohols/1-dodecanol/p/9853.3 (accessed on 13 June 2024).
41. Krzemiński, A.; Jaworski, A.; Kuszewski, H.; Woś, P. A comparative study on selected physical properties of diesel–ethanol–dodecanol blends. *Combust. Engines* **2024**, *196*, 99–105. [CrossRef]
42. *Przewodnik Głównego Urzędu Miar: Wyrażanie Niepewności Pomiaru (Guide of the Central Office of Measures: Expressing Measurement Uncertainty)*; Central Office of Measures: Warszawa, Poland, 1999.
43. Zięba, A. *Pracownia Fizyczna Wydziału Fizyki i Techniki Jądrowej*; AGH: Kraków, Poland, 2002.

Disclaimer/Publisher's Note: The statements, opinions and data contained in all publications are solely those of the individual author(s) and contributor(s) and not of MDPI and/or the editor(s). MDPI and/or the editor(s) disclaim responsibility for any injury to people or property resulting from any ideas, methods, instructions or products referred to in the content.

Article

The Oxidation Performance of a Carbon Soot Catalyst Based on the Pt-Pd Synergy Effect

Diming Lou, Guofu Song, Kaiwen Xu, Yunhua Zhang * and Kan Zhu

School of Automotive Studies, Tongji University, Shanghai 201800, China; loudiming@tongji.edu.cn (D.L.); 2180073@tongji.edu.cn (G.S.); xukaiwen@tongji.edu.cn (K.X.); zhuk_74@tongji.edu.cn (K.Z.)
* Correspondence: zhangyunhua@tongji.edu.cn

Abstract: Pt-Pd-based noble metal catalysts are widely used in engine exhaust aftertreatment because of their better carbon soot oxidation performance. At present, the synergistic effect of Pt and Pd in CDPFs, which is the most widely used and common doping method, in catalyzing the combustion of carbon smoke has not been reported, and it is not possible to give an optimal doping ratio of Pt and Pd. This paper investigates the carbon soot oxidation performance of different Pt/Pd ratios (Pt/Pd = 1:0, 10:1, 5:1, 1:1) based on physicochemical characterization and particle combustion kinetics calculations, aiming to reveal the Pt-Pd synergistic effect and its carbon soot oxidation law. The results show that Pt-based catalysts doped with Pd can improve the catalyst dispersion, significantly increase the specific surface area, and reduce the activation energy and reaction temperature of carbon soot reactions, but excessive doping of Pd leads to the enhancement of the catalyst agglomeration effect, a decrease in the specific surface area, and an increase in the activation energy and reaction temperature of the carbon soot reaction. The specific surface area and pore capacity of the catalyst are the largest, and the activation energy of particle oxidation and the pre-exponential factor are the smallest (203.44 kJ·mol^{-1} and 6.31 × 10^7, respectively), which are 19.29 kJ·mol^{-1} and 4.95 × 10^8 lower than those of pure carbon soot; meanwhile, the starting and final combustion temperatures of carbon soot (T_{10} and T_{90}) are the lowest at 585.8 °C and 679.4 °C, respectively, which are 22.1 °C and 20.9 °C lower than those of pure carbon soot.

Keywords: Pt-Pd synergy; carbon soot; physicochemical properties; combustion kinetics; oxidation properties

Citation: Lou, D.; Song, G.; Xu, K.; Zhang, Y.; Zhu, K. The Oxidation Performance of a Carbon Soot Catalyst Based on the Pt-Pd Synergy Effect. *Energies* **2024**, *17*, 1737. https://doi.org/10.3390/en17071737

Academic Editors: Kazimierz Lejda, Artur Jaworski and Maksymilian Mądziel

Received: 2 March 2024
Revised: 24 March 2024
Accepted: 26 March 2024
Published: 4 April 2024

Copyright: © 2024 by the authors. Licensee MDPI, Basel, Switzerland. This article is an open access article distributed under the terms and conditions of the Creative Commons Attribution (CC BY) license (https:// creativecommons.org/licenses/by/ 4.0/).

1. Introduction

Diesel engines with low fuel consumption, high torque, high reliability, and other advantages are widely used in road transport, marine transport, national defense equipment, non-road mobile machinery (construction machinery, agricultural machinery, generator sets, etc.), and other fields. All kinds of trucks and medium and large buses mainly use diesel engines as their power source. China has a vast territory and a large mountainous area, and both passenger and cargo transport mainly involve road transport. In 2021, road transport dominated by diesel vehicles bore about 61.2% of passenger transport and 73.8% of cargo transport [1], and for a long time, diesel power will still be the backbone of national defense equipment, economic development, and infrastructure construction. Compared with gasoline engines, the emission of pollutants has been one of the important factors that trouble the wide popularization of diesel engines. As a result of the growing number of motor vehicles and industrialization, air pollution has become an important environmental problem faced by all parts of the world. Diesel engines, as one of the important driving forces in modern society and an irreplaceable part of the human production process, are widely used in various domains such as transportation, industry, and agricultural production, and are also one of the main contributors to the air pollution problem. As emission regulations are constantly upgraded, emission requirements are also becoming stricter and stricter. Diesel National VI emission regulations are more stringent compared to the National V emission limits and test methods. In diesel engine emissions, particulate matter (PM) emission is of concern. Diesel exhaust particles are mainly composed of non-volatile (insoluble) and volatile (soluble) parts. Volatile compounds are composed of organic carbon, sulfate, and nitrate compounds, while the non-volatile portion is composed of the carbonaceous (soot) portion and ash [2]. Particulate matter can be suspended in the air for a long time, polluting the environment and affecting human physical and mental health. The size of 70% of particles is less than 0.3 μm, mainly composed of carbon smoke particles and soluble organic fraction (SOF) with hydrocarbon droplets and sulfuric acid droplets adsorbed on the surface. These fine carbon smoke particles are easily inhaled by the human body in the alveoli and are difficult to eliminate. SOF and other chemicals adsorbed on the particles in particular will lead to toxic metals such as lead, cadmium, and zinc being absorbed into the blood through the respiratory tract and cause hypersensitivity and allergic reactions, which may also lead to bacterial and fungal infections in living organisms, as well as cancer, pulmonary fibrosis, and mucosal irritation [3]. At the same time, PM emitted by engine exhausts into the atmosphere becomes a pollutant and disturbs the atmospheric radiation balance, thus aggravating the effect of global warming [4].

For this reason, emission regulations strictly control particulate matter emissions from diesel vehicles. As a result, the Diesel Particulate Filter (DPF) is an essential device in diesel exhaust treatment systems, capturing particles through a mixture of surface and internal filters using methods such as such as diffusion precipitation, inertial precipitation, and linear interception. DPFs can effectively purify 65–90% of the particulates in exhaust gas, making it one of the most effective and direct methods to purify the particulates produced by diesel vehicles [5]. However, with the accumulation of DPF running time and mileage, a large number of particles will accumulate and block the DPF, which will increase the exhaust back pressure and lead to the deterioration of engine power performance and economic performance. Therefore, it is necessary to burn off the particles gathered on the surface of the DPF in a timely manner and carry out a regeneration treatment on the DPF. There are two regeneration methods for DPFs [6], namely active regeneration and passive regeneration. Because the combustion of particulate matter requires a very high temperature, much higher than the temperature of diesel engine exhausts, passive regeneration usually involves coating a catalyst on the surface of DPF carriers to reduce the combustion temperature of particulate matter so that it can be ignited and burned at a lower exhaust temperature.

The catalyst can participate in the process of the combustion reaction, reduce the activation energy of the chemical reaction, and accelerate the reaction rate. In the passive regeneration process of Catalyzed Diesel Particulate Filters (CDPFs), the catalyst plays a key role. The effect of passive regeneration often depends on the quality of the catalyst.

There are two types of catalysts currently used in CDPFs. The first type is a non-noble-metal catalyst without noble metal doping. Catalysts based on non-noble metals can be divided into the following main types: perovskite-type catalysts [7], spinel-type catalysts [8], and transition metal oxide catalysts. Among them, cerium dioxide-based catalysts [9,10] and other metal oxide catalysts [11–13] are represented. The other type of catalyst is the noble-metal-doped catalyst. The noble-metal catalyst is the most widely used CDPF catalyst, as it has the highest catalytic efficiency and best performance among catalysts. At present, the noble metals used mainly include Pt, Pd, Rh, and other elements. Usually, metal oxides such as alumina and titanium dioxide are used as the carriers of noble-metal catalysts. The carbon smoke particles on CDPFs can be effectively removed under the joint action of noble-metal elements and carriers. Moreover, relevant studies showed that [14] the activity of Pt supported on TiO_2-SiO_2 (Pt/TiO_2-SiO_2) was higher than those of other Pt/MOx systems (MOx = TiO_2, ZrO_2, SiO_2, MOX). Al_2O_3, TiO_2-ZrO_2, TiO_2-Al_2O_3, ZrO_2-SiO_2, ZrO_2-Al_2O_3, and SiO_2-Al_2O_3 had higher activity. The form of carbon particle oxidation catalyzed by noble-metal catalysts is generally considered to be gas-phase catalysis [15]. Pt indirectly promotes the oxidation of soot; that is, it catalyzes the oxidation of NO in the exhaust gas to NO_2 and then oxidizes the soot to CO_2 [16,17].

Noble-metal (Ag, Pt, and Pd) catalysts using CeO_2 and TiO_2 as carriers showed higher particulate matter (PM) catalytic oxidation activity for Ag than for Pt and Pd under tight and loose contact conditions [18]. The thermal stability and catalytic activity of a Pt/Al_2O_3 catalyst modified with Pd became better due to the addition of Pd [19]. A comparison of the isothermal oxidation kinetic properties (rate and activation energy) of soot with the physicochemical properties of diesel, charcoal, and commercial (Printex U, Vulkan XC-72, Mumbai, India) soot samples showed that the oxidative reactivity of soot depended on many parameters. The oxidative reactivity of soot on the surface of $Pt-Pd/MnOx-Al_2O_3$ oxidation catalysts with low Pt-Pd content aimed at removing diesel soot depended on many parameters, but mainly on the nanostructure, microstructure, and composition of surface functional groups [20].

Worldwide, scholars have confirmed that DPF catalysts doped with Pt and other noble-metal elements have more obvious catalytic effects on soot combustion, and that such catalytic reactions are related to the microstructure of the catalysts, the functional groups on the surface of the catalysts, and the use of different elements in the doping. However, the synergistic effect of Pt and Pd in CDPFs catalyzing the combustion of soot, which is the most widely used and common doping method, has not yet been reported, and the optimal doping ratio of Pt and Pd could not be given.

Therefore, in this paper, four groups of DPF catalysts with different Pt and Pd content ratios are studied, the microscopic characteristics of the catalyst surface are characterized and analyzed, and the activation energy (E) and predigital factor (A) of the reaction catalyzed by soot are calculated using the Coats–Redfern integral method. At the same time, the influence of different Pt-Pd catalysts on the soot catalytic combustion of diesel engines is analyzed based on the key temperature of the soot combustion reaction.

2. Materials and Methods

Through scanning electron microscope (SEM) imaging and a specific surface area test (BET test), the micro photos and physical characteristics of the sample were obtained. Derivative thermogravimetric (DTG) curves, as well as the curves of mass and temperature (TG) and differential scanning calorimetry (DSC) were obtained by thermogravimetric testing and a differential scanning calorimetry test to describe the combustion kinetic characteristics of the sample.

Test Materials and Equipment

According to the different content ratio of Pt and Pd in the catalyst, the test samples were divided into 4 types, as shown in Table 1. The soot used is 99% pure carbon powder. The mixed state of the catalyst and soot particles can be divided into tight contact and loose contact, which will have a great influence on the thermogravimetric results. The loose contact state was more consistent with the actual state on the DPF, and the loose contact mode was used in this study. The total mass of each test sample was 6 mg, the mass ratio of catalyst to carbon smoke was 1:1 (that is, the mass fraction of carbon smoke was 50%), and they were gently stirred evenly to ensure that the two were in loose contact (stirring for at least 3 min).

Table 1. Test samples.

Serial No.	Catalyst Proportioning
1	Pt/Pd = 1:0
2	Pt/Pd = 10:1
3	Pt/Pd = 5:1
4	Pt/Pd = 1:1

The instrument used for SEM analysis in this experiment was a Thermo Fisher Apreo2C scanning electron microscope (Waltham, MA, USA). The scanning electron microscope has a resolution of 1.2 nm at 1 kV, a landing energy range of 20 eV~30 keV, and a maximum beam current of 50 nA. The BET data were determined using a NOVA-1200 spectrophotometer manufactured by Quantchrom (Boynton Beach, FL, USA). The instrument has a minimum measurable specific surface area value of 0.01 m^2/g, a measurable pore size range of 0.35 nm to >400 nm, and a minimum pore volume of 0.0001 cc/g. Thermogravimetric testing was carried out with a TG209F1 Thermogravimetric Analyzer from NETZSCH (Selb, Bavaria, Germany). The thermogravimetric analyzer has a maximum sample mass of 2 g, a maximum temperature of 1100 °C, an adjustable heating rate between 0.001 K/min and 200 K/min, and a TGA resolution of 0.1 µg. DSC tests were performed with a Synchronized Thermal Synthesis Analyzer (STA449F3) from NETZSCH (Selb, Bavaria, Germany). The analyzer has a temperature range of −150~2400 °C, a temperature increase rate of up to 50 K/min, a maximum test sample mass of 35 g, and a test resolution of 0.1 µg.

In order to simulate the actual oxygen content of diesel exhaust gas, the set working atmosphere of the thermogravimetric test was 13% O_2 and 87% N_2 syngas. The working gas flow rate was 50 mL/min. The temperature rise rate was 10 °C/min, and the temperature rise range was 40~800 °C. The TG curve, DTG curve, and DSC curve were obtained.

3. Results and Discussion

3.1. Analysis of Catalyst Characterization Results

3.1.1. Analysis of SEM Results

SEM images of the four catalysts are shown in Figure 1, and the surfaces of the catalysts of the four proportioning schemes showed good granularity. The catalyst surface had fine particles, fluffy organization, and a lot of pores, which increase the surface area of the catalyst, making the catalyst have a sufficient contact area with the carbon fume, and the catalytic effect of the catalyst is maximized.

The catalyst particles in the Pt/Pb = 1:0 scheme show a small state. When Pb is added, Pt/Pb = 10:1; the catalyst particles have a small increase, but at the same time, the particles are more loose from each other, and the pores are obviously increased. When Pb is further added, Pt/Pb = 5:1; the catalyst particles have an obvious effect of increasing, and the agglomeration effect occurs, which makes the surface area experience a small decrease. When Pt/Pb = 1:1, the catalyst particles further increase, resulting in a significant decrease in surface area.

When a small amount of the Pd element is added (Pt/Pb = 10:1), the particles on the surface of the catalyst are more dispersed, the granularity is more obvious, and the pores are more adequate, with a better synergistic effect. With the increase in the added Pd content, the particles on the catalyst surface have a tendency to re-colonize and become larger.

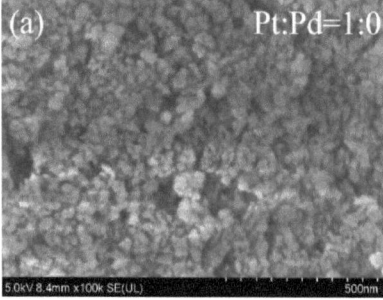 

Figure 1. Cont.

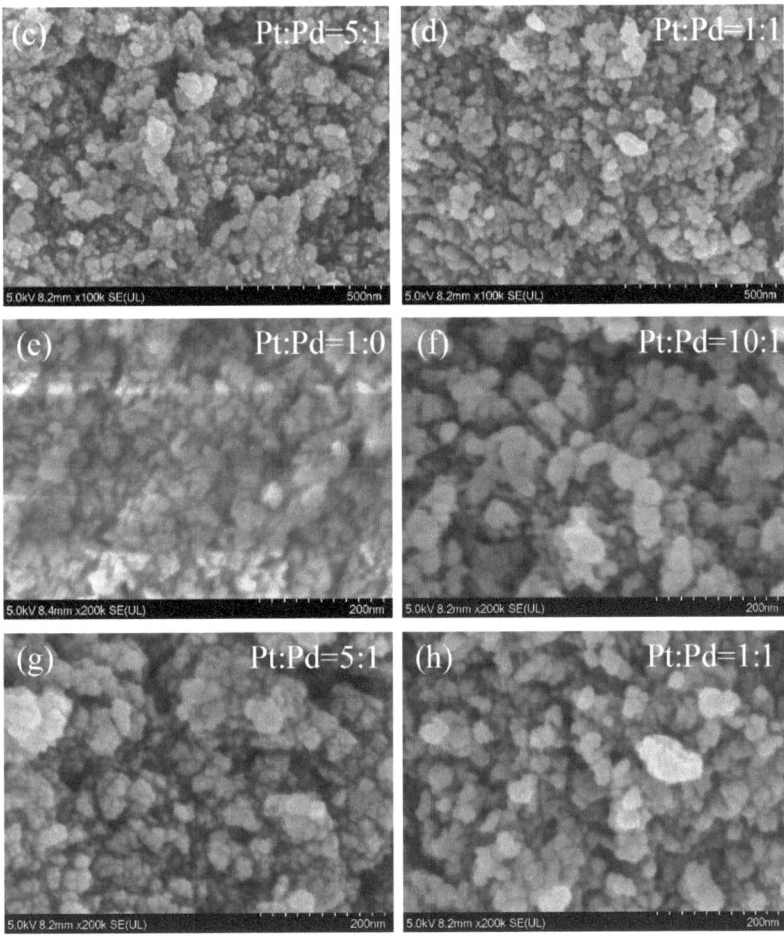

Figure 1. SEM photos of the four catalysts.

3.1.2. Analysis of BET Results

The specific surface area and pore volume parameters of the four catalysts obtained after BET tests are shown in Table 2. The specific surface areas of the four catalyst schemes were more than 110 m^2/g, among which the Pt/Pd = 10:1 ratio scheme had the largest specific surface area of 140.366 m^2/g; the Pt/Pd = 1:0 ratio scheme had the smallest specific surface area of 113.842 m^2/g. The pore volumes of the four catalyst schemes were the largest for the Pt/Pd = 10:1 ratio scheme at 0.5359 cm^3/g; the Pt/Pd = 1:1 ratio scheme has the smallest pore volume of 0.3644 cm^3/g.

Table 2. BET results for the four proportioning schemes.

Catalyst	SBET (m^2/g)	Vp (cm^3/g)
Pt/Pd = 1:0	114	0.44
Pt/Pd = 10:1	140	0.54
Pt/Pd = 5:1	132	0.40
Pt/Pd = 1:1	116	0.36

According to the results in Table 2, a small amount of added Pd can increase the specific surface area and pore volume of the catalyst, but with the increase in Pd addition, both the specific surface area and pore volume decrease to different extents, and the pore volume decreases to lower than that of the un-added Pd solution after the ratio of Pt and Pd reaches or exceeds 5:1.

3.2. Analysis of Test Results

The TG curves, DTG curves, and DSC curves of the four catalyst ratios obtained from the experiments are shown in Figure 2, and the downward direction in the DSC curves is the exothermic direction. The maximum exothermic rates of each scheme are close to each other, among which the Pt/Pd = 5:1 scheme has the lowest maximum exothermic rate of 11.5 mW/mg and the Pt/Pd = 10:1 scheme had the highest maximum exothermic rate of 14.3 mW/mg.

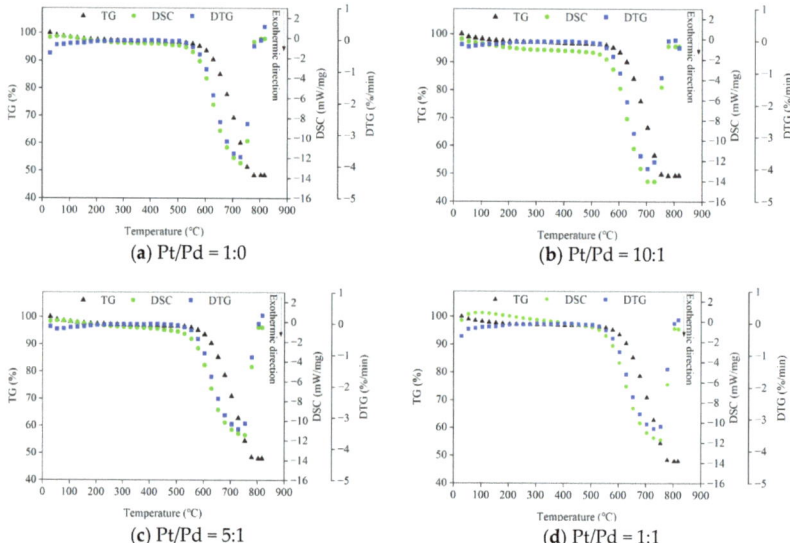

Figure 2. TG-DTG-DSC test curve.

3.3. Calculation of Kinetic Parameters and Combustion Temperature

3.3.1. Analysis of SEM Results

In this paper, the thermodynamic parameters, such as the activation energy (E) and pre-exponential factor (A) of carbon black, were calculated using the Coats–Redfern integral method [21], and the effects of different Pt/Pd ratios on the catalytic oxidation characteristics and thermodynamic parameters of diesel particulate matter were analyzed based on the results.

The initial mass of the test sample is set to be m_0; the sample is burned and consumed during the test, and the mass changes to m at time t. The final residual sample mass is m_e. Therefore, the combustion decomposition rate of the sample is

$$d\alpha/dt = kf(\alpha) \tag{1}$$

In Equation (1), α is the sample conversion rate; $\alpha = [(m_0 - m)/(m_0 - m_e)] \times 100\%$; t is the response time; k is the Arrhenius reaction rate constant; $k = A\exp(-E/RT)$; A is the pre-exponential factor of an index; E is the activation energy; $R = 8.314$ J/(mol · K) is the molar gas constant; T is the thermodynamic temperature; and $f(\alpha)$ is the differential form of the reaction mechanism function. The reaction temperature is ramped up at a certain rate of increase, β; the expression is $\beta = dT/dt$. Substituting into Equation (1) gives the kinetic equation as

$$\frac{d\alpha}{dT} = \frac{A}{\beta} \exp(-\frac{E}{RT})(1-\alpha)^n \qquad (2)$$

The Coats–Redfern equation is obtained by integrating both sides of Equation (2) and taking logarithmic treatment where the number of reaction stages (n) on the right-hand side of Equation is taken as 1, which simplifies to

$$\ln\left[\frac{-\ln(1-\alpha)}{T^2}\right] = \ln\left[\frac{AR}{\beta E}(1 - \frac{2RT}{E})\right] - \frac{E}{RT} \qquad (3)$$

In particular, it is known from the trial analysis that the first term on the right-hand side of the equation in Equation (3) varies very little in value over the temperature increase interval selected for conventional thermogravimetric experiments, and it can be regarded as a constant. Therefore, Equation (3) can be simplified to a one-dimensional linear equation in the form of

$$Y = b + kX \qquad (4)$$

If we make drawings with $Y = \ln\left[\frac{-\ln(1-\alpha)}{T^2}\right]$ and $X = \frac{1}{T}$, a straight line with slope $k = -\frac{E}{R}$ can be obtained. The reaction activation energy (E) can be calculated from the slope (k). The calculated activation energy and the fitted intercept can be brought into Equation (3) to calculate the pre-exponential factor (A).

3.3.2. Analysis of SEM Results

Figure 3 shows the conversion rate curve of carbon black with temperature change when Pt/Pd = 1:1 in the catalyst. From the curve, the decomposition rate between 0 and 10% is almost unchanged with the increase in temperature, the combustion rate is very slow, and the mass consumed at this time is the mass of other impurities such as water in the carbon black. After the decomposition rate rises to 10%, it rises rapidly with the temperature increase. After the decomposition rate reaches 90%, most of the carbon black is already consumed, so the rising trend becomes slower. To make the parameter analysis more reasonable, thermodynamic parameters such as the reaction activation energy (E) and pre-exponential factor (A) were calculated using data with conversion rates in the range of 10% to 90%.

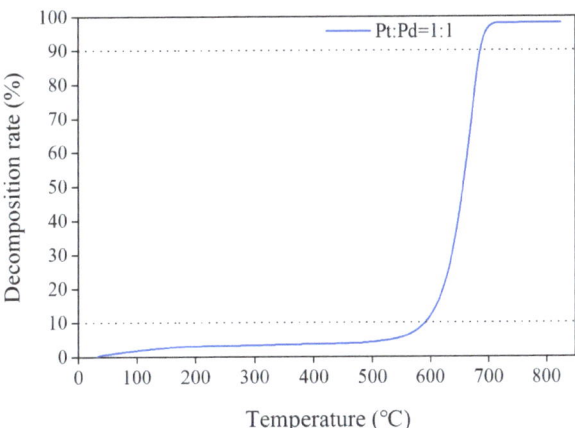

Figure 3. Decomposition rate of carbon black with temperature change.

Figure 4 shows the curves processed and fitted to the experimental data using the Coats–Redfern integration method for the Pt/Pd = 1:1 scheme and the pure carbon smoke, while using the same method to process the data for the other experimental schemes to obtain the kinetic parameters for each experimental scheme, as shown in Table 3, where logarithmic treatment is taken for the pre-exponential factor.

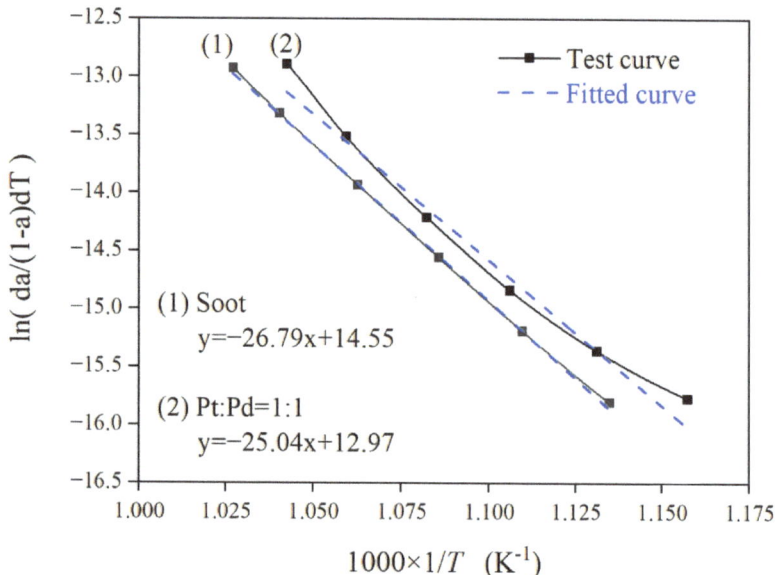

Figure 4. Test curves and corresponding fitted curves.

Table 3. Kinetic parameters of the programs.

Serial Number	Programmatic	Fitted Curve	Activation Energy, E (kJ·mol^{-1})	Pre-Exponential Factor, A (min^{-1})	ln(A)
1	Pt/Pd = 1:0	y = −24.91x + 12.94	207.10	1.04×10^8	18.46
2	Pt/Pd = 10:1	y = −24.47x + 12.46	203.44	6.31×10^7	17.96
3	Pt/Pd = 5:1	y = −24.89x + 12.88	206.94	9.77×10^7	18.40
4	Pt/Pd = 1:1	y = −25.04x + 12.97	208.18	1.08×10^8	18.49
5	Soot	y = −26.79x + 14.55	222.73	5.58×10^8	20.14

3.3.3. Comparison of Kinetic Parameters and Combustion Temperatures

Figure 5 shows the values of the reaction activation energy and pre-exponential factor for the four Pt/Pd ratio catalyst schemes and the pure carbon fume, where the values of the corresponding parameters for the pure carbon fume are indicated by dashed lines. Reaction activation energy refers to the energy required for the molecules to change from the normal state to the active state in which chemical reactions can easily take place; the lower the activation energy is, the faster the reaction rate is. Therefore, lowering the activation energy will effectively promote the reaction. The pre-exponential factor is the factor representing the total number of effective collisions of activated molecules; the higher the value of the pre-exponential factor, the greater the number of effective collisions of activated molecules, the more easily the reaction is carried out, and the more intense is the degree of the reaction.

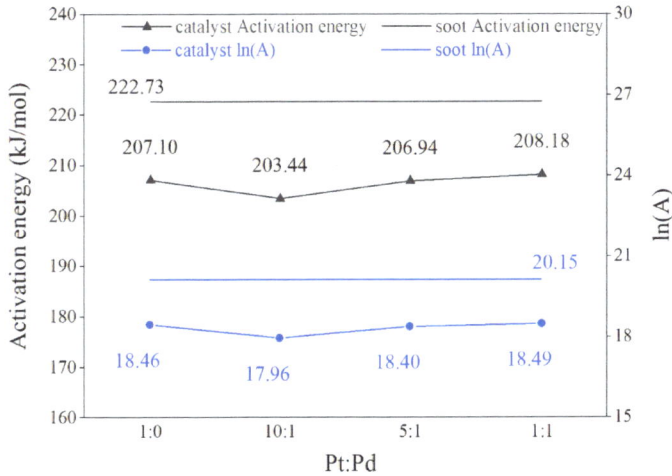

Figure 5. Kinetic parameters of different Pt/Pd ratio schemes.

The activation energies (E) of each scheme are in the range of 203.44–208.18 kJ/mol, and the logarithmic pre-exponential factors (ln(A)) are in the range of 17.96–18.49. Compared with the pure carbon fume, the reaction kinetic parameters of each scheme after doping catalysts are decreased to different extents with the same trend. Among them, the activation energy and logarithmic pre-exponential factor of the Pt/Pd = 10:1 scheme are the lowest at 203.44 kJ/mol and 18.0, respectively, which are 19.29 kJ/mol and 2.19 lower than those of the pure carbon fume.

As can be seen from Figure 4, the addition of a small amount of Pd to the Pt-containing DPF catalyst can rapidly reduce the reaction activation energy of carbon smoke combustion. With the increase in the Pd content, the reaction activation energy of carbon smoke rises and the catalytic effect deteriorates, but the catalytic effect is still obvious compared with that of pure carbon smoke.

Figure 6 shows the starting temperature (T_{10}), the final combustion temperature (T_{90}), and the peak combustion temperature (T_P) of four different Pt/Pd ratios, and T_P is the temperature of the fastest combustion rate during the combustion of carbon smoke.

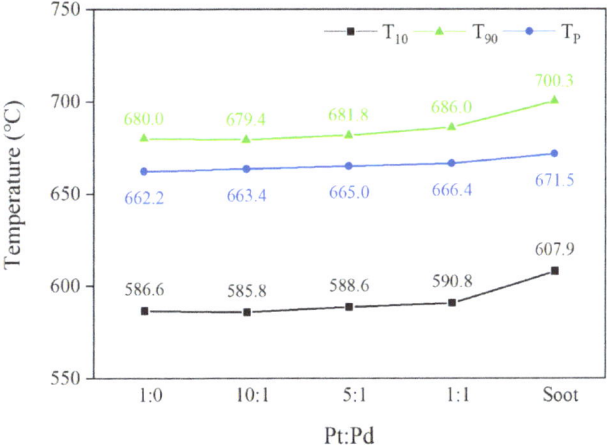

Figure 6. Critical temperature points for different Pt/Pd ratio schemes.

The trends of T_{10} and T_{90} for the four catalyst schemes are approximately the same as those of the activation energy, with the lowest T_{10} and T_{90} for the Pt/Pd = 10:1 scheme at 585.8 °C and 679.4 °C, respectively, and compared with the pure carbon smoke combustion without the catalyst included, both T_{10} and T_{90} are drastically reduced by 22.1 °C and 20.9 °C, respectively. The use of Pt- and Pd-doped catalysts can effectively reduce the ignition temperature of carbon smoke and promote the passive regeneration of DPF. The T_P of the four scenarios does not differ much from that of the pure carbon smoke case, with only a slight decrease, among which the lowest T_P scenario is Pt/Pd = 1:0 with a temperature of 662.2 °C, which is 9.3 °C lower than that of the pure carbon smoke case.

4. Conclusions

This paper focuses on analyzing the characterization of different Pt-Pd catalysts and their catalytic performance on soot. Several conclusions are as follows:

1. Pt-Pd has a good synergistic effect, and the doping of Pt-based catalysts with Pd can improve the dispersion of the catalysts, significantly increase the specific surface area, and reduce the activation energy and reaction temperature of the soot reaction, but excessive doping of Pd will lead to the enhancement of the agglomeration effect of the catalysts, a reduction in the specific surface area, and an increase in the activation energy and reaction temperature of the soot reaction;
2. When the ratio of Pt/Pd was 10:1, the specific surface area and the pore volume of the catalyst were the largest, and the activation energy of combustion and the pre-exponential factor of carbon soot particles were the smallest, which were 203.44 kJ/mol and 6.31×10^7, respectively, and were 19.29 kJ/mol and 4.95×10^8 lower than those of pure carbon soot. At the same time, the starting temperature of the soot, T_{10}, and the final temperature of the carbon soot, T_{90}, were the lowest, which were 585.8 °C and 679.4 °C, respectively: 22.1 °C and 20.9 °C lower those that of pure carbon smoke;
3. The activation energies of the four different Pt-Pd catalysts ranges from 203 kJ/mol to 209 kJ/mol, with the pre-exponential factor between 6.31×10^7 and 1.08×10^8, and the temperature range of 580 °C to 690 °C is the main temperature range for the catalytic combustion of carbon smoke.

From the comprehensive results of the full analysis, the ratio scheme of Pt/Pd = 10:1 has the lowest activation energy, T_{10}, and T_{90} for the soot combustion reaction, the highest exothermic rate, and the most active reaction, which is the best scheme for the passive regeneration catalyst of DPFs. Further studies on different doping modes of other elements, including non-noble metals, will be conducted in the future.

Author Contributions: Conceptualization, D.L., Y.Z. and K.X.; methodology, K.X.; software, K.X. and G.S.; validation, K.X.; formal analysis, K.X.; investigation, Y.Z. and G.S.; writing—original draft preparation, K.X. and K.Z.; writing—review and editing, K.Z.; supervision, D.L.; project administration, Y.Z. All authors have read and agreed to the published version of the manuscript.

Funding: This research was funded by the National Key R&D Program of China, grant number 2022YFE0100100; the National Natural Science Foundation of China, grant number 52206167.

Data Availability Statement: The raw data supporting the conclusions of this article will be made available by the authors on request.

Conflicts of Interest: The authors declare no conflicts of interest.

References

1. Ministry of Ecology and Environment of the People's Republic of China. 2022 Annual Report on Environmental Management of Mobile Sources in China. Available online: https://www.mee.gov.cn/hjzl/sthjzk/ydyhjgl/202212/t20221207_1007111.shtml (accessed on 7 December 2022).
2. Khobragade, R.; Singh, S.K.; Shukla, P.C.; Gupta, T.; Al-Fatesh, A.S.; Agarwal, A.K.; Labhasetwar, N.K. Chemical composition of diesel particulate matter and its control. *Catal. Rev.* **2019**, *61*, 447–515. [CrossRef]

3. Niessner, R. The many faces of soot: Characterization of soot nanoparticles produced by engines. *Angew. Chem. Int. Ed.* **2014**, *53*, 12366–12379. [CrossRef] [PubMed]
4. Bond, T.C.; Doherty, S.J.; Fahey, D.W.; Forster, P.M.; Berntsen, T.; DeAngelo, B.J.; Flanner, M.G.; Ghan, S.; Kärcher, B.; Koch, D.; et al. Bounding the role of black carbon in the climate system: A scientific assessment. *J. Geophys. Res. Atmos.* **2013**, *118*, 5380–5552. [CrossRef]
5. Zhong, H.; Tan, J.W.; Wang, Y.L.; Tian, J.L.; Hu, N.T.; Cheng, J.; Zhang, X.M. Effects of a diesel particulate filter on emission characteristics of a China II non-road diesel engine. *Energy Fuel* **2017**, *31*, 9833–9839. [CrossRef]
6. Wang, Z.B.; Liu, P.; Li, H.M.; Li, R.; Pan, X.B.; Zhao, Y. The development of diesel particulate filter technology. *IOP Conf. Ser. Earth Environ. Sci.* **2021**, *632*, 032012. [CrossRef]
7. Feng, N.J.; Meng, J.; Wu, Y.; Chen, C.; Wang, L.; Gao, L.; Wan, H.; Guan, G.F. KNO_3 supported on three-dimensionally ordered macroporous $La_{0.8}Ce_{0.2}Mn_{1-x}Fe_xO_3$ for soot removal. *Catal. Sci. Technol.* **2016**, *6*, 2930–2941. [CrossRef]
8. Li, Q.; Meng, M.; Xian, H.; Tsubaki, N.; Li, X.G.; Xie, Y.N.; Hu, T.D.; Zhang, J. Hydrotalcite-Derived $Mn_xMg_{3-x}AlO$ Catalysts Used for Soot Combustion, NOx Storage and Simultaneous Soot-NOx Removal. *Environ. Sci. Technol.* **2010**, *44*, 4747–4752. [CrossRef]
9. Lee, J.H.; Lee, S.H.; Choung, J.W.; Kim, C.H.; Lee, K.Y. Ag-incorporated macroporous CeO_2 catalysts for soot oxidation: Effects of Ag amount on the generation of active oxygen species. *Appl. Catal. B Environ.* **2019**, *246*, 356–366. [CrossRef]
10. Lin, X.T.; Li, S.J.; He, H.; Wu, Z.; Wu, J.L.; Chen, L.M.; Ye, D.Q.; Fu, M.L. Evolution of oxygen vacancies in MnO_x-CeO_2 mixed oxides for soot oxidation. *Appl. Catal. B Environ.* **2018**, *223*, 91–102. [CrossRef]
11. Qi, B.Y.; Li, Z.G.; Lou, D.M.; Zhang, Y.H. Experimental investigation on the effects of DPF Cs-V-based non-precious metal catalysts and their coating forms on non-road diesel engine emission characteristics. *Environ. Sci. Pollut. Res.* **2023**, *30*, 9401–9415. [CrossRef] [PubMed]
12. Wagloehner, S.; Baer, J.N.; Kureti, S. Structure–activity relation of iron oxide catalysts in soot oxidation. *Appl. Catal. B Environ.* **2014**, *147*, 1000–1008. [CrossRef]
13. Ji, F.; Men, Y.; Wang, J.G.; Sun, Y.L.; Wang, Z.D.; Zhao, B.; Tao, X.T.; Xu, G.J. Promoting diesel soot combustion efficiency by tailoring the shapes and crystal facets of nanoscale Mn_3O_4. *Appl. Catal. B Environ.* **2019**, *242*, 227–237. [CrossRef]
14. Oi-Uchisawa, J.; Wang, S.D.; Nanba, T.; Ohi, A.; Obuchi, A. Improvement of Pt catalyst for soot oxidation using mixed oxide as a support. *Appl. Catal. B Environ.* **2003**, *44*, 207–215. [CrossRef]
15. Wang, K.X. Low-Temperature Activity and Sulfur Resistance of Non-Precious Metal CDPF Catalysts for the Catalytic Oxidation of Carbonaceous Smoke. Master's Thesis, Shanghai Jiao Tong University, Shanghai, China, 2019.
16. Oi-Uchisawa, J.; Obuchi, A.; Wang, S.; Nanba, T.; Ohi, A. Catalytic performance of Pt/MO_x loaded over SiC-DPF for soot oxidation. *Appl. Catal. B Environ.* **2003**, *43*, 117–129. [CrossRef]
17. Zhang, Y.H.; Lou, D.M.; Tan, P.Q.; Hu, Z.Y.; Fang, L. Effect of catalyzed diesel particulate filter and its catalyst loading on emission characteristics of a non-road diesel engine. *J. Environ. Sci.* **2023**, *126*, 794–805. [CrossRef] [PubMed]
18. Lim, C.B.; Kusaba, H.; Einaga, H.; Teraoka, Y. Catalytic performance of supported precious metal catalysts for the combustion of diesel particulate matter. *Catal. Today* **2011**, *175*, 106–111. [CrossRef]
19. Kaneeda, M.; Iizuka, H.; Hiratsuka, T.; Shinotsuka, N.; Arai, M. Improvement of thermal stability of NO oxidation Pt/Al_2O_3 catalyst by addition of Pd. *Appl. Catal. B Environ.* **2009**, *90*, 564–569. [CrossRef]
20. Yashnik, S.A.; Ismagilov, Z.R. Pt–Pd/MnO_x–Al_2O_3 oxidation catalysts: Prospects of application for control of the soot emission with diesel exhaust gases. *Kinet. Catal.* **2019**, *60*, 453–464. [CrossRef]
21. Meng, Z.W.; Yang, D.; Yan, Y. Comparison of methods for analyzing the kinetic response of diesel particles to oxidation. *J. Xihua Univ. (Nat. Sci. Ed.)* **2013**, *32*, 51–55.

Disclaimer/Publisher's Note: The statements, opinions and data contained in all publications are solely those of the individual author(s) and contributor(s) and not of MDPI and/or the editor(s). MDPI and/or the editor(s) disclaim responsibility for any injury to people or property resulting from any ideas, methods, instructions or products referred to in the content.

Article

The Effectiveness of HEVs Phase-Out by 2035 in Favor of BEVs with Respect to the Production of CO_2 Emissions: The Italian Case

Francesca Maria Grimaldi * and Pietro Capaldi

CNR-STEMS, Via G.Marconi, 4, 80125 Naples, Italy
* Correspondence: francescamaria.grimaldi@stems.cnr.it or francesca.grimaldi@gmail.com

Abstract: The EU has planned the phase-out of new vehicles based on internal combustion engines in favor of high-efficiency battery electric vehicles (BEV) by 2035 (Fit for 55 package). However, many doubts remain about the effectiveness of this choice for each country of the Union in terms of CO_2 emissions reduction, as each State is characterized by a different carbon intensity related to the production of electricity needed to manufacture and recharge vehicles. This study seeks to explore the Italian case. To this aim, carbon intensities related to electricity production were calculated considering both the Italian electricity mix production in 2022 and those envisaged in 2035, considering two energy scenarios based on different introductions of renewable energy sources (RES). Afterward, the values obtained were adopted for determining the CO_2 emissions related to the whole production process of battery systems in Italy (emissions from mining and refining, scrap materials, and final assembly included) by comparing some of the most up-to-date Life-Cycle Assessment (LCA) analyses related to the manufacturing cycle of the batteries. Finally, the results were adopted to calculate the starting carbon debit for A, B, C, and M car segments for Mild Hybrid, Full Hybrid, and Full Electric powertrains. At the same time, statistical road fuel/electricity consumption data were collected and overall CO_2 emissions were calculated for the same vehicles adopting a dynamic approach and plotted for a defined distance, so as to determine break-even points with respect to the cumulative (i.e., from battery and road) carbon emissions. The results showed that advantages related to electric vehicles are significant only if a low carbon intensity related to electricity production is reached by means of a very high introduction of RES, thus keeping the door open for innovative hybrid powertrain technologies, if fed with low carbon fuels.

Keywords: ICE-based vehicles 2035 phase-out; BEV; HEV; battery; carbon intensity

Citation: Grimaldi, F.M.; Capaldi, P. The Effectiveness of HEVs Phase-Out by 2035 in Favor of BEVs with Respect to the Production of CO_2 Emissions: The Italian Case. *Energies* **2024**, *17*, 961. https://doi.org/10.3390/en17040961

Academic Editors: Maksymilian Mądziel, Kazimierz Lejda and Artur Jaworski

Received: 12 December 2023
Revised: 15 February 2024
Accepted: 16 February 2024
Published: 19 February 2024

Copyright: © 2024 by the authors. Licensee MDPI, Basel, Switzerland. This article is an open access article distributed under the terms and conditions of the Creative Commons Attribution (CC BY) license (https://creativecommons.org/licenses/by/4.0/).

1. Introduction

In order to contain—or rather avoid—the imminent effects of climate change, rapid decarbonization of all economic sectors is required [1]. In particular, a fast decrease in emissions production within the vehicular transport sector is necessary, as it alone emits 23.2% [2] of total CO_2 emissions (2022) and, in general, air pollution [3,4]. To this end, a wide variety of technologies and policies has been suggested and implemented in many countries so far, along with diversified ways to achieve the targets needed to meet mitigation in CO_2 production [5]. Among these, electrification of transport is considered one of the most noticeable actions to be adopted, above all in Italy where the fleet is constituted by a large number (39.8 million) of quite old (12 years on average) and polluting units [6]. In fact, it is well known that Battery Electric Vehicles (BEVs) have significant advantages, as highlighted in various papers [7–9], such as no tail-pipe emissions and an overall greater efficiency than internal combustion engine (ICE)-based vehicles in terms of energy and emissions, even when electricity production is taken into account. However, BEVs cannot be considered as zero-emission vehicles, as they are characterized by significant values of cumulative CO_2 emissions over their entire life cycle, especially when the Life

Cycle Assessment (LCA) analysis methodology is applied [10,11]. The latter represents the most comprehensive approach to be adopted for evaluating the closest-to-reality CO_2 emissions related to vehicles. However, it is not the only one that could be found in the literature. In fact, many papers, such as [12,13], applied instead a Well-to-Wheel (WTW) methodology, which is a simplified LCA and therefore provides a limited quantity of the effective emissions produced by vehicles. For example, it does not consider the hardware construction and decommissioning of vehicles, including materials cycles [12]. An even more misleading approach is the one adopted by the European Community (EC), which, implementing the Tank-to-Wheel (TTW) methodology [14], takes into account only the emissions produced during the road use-phase in order to determine the vehicle carbon footprint, thus considering BEVs as zero-emission vehicles [15]. Therefore, as the European Union has recently guided its member States to adopt a phase-out in the production of new conventional ICE vehicles and hybrid vehicles (HEV) by 2035 in favor of BEVs (Fit for 55 package) [16], mostly due to their reduced carbon footprint, this paper seeks to determine the real advantage of BEVs over Mild Hybrid (M-HEV) and Full Hybrid (F-HEV) cars in terms of CO_2 emissions. In detail, the authors focused their analysis on the energy aspect of the problem, as they believe that the real advantage of adopting BEVs strictly depends on the future value of the carbon intensity of the electricity (c.i.e.) produced in the country where vehicles, and especially battery systems, have to be manufactured and, finally, used, along with the evolution of the c.i.e. value over time. This aspect has to be taken into major account if considering the Regulation (EU) 2023/1542 [17], which implicitly states that future batteries have to be produced in Europe in order to promote ethical and transparent sourcing/processing of raw materials, favor security of supply, and support the re-use, repurposing, and recycling of batteries. Since Italy has a long tradition as a carmaker, the analysis is focused on the comparison between BEVs and HEVs in the eventuality that these vehicles, and therefore batteries, are produced and adopted in the country.

To determine the potential advantage cited above, the cumulative CO_2 emissions, related to some of the most popular and advanced BEVs, M-HEVs, and F-HEVs available on the market (Best in Class), were preliminary calculated adopting a dynamic approach, which foresees the evolution of the Italian grid mix composition over time (2022–2034) according to two energy scenarios toward the Fit for 55 objectives. Afterward, the cumulative CO_2 emissions were plotted versus distance throughout the mean service life of the vehicle, and break-even points in terms of cumulative CO_2 emissions were highlighted for all the vehicles along with percentage variations in carbon emissions between HEVs and BEVs. The same methodology was, then, employed in order to predict the cumulative CO_2 emissions in 2035 and their evolution over time (2035–2047) according to the same energy scenarios but extended up to 2047 toward the decarbonization pledges. In this context, both technology development of the vehicles in terms of running efficiency and energy battery density were taken into account. Furthermore, even in this case, break-even points in terms of cumulative CO_2 emissions were highlighted for all the vehicles, and percentage variations in carbon emissions between HEVs and BEVs were calculated.

The study presented shows that a significant advantage related to BEVs deployment in terms of CO_2 reduction could be obtained only if a much lower c.i.e. than the one in 2022 is reached. In fact, if the introduction of newly installed RES capacity were at a standstill or proceeded at the same pace as the last past years (2016–2021), it could be reasonable to think about HEVs—in particular, new generation HEVs—as a valid alternative, especially if fed with low-carbon fuels (i.e., renewable fuels [18], advanced biofuels [19], e-fuels [20]), as foreseen by RED II [21] and RED III [22] for the future. Moreover, the study suggests that the advantage of BEV deployment in terms of CO_2 emissions reduction differs significantly when it is compared to those found in other European countries characterized by a much lower c.i.e. than the Italian one.

2. Materials and Methods

In order to assess the advantage in phasing out the production of new HEV powertrains (Mild or Full) in 2035, 12 vehicles belonging to A, B, C, and M segments, which represent around 85% of the Italian fleet, were initially selected (Table 1). Each of them was characterized by one of the three recalled powertrains, was a model year 2022, and was identified by authors after a market inquiry as "Best in Class" models in terms of efficiency. They were currently manufactured by the following Brands: Suzuki, Mazda, Ford, Honda, Toyota, Hyundai, Volkswagen, Nissan, Tesla, and Opel.

Table 1. The chosen vehicles according to the segment and powertrain.

Car Segment	Powertrain		
	M-HEV	F-HEV	BEV
A [1]	Suzuki Ignis	Honda Jazz	VW UP
B [2]	Mazda 2	Toyota Yaris	Nissan Leaf
C [3]	Ford Focus	Hyundai Ioniq	Tesla Mod.3
M [4]	Suzuki Vitara	VW Tiguan	Opel Mokka

[1] A = city cars with a full length up to 3.7 m; [2] B = small cars with a full length between 3.7 m and 4.2 m; [3] C = small family cars with a full length between 4.2 m and 4.6 m; [4] D = compact Multi-Purpose Vehicle with a full length up to 4.5 m.

Afterward, the values related to the cumulative CO_2 emissions produced by the aforementioned vehicles were calculated by summing up the emissions coming from (a) the use-phase of the vehicle; (b) the vehicle production phase.

2.1. The Use-Phase Emissions of the Vehicle

The use-phase emissions can be obtained multiplying the specific use-phase emissions produced by the vehicles (Table 1) during their functioning on the road (expressed in terms of $kgCO_2/km$) for the distance run by the vehicles throughout their mean service life. Moreover, these emissions depend on the energy/fuel consumption data, which, in this case, are not homologation data but real statistical mean records. In particular, in the case of BEVs, the CO_2 emissions related to the use-phase depend on the c.i. of the electricity consumed to power vehicles, with the latter calculated by taking into account all the contributions related to the electricity production and transformation (from power plant to the battery). Along with this, they depend on the c.i.e evolution over time, the latter determined by means of a dynamic approach throughout the mean life service of the vehicle. The dynamic approach, described in detail elsewhere [11], foresees the evolution of the grid mix composition over time according to two energy scenarios: one conservative, i.e., Business-As-Usual (BAU), and one more aggressive, which takes into account a significant future increase in newly installed RES capacity.

In this work, the CO_2 emissions related to the use-phase of BEVs were preliminary calculated in 2022 according to the 2022 c.i.e, calculated as mean value over the period 2016–2021, in order to obtain reference data. Afterward, their evolution over time (2022–2034) was dynamically determined according to two different energy scenarios, the BAU@2035 and the Fit for 55 (FF55) @2035, described as follows:

- The BAU@2035: is a conservative scenario, i.e., Business-As-Usual. It was constructed considering a consolidated average growth rate (CAGR) of newly installed RES capacity over the period 2016–2021, which was afterward extended until obtaining a certain value of RES in 2035;
- The FF55@2035: is a more aggressive and binding scenario. It was constructed by taking into account the same growth rate of newly installed RES capacity as stated to fulfill the Fit for 55 package objectives by 2030, but extended until obtaining a certain value of RES in 2035

The same methodology was, then, employed in order to predict the CO_2 emissions related to the use-phase of BEVs at 2035 by means of the previously calculated c.i.e. for BAU@2035 and FF55@2035 scenarios at 2035. Furthermore, the evolution of the CO_2 emissions related to the use-phase of BEV was dynamically determined in the timeframe 2035–2047 adopting the same energy scenarios and, therefore, the same CAGR, but extended up to 2047 toward the decarbonization pledges. In this context, both technology development of the vehicles in terms of running efficiency and energy battery density were taken into account. The obtained emissions for BEV, which are dependent on the c.i.e., can be defined as Power Plant-to-Wheel (PTW) emissions and can be considered equivalent to the Well-to-Wheel (WTW) emissions related to liquid fuels adopted for HEVs. Despite this, in this paper, we chose not to adopt a dynamic but rather a static approach for liquid fuels for HEVs, as the effective degree of reduction in carbon intensity due to the future introduction of low-carbon fuels (renewable fuels, advanced biofuels, and e-fuels), as stated by the RED II and RED III, is still uncertain.

2.2. The Vehicle Production Phase Emissions

As regards the CO_2 emissions related to the vehicle production phase, it was assumed that battery production phase emissions were the only significant ones to be taken into account when comparing the selected vehicles (Table 1) and their relative powertrains. This approach derives from the analysis of some papers [9,11,23–26], in which the CO_2 emissions produced during the material extraction and transformation processes (mining and refining), the vehicle production (manufacturing, battery production, etc.), and the final vehicle dismantlement (disposal phase) are reported. In fact, the aforementioned emissions, which depend on the carbon intensity of the manufacturing country at the time of construction and are therefore fixed, could be partly deconstructed into their own elements and later clustered again into the following three groups: (1) car glider group (body, suspensions, wheels, interiors), (2) powertrain system group (engines/motors, inverters and transmissions), (3) energy storage system group (power battery, tanks). As regards the first group, the CO_2 emissions related to the construction of the whole vehicle body were found to be very similar for both HEVs and BEVs, when considering vehicles belonging to the same car segment, thus making the CO_2 emissive differences quite negligible in the comparison between these two vehicle categories. As regards the CO_2 emissions related to the construction of the two powertrains for HEVs and BEVs, it was considered that HEVs in general—and, in particular, M-HEVs—are characterized by an electric powertrain of quite limited power. In fact, HEV powertrain cannot even ensure a full electric running condition except for very short periods and at low speeds. This aspect limits the contribution of CO_2 production by the electric elements (motor, inverters, and other auxiliary elements) to be added to the CO_2 emissions related to the internal combustion engine. For this reason, the total CO_2 emissions for HEVs powertrains are expected to be higher than those of a conventional ICEV powertrain, but slightly lower than those of a BEV powertrain [27], the latter being characterized by a significantly bigger electric motor and power control system, along with a high content of critical materials. This result is confirmed by ECF [28], which suggests that BEVs powertrains have higher emissions than HEVs when larger and more powerful vehicles are taken into account. On the contrary, other papers, such as [29], report that HEV small/medium car powertrains lead to slightly higher CO_2 emissions, if compared to BEVs. As these differences are due to many aspects, which are specific for each powertrain characteristic and which cannot be extended to all vehicles, it was decided to consider as fully comparable the total CO_2 emissions for HEV and BEV powertrains. It must be pointed out that this approach cannot be adopted in the case of PHEVs (plug-in hybrids) [10], where thermal and electric powertrains are both significant in terms of power and can perform independently. The above-seen considerations regarding the CO_2 emissive content related to the car glider group and powertrain group enabled us to simplify the comparison between HEVs and BEVs, as they allowed us to assume the CO_2 emissions generation related to these two groups as being fully comparable. For this reason, the car glider group and powertrain group emissive contributions will not be further considered in the comparison, allowing us to focus the analysis only on the emissions from energy storage systems, notably batteries, since fuel tank emissions for HEVs are not significant. In fact, batteries are considered the most significant and distinctive element to be measured in the comparison between HEVs and BEVs. In particular, in this work, we chose to focus the analysis on the battery production phase (mining and refining processes along with scrap materials production included) and on the determination of the consequent CO_2 emissions when batteries are produced in Italy. To this aim, the CO_2-specific emissions value, expressed in terms of $kgCO_2/kWh$ of battery capacity, was determined by performing a critical comparison between six of the most recent and relevant LCA analyses available in the literature [7,23–25,30,31], so to define a specific reference value of battery CO_2 emissions when the Italian c.i.e. at the date of the vehicle construction (2022) is considered. The cross-comparison of these analyses was necessary as the LCA results reported in the literature, even if calculated by means of important and detailed databases, are commonly presented as local values of c.i.e., which could strongly differ from each other and from the Italian one. In fact, unlike the Italian case, they could depend on coal or nuclear energy for electricity production. Afterward, the specific CO_2 emission value related to the battery production was multiplied for the battery capacity of each aforementioned vehicle and, then, summed to the relative total use-phase emissions, thus determining the final cumulative CO_2 emissions value for each car over time. Lastly, the obtained cumulative CO_2 emissions were plotted versus distance throughout the mean service life of the vehicle (120,000 km in 12 years [32]), and break-even points were highlighted for all the vehicles along with percentage variations in carbon emissions between HEVs and BEVs. The comparison was carried out for the three different powertrains and for each of the four vehicle segments taking into account the above-mentioned scenarios: 2022, BAU@2035, and FF55@2035. It has to be specified that while in the case of 2022 scenario, the c.i. that referred to battery production was determined by taking into account the c.i.e. in 2022, in the case of BAU@2035 and FF55@2035 scenarios, the c.i. for battery production was obtained considering the different c.i.e. in 2035 according to the above-mentioned scenarios. Moreover, it has to be pointed out that the analysis performed was based not only on the 2022 total consumption data, which considers the electrical and/or thermal powertrain efficiencies, battery energy capacity, etc., but also on their future development until 2035 in terms of technical evolution and enhanced performances of the recalled parameters [33]. For this reason, in 2035 energy scenarios, an increase in performance regarding battery capacity and better running efficiency were considered for BEVs. At the same time, a lower fuel consumption was considered for M-HEV and F-HEV, due to an enhanced matching between electric powertrain and ICE, which could reduce even more the low-efficiency part-load operation.

3. Results

In this Section, the value of the Italian c.i.e. in 2022 was first determined and reported (§3.1). Then, two different predictions of c.i.e. in 2035 were made by taking into account the evolution of the parameter over time (2022–2034) according to a conservative (BAU@2035) and a binding scenario (FF55@2035) (§3.2). These scenarios differ from each other above all for the different future electricity generation mix values and, in particular, for the different potential RES installed capacities. The obtained c.i.e. in 2022 and 2035 was newly determined, in comparison to a previous paper [34], by taking into account some updated values of the CO_2-specific emissions, expressed in terms of gCO_2/kWh, regarding renewables (e.g., new technologies for the production of photovoltaic panels, etc.) and fossil fuels (e.g., more efficient CCGT power plants), according to the American National Renewable Energy Laboratories (NREL) [35]. Furthermore, it was considered that the contribution coming from the CO_2-specific emissions related to the imported electricity from other countries. In particular, the mean European c.i.e. value was taken into account according to the 2022 and future (2035) scenarios. Lastly, the CO_2 emissions values relating both to the battery production (§3.3) and to the use-phase of the vehicle (§3.4) were calculated. The resulting data, which correspond to the cumulative value of CO_2 emissions for each vehicle, were then dynamically plotted versus distance until the vehicle's end of life (120,000 km in 12 years) according to the evolution of the 2022 scenario and to the four vehicle segments (respectively, A, B, C, and M). In particular, in all the four plots obtained, a comparison between the three different powertrains was made both highlighting the break-even points in terms of CO_2 emissions and calculating the percentage variations in carbon emissions produced by HEVs, if compared to BEVs. Afterward, by taking into account the c.i.e. in 2035 for both the scenarios and the evolution of the vehicles in terms of running efficiency and energy battery density, a new set of data, corresponding to the future cumulative CO_2 emissions value over time, was dynamically plotted according to the BAU@2035 and FF55@2035 scenarios (extended up to 2047) and to the four vehicle segments. In particular, in all the eight plots obtained, a comparison between the three different powertrains was made both highlighting the break-even points in terms of CO_2 emissions and calculating the percentage variations in carbon emissions (expressed as percentages) produced by HEVs, if compared to BEVs.

3.1. The Italian c.i. at 2022

In general, the mean c.i. of the electricity produced in a country depends on the following features:

- The primary fuel used and the thermodynamic cycle allowable by the fuel.
- The contribution of RES, whose total incidence varies according to the season and to the different weather conditions.
- The net energy flux imported and exported through the international exchange.

Regarding the Italian situation (updated to the end of 2021, as 2022 was considered statistically not significant because of the Russia–Ukraine conflict and the exceptional drought), the energy consumption was found to be around 320 TWh, with an electricity mix depending on the deployment of thermoelectric non-renewable plants based on natural gas (about 50% of the whole electricity production), oil (6.1%), and coal (6.9%), along with a significant percentage of RES (about 37%), mainly hydroelectric (15.7%), and photovoltaic and wind (16.1%) [36–38]. According to this energy panorama, the total electricity produced in the country was found to be 278 TWh, 116 TWh of which came from renewable sources and 162 TWh from fossil fuels (Table 2). Instead, the net electricity import was approximately 43 TWh [36] and the curtailment/storage losses accounted for about 1 TWh [39]. Therefore, based on the specific c.i. for each technology production as reported by NREL [35] and considering a mean European c.i. related to the net imported electricity (287 gCO$_2$/kWh in 2022) [40], a quite accurate c.i. value of 293 gCO$_2$/kWh was determined by means of the following equation:

$$C.I. E. (2022) = \sum gCO_2/kWh_{res} \times (TWh_{res}/TWh_{tot}) + \sum gCO_2/kWh_{fos} \times (TWh_{fos}/TWh_{tot}) + \sum gCO_2/kWh_{I} \times (TWh_{I}/TWh_{tot}) \quad (1)$$

Table 2. The Italian electricity mixes (TWh) and their relative c.i.e. (gCO$_2$/kWh) according to current, BAU@2035, and FF55@2035 scenarios. The c.i.e. reported for 2022 and future (2035) scenarios have to be considered as gross values.

	NG (TWh)	OTHER NON-RES (TWh)	HYDRO (TWh)	PV (TWh)	WIND (TWh)	OTHER RES (TWh)	CURTAILMENT/ STORAGE LOSSES (TWh)	Net I/E (TWh)	c.i.e. (gCO$_2$/kWh)
2022	140	22	47	25	21	23	−1	43	293
2035 BAU	155	7	51	35	30	23	−4	53	249
2035 FF55	57	3	51	136	90	23	−26	53	110

3.2. The Italian c.i. at 2035

In order to calculate the c.i.e. in 2035, two different scenarios were defined by the authors: the BAU@2035 and the FF55@2035. These were outlined by the comparison and merger of different scenarios defined in several analyses performed by Terna, Snam, and the Institute for Sustainable Future (ISF) for 2030 and 2040 [39,41,42]. The BAU@2035 is a conservative scenario that takes into account the evolution of the Italian installed RES capacity over the period 2022–2035. In particular, it considers the trend of newly installed RES capacity in the timeframe 2016–2021 extending up to 2035, along with a constant capacity of fossil-fuel-based power plants, if compared to 2022 values. The latter is constant as a whole, but foresees an increase in the deployment of natural gas and a decrease in other non-RES (mainly coal and oil); this is to accomplish the decarbonization pledges. In detail, the BAU@2035 scenario is characterized by a consolidated compound annual growth rate (CAGR) of newly installed RES equal to 1.0 GW/year (corresponding to an energy growth rate of about 1.35 TWh/year), a value which was calculated over the period 2016–2021 (2020 excluded because of COVID-19) and fully confirmed in the 2023 IEA report about Italy [37,38]. The trend depicted in the BAU@2035 scenario reflects the condition of the minimum introduction of RES, therefore representing the lowest trend of decarbonization. This situation reflects the limiting conditions that exist in the Italian panorama, such as cumbersome and time-consuming authorization processes for new power plants, inadequate grid characteristics [43], high costs of logistics, and critical procurement of base materials, which influence the success of the auctions. Last, but not least, the increasing opposition from local populations due to the NIMBY syndrome should be underlined. According to this scenario, the total electricity produced in the country in 2035 would be approximately 301 TWh, 139 TWh of which from renewable sources and 162 TWh from conventional power plants based on fossil fuels (Table 2). Furthermore, the total electricity consumption would be 350 TWh, and the annual electricity import from neighboring countries would increase to 53 TWh [36]. In this energy panorama, the curtailment/storage losses would be 4 TWh [id.]. If the above-seen Equation (1) is taken for granted, and if the specific c.i. for each energy source is considered, then the final c.i.e. value would almost be equal to 249 gCO_2/kWh, when a mean European c.i.e. value reduced to 100 gCO_2/kWh, due to increased decarbonization, is taken into account [40,41,44]. The FF55@2035, instead, is a binding scenario strongly based on RES, whose trend is outlined considering the same CAGR of newly installed RES planned in 2019 to achieve the Fit for 55 package pledges by 2030 (i.e., 7 GW/y), although extended up to 2035 through extrapolation. This scenario leads to an electricity generation of 413 TWh, with 300 TWh of RES, 60 TWh of fossil energy (mainly natural gas), and 53 TWh of imported energy. Because of the inconstancy of RES (mainly based on photovoltaic systems), which can cause great difficulty in following the real electric loads and can require large and expensive storage systems, significant energy curtailment and storage losses are expected (around 26 TWh), thus making the total national electric availability decrease to 387 TWh. The resulting trend depicted in this scenario reflects the condition of the maximum introduction of RES and therefore represents the highest trend of decarbonization. In this scenario, still considering equation (1) and a mean European c.i.e. of 100 gCO_2/kWh [42,44], the c.i.e. would be almost equal to 110 gCO_2/kWh. In the following Table 2, the 2022 and 2035 electricity mixes and c.i.e. are reported.

However, it has to be pointed out that the c.i.e. values reported in Table 2 have to be considered as gross values due to the fact that they are calculated at the power plants. Therefore, grid efficiency should also be taken into account in order to consider the real c.i.e. of the delivered energy. The latter may be generally available for production plants and industries (such as a battery production plant) at medium voltage (about 20 kV) with an average transmission efficiency of 95%. At the same time, electricity may be available at low voltage (380 V) for non-industrial uses (e.g., electric socket) after more transformation stages and consequent losses, with a final transmission efficiency of 92%. In the following Table 3, the resulting c.i.e. according to the different grid transmission efficiencies is provided with respect to scenarios and final applications.

Table 3. Specific carbon intensities (c.i.e.) (kgCO$_2$/kWh) related to the energy scenarios and delivery points.

Energy Scenario	Electric C.I. (p.p.) [1] (kgCO$_2$/kWh)	Electric C.I. (m.v.) [2] (kgCO$_2$/kWh)	Electric C.I. (l.v.) [3] (Socket c.i.) (kgCO$_2$/kWh)
2022	0.293	0.308	0.318
2035 BAU	0.249	0.262	0.270
2035 FF55	0.110	0.116	0.120

[1] p.p. = electricity at power-plant; [2] m.v. = medium voltage; electricity for production plants and industries (battery production plants) at medium voltage (20 kV) with 95% grid transmission efficiency; [3] l.v. = low voltage; electricity for non-industrial uses (e.g., electric socket) at low voltage (380 V) with 92% grid transmission efficiency.

3.3. The Carbon Footprint Related to the Battery Production (c.i.b)

In order to calculate the specific c.i. related to the production of the batteries (c.i.b) in Italy, six life cycle assessment analyses (LCAs) that referred to the production of BEVs or batteries (cells or packs) were considered (Table 4). The latter were chosen from the literature due to their relevance, their being up-to-date, and because they contained, implicitly or explicitly, values of c.i.b related to defined energy contexts (e.g., nuclear-based or coal-based ones). Each of the reported values related to the c.i.b was, then, singularly highlighted or extrapolated from each LCA (Table 4), although the energy contexts they were related to were quite different, when compared, to the 2022 Italian one (Table 2). Along with them, the CO$_2$ emissions related both to the mining and refining processes (from brine or ore) and to scrap materials production in a battery factory, when available, were taken into account. As regards the latter, the authors decided to completely allocate them to the battery, following the methodology of the cut-off approach according to the Environmental Product Declaration (EPD) system [45]. For this reason, no system expansion was applied to this analysis, as no credits were recognized for material recycling and/or energy recovery from the same recycling.

Table 4. CO$_{2e}$ emissions (kgCO$_2$/kWh) related to (a) the battery cell production, (b) mining and refining processes, and (c) the scrap materials production found in the literature.

Authors	CO$_2$ Specific Emissions Related to the Battery Cell (kgCO$_2$/kWh)	CO$_{2e}$ Emissions Related to Mining and Refining Processes (kgCO$_2$/kWh)	CO$_{2e}$ Emissions from Scrap Materials (kgCO$_2$/kWh)
Cox et al. (2017) [22]	100	N.C. [1]	N.C.
IVL (2019) [30]	80	N.C.	N.C.
Helmers et al. (2020) [23]	90	N.C.	N.C.
Woody et al. (2021) [7]	90	N.C.	N.C.
Volvo [2] (2021) [24]	90	50	15
Northvolt (2022) [31]	130 [3]	//	//

[1] N.C. = not considered; [2] Total emissions referred to the whole battery pack; [3] Global "State-of-the-Art" value also considering mining, refining, and emissions from un-recycled scrap materials.

Afterward, the aforementioned CO$_2$ emissions values that related only to the production phase of the battery cells (all reports but Volvo, whose study referred to the whole battery pack) were compared in order to obtain a law of dependence of the specific CO$_2$

emissions related to the production of the batteries with the electricity mix available at the production site, as indicated in Table 3. Once determined, the CO_2 emissions values related to the production of the batteries were scaled based on the 2022 and future (2035) Italian c.i.e. (§ 3.1, 3.2), with the aim of obtaining a reliable mean specific CO_2 emissions value that related to the production of the whole battery pack in Italy. Furthermore, for the BAU@2035 and FF55@2035 scenarios, an assumption was made regarding the evolution of the battery system in terms of energy density. In fact, by taking into account the energy increase in recent years, a 20% increase in battery capacity at constant weight was assumed [46]. Moreover, the authors did not consider solid-state cells, whose industrial applicability is not yet clearly defined. Therefore, 74 kgCO_2/kWh was obtained as the CO_2 emissions value for the production of the battery in Italy when considering the 2022 energy scenario, while 50 kgCO_2/kWh and 23 kgCO_2/kWh were determined in the case of the BAU@2035 and FF55@2035 scenarios, respectively (Table 5). Instead, the CO_2 emissions related to the mining and refining processes resulted in 45 kgCO_2/kWh in 2022 and 40 kgCO_2/kWh in both the 2035 scenarios (Table 5). In fact, they were scaled only by 10%, as these processes are not always supported by publicly available data on energy consumption [47] and generally take place in non-EU countries where carbon-free electricity will not be widely accessible even in the near future. Instead, the contribution of scrap materials (in terms of CO_2 emissions) during the production of batteries was calculated considering that it will follow the same c.i.e. reduction trends by 2035. In fact, after calculation, the related specific CO_2 emissions resulted in 15 kgCO_2/kWh in 2022, which are expected to become, respectively, 14 kgCO_2/kWh and 10 kgCO_2/kWh in the BAU@2035 and FF55@2035 scenarios (Table 5). Lastly, 20 kgCO_2/kWh was added to the obtained c.i.b values in order to determine the CO_2 content related to the construction of the whole battery pack, thus taking into account the significant contribution of the assembly process due to the presence of hundreds of welded connections, frame structures, containers, etc. [31]. As with the previous terms, the CO_2 emissions related to the battery-pack assembly process were also calculated and were expected to decrease over time, from 20 to 18 or 14 kgCO_2/kWh, according to the BAU@2035 scenario or FF55@2035 scenario, respectively (Table 5).

Table 5. CO_2 emissions (kgCO_2/kWh) related to (a) mining and refining processes, (b) battery cell production, (c) scrap materials production, (d) battery pack assembly process, (e) total emissions related to battery pack production in the Italian case study, and (f) total emissions related to battery pack production in the Italian case study in case of enhanced battery efficiency. As b–f are dependent on the same c.i.e., the latter is reported for all the energy scenarios at the battery production plant.

Energy Scenario	Electric C.I. (Battery Production Plant) (kgCO_2/kWh)	(a) CO_2 for Mining and Refining (kgCO_2/kWh)	(b) CO_2 for Battery Cell Production (kgCO_2/kWh)	(c) CO_2 for Scrap Materials Production (kgCO_2/kWh)	(d) CO_2 for Battery Pack Assembly (kgCO_2/kWh)	(e) Total CO_2 for Battery Pack Production (kgCO_2/kWh)	(f) Total CO_2 for Enhanced Battery @2035 (+20% Energy) (kgCO_2/kWh)
2022	0.308	45	74	15	20	154	//
2035 BAU	0.262	40	50	14	18	122	97
2035 FF55	0.116	40	23	10	14	87	70

3.4. The c.i. of the Vehicles' Use-Phase

As is well-known, the use-phase CO_2 emissions of any vehicle are determined by two factors: (1) the fuel/electricity consumption of the vehicle, and (2) the c.i. of the fuel/electricity used to feed the vehicle. In order to determine the first one, the authors did not consider simulation models that calculate the vehicle fuel/electricity consumption according to different vehicle's characteristics and operation points, nor WLTP homologation procedures, the implementation of which always provides values of the vehicle's fuel requirement that are quite far from the real ones (−15% on average). Indeed, in order to determine the real fuel/electricity consumption of vehicles, in this paper, only significant statistical records relating to fuel needs were taken into account [48]; these were eventually compared with other consumption data delivered by independent institutions that had carried out road tests [49]. On the other hand, in order to determine the c.i. of the fuel/electricity used to feed the vehicle, a "Well-to-Wheel" (WTW) approach was adopted, both for HEVs and BEVs. Therefore, the CO_2 emissions produced during both the "Well-to-Tank" (WTT) phase, which also includes the refinery and transportation contributions, and the "Tank-to-Wheel" (TTW) phase, also called the "Battery-to-Wheel" (BTW) phase for BEVs, were taken into account. In particular, for HEVs fed with gasoline, the values reported in the following two references were considered:

- JEC Consortium (JRC, EUCAR, CONCAWE) [50];
- ARGONNE Laboratories (adopting GREET-1 model) [51].

After having analyzed these data, a mean value of 470 gCO_2/dm3 for the WTT phase was determined along with a mean value of 2330 gCO_2/dm3 for the TTW phase. Therefore, a resulting total value of 2800 gCO_2/dm3 for the WTW phase was obtained, which was later used in the computation of the real total CO_2 emissions related to vehicle fuel consumption. It has to be pointed out that the obtained value was not supposed to change over time, as it is still uncertain which will be the effective future degree in carbon intensity reduction, and therefore in CO_2 reduction, linked to the introduction of renewable fuels (renewable fuels, advanced biofuels, e-fuels) over time. Instead, as regards the determination of the c.i. of the electricity used to feed BEVs in the road use-phase, the values reported in Table 3 regarding the c.i.e. at low voltage (or socket c.i.) in the three scenarios were considered as references, to be afterward referred to as the efficiency of the whole charging phase. In fact, in order to determine the c.i. of the electricity used to feed BEVs as close to real conditions as possible, several losses should be considered, starting from the charging system, whose mean efficiency during charging can be sensibly lower than the nominal value, passing through cables and connections up to the battery itself. In particular, the global charging phase can be characterized by different efficiencies, ranging from 83%, in the case of an AC single-phase home apparatus, up to 87%, which is typical of an AC three-phase road charging tower (up to 22 kW of power) [49,52]. As regards DC systems, which can reach up to 300 kW of power and are normally adopted for fast charging on highways, they are characterized by a higher charging efficiency (around 91%) [53], when compared to AC systems. However, they were not considered in this analysis because of their much lower diffusion due to high costs. In conclusion, a mean efficiency value of about 85%, fully confirmed by [49], was chosen for the existing (2022) AC charging systems, while for the future (2035) ones, a significant improvement was considered by the authors, fixing the global mean charging efficiency at 88%. It must be underlined that the charging phase represents the most inefficient stage of the whole energy transmission process (from energy production sites up to car batteries) and, therefore, that the energy effectively stored in the battery can be significantly lower than the energy provided by the grid. For this reason, in order to determine the effective BEV efficiency in terms of CO_2 emissions, the global mean AC charging efficiency value according to the different scenarios should be adopted. The latter values are reported in the following Table 6 along with the c.i.e. at low voltage and the resulting charging c.i. (c.i.c.) values for BEVs. It has to be pointed out that the c.i. of the electricity used to feed BEVs could be considered equivalent to the WTT c.i. previously reported for gasoline; alternatively, it could be called Plant-to-Battery efficiency (PTB). However, unlike the c.i. related to the fuel, which was considered to be constant over time throughout the mean life service of the vehicle, the c.i. of the electricity used to feed BEVs was assumed to dynamically change according to the scenarios previously described.

Table 6. Socket carbon intensity (kgCO$_2$/kWh), charging efficiency, and vehicle battery carbon intensity (kgCO$_2$/kWh), for the 2022 scenario and 2035 scenarios.

Energy Scenario	C.I.E. @Socket	AC Charging Phase Efficiency	C.I.C. @Vehicle Battery (PTB)
2022	0.318	0.85	0.374
BAU@2035	0.270	0.88	0.307
FF55@2035	0.120	0.88	0.137

As stated before, the fuel/electricity consumptions of the vehicles taken into account in this investigation were gained from statistical records from customers, along with other real-world consumption tests performed by independent organizations [48,49]. Afterward, all the obtained data were compared, thus determining a mean fuel consumption expressed in terms of dm^3/km for gasoline, in the case of M-HEVs and F-HEVs, or in terms of kWh/km for electricity, in the case of BEVs. Also, the battery capacity of each vehicle was considered and then multiplied for the total value of the battery-specific CO$_2$ emissions in 2022 (i.e., date of vehicle construction reported in Table 5), thus determining the starting CO$_2$ emissions for each model, to be summed up to the road use-phase emissions (Table 7).

Table 7. Fuel/energy consumption, on-board battery capacity, and battery production CO$_2$ emissions by the 12 vehicles taken into account.

Mild HEV Car Segment	Model	Fuel Consumption (dm^3/km)	Battery Capacity (kWh)	CO$_2$ for Battery Production (kgCO$_2$)
A	Suzuki Ignis	0.054	0.13	20
B	Mazda 2	0.046	0.90	138
C	Ford Focus	0.058	1.20	185
M	Suzuki Vitara	0.062	1.10	169
Full HEV Car Segment	**Model**	**Fuel Consumption (dm^3/km)**	**Battery Capacity (kWh)**	**CO$_2$ for Battery Production (kgCO$_2$)**
A	Honda Jazz	0.046	0.75	115
B	Toyota Yaris	0.045	0.90	138
C	Hyundai Ioniq	0.048	1.6	200
M	VW Tiguan	0.050	1.20	185
BEV Car Segment	**Model**	**Energy Consumption (kWh/km)**	**Battery Capacity (kWh)**	**Battery Carbon Intensity (kgCO$_2$)**
A	VW UP	0.141	37	5700
B	Nissan Leaf	0.162	60	9240
C	Tesla Mod.3	0.184	60	9240
M	Opel Mokka	0.177	50	7700

Afterward, fuel and energy consumptions were multiplied respectively for the previously calculated fuel c.i. and the c.i.c., the latter dynamically updated to the period 2022–2034 by taking into account c.i.c. reduction according to BAU@2035 and FF55@2035, in order to determine the cumulative CO$_2$ emissions for each vehicle. To this aim, a graphical comparison, in terms of CO$_2$ emissions versus distance, was performed between the three powertrains when these were applied to the four vehicle segments according to the 2022 scenario in the timeframe 2022–2034 (Figure 1). In particular, two different curves were graphed for BEVs, which report their cumulative CO_2 emission trends over distance when two different evolutions for c.i.c. are considered according to the BAU@2035 and FF55@20335 scenarios, respectively. The intersections between the green lines and blue line or red line, respectively, represent the break-even points of BEVs with respect to M-HEVs and F-HEVs according to the two scenarios.

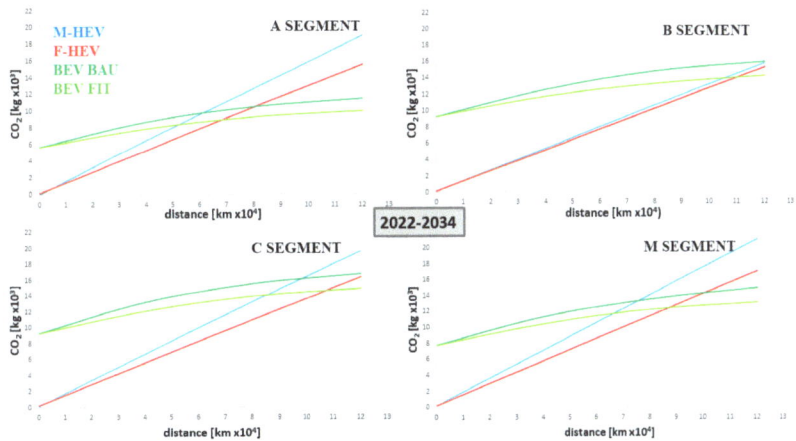

Figure 1. The comparison (in terms of CO_2 emissions versus distance) between the three powertrains when applied to the four vehicle segments according to the 2022 scenario.

By analyzing Figure 1, it can be seen that parity between BEVs and HEVs is obtained after significant distances in most of the cases, above all, when F-HEVs are considered, mainly due to their higher global efficiency. In fact, only in one case did a M-HEV (i.e., Mazda 2 provided with Skyactive-G engine family) show similar CO_2 emissions with respect to a F-HEV, mostly because of the very high efficiency of its internal combustion engine even at partial loads [54]. Also, the significant role played by battery capacity is quite evident, as it increases considerably both the vehicle starting carbon debit and the energy consumption related to a greater moving mass (increasing the green line slope).

Lastly, in order to make predictions for 2035, the above-mentioned vehicles were again considered, although with upgraded performances in terms of global efficiency. In fact, a reduction in fuel consumption of about 5% and 10% for M-HEVs and F-HEVs, respectively, was considered, because of the higher potential expressed by the second ones in reducing inefficient partial load operations due to higher electric/thermal ratios. At the same time, a 5% improvement in efficiency was considered for BEVs, due to the already very high level achieved by each single component, which will be very difficult to improve further. Instead, as regards the battery, a 20% higher specific energy density was taken into account along with a lower specific c.i.b in the BAU@2035 and FF55@2035 scenarios, as reported in Table 5. In the following Figures 2 and 3, graphical comparisons, in terms of CO_2 emissions versus distance, of the three powertrains, when these are applied to the four vehicle segments according to the BAU@2035 and the FF55@2035 scenarios, are shown. Even in these cases, a dynamic scenario was adopted in order to take into account the decrease in c.i.e. occurring during the period 2035–2047. The intersections between lines represent the break-even points in terms of CO_2 emissions between BEVs and HEVs.

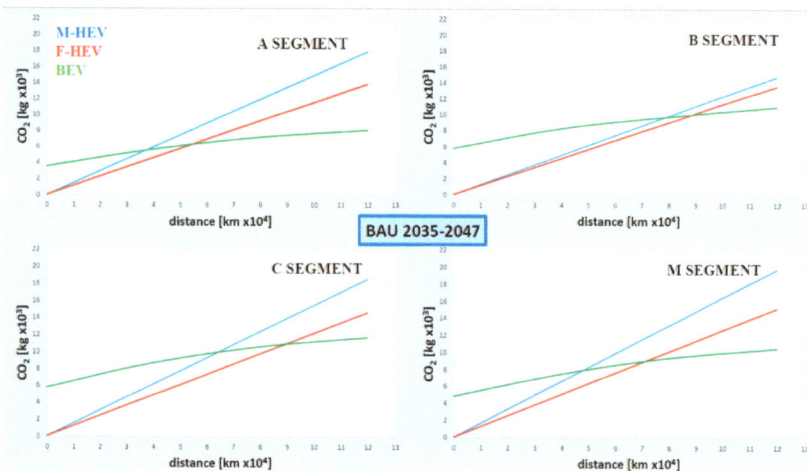

Figure 2. A comparison (in terms of CO_2 emissions versus distance) between the three powertrains when applied to the four vehicle segments according to the BAU@2035 scenario.

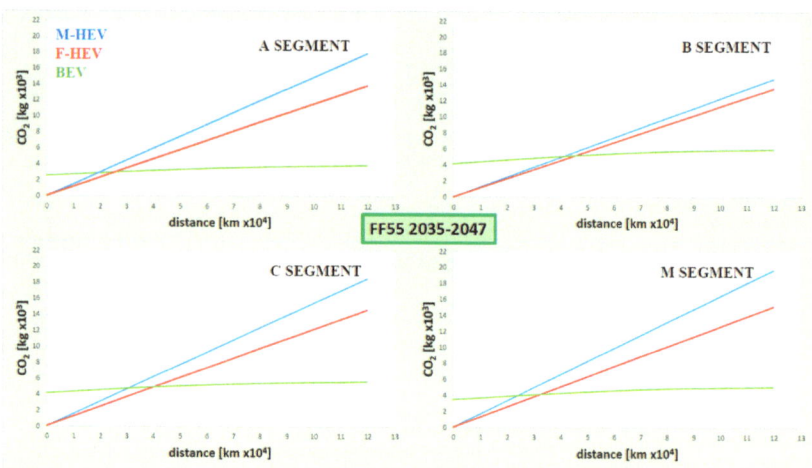

Figure 3. A comparison (in terms of CO_2 emissions versus distance) between the three powertrains when applied to the four vehicle segments according to the FF55@2035 scenario.

By analyzing Figures 2 and 3, it can be easily seen that the advantage of adopting BEVs is disruptive only in the case of the FF55@2035 scenario and when the whole life of the vehicle is considered. On the contrary, the break-even points are significantly higher for most of the vehicles in the BAU@2035 scenario, leading to a lower advantage for BEVs in terms of avoided CO_2 emissions when HEVs are considered. In the following Table 8, the variations in CO_2 emissions for M-HEVs and F-HEVs compared to BEVs are reported in terms of percentages for each of the four vehicle segments and for the scenarios considered. It has to be highlighted that the values reported are related to the mean statistical lifespan, which corresponds to the average running range for vehicles in Europe (120,000 km [32]).

Table 8. Variations in CO_2 emissions for M-HEVs and F-HEVs compared to BEVs according to the energy scenarios considered.

SCENARIO	CAR SEGMENT	POWERTRAIN		
		BEV	Full HEV	Mild HEV
2022@BAU	A	Ref.	+35%	+66%
	B	"	−4%	−1%
	C	"	−2%	+17%
	M	"	+14%	+40%
2022@FF55	A	"	+55%	+90%
	B	"	+8%	+11%
	C	"	+10%	+32%
	M	"	+29%	+60%
BAU@2035	A	"	+71%	+121%
	B	"	+23%	+34%
	C	"	+25%	+58%
	M	"	+44%	+88%
FF55@2035	A	"	+259%	+364%
	B	"	+126%	+147%
	C	"	+160%	+229%
	M	"	+197%	+286%

4. Discussion

As can be easily seen from Figure 1, the 2022 Italian electricity mix is not able to bring a significant advantage to BEVs over HEVs in reducing cumulative CO_2 emissions, if the 2022@BAU scenario is taken into account. In fact, the necessary distance for reaching the break-even point, therefore, the carbon emission parity, is significant, and the advantages in terms of avoided emissions over the mean statistical lifespan @120,000 km can be negative in some cases (Table 8). This issue is evident for most of the vehicle segments, especially when comparing BEVs to F-HEVs, the latter characterized by a higher efficiency powertrain in comparison to M-HEVs. On the contrary, a better result can be achieved for BEVs, if a significant introduction of RES takes place. Indeed, an increase in CO_2 can be the result for M-HEVs (around +90%) in the case of city cars (Segment-A) when these are compared to the equivalent BEVs, the latter characterized by the presence of a small battery pack that determines a lower carbon debit and a better running efficiency due to the reduced mass. As regards the 2035 objectives imposed by the Fit for 55 package within the Green Deal, the advantage of phasing out new HEVs will depend on the future introduction of RES. In fact, in the case of the BAU@2035 scenario, HEVs would see a major increase in CO_2 emissions in comparison to BEVs (up to +121%, in the case of Segment-A), whereas, in the case of the FF55@2035 scenario, BEVs would achieve an impressive advantage over HEVs (Table 8). However, the latter would be very difficult to reach in Italy because of the very low CAGR related to the newly installed RES capacity connected to the grid in 2022. Moreover, political uncertainty, e.g., linked to the Russia–Ukraine conflict, could play a troublesome action in future scenarios, as confirmed by the negative trends of c.i.e. reduction recorded during the last period in Italy [44,55] and in some parts of Europe [44,56].

The performed analysis has also underlined that it will be hard for Italy to give up using natural gas power plants because of its dependence on photovoltaic energy, which is characterized by a very strong seasonal fluctuation in power production. In fact, it will be hard for energy compensation to be resolved by means of storage systems based on hydrogen or battery systems, which would have enormous consequences, at least at the moment, in terms of capacity and high cost [36,57]. For these reasons, it is improbable that the Italian energy setup will soon reach a very low-carbon intensity value in electricity production like other countries can provide, for instance, with nuclear power plants and more stable RES, the latter based on wind turbines and hydroelectric power plants. For this reason, it is expected that a feasible c.i.e. value in 2035 would fall between the two values reported in the BAU@2035 and FF55@2035 scenarios, resulting in higher-than-expected specific CO_2 emissions from cars and in limited global savings of total carbon emissions. Indeed, this perspective would represent quite a different situation from what the EU has planned in phasing out new ICE-based vehicles by 2035.

The research has also shown other issues that could reduce the advantages of BEVs over HEVs in terms of CO_2 emissions, at least for the moment. the main issue still remains, namely, the high specific c.i. related to battery production, which was found to be quite higher than the ones assessed in other papers., To explain this underestimation, the authors have highlighted the following reasons: (1) mining and refining of base materials are often not taken into account; (2) CO_2 emissions related to batteries refer to a single cell and not to the battery pack, thus excluding the emissive content related to the assembly phase; (3) emissions related to scrap materials are not allocated to the battery; and (4) the c.i.e. taken into account for the production of batteries is often theoretical. In fact, the latter could be referred to as the proper average value of a very large region and not the local one available at the production site or, in contrast, it could be distinctive of a specific country characterized by a very low c.i.e. On the contrary, if strictly applying recognized LCA guidelines and considering only the effective c.i.e. of the manufacturing country (Italy, in this case), higher specific carbon emissions than average could be obtained for the whole production process. Lastly, it must be underlined that the carbon footprint could be further reduced in the future through the adoption of electricity with a lower c.i. for cell manufacturing and assembly. However, the emissive content from mining and refining processes will not easily decrease, as these processes, often not supported by publicly available data on energy consumption [47], generally take place in non-EU countries where carbon-free electricity will not be widely accessible even in the near future. All these aspects together will keep the total carbon footprint value for battery production significant, above all in countries, like Italy, which will not easily reach very low c.i. values for electricity production.

Furthermore, other important issues regarding the effective advantages of BEVs over HEVs are set out below. The first one regards the effectiveness of the transmission of electricity from power plant to battery; in particular, the battery charging phase. In fact, the latter is usually found to be the most critical one, as it contributes to reducing BEVs' global energy efficiency. Instead, the second issue regards the size of the battery pack and the consequences it could determine on-road efficiency and relative CO_2 emissions. In fact, BEVs provided with larger battery packs lead to higher CO_2 emissions during the battery production phase and to a lower global efficiency due to a higher vehicle mass. Instead, Segment-A (city cars) BEV vehicles show better performances in comparison to HEVs because of the smaller battery pack and the consequent lower mass, although this involves a shorter running range. It must be underlined that, even with cars provided with large battery systems, the current distance may be much shorter than the one achievable by HEVs. This aspect is still considered crucial by customers, making the trend toward bigger batteries quite unavoidable. On the other hand, the analysis has shown that 2022 HEVs, provided with high-efficiency right-sized combustion engines, could introduce a significant evolution in terms of fuel consumption and CO_2 emissions production, as they were expressly designed to match with electric powertrains, thus achieving better global efficiency, when compared to past models. Moreover, HEVs could fully take advantage of the availability of low-carbon fuels (LCF), thus giving impressive attractiveness in the reduction in CO_2 emissions. In fact, these fuels could even be produced in a sustainable and affordable way thanks to the availability of low-cost electricity instead of performing

RES curtailment (quite probable in the case of the FF55@2035 scenario). This could lead to the availability of a low-c.i. energy source for the automotive sector, making HEVs again competitive with BEVs in many situations.

Lastly, the research has shown that the aforementioned issues along with the problems related to the industrial feasibility of BEVs (availability of critical materials in battery production, power control, motors/generators, etc.) could make the industrial transformation to electric vehicles very challenging for Italy. In fact, in these conditions, it could be quite impossible to create a production context capable of being competitive on the global market with other countries, characterized by lower c.i.e. and cost of energy (i.e., with a nuclear base production and wind/hydro renewables). Also, the resulting uncertainty could even have a negative impact on the country, reducing its potential and any related economic benefits. This panorama could make unworthy the huge economic efforts necessary to prepare adequate infrastructure for vehicle recharging without yielding substantial results in terms of CO_2 emissions reduction.

5. Conclusions

The final aim of this paper was an evaluation of the effectiveness of phasing out hybrid vehicles (HEVs) in favor of battery electric vehicles (BEVs) by 2035 when considering the Italian energy scenario. To this aim, a preliminary analysis of the 2022 Italian electricity mix production was performed and a reliable value, comprehensive of the transformation losses, of the carbon intensity of the electricity (c.i.e) was obtained. The same methodology was adopted for predicting two different c.i.e. values in 2035 when taking into account, respectively, a conservative (BAU@2035) and a more aggressive (FF55@2035) scenario, characterized by a different CAGR in RES. Afterward, the obtained data were employed in order to evaluate the total carbon footprint of BEVs with respect to the HEVs' (M-HEVs and F-HEVs) one, for all the selected vehicles and according to all the energy scenarios considered. In particular, the c.i.e was adopted for determining, at first, the emissions related to battery production (emissions from mining and refining, scrap materials, and final assembly included), then those related to the road use-phase over time adopting a dynamic approach. Afterward, both the CO_2 emissive contributions, defined as "cumulative CO_2 emissions", were plotted over a defined distance throughout the mean service life of the vehicle (120,000 km in 12 years) in order to determine break-even points with respect to the cumulative carbon emissions. Furthermore, percentage variations in CO_2 emissions for M-HEVs and F-HEVs with respect to BEVs were calculated for all the vehicles and energy scenarios. The obtained results show that, when the BAU@2035 scenario is considered over the timeframe 2022–2034, the Italian electricity mix does not manage to bring a significant advantage to BEVs over HEVs in reducing cumulative CO_2 emissions. In fact, in most cases, break-even points are reached only after a significant distance, while percentage variations in CO_2 emissions between HEVs over BEVs can be even negative when the B- and C-segments are taken into account. On the contrary, if, in the same timeframe, the FF55@2035 scenario is taken into account, break-even points are also reached after a not negligible distance, although percentage variations in CO_2 emissions between HEVs over BEVs are found to be more accentuated, especially in case of A- segment M-HEV (+90%). As regards the 2035 objectives imposed by the Fit for 55 package, the advantages of phasing out new HEVs will depend on the future introduction of RES. In fact, when considering the BAU@2035 scenario in the timeframe 2035–2047, break-even points would be reached after a short distance (38,000 km) only when A-segment BEV is compared to A-segment M-HEV, while achieving minor advantages in the case of F-HEV and of the other segments, especially B- and C-segments F-HEV. Instead, in the case of the FF55@2035 scenario, BEVs would achieve an impressive advantage over HEVs (Table 8), both in terms of break-even points, which are found to be always < 50,000 km, and of percentage variations in CO_2 emissions between HEVs and BEVs (from +126% to +364%). However, it would be very difficult to reach the latter condition in Italy because of the 2022 very low CAGR in newly installed RES capacity connected to the grid. For this reason, if the introduction of newly installed

RES capacity was at a standstill or proceeded at the same pace as past years (2016–2021), it could be reasonable to think about HEVs—in particular, new generation HEVs—as a valid alternative, especially if fed with low-carbon fuels. Otherwise, an intermediate situation between those outlined in the timeframe 2035–2047 for the BAU@2035 and FF55@2035 scenarios would be expected, which will involve important but not disruptive results as those expected, as a starting point, by the EU for the phase-out of new HEVs by 2035 (Fit for 55 package). The same assumption could be made for all the other European countries whose c.i.e. value is similar to or exceeds the Italian one, while it could represent an absolute revolution in terms of efficiency and CO_2 emissions reduction for countries already characterized by a very low c.i.e. value.

As this transformation could profoundly affect the Italian automotive industry, policymakers should be careful not to move too rapidly toward an electrification roadmap, which could definitively damage the significance of the car industry, which represents one of the most important Italian economic pillars. For all the above-mentioned reasons, the authors suggest that the phase-out of thermal engines has to be postponed until the achievement of a real reduction in CO_2 electricity emissions, making battery production and end-use phase less critical for customers from both an economic and environmental point of view.

The latter concept finds confirmation in the small percentage of BEVs sold in Italy to mid-2023 (well below the European average), which, therefore, leads to a negligible percentage of BEVs in circulation and a small consequent impact on total CO_2 emissions.

Author Contributions: Conceptualization, F.M.G. and P.C.; methodology, P.C.; investigation, F.M.G. and P.C.; resources, F.M.G. and P.C.; data curation, P.C.; writing—original draft preparation, F.M.G. and P.C.; writing—review and editing, F.M.G.; visualization, F.M.G. and P.C. All authors have read and agreed to the published version of the manuscript.

Funding: This research received no external funding.

Data Availability Statement: The original contributions presented in the study are included in the article, further inquiries can be directed to the corresponding author.

Acknowledgments: We warmly thank Antonio Di Meo for the precious help.

Conflicts of Interest: The authors declare no conflicts of interest.

References

1. Abdul-Manan, A.F.N.; Zavaleta, V.G.; Agarwal, A.K.; Kalghatgi, G.; Amer, A.A. Electrifying passenger road transport in India requires near-term electricity grid decarbonisation. *Nat. Commun.* **2022**, *13*, 2095. [CrossRef]
2. Eurostat. *Shedding Light on Energy*; 2023 ed.; Eurostat: Luxembourg, 2023. Available online: https://ec.europa.eu/eurostat/web/interactive-publications/energy-2023 (accessed on 27 October 2023).
3. McBain, S.; Teter, J. *Tracking Transport 2021*; IEA: Paris, France, 2021. Available online: https://www.iea.org/reports/tracking-transport-2021 (accessed on 13 September 2023).
4. International Energy Agency (IEA). Data Browser. Available online: https://www.iea.org/countries/italy (accessed on 15 November 2023).
5. International Energy Agency (IEA). *Clean Energy Innovation*; IEA: Paris, France, 2020. Available online: https://www.iea.org/reports/clean-energy-innovation (accessed on 15 October 2023).
6. The European Automobile Manufacturers' Association (ACEA). *Vehicles in Use, Europe*; ACEA: Brussels, Belgium, 2023. Available online: https://www.acea.auto/publication/report-vehicles-in-use-europe-2023/ (accessed on 4 October 2023).
7. Woody, M.; Vaishan, P.; Keoleian, G.A.; De Kleine, R.; Kim, H.C.; Anderson, J.E.; Wallington, T.J. The role of pickup truck electrification in the decarbonisation of light-duty vehicles. *Environ. Res. Lett.* **2022**, *17*, 034031. [CrossRef]
8. The European Automobile Manufacturers' Association (ACEA). *Fuel Types of New Cars: Battery Electric, 9.1.%.; Hybrid, 1.9.6.%.; Petrol. 400% Market Share Full-Year 2021*; ACEA: Brussels, Belgium, 2021. Available online: https://www.acea.auto/files/202202 02_PRPC-fuel_Q4-2021_FINAL.pdf (accessed on 13 September 2023).
9. Xia, X.; Li, P. A review of the Life-cycle Assessment of electric vehicles: Considering the influence of batteries. *Sci. Total Environ.* **2022**, *814*, 152870. [CrossRef] [PubMed]
10. Zeng, D.; Dong, Y.; Cao, H.; Li, Y.; Wang, J.; Li, Z.; Hauschild, M.Z. Are the electric vehicles more sustainable than the conventional ones? Influences of the assumptions and modeling approaches in the case of typical cars in China. *Resour. Conserv. Recycl.* **2021**, *167*, 105210. Available online: https://www.sciencedirect.com/science/article/abs/pii/S0921344920305279 (accessed on 30 January 2024). [CrossRef]

11. European Commission, Directorate-General for Climate Action; Hill, N.; Amaral, S.; Morgan-Price, S.; Nokes, T.; Bates, J.; Helms, H. Determining the Environmental Impacts of Conventional and Alternatively Fuelled Vehicles through LCA—Final Report, Publications Office of the European Union. 2020. Available online: https://data.europa.eu/doi/10.2834/91418 (accessed on 26 December 2023).
12. Moro, A.; Lonza, L. Electricity carbon intensity in European Member States: Impacts on GHG emissions of electric vehicles. *Transp. Res. Part D Transp. Environ.* **2018**, *64*, 5–14. [CrossRef] [PubMed]
13. Albatayneh, A.; Assaf, M.N.; Alterman, D.; Jaradat, M. Comparison of the overall energy efficiency for internal combustion engine vehicles and electric vehicles. *Environ. Clim. Technol.* **2020**, *24*, 669–680. Available online: https://sciendo.com/it/article/10.2478/rtuect-2020-0041 (accessed on 27 October 2023). [CrossRef]
14. European Council. 'Fit for 55': Council Adopts Regulation on CO_2 Emissions for New Cars and Vans. Available online: https://www.consilium.europa.eu/it/press/press-releases/2023/03/28/fit-for-55-council-adopts-regulation-on-co2-emissions-for-new-cars-and-vans/ (accessed on 12 September 2023).
15. European Commission. On the Path to a Climate-Neutral Europe by 2050. Available online: https://commission.europa.eu/strategy-and-policy/priorities-2019-2024/european-green-deal/delivering-european-green-deal_en (accessed on 12 September 2023).
16. European Council. Fit for 55. Available online: https://www.consilium.europa.eu/en/policies/green-deal/fit-for-55-the-eu-plan-for-a-green-transition/ (accessed on 1 May 2023).
17. European Parliament, Council of the European Union. REGULATION (EU) 2023/1542 of the European Parliament and of the Council of 12 July 2023 Concerning Batteries and Waste Batteries, Amending Directive 2008/98/EC and Regulation (EU) 2019/1020 and Repealing Directive 2006/66/EC. *Off. J. Eur. Union* **2023**, *66*. Available online: http://data.europa.eu/eli/reg/2023/1542/oj (accessed on 1 December 2023).
18. European Commission. Biofuels. Available online: https://energy.ec.europa.eu/topics/renewable-energy/bioenergy/biofuels_en (accessed on 3 January 2024).
19. Gustafsson, M.; Svensson, N.; Eklund, M.; Möller, B.F. Well-to-wheel climate performance of gas and electric vehicles in Europe. *Transp. Res. Part D Transp. Environ.* **2021**, *97*, 102911. [CrossRef]
20. Efuel Alliance. What Are Efuels. Available online: https://www.efuel-alliance.eu/efuels/what-are-efuels (accessed on 15 September 2023).
21. European Commission. Renewable Energy—Recast to 2030 (RED II). Available online: https://joint-research-centre.ec.europa.eu/welcome-jec-website/reference-regulatory-framework/renewable-energy-recast-2030-red-ii_en (accessed on 15 September 2023).
22. European Parliament. Legislative Train Schedule. Revision of the Renewable Energy Directive. In "A European Green Deal". Available online: https://www.europarl.europa.eu/legislative-train/package-fit-for-55/file-revision-of-the-renewable-energy-directive (accessed on 27 December 2023).
23. Cox, B.; Bauer, C.; Beltran, A.M.; van Vuuren, D.P.; Mutel, C.L. Life cycle environmental and cost comparison of current and future passenger cars under different energy scenarios. *Appl. Energy* **2020**, *269*, 115021. [CrossRef]
24. Helmers, E.; Johannes, D.; Weiss, M. Sensitivity Analysis in the Life-Cycle Assessment of Electric vs. Combustion Engine Cars under Approximate Real-World Conditions. *Sustainability* **2020**, *12*, 1241. [CrossRef]
25. Evrard, E.; Davis, J.; Hagdahl, K.-H.; Palm, R.; Lindholm, J.; Dahlöf, L. *Carbon Footprint Report—Volvo C40 Recharge*; Volvo: Goteborg, Sweden, 2021. Available online: https://www.volvocars.com/us/news/sustainability/Wanted-clean-energy-for-full-climate-benefits/ (accessed on 15 August 2023).
26. Petrauskienė, K.; Galinis, A.; Kliaugaitė, D.; Dvarionienė, J. Comparative environmental life cycle and cost assessment of electric, hybrid, and conventional vehicles in Lithuania. *Sustainability* **2021**, *13*, 957. [CrossRef]
27. Onat, N.C.; Kucukvar, M.; Tatari, O. Conventional, hybrid, plug-in hybrid or electric vehicles? State-based comparative carbon and energy footprint analysis in the United States. *Appl. Energy* **2015**, *150*, 36–49. Available online: https://www.sciencedirect.com/science/article/abs/pii/S0306261915004407 (accessed on 15 December 2023). [CrossRef]
28. Schuller, A.; Stuart, C. From Cradle to Grave: E-Mobility and the Energy Transition. From Cradle to Grave: E-Mobility and the Energy Transition. 2018. Available online: https://europeanclimate.org/wp-content/uploads/2018/09/From-cradle-to-grave-e-mobility-and-the-energy-transition_IT_SP_UK_EU.pdf (accessed on 31 December 2023).
29. Pipitone, E.; Caltabellotta, S.; Occhipinti, L. A life cycle environmental impact comparison between traditional, hybrid, and electric vehicles in the European context. *Sustainability* **2021**, *13*, 10992. [CrossRef]
30. Emilsson, E.; Dahllöf, L. *Lithium-Ion Vehicle Battery Production, Status 2019 on Energy Use, CO_2 Emissions, Use of Metals, Products Environmental Footprint, and Recycling*; IVL: Stockholm, Sweden, 2019. Available online: https://www.ivl.se/download/18.14d7b12e16e3c5c36271070/1574923989017/C444.pdf (accessed on 8 September 2023).
31. Northvolt. *Sustainability and Annual Report, 2022*; Northvolt: Stockholm, Sweden, 2023. Available online: https://northvolt.com/environment/report2022/ (accessed on 1 November 2023).

32. UNRAE. *Per Quattroruote L'automobile: Italiani a Confronto*; UNRAE: Roma, Italy, 2022. Available online: https://unrae.it/files/Studio%20UNRAE%20L_automobile%20Italiani%20a%20confronto_6336ab4170253.pdf (accessed on 15 October 2023).
33. Islam, E.; Vijayagopal, R.; Moawad, A.; Kim, N.; Dupont, B.; Prada, D.N.; Rousseau, A. *A Detailed Vehicle Modeling & Simulation Study Quantifying Energy Consumption and Cost Reduction of Advanced Vehicle Technology through 2050*; Argonne National Lab (ANL): Argonne, IL, USA, 2021. [CrossRef]
34. Capaldi, P.; Grimaldi, F.M. The effectiveness of ICEV phase-out at 2035 in terms of CO_2 emission reduction in the Italian scenario. In Proceedings of the 6th International Conference on Energy Harvesting, Storage, and Transfer (EHST'22), Niagara Falls, ON, Canada, 8–10 June 2022. Available online: https://avestia.com/EHST2022_Proceedings/files/paper/EHST_146.pdf (accessed on 27 October 2023).
35. Life Cycle Greenhouse Gas Emissions from Electricity Generation: Update. Available online: https://www.nrel.gov/docs/fy21osti/80580.pdf (accessed on 10 October 2023).
36. Ministero della Transizione Ecologica—Dip.Energia—Direzione Generale Infrastrutture e Sicurezza. La Situazione Energetica Nazionale Nel 2021. Italy. Available online: https://dgsaie.mise.gov.it/pub/sen/relazioni/relazione_annuale_situazione_energya_nazionale_dati_2021.pdf (accessed on 27 September 2023).
37. ANIE Federazione. Sala Stampa—Comunicati Stampa. Osservatorio FER. Available online: https://anie.it/sala-stampa/comunicati-stampa/ (accessed on 10 October 2023).
38. International Energy Agency (IEA). *Italy 2023—Energy Policy Review*; IEA: Paris, France, 2023. Available online: https://www.iea.org/reports/italy-2023 (accessed on 27 October 2023).
39. Terna—Snam. *Documento di Descrizione Degli Scenari 2022*; Terna/Snam: Roma, Italy, 2022. Available online: https://www.terna.it/it/sistema-elettrico/rete/piano-sviluppo-rete/scenari (accessed on 10 October 2023).
40. EMBER European Union. The EU Accelerates Electricity Transition in the Wake of Crisis. Explore Open Data (Emissions Intensity). Available online: https://ember-climate.org/countries-and-regions/regions/european-union/ (accessed on 2 October 2023).
41. Terna. *Evoluzione Rinnovabile 2021*; Terna: Roma, Italy, 2021. Available online: https://download.terna.it/terna/Evoluzione_Rinnovabile_8d940b10dc3be39.pdf (accessed on 27 October 2023).
42. Teske, S.; Morris, T.; Nagrath, K. 100% Renewable Energy: An Energy [R] Evolution for ITALY. Report Prepared by ISF for Greenpeace Italy. 2020. Available online: https://www.uts.edu.au/sites/default/files/article/downloads/100-Percent-Renewable-Energy-Italy-report.pdf (accessed on 1 May 2023).
43. International Energy Agency (IEA). *Electricity Grids Secure Energy Transition*; IEA: Paris, France, 2023. Available online: https://www.iea.org/reports/electricity-grids-and-secure-energy-transitions (accessed on 21 October 2023).
44. International Energy Agency (IEA). *Electricity 2024, Analysis and Forecast to 2026*; IEA: Paris, France, 2024. Available online: https://www.iea.org/reports/electricity-2024 (accessed on 24 January 2024).
45. The International EPD System Documentation Overview for, Environmental Product Declaration; PCR. Available online: https://environdec.com/resources/documentation#generalprogrammeinstructions (accessed on 1 November 2023).
46. Link, S.; Neef, C.; Wicke, T. Trends in Automotive Battery Cell Design: A Statistical Analysis of Empirical Data. *Batteries* **2023**, *9*, 261. [CrossRef]
47. Maus, V.; Werner, T.T. Impacts for half of the world's mining areas are undocumented. *Nature* **2024**, *625*, 26–29. [CrossRef] [PubMed]
48. Spritmonitor.de. MPG and Cost Calculator and Tracker. Available online: https://www.spritmonitor.de/en/ (accessed on 21 October 2023).
49. ADAC. Elektroautos im Test: So Hoch ist die Reichweite Wirklich. Available online: https://www.adac.de/rund-ums-fahrzeug/elektromobilitaet/tests/stromverbrauch-elektroautos-adac-test/ (accessed on 10 October 2023).
50. Edwards, R.; Larivé, J.-F.; Rickeard, D.; Weindorf, W. *JEC, Well-to-Wheels Analysis, Well-to-Wheels Analysis of Future Automotive Fuels and Powertrains in the European Context"*, Concawe-EUCAR-JRC; JRC: Bruxelles, Belgium, 2014. Available online: https://publications.jrc.ec.europa.eu/repository/bitstream/JRC85326/wtt_report_v4a_april2014_pubsy.pdf (accessed on 10 September 2023).
51. Argonne National Laboratory. Energy Systems and Infrastructure Analysis, GREET Model—The Greenhouse Gases, Regulated Emissions, and Energy Use in Technologies Model. Available online: https://greet.es.anl.gov/ (accessed on 1 October 2023).
52. Reick, B.; Konzept, A.; Kaufmann, A.; Stetter, R.; Engelmann, D. Influence of charging losses on energy consumption and CO_2 emissions of battery-electric vehicles. *Vehicles* **2021**, *3*, 736–748. [CrossRef]
53. Genovese, A.; Ortenzi, F.; Villante, C. On the energy efficiency of quick DC vehicle battery charging. *World Electr. Veh. J.* **2015**, *7*, 570–576. [CrossRef]
54. Goto, T.; Isobe, R.; Yamakawa, M.; Nishida, M. The New Mazda Gasoline Engine Skyactiv-G. *MTZ Worldw. eMag.* **2011**, *72*, 40–47. [CrossRef]
55. EMBER Italy. Fossil Gas still Dominates Italy's Electricity but Renewable Power Targets Have Increased. Available online: https://ember-climate.org/countries-and-regions/countries/italy/ (accessed on 24 November 2023).

56. EMBER Europe. Uneven Progress Towards Clean Electricity. Available online: https://ember-climate.org/countries-and-regions/regions/europe/ (accessed on 17 December 2023).
57. SNAM; The European House-Ambrosetti. *H2 Italy 2050. A National Hydrogen Value Chain for the Growth and Decarbonization of Italy*; SNAM, The European House-Ambrosetti: Milano, Italy, 2020. Available online: https://www.ambrosetti.eu/en/community-innotech/hydrogen-community/ (accessed on 30 September 2023).

Disclaimer/Publisher's Note: The statements, opinions and data contained in all publications are solely those of the individual author(s) and contributor(s) and not of MDPI and/or the editor(s). MDPI and/or the editor(s) disclaim responsibility for any injury to people or property resulting from any ideas, methods, instructions or products referred to in the content.

Article

The Influence of the Type and Condition of Road Surfaces on the Exhaust Emissions and Fuel Consumption in the Transport of Timber

Andrzej Ziółkowski [1,*], Paweł Fuć [1], Piotr Lijewski [1], Maciej Bednarek [1], Aleks Jagielski [1], Władysław Kusiak [2] and Joanna Igielska-Kalwat [3]

1. Faculty of Civil Engineering and Transport, Poznan University of Technology, ul. Piotrowo 3, 60-695 Poznan, Poland; pawel.fuc@put.poznan.pl (P.F.); piotr.lijewski@put.poznan.pl (P.L.); maciej.bednarek@doctorate.put.poznan.pl (M.B.); aleks.jagielski@doctorate.put.poznan.pl (A.J.)
2. Faculty of Forestry and Wood Technology, Poznan University of Life Sciences, ul. Wojska Polskiego 71C, 60-625 Poznan, Poland; wladyslaw.kusiak@up.poznan.pl
3. Faculty of Cosmetology, University of Education and Therapy, Grabowa 22, 61-473 Poznan, Poland; joanna.igielska-kalwat@symbiosis.pl
* Correspondence: andrzej.j.ziolkowski@put.poznan.pl

Abstract: Owing to society's growing ecological awareness, researchers and car manufacturers have increasingly been focusing on the adverse impact of transport on the environment. Many scientific publications have been published addressing the influence of a variety of factors on the exhaust emissions generated by vehicles and machinery. In this paper, the authors present an analysis of the exhaust emissions of components such as CO, THC, and NO_x in relation to the type and condition of the road surface. The analysis was performed on a heavy-duty truck designed for carriage of timber. The investigations were carried out with the use of the PEMS equipment (portable emission measurement system) on bitumen-paved roads and unpaved forest access roads. The portable measurement system allowed for an accurate determination of the influence of the road conditions on the operating parameters of the vehicle powertrain and its exhaust emissions. Additionally, the authors present the influence of the type of road surface on the vehicle fuel consumption calculated based on the carbon balance method.

Keywords: fuel consumption; emission; wood transportation; PEMS

1. Introduction

The COVID pandemic has led to a substantial slowdown in economic production. Transport has also found itself among the affected sectors, which was consequential to the imposed lockdown restrictions. According to the EUROSTAT data, in the first months of the pandemic, a significant drop in the use of crude oil-based fuels was observed. It is noteworthy, however, that despite such a drop, in 2020 transport was still the main sector responsible for the consumption of crude oil (approx. 62%), 48% of which was attributed to road transport [1]. The above hints to the fact that transport is one of the most fossil fuel-dependent sectors of the economy.

The main negative effect of the combustion of fuel in vehicle engines are the exhaust emissions. As presented in the report of [2], transport is one of the main factors responsible for air pollution. The exhaust components contributing to the degradation of the Earth's atmosphere are nitrogen oxides (NO_x), carbon monoxide (CO), and hydrocarbons (HCs). These components also have a negative impact on human health, leading to a multitude of health conditions and ailments. Among the exhaust components released to the atmosphere, we can also find highly hazardous particulate matter (PM) [3]. Particles of a small size, 30% of the emission of which is attributed to transport, are especially hazardous to human health. As has been confirmed, these components deeply penetrate the lungs, contributing to the occurrence of cardio-pulmonary disorders. PM is also considered to be carcinogenic [4]. When analyzing the impact of transport on the natural environment, one should not neglect the emission of carbon dioxide. According to the data obtained from the European Environment Agency, in 2020 the transport sector generated over 700 million tons of CO_2. It is noteworthy that 77% of these emissions were attributed to road transport. However, as the EEA suggests, the share of the CO_2 emissions from transport in the EU will continually drop, provided the member states continue to implement efficient actions aiming at their reduction [5].

When taking a closer look at the problem of reduction in vehicle exhaust emissions, one should consider the complexity of the factors determining their amount. There are many publications whose main topic is the identification of the main factors determining the level of emissions of individual exhaust components [6–9]. The investigations presented in the papers [10,11] discuss the problem of the influence of the engine thermal state on the formation of the emission components and exhaust emissions. In the analysis found in [12], the authors analyzed the exhaust emissions from vehicles during cold starts at low ambient temperatures. The main topic of article [13] was the influence of the vehicle air conditioning system on its exhaust emissions and fuel consumption (and in the case of an electric vehicle, on its range). In review [14], the influence of the SCR system on the formation of particulate matter was estimated. In the work by Giechaskiel et al., the authors undertook the objective to determine the influence of extreme ambient temperatures (-30–$50\ °C$) on the emission of carbon dioxide and fuel consumption. An important part of the said investigations was also the influence of the road gradient on the exhaust emissions. In this case, the investigations clearly showed a linear proportionality of 12–14% increase in the emission of CO_2 for each gradient degree. The literature studies allow for the conclusion that researchers present a multitude of approaches toward estimating the emission level from vehicles.

Currently, the most accurate method of determination of exhaust emissions from road vehicles is the measurement of the concentration of exhaust components using analyzers while the vehicle is in operation. Advanced PEMS equipment (portable emission measurement systems) allows this to happen under almost any type of conditions of operation and almost any type of machinery fitted with a combustion engine. Therefore, portable analyzers are the most useful group of research equipment serving the purpose of determining the relations between the exhaust emissions and external factors and are widely applied in scientific research, as explained further in this section.

In paper [15], the authors attempted to estimate the influence of the driving style on the vehicle exhaust emissions. Investigations of such a complex factor such as driving style were carried out using the above-mentioned equipment. The analysis performed by the authors confirmed the impact of the driving style on the exhaust emissions, particularly on such exhaust components as CO_2 and NO_x. The significance of the driving style (eco-driving) on the emissions and fuel consumption was also confirmed in [16].

The authors of the work [17] attempted to determine the relation between real-world traffic and the emission of nitrogen oxides, carbon dioxide, PM number, and DPF active regeneration performed in laboratory tests. The investigations performed on eight Euro 5 and Euro 6 vehicles showed a significant influence of the DPF regeneration on the level of exhaust emissions of the above-mentioned exhaust components, confirming, inter alia, the elevated emission of CO_2. The problem of the influence of DPF regeneration on the vehicle exhaust emissions was also discussed in [18]. The investigations of an LDV vehicle in real-world traffic conditions has shown that the DPF regeneration process may contribute to the increased emissions of hydrocarbons and carbon monoxide. As the authors of the publication conclude, this is the effect of oxygen deficit during the regeneration process.

In article [19], the authors analyzed the factors influencing the emission of PN and PM from a diesel engine under real-world city traffic conditions. The problem of the emission of particulate matter and factors that have an impact on these emissions was also discussed in [20]. Based on rather extensive research covering HDV (heavy duty vehicle), LDV (light duty vehicle), as well as PC (passenger car) vehicles under laboratory and real-world conditions, the authors observed that sudden accelerations and cold starts result in increased particle number. This is, inter alia, the effect of the too-low temperature of the aftertreatment systems, which contributes to their reduced initial efficiency. Additionally, as is worth noting, the authors compared the results obtained under laboratory and real-world traffic conditions, showing considerable differences between them, which may be the effect of factors such as ambient temperature, driving style, or the dynamometer settings.

In paper [21], the main research topic was the determination of the influence of the DPF support element on the exhaust emissions. The same authors in [22] investigated the influence of the application of GPF in a gasoline engine on the exhaust emissions under real-world traffic conditions. Sarkan et al. [23] compared the exhaust emissions from vehicles fueled with gasoline and LPG. In this case, the real-world tests (RDEs: real driving emissions) showed that, upon changing the fuel type from gasoline to LPG, the vehicle generated less carbon monoxide and carbon dioxide (by approx. 0.03 g/km CO and approx. 3.6 g/km CO_2) but more hydrocarbons and nitrogen oxides (0.0003 g/km and 0.02 g/km, respectively). The problem of differences in the exhaust emissions from vehicles fueled with gasoline or LPG under real-world conditions was also discussed in [24].

In review [25], the authors attempted to determine the influence of converting a traffic lights-controlled intersection into a traffic circle. Based on more than 300 exhaust emission trials using PEMS, the authors observed that replacing a conventional intersection with a traffic circle results in a reduction in the emission of CO_2 and NO_x. This research shows that PEMS is also suitable for studying the impact of infrastructure on exhaust emissions from vehicles of different types.

Owing to the advancement of exhaust emission measurement systems, scientists can accurately determine the impact of transport on the natural environment. As has been proven above, the authors of scientific research investigate a multitude of factors that may affect exhaust emissions. There is, however, an insufficient number of research works determining the influence of the type of road surface on the exhaust emissions. A mere study of the literature allows for the conclusion that all sorts of vehicles are operated not only on bitumen-paved roads but also on unpaved off-road ones [26–28].

Among the common features of the said publications is the fact that the operation models described there are characteristic of utility vehicles, including heavy-duty (HDV) ones. Therefore, the authors of this paper decided to assess the influence of the condition and type of road surfaces on the exhaust emissions based on the investigations of a heavy-duty vehicle used in the transport of timber. The investigated vehicle is regularly operated on bitumen-paved roads and forest access roads, the condition of the latter being heavily dependent on the time of year and ambient conditions. In the generally available literature, there are not many similar analyses. The problem was mentioned by the authors in [29] who also drew attention to the question of increased exhaust emissions under conditions diverging from the RDE standards.

2. Materials and Methods

2.1. Research Objects

For the investigations, the authors used a heavy-duty vehicle designed for the carriage of timber (Figure 1). The operating model of the research object is characterized by highly variable operating conditions as the process of timber extraction requires the use of bitumen-paved roads (urban and rural) and unpaved forest access roads. Therefore, the authors consider the vehicle appropriate for the investigations of the influence of the type and condition of the road surface on the exhaust emissions. The investigated object consisted of a tractor unit and a timber trailer. The vehicle was fitted with a Euro VI 353 kW 6-cylinder diesel engine. The combustion engine is manufactured by MAN. The vehicle was fitted with a series of aftertreatment systems (SCR, DPF, DOC). A detailed specification of the investigated object is presented in Table 1.

Figure 1. Research object.

Table 1. Specification of the research objects.

Parameter	Research Object
Engine type	Diesel
Displacement	12.4 dm^3
Maximum power output @1900 rpm	353 kW
Number of cylinders	6
Emission standard	Euro VI
Aftertreatment	SCR, DPF, DOC

2.2. Measurement Equipment

For the investigations, the authors used Axion RS+, a portable emission measurement system (PEMS). Axion RS+ (Figure 2) allows for the measurement of the concentrations of gaseous exhaust components and particle emissions. It is equipped with a GPS system, an on-board weather station, and a module pulling data from the on-board diagnostic system. For the measurement of the emissions of carbon monoxide, carbon dioxide, and hydrocarbons, the analyzer utilizes an infrared non-dispersive method (NDIR: non-dispersive infra-red). The electrochemical analyzer is used for the measurement of the content of nitrogen oxides and oxygen in the exhaust gas. Particulate matter is measured based on the phenomenon of laser dispersion [30]. Detailed specifications of the measurement equipment are shown in Table 2.

Table 2. Axion RS+ technical specifications.

Parameter	Measurement Method	Measurement Range	Measurement Accuracy
CO_2	Non-dispersive infrared	0–16%	±3%
CO	Non-dispersive infrared	0–10%	±3%

Table 2. Cont.

Parameter	Measurement Method	Measurement Range	Measurement Accuracy
HC	Non-dispersive infrared	0–2000 ppm	±4%
O_2	Electrochemical	0–25%	±1%
NO_x	Electrochemical	0–5000 ppm	±1%
PM	Laser dispersion	0–250 g/m^3	±2%

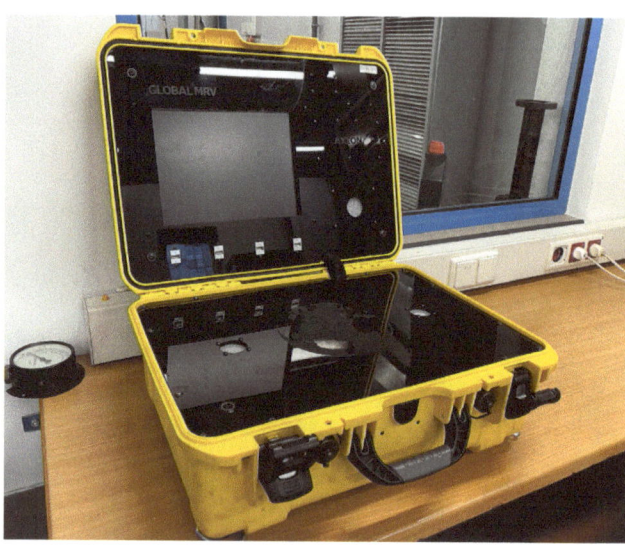

Figure 2. Portable emission measurement system Axion RS+.

2.3. The Test Route

The investigations were performed under real-world operating conditions. In order to determine the actual influence of the road condition and type on the exhaust emissions, the authors selected a heavy-duty truck used for the transport of timber. This selection was related to the nature of the transport process of timber, which is characterized by driving on roads of different types. The route on which the tests were carried out is representative of the process of timber extraction. It led through unpaved forest access roads and bitumen-paved public roads (urban and rural). The condition of the unpaved roads is heavily dependent on, inter alia, the ambient conditions. In order to allow for the variability of the road surface and its impact on the fuel consumption and exhaust emissions, the tests were performed in January, shortly after rainfall. This allowed the authors to obtain very arduous and variable conditions on the forest roads. The authors divided the test route into several stages of different arduousness. Stage one was marked blue, stage two magenta, stage three yellow, and stage four (covering the bitumen-paved road) red. The stages were properly marked with colors on the images (Figure 3). The route was divided according to the condition of the dirt road, which are shown in Figure 4. The proposed transport cycle corresponded to a regular timber transport process; hence, the authors could determine the emission level of the entire transport cycle in relation to the road conditions. The test run from point A to point B was realized on unpaved forest roads. Points B–C correspond to the test run on bitumen-paved roads. The length of the timber extraction route was 20.7 km. The unpaved road constituted approx. 20% of the test run.

Figure 3. Timber extraction route.

Figure 4. Condition of the unpaved roads: (**a**)—stage one (road portion marked blue), (**b**)—stage two (road portion marked magenta), (**c**)—stage three (road portion marked yellow).

3. Results and Discussion

3.1. Analysis of Vehicle Movement and Drive Unit Performance Parameters

Based on the obtained data, the authors carried out an analysis of the vehicle speed on each of the individual stages of the test (Figure 5). The test run on the paved roads was realized with an average speed of approx. 48 km/h. The transport of timber on the forest roads, whose technical condition caused considerable motion resistance, was almost four times slower. This confirms the influence of the road condition on the efficiency of the transport process. This is particularly the case for the unpaved roads. The operation of heavy-duty trucks combined with bad weather conditions renders these roads almost impassable. In the tests performed by the authors, the road portion characterized by the highest variability of the road profile and the occurrence of the deepest ruts (Figure 4a) was covered with the speed of approx. 3 km/h. The portion marked purple (Figure 4b) was covered with the speed of 7 km/h and the portion characterized with the relatively best condition was covered with the speed of 14.6 km/h. Even though the drive on the unpaved roads constituted only 20% of the entire test, it took approx. half of the time for the entire procedure.

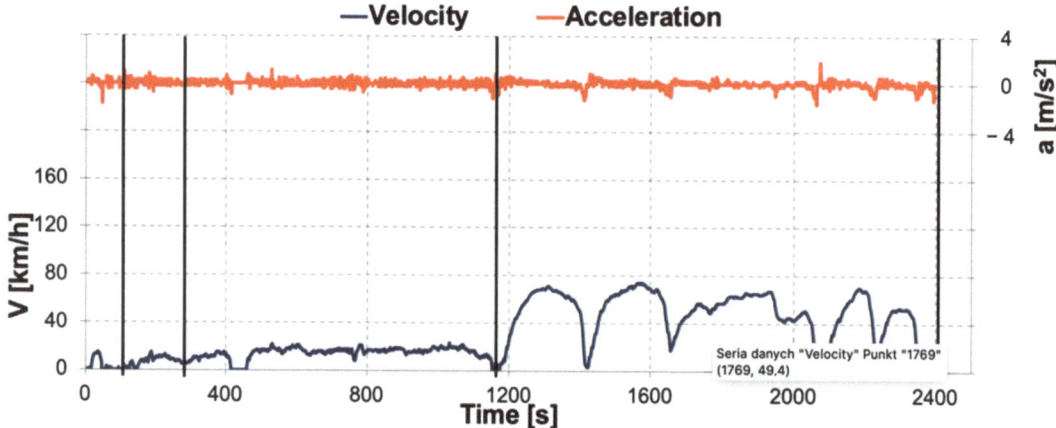

Figure 5. Vehicle velocity profile. (Vertical lines mark the end of the subsequent stages of the study).

In order to visualize the influence of the type and condition of the road surface on the vehicle driving, the authors prepared operating time characteristics of the tested object as a function of speed and acceleration (Figure 6). The test run in the first stage of the test was characterized by a significant an approximative 40% share of stops, which directly resulted from the motion resistance and wheel slippage. The second stage was characterized by the greatest share of the vehicle in motion at the lowest speed values in the range of 4.8 m/s and accelerations from -0.6 m/s^2 to 0.6 m/s^2. The final stage of the drive on the forest roads was covered in a similar way to the previous one. In that case, the main share of the speed intervals was one order of magnitude higher. The most diverse characteristic was observed for the final part of the entire test, i.e., the test run on the bitumen-paved roads. The main area of operation in this part of the test was described by speeds in the range of 12 m/s; 20 m/s and accelerations in the range of -0.6 m/s^2; 0.6 m/s^2. When analyzing the obtained characteristics, one may observe a significant share of operation at lower speeds and at the entire range of accelerations. Such a diverse nature of the test run results from driving on the roads in the urban area as well as rural area with greater traffic intensity.

Recording the engine parameters allows for developing the operating time share characteristics as a function of engine load and speed (Figure 7). This allows for observing the main areas of operation in individual stages of the test. When analyzing the obtained graphs, one should observe that stage one of the test had different characteristics (Figure 7a). In this stage, we may distinguish three significant areas of the engine operation in the engine speed intervals of (600 rpm; 1200 rpm), (1400 rpm; 1800 rpm), (4000 rpm; 4400 rpm), and torque in the range of 0–1200 Nm. In this stage, the test run was performed on a highly uneven and slushy road, the effect of which was a significant wheel slippage. In the subsequent stages of the test, when the road condition improved, we may observe a single dominating operating area on the characteristics (Figure 7b,c). During the second stage, the engine operated in the speed range of (600 rpm; 1800 rpm) and torque of up to 800 Nm as well as (1400 rpm–2400 rpm) and torque of 800–1200 Nm. For the test drive on the bitumen-paved roads, we may observe an increase in the operating time share at the lowest load and engine speeds of 600–1800 rpm. The highest operating time shares in this case were observed for the load from 400 Nm to 1600 Nm and engine speed (1400 rpm–1800 rpm).

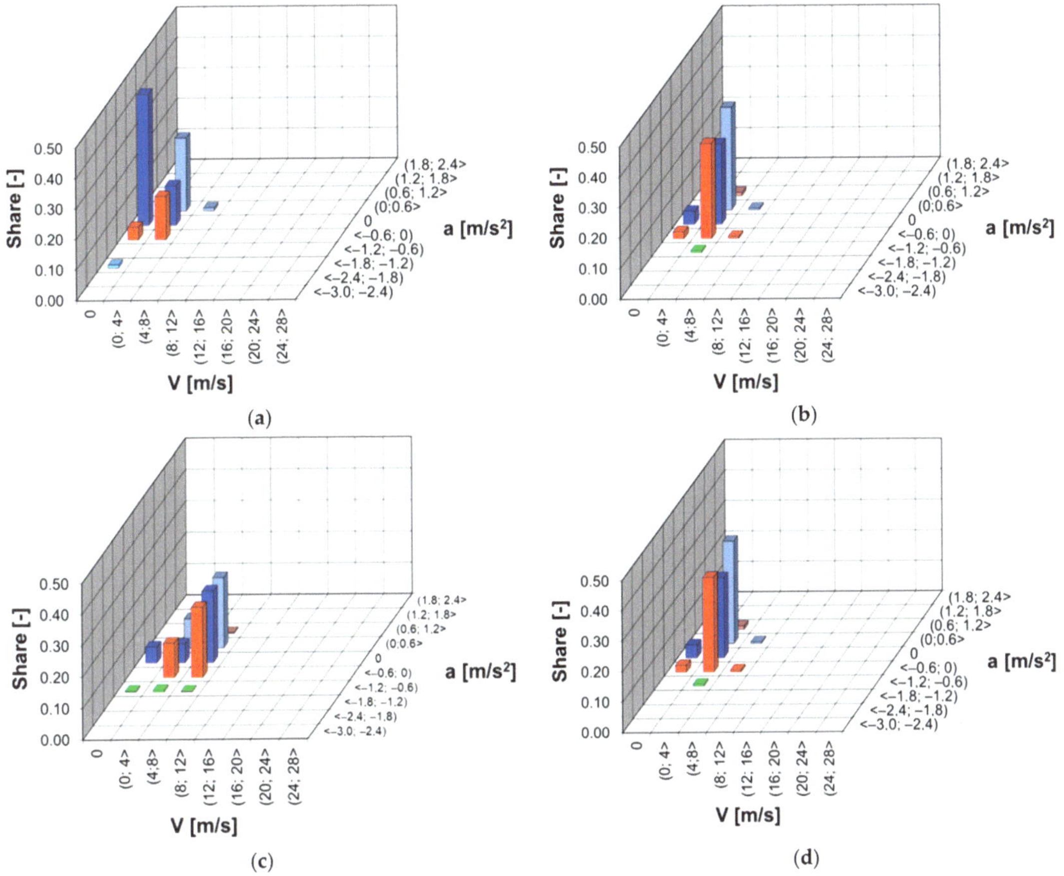

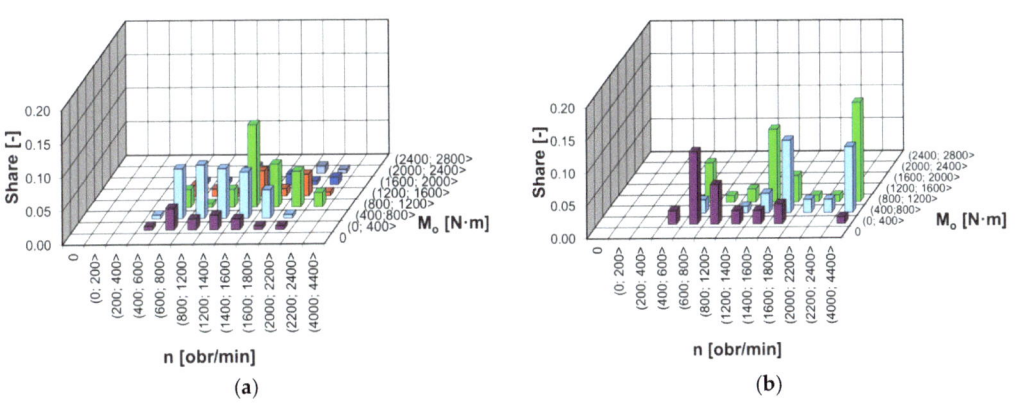

Figure 6. Operating time share as a function of speed and acceleration during individual test stages, (**a**) stage one, (**b**) stage two, (**c**) stage three, (**d**) stage four—bitumen-paved road.

Figure 7. Cont.

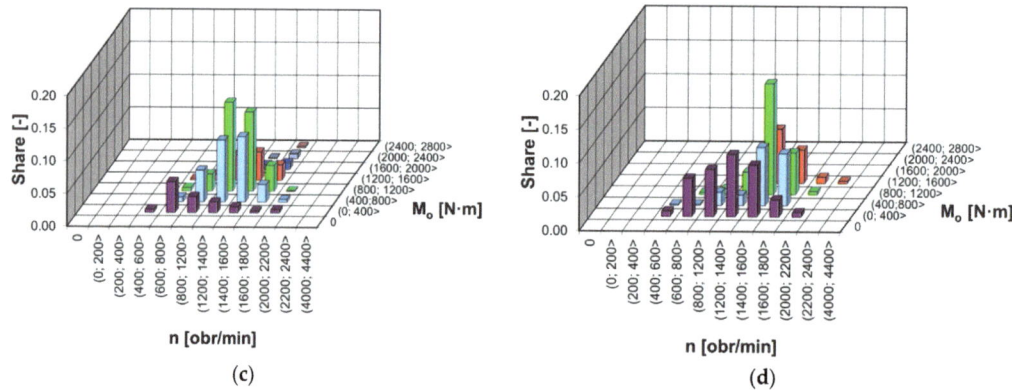

(c)   (d)

Figure 7. Operating time share as a function of engine load and engine speed during individual test stage, (**a**) stage one, (**b**) stage two, (**c**) stage three, (**d**) stage four—bitumen-paved road.

3.2. Analysis of the Individual Steps in the Timber Transport Process

Based on the parameters obtained from the GPS system, the authors recorded accurate data related to the course of each of the test stages. In the timber extraction process, the greatest share corresponds to the drive on paved roads. In the proposed cycle, this stage constituted approx. 80% of the entire test. The stages of the most difficult conditions constituted only a small part of the investigations, namely −0.5% for stage one and almost 2% for stage two (Figure 8).

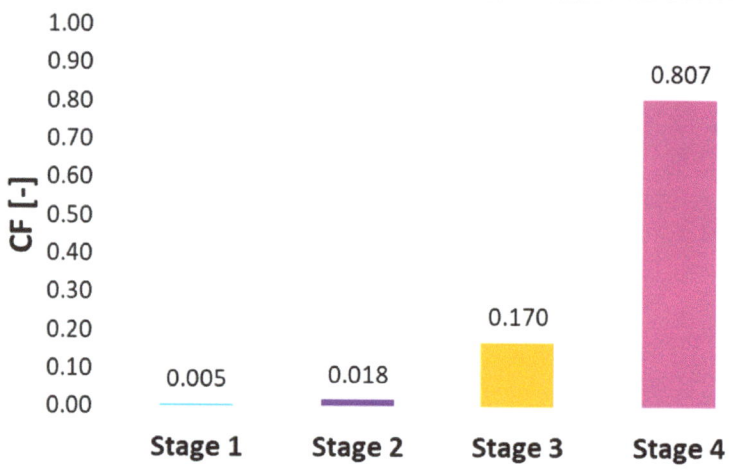

Figure 8. Coefficient of road type share on individual stages of the test.

The authors performed a similar analysis for the vehicle engine. In the entire research cycle, the vehicle engine performed work of over 67.6 kWh. Stage four of the test had the greatest share in the work performed by the engine. Compared to the share of the covered distance, the difference among the stages is not as significant (Figure 9). The share of the least arduous conditions (stage 3 and 4 of the test) referred to the performed work was 86% collectively. The authors also observed much greater shares of the first two stages of the test in the work performed by the engine (Figure 9).

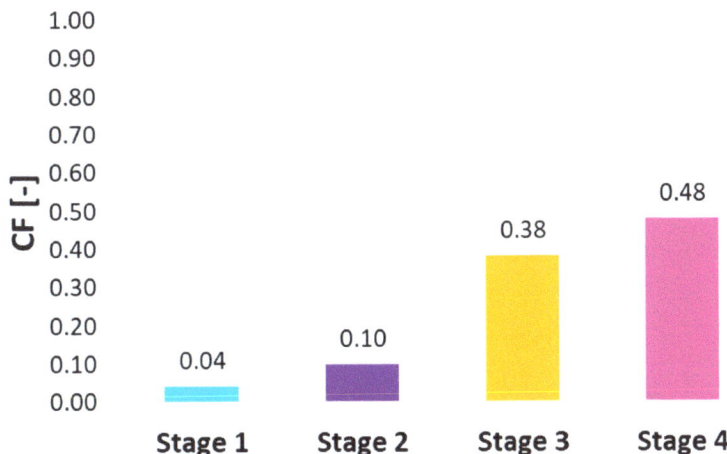

Figure 9. Coefficient of the share of work in individual stages.

3.3. Analysis of Emissions of Harmful Compounds in the Process of Timber Transport

Tests performed under real-world conditions using the PEMS equipment allow for the recording of detailed emission-related data on each stage of the test. This allowed for calculating the basic values of the exhaust emissions (road and specific emissions—Table 3) and estimating the share that each of the road conditions distinguished in the test had in the emission of individual exhaust components. Therefore, the authors were able to propose a new emission assessment method by defining the contribution factor (CF):

$$CF = \frac{\text{emissions in the subsequent stages of the test } \left[\frac{g}{km} \text{ or } \frac{g}{kWh}\right]}{\text{emissions in the entire test } \left[\frac{g}{km} \text{ or } \frac{g}{kWh}\right]} \quad (1)$$

Table 3. Comparison of the emissions in the test.

Index	Exhaust Component			
	CO	THC	NO_x	CO_2
Road emission	2.98 g/km	0.10 g/km	0.14 g/km	1557.29 g/km
Specific emission	0.91 g/kWh	0.03 g/kWh	0.04 g/kWh	477 g/kWh

Stage four of the test had the greatest share in the mass of the generated exhaust emissions. The mass of carbon dioxide and unburnt hydrocarbons in this stage was approx. 60% of the entire mass of produced emissions. As for stage three, for the above-mentioned exhaust components, this share was approx. 30%. The first two stages of the tests were responsible for the mass emission of approx. 5% CO_2 and THC and 7% CO. The results are directly related to the specificity of the timber transport process. Stages three and four were the longest road portions. An opposite trend was observed for nitrogen oxides (Figure 10). In the case of nitrogen oxides, over 70% of the mass emission occurred in stage one. This is the effect of the low efficiency of the selective catalytic reduction system, which was not yet up to temperature.

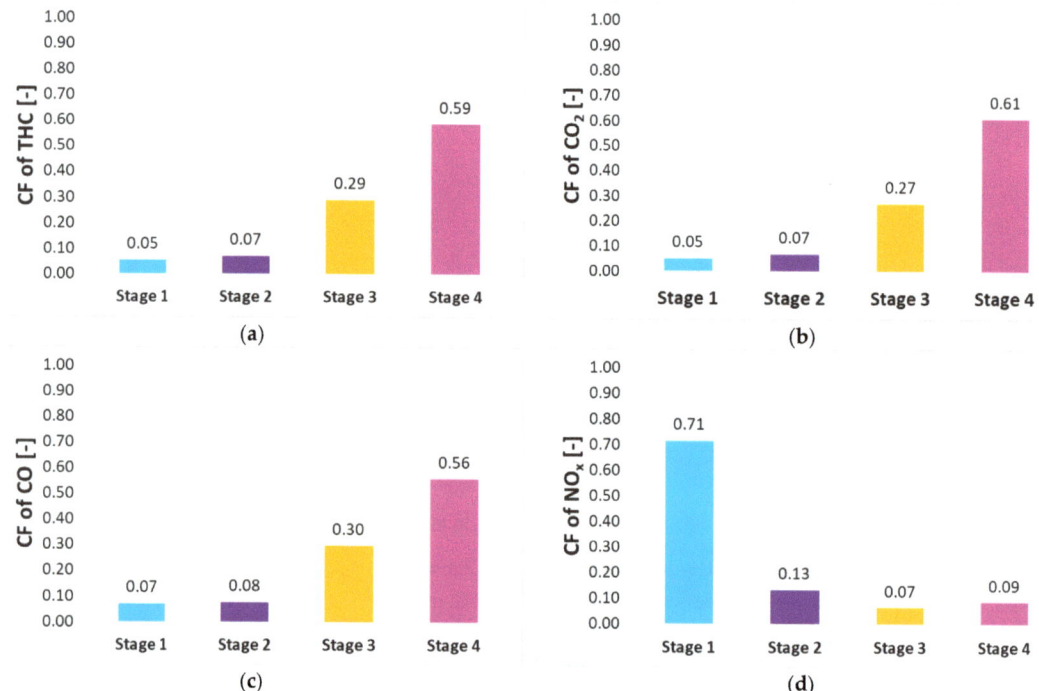

Figure 10. Coefficients of mass share of the emission components in individual stages of the test (**a**) THC, (**b**) CO_2, (**c**) CO, and (**d**) NO_x.

Due to the nature of the test route and the varied transport models, the mass of the emitted exhaust components does not constitute a universal indicator providing valuable information regarding the vehicle exhaust emission level. Hence, the authors performed an analysis of the contribution factors of individual test stages for the road and specific emissions. In the case of the road emissions, the greatest share was observed for the first stage of the test. This results from the fact that this stage was the shortest (approx. 100 m.). It is noteworthy that the contribution factor for nitrogen oxides for the first stage significantly diverges from the contribution factors of the other analyzed exhaust components. For the first stage, it reached almost 150. Additionally, for the subsequent stages of the test, it dramatically decreased to reach 0.1 in the final stage. This relation confirms the assumption that, in the beginning of the test, the emission of NO_x was high and, as the following test stages took place, the SCR system (selective catalyst reduction) became more efficient, thus reducing the said emission (Figure 11). When analyzing the obtained results, one may observe that the contribution factors of CO, THC, and CO_2 for each stage of the test decreased with the progression of the test.

In the homologation tests of heavy-duty vehicles, the exhaust emissions are referred to the work performed by the engine. Therefore, in the case of the tested object, it was more important to determine the specific emission share coefficients for individual stages of the test. The obtained results are shown in Figure 12. As for the emission coefficients for CO_2, THC, and CO, it is noteworthy that the greatest values were observed for stage one and stage four. In the case of the first two components, the values of the obtained share coefficients were very similar. The lowest values were recorded for stage two and the last stage. For the above-mentioned exhaust components, the share coefficients were almost the same and amounted to 0.71–0.78. Analogously to the analysis of the road emission share coefficients, in the analysis of the specific emission share coefficients, nitrogen oxides also had a different trend. The greatest share coefficient was recorded for stage one—its value was over 17. The share of the subsequent stages in the specific emission during the test was negligible. The test and the performed analysis confirmed that the greatest share of the emission of nitrogen oxides occurred in stage one. This stage was characterized by the shortest distance and the most arduous conditions. The higher emission was influenced by a high engine load that generated excess NO_x. Additionally, the cold start and low exhaust gas temperature may have rendered the SCR system inactive. The influence of the temperature on the operation of the SCR system has been proven in many publications [31–33]. The research in paper [34] confirms that the operation of an engine that is not up to temperature increases the emission of nitrogen oxides by over 50%. The investigations presented in work [35] covered the measurement of the emission of nitrogen oxides in a laboratory WHTC test. They confirm an increased emission of nitrogen oxides during the cold phase of the test. The fact that the temperature of this aftertreatment system plays a key role is also confirmed by the research presented in [36]. The authors have proven that the low temperature of the system results in an insufficient injection of the water solution of urea.

The fuel demand was calculated based on the carbon balance method. This method is currently the most accurate way to determine fuel consumption. The transport of timber in the proposed cycle resulted in a fuel consumption of 59 dm^3/100 km. Knowing the fuel mileage in individual stages of the test, the authors could calculate the share coefficients presented in Figure 13. The greatest coefficient was recorded for stage one, which directly results from the length of this road portion and the specificity of the test run. This stage included negotiating an extremely requiring turn in terms of the road conditions (Figure 14). A slushy and muddy road surface caused the vehicle to move at a very low speed, interrupted with frequent stops. These factors caused high fuel consumption that translated into a high value of fuel mileage of stage one compared to the entire research cycle. The fact that the condition and type of the road surface plays a key role in the fuel consumption has been proven in many publications presenting both simulations and real-world trials [27,37,38]. Fuel consumption constitutes the main indicator of the economy of a vehicle or a transport process. Hence, many research works focus on this aspect or are supplemented with this particular fuel mileage indicator. For example, during the testing of a heavy-duty truck [39] in real-world traffic on bitumen-paved roads, the fuel mileage reached a little over 28 dm^3/100 km.

The test was performed based on a cycle designed so as to reflect the actual timber transport process. The test covered the extraction of 24 m^3 pine timber cut in October 2021. The transport process took place in January 2022. The length of timber was from 6.2 to 13.8 m, and its volume was in the range of 0.4–1.05 m^3. The density of the transported timber was approx. 1000 kg/m^3. Based on the above data, the authors determined the coefficient of energy demand (understood as fuel demand) referred to the amount of transported timber and length of the route depending on the road conditions. The obtained results are shown in Table 4.

Table 4. Fuel demand in timber transport.

Stage	Fuel Consumption	
	dm^3/100 km × 1 m^3	dm^3/1 km × 1 m^3
Stage 1	26.57	1.100
Stage 2	9.34	0.093
Stage 3	3.91	0.039
Stage 4	1.85	0.018
Entire test	2.45	0.024

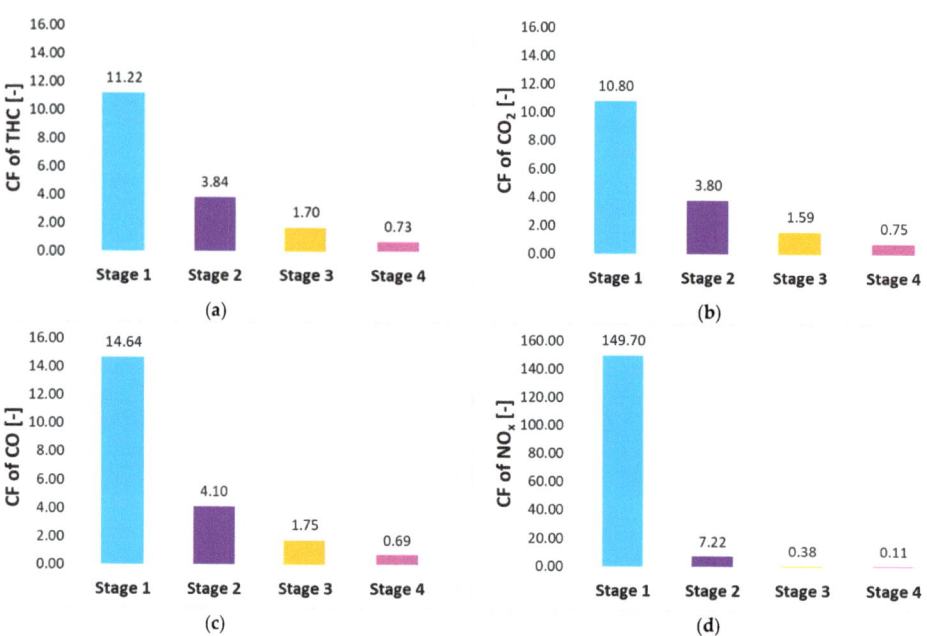

Figure 11. Road emission share coefficients in individual stages of the test for (**a**) THC, (**b**) CO_2, (**c**) CO, and (**d**) NO_x.

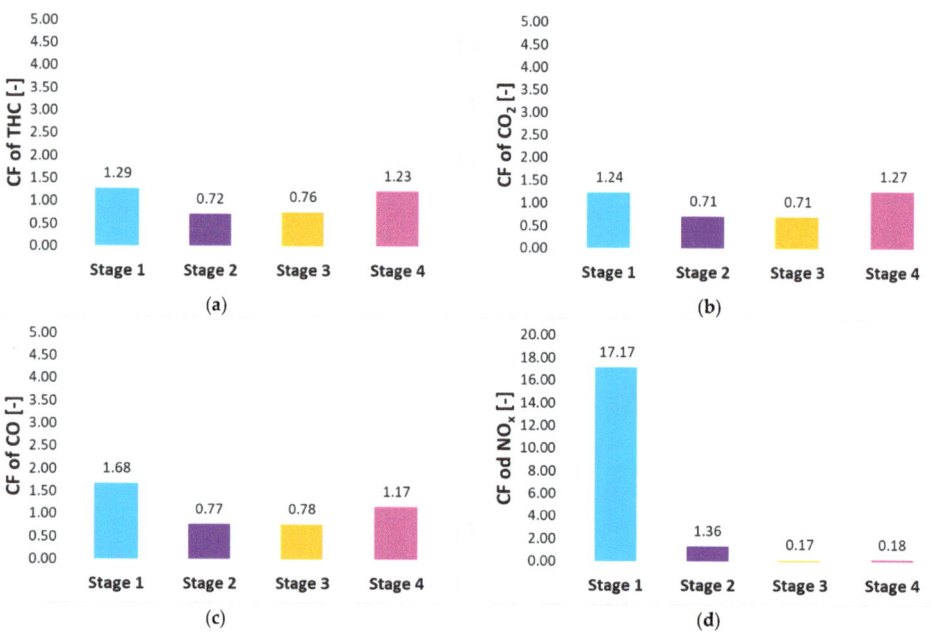

Figure 12. Specific emission share coefficients in individual stages of the test for (**a**) THC, (**b**) CO_2, (**c**) CO, and (**d**) NO_x.

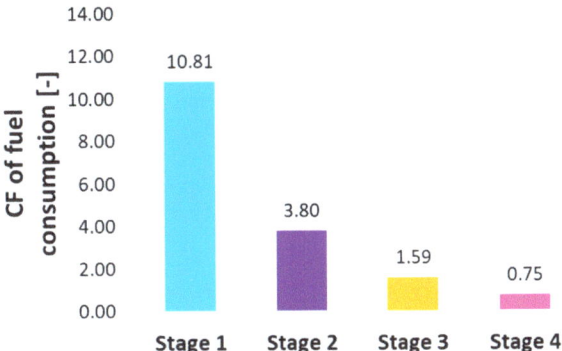

Figure 13. Fuel mileage share coefficients in individual stages.

Figure 14. Stage one of the research cycle—a turn on the forest road.

4. Conclusions

Currently, a variety of actions are being undertaken to reduce the negative impact of all types of industry on the natural environment. In the case of transport, the majority of the actions focus on the reduction in the exhaust emissions through aftertreatment systems, improvements in the combustion process, or the application of alternative powertrains. However, it is important to note that the emissions are influenced by a variety of factors and one of the methods of their reduction is a skillfully designed transport process and road infrastructure. The transport of timber is a great example of a transport process taking place under varied conditions. The main objective of this article was to find the relationship between the type of road surface and the emission of harmful compounds in the process of timber transport. The real-world investigations using portable emission measurement systems allowed for a detailed analysis of the influence of a given condition on the exhaust emissions and fuel consumption. The authors' analysis confirmed the fundamental impact of the road conditions on the exhaust emissions. Stage one, characterized by the most arduous road conditions, had the greatest coefficients of both road and specific emissions. As the condition of the unpaved roads improved in the course of the test, the coefficients of emission dropped. The investigations also confirmed the key role of the aftertreatment systems and the technique of realization of the transport process in the vehicle's environmental performance. This is confirmed by the coefficients related to the emission of nitrogen oxides. The question of fuel consumption is important in the performed analysis. From the investigations, it results that the operation of a vehicle under difficult conditions (significant wheel slippage) heavily influences the fuel consumption. The fuel demand coefficients developed by the authors allow for estimating the fuel consumption for the transport of timber depending on the length of the extraction route and the volume of the cargo, which have a clear impact on both emissions and fuel consumption. The harshest conditions resulted in a more than 14 times higher fuel consumption relative to driving on a paved road despite being the shortest stage in terms of length. Equally significantly, higher specific emissions were observed in the first stage relative to the last stage. For THC, specific emissions were 1.04 times higher, for CO almost 1.4 times higher, and for NOx more than 95 times higher.

The authors of the publication plan to carry out a number of studies in the near future, taking into account other road conditions (including rain, snow). For control purposes, it is planned to test emissions for a process using paved forest roads. These tests will be a continuation of the amalgamation presented in this article.

It is noteworthy that the vehicle used for the investigations belongs to the HDV category. Such vehicles carry out transport processes under varied conditions on construction sites, in forests, or mountainous areas. Importantly, vehicles operating under different conditions are homologated to the same standards, and laboratory and road tests do not allow for such varied conditions, which can significantly affect engine operating range and emissions.

Author Contributions: Conceptualization, A.Z., P.F., P.L., A.J. and M.B.; methodology, A.Z., P.F., P.L., A.J., M.B. and W.K.; software, A.Z., P.F. and W.K.; formal analysis, A.Z., P.F. and A.J.; writing, M.B., A.J. and A.Z.; writing—review and editing, A.Z. and A.J.; visualization, M.B. and J.I.-K.; supervision, A.Z. and P.F.; project administration, A.Z.; funding acquisition, P.F. All authors have read and agreed to the published version of the manuscript.

Funding: The study presented in this article was performed within the statutory research (contract No. 0415/SBAD/0326).

Data Availability Statement: Not applicable.

Conflicts of Interest: The authors declare no conflict of interest.

References

1. Available online: https://ec.europa.eu/eurostat/web/main/data/database (accessed on 6 December 2022).
2. Joint Research Centre, European Commission. *The Future of Road Transport*; EUR 29748 EN; Publications Office of the European Union: Luxembourg, 2019. [CrossRef]
3. Merkisz, J.; Pielecha, J. *Emisja Cząstek Stałych ze Źródeł Motoryzacyjnych*; Wydawnictwo Politechniki Poznańskiej: Poznań, Poland, 2014.
4. Leikauf, G.D.; Kim, S.H.; Jang, A.S. Mechanizmy skutków zdrowotnych układu oddechowego wywołanych najdrobniejszymi cząstkami. *Exp. Mol. Med.* **2020**, *52*, 329–337. [CrossRef] [PubMed]
5. Available online: https://www6.eea.com (accessed on 6 December 2022).
6. Aminzadegan, S.; Shahriari, M.; Mehranfar, F.; Abramović, B. Factors affecting the emission of pollutants in different types of transportation: A literature review. *Energy Rep.* **2022**, *8*, 2508–2529. [CrossRef]
7. Grieshop, A.P.; Lipsky, E.M.; Pekney, N.J.; Takahama, S.; Robinson, A.L. Fine particle emission factors from vehicles in a highway tunnel: Effects of fleet composition and season. *Atmos. Environ.* **2006**, *40* (Suppl. S2), 287–298. [CrossRef]
8. André, M.; Hammarström, U. Driving speeds in Europe for pollutant emissions estimation. *Transp. Res. Part D Transp. Environ.* **2000**, *5*, 321–335. [CrossRef]
9. Kean, A.J.; Harley, R.A.; Kendall, G.R. Effects of Vehicle Speed and Engine Load on Motor Vehicle Emissions. *Environ. Sci. Technol.* **2003**, *37*, 3739–3746. [CrossRef]
10. Zare, A.; Nabi, M.N.; Bodisco, T.A.; Hossain, F.M.; Rahman, M.M.; Van, T.C.; Ristovski, Z.D.; Brown, R.J. Diesel engine emissions with oxygenated fuels: A comparative study into cold-start and hot-start operation. *J. Clean. Prod.* **2017**, *162*, 997–1008. [CrossRef]

11. Leblanc, M.; Noël, L.; R'Mili, B.; Boréave, A.; D'Anna, B.; Raux, S. Impact of Engine Warm-up and DPF Active Regeneration on Regulated & Unregulated Emissions of a Euro 6 Diesel SCR Equipped Vehicle. *J. Earth Sci. Geotech. Eng.* **2016**, *6*, 29–50.
12. Weilenmann, M.; Favez, J.-Y.; Alvarez, R. Cold-Start Emissions of Modern Passenger Cars at Different Low Ambient Temperatures and Their Evolution over Vehicle Legislation Categories. *Atmos. Environ.* **2009**, *43*, 2419–2429. [CrossRef]
13. Farrington, R.; Rugh, J. *Impact of Vehicle Air-Conditioning on Fuel Economy, Tailpipe Emissions, and Electric Vehicle Range*; National Renewable Energy Lab. (NREL): Golden, CO, USA, 2000.
14. Amanatidis, S.; Ntziachristos, L.; Giechaskiel, B.; Bergmann, A.; Samaras, Z. Impact of Selective Catalytic Reduction on Exhaust Particle Formation over Excess Ammonia Events. *Environ. Sci. Technol.* **2014**, *48*, 11527–11534. [CrossRef]
15. Gallus, J.; Kirchner, U.; Vogt, R.; Benter, T. Impact of driving style and road grade on gaseous exhaust emissions of passenger vehicles measured by a Portable Emission Measurement System (PEMS). *Transp. Res. Part D Transp. Environ.* **2017**, *52*, 215–226. [CrossRef]
16. Merkisz, J.; Andrzejewski, M.; Pielecha, J. The effect of applying the eco-driving rules on the exhaust emissions. *Combust. Engines* **2013**, *155*, 66–74. [CrossRef]
17. Dimaratos, A.; Giechaskiel, B.; Clairotte, M.; Fontaras, G. Impact of Active Diesel Particulate Filter Regeneration on Carbon Dioxide, Nitrogen Oxides and Particle Number Emissions from Euro 5 and 6 Vehicles under Laboratory Testing and Real-World Driving. *Energies* **2022**, *15*, 5070. [CrossRef]
18. Huang, Y.; Ng, E.C.Y.; Surawski, N.C.; Zhou, J.L.; Wang, X.; Gao, J.; Lin, W.; Brown, R.J. Effect of diesel particulate filter regeneration on fuel consumption and emissions performance under real-driving conditions. *Fuel* **2022**, *320*, 123937. [CrossRef]
19. Barrios, C.C.; Domínguez-Sáez, A.; Rubio, J.R.; Pujadas, M. Factors influencing the number distribution and size of the particles emitted from a modern diesel vehicle in real urban traffic. *Atmos. Environ.* **2012**, *56*, 16–25. [CrossRef]
20. Giechaskiel, B.; Riccobono, F.; Vlachos, T.; Mendoza-Villafuerte, P.; Suarez-Bertoa, R.; Fontaras, G.; Bonnel, P.; Weiss, M. Vehicle emission factors of solid nanoparticles in the laboratory and on the road using portable emission measurement systems (PEMS). *Front. Environ. Sci.* **2015**, *3*, 82. [CrossRef]
21. Sokolnicka-Popis, B.; Szymlet, N.; Siedlecki, M.; Gallas, D. The impact of particulate filter substrate type on the gaseous exhaust components emission. *Combust. Engines* **2020**, *183*, 58–62. [CrossRef]
22. Siedlecki, M.; Szymlet, N.; Sokolnicka, B. Influence of the Particulate Filter Use in the Spark Ignition Engine Vehicle on the Exhaust Emission in Real Driving Emission Test. *J. Ecol. Eng.* **2020**, *21*, 120–127. [CrossRef]
23. Šarkan, B.; Jaśkiewicz, M.; Kubiak, P.; Tarnapowicz, D.; Loman, M. Exhaust Emissions Measurement of a Vehicle with Retrofitted LPG System. *Energies* **2022**, *15*, 1184. [CrossRef]
24. Jaworski, A.; Lejda, K.; Lubas, J.; Mądziel, M. Comparison of exhaust emission from Euro 3 and Euro 6 motor vehicles fueled with petrol and LPG based on real driving conditions. *Combust. Engines* **2019**, *178*, 106–111. [CrossRef]
25. Meneguzzer, C.; Gastaldi, M.; Rossi, R.; Gecchele, G.; Prati, M.V. Comparison of exhaust emissions at intersections under traffic signal versus roundabout control using an instrumented vehicle. *Transp. Res. Procedia* **2017**, *25*, 1597–1609. [CrossRef]
26. Ziółkowski, A.; Fuć, P.; Lijewski, P.; Jagielski, A.; Bednarek, M.; Kusiak, W. Analysis of Exhaust Emissions from Heavy-Duty Vehicles on Different Applications. *Energies* **2022**, *15*, 7886. [CrossRef]
27. Lijewski, P.; Merkisz, J.; Fuć, P.; Ziółkowski, A.; Rymaniak, Ł.; Kusiak, W. Zużycie paliwa i emisja spalin w procesie zmechanizowanego pozyskiwania i transportu drewna. *Eur. J. Za. Rez.* **2017**, *136*, 153–160. [CrossRef]
28. Hong, B.; Lü, L. Assessment of Emissions and Energy Consumption for Construction Machinery in Earthwork Activities by Incorporating Real-World Measurement and Discrete-Event Simulation. *Sustainability* **2022**, *14*, 5326. [CrossRef]
29. Triantafyllopoulos, G.; Dimaratos, A.; Ntziachristos, L.; Bernard, Y.; Dornoff, J.; Samaras, Z. A study on the CO_2 and NO_x emissions performance of Euro 6 diesel vehicles under various chassis dynamometer and on-road conditions including latest regulatory provisions. *Sci. Total Environ.* **2019**, *666*, 337–346. [CrossRef]
30. Press Materials. Available online: https://www.globalmrv.com (accessed on 13 January 2023).
31. Giechaskiel, B.; Gioria, R.; Carriero, M.; Lähde, T.; Forloni, F.; Perujo, A.; Martini, G.; Bissi, L.M.; Terenghi, R. Emission Factors of a Euro VI Heavy-duty Diesel Refuse Collection Vehicle. *Sustainability* **2019**, *11*, 1067. [CrossRef]
32. Zhang, Y.; Lou, D.; Tan, P.; Hu, Z.; Fang, L. Effect of SCR downsizing and ammonia slip catalyst coating on the emissions from a heavy-duty diesel engine. *Energy Rep.* **2022**, *8*, 749–757. [CrossRef]
33. Leskovjan, M.; Kočí, P.; Maunula, T. Simulation of diesel exhaust aftertreatment system DOC—pipe—SCR: The effects of Pt loading, PtOx formation and pipe configuration on the deNOx performance. *Chem. Eng. Sci.* **2018**, *189*, 179–190. [CrossRef]
34. Varella, R.; Faria, M.; Mendoza-Villafuerte, P.; Baptista, P.C.; Sousa, L.; Duarte, G.O. Assessing the influence of boundary conditions, driving behavior and data analysis methods on real driving CO_2 and NO_x emissions. *Sci. Total Environ.* **2019**, *658*, 879–894. [CrossRef]
35. Sala, R.; Krasowski, J.; Dzida, J. *Selective Catalytic Reduction of Nitrogen Oxides in Vehicles Complaint to Euro VI Emission Limit, Autobusy: Technika, Eksploatacji, Systemy Transportwoe*; Instytut Naukowo-Wydawniczy Spatium: Radom, Poland, 2017; pp. 1052–1056.
36. Li, P.; Lü, L. Evaluating the Real-World NO_x Emission from a China VI Heavy-Duty Diesel Vehicle. *Appl. Sci.* **2021**, *11*, 1335. [CrossRef]
37. Drywa, M. Wpływ nawierzchni jezdni na zużycie paliwa. Autobusy. *Tech. Eksploat. Syst. Transp.* **2014**, *10*, 10.

38. Kokoszka, S.; Kuboń, M. *A Comperative Analysis of the Roads for Agriculture Transport on Plain and Piedmont Areas*; Inżynieria Rolnicza: Cracov, Poland, 2004; pp. 253–261.
39. Merkisz, J.; Kozak, M.; Molik, P.; Nijak, D.; Andrzejewski, M.; Nowak, M.; Rymaniak, Ł.; Ziółkowski, A. The analysis of the emission level from a heavy-duty truck in city traffic. *Silniki Spalinowe* **2012**, *150*, 80–88.

Disclaimer/Publisher's Note: The statements, opinions and data contained in all publications are solely those of the individual author(s) and contributor(s) and not of MDPI and/or the editor(s). MDPI and/or the editor(s) disclaim responsibility for any injury to people or property resulting from any ideas, methods, instructions or products referred to in the content.

Article

Application of Machine Learning to Classify the Technical Condition of Marine Engine Injectors Based on Experimental Vibration Displacement Parameters

Jan Monieta * and Lech Kasyk

Maritime University of Szczecin, Wały Chrobrego 1-2, 70-500 Szczecin, Poland; l.kasyk@pm.szczecin.pl
* Correspondence: j.monieta@pm.szczecin.pl; Tel.: +48-91-48-09-415

Abstract: The article presents the possibility of using machine learning (ML) in artificial intelligence to classify the technical state of marine engine injectors. The technical condition of the internal combustion engine and injection apparatus significantly determines the composition of the outlet gases. For this purpose, an analytical package using modern technology assigns experimental test scores to appropriate classes. The graded changes in the value of diagnostic parameters were measured on the injection subsystem bench outside the engine. The influence of the operating conditions of the fuel injection subsystem and injector condition features on the injector needle vibration displacement waveforms was subjected to a neural network (NN) ML process and then tested. Diagnostic parameters analyzed in the amplitude, frequency, and time–frequency domains were subjected after a learning process to recognize simulated various regulatory and technical states of suitability and unfitness with single and complex damage of new and worn injector nozzles. Classification results were satisfactory in testing single damage and multiple changes in technical state characteristics for unfitness states with random wear injectors. Testing quality reached above 90% using selected NNs of Statistica 13.3 and MATLAB R2022a environments.

Keywords: marine engines; injectors; experimental states; machine learning; state classification

1. Introduction

One of the most unreliable components of a marine engine is the injector, which is why it is often included in diagnoses [1]. In operation, studies of wear and damage in fuel apparatuses have shown that damage is often misdiagnosed and entails unnecessarily extensive engine repair [2]. Current diagnostic systems do not always allow for the detection of all engine deficiencies without performing at least partial disassembly.

One of the critical problems in diagnosing marine internal combustion engines is finding out in what technical state the component under examination is but also what, where, and how it became damaged (Zoltowski and Cempel [3]). This is of interest primarily to the user and when developing scientific diagnostic procedures.

The technical condition of the engine and injection apparatus significantly determines the exhaust gas composition. Since injectors are an important assembly and significantly impact marine engine performance values and environmental characteristics, as dictated by the stringent International Maritime Organization (IMO) requirements [4], they are one object of study in this work. The IMO has developed mandatory or optional instruments to reduce emissions of nitrogen oxides (NO_x), sulfur oxides (SO_x), and carbon dioxide on ships on international voyages. Carbon dioxide (CO_2) emissions from a ship's stack are directly related to the oxidation of carbon chains present in fuels and used on board for combustion. At high temperatures, the combustion products CO_2 and H_2O dissociate to CO and H_2 in the exhaust gas. For injectors, extensive studies have been conducted to measure technical state features and wear and damage processes [5,6].

In addition to diagnosis, modern diagnostic systems enable prognosis and genesis [3]. Such systems allow for the identification of the type of state in addition to fault localization. Identifying elements is assigning information about their properties [7]. Knowledge of the permissible and inoperable states that occur, as well as their characteristic symptoms, should allow for the development of various diagnostic methods that make it possible to identify a specific state in the functioning of a given object.

The NN methods were used extensively in monitoring technical states by identifying dynamic models but with simplifying assumptions. Panetelas et al. [8] presented a method for modeling the selected faults of large self-ignition marine engines. This method includes measurements of vibro-acoustics signals recorded by the operating engine and uses wavelets to an NN classification system toward marine combustion engines' damage.

In this article, therefore, an attempt is made to identify the unfitness of the most unreliable engine component—the injector [1]—based on simulated states in the off-engine injection subsystem test table of technical state changes and the impact of these changes on the waveform of vibration displacement signals in the injector in various signal analysis domains. Recognition of single and combined selected changes in technical state features was undertaken using artificial neural networks (ANNs).

2. State of Knowledge

2.1. About Neural Networks in General

ANNs are various systems of nonlinear signal processing created from observing the functioning of the nervous system of living organisms. They are an effective tool for searching for new technological solutions, using phenomena occurring in nervous systems [9–12]. In many problems, NNs are used to approximate the functions of many variables. They represent nonlinear maps of the form $y = f(x)$, in which the input vector (x) corresponds to a vector of multivariate functions (y) [11]. ANNs are mainly used to analyze regression or classification problems. In the process of training the network, two sets of vectors are provided: the input data vector (x) and the corresponding output data vector (y). It is a type of learning network with a teacher.

Among the many types of the NNs architecture, two basic types can be distinguished: a multilayer perceptron with a sigmoidal activation function (called the MLP—multilayer perceptron network) and a network using finite support base functions (called the RBF network—radial basis functions). Both types of networks are used both as a predictor and a classifier.

The MLP type is one of the simplest and most frequently used types of the ANN, often used and described in various publications [9,11]. The operating diagram of such a network, which has two hidden layers, is shown in Figure 1.

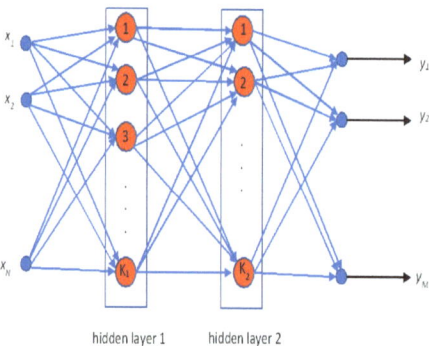

Figure 1. Schematic of an MLP network with two hidden layers.

Although the number of hidden layers can be arbitrary, two layers are usually sufficient to precisely map the input represented by the vector x ($x \in R^n$) to the output data represented by the vector y ($y \in R^m$) according to the formula $y = f(x)$. Individual neurons of the network implement nonlinear mapping, which, depending on the type of neuron, has a different form. In this study, sigmoidal neurons were used.

The output vector is described by the formula below

$$y(t) = f\left(\sum_{i=0}^{N} w_i(\tau) x_i(\tau)\right) \qquad (1)$$

where w_i creates a vector of weights w, x_i creates a vector of input values x, f is an activation function, and τ is a number of the next iteration of the algorithm.

Activation functions are an important element of the ANN. The activation function $f(u)$ in a perceptron network has a sigmoidal character of an unipolar or a bipolar type, where β is a numerical coefficient with a value usually equal to 1. A characteristic feature of the MLP-type network is the sigmoidal activation function [11].

This study uses networks with one-way information flow—from input to output. The aim of learning is to determine the weight values of all network layers so that, given the input vector x, the output values of the y vector correspond with appropriate accuracy to the given values of the y_0 vector. It boils down to minimizing the error function, which in the case of the discussed sigmoid neuron has the form [13]:

$$Q(w) = \frac{1}{2}\left[d - f\left(\sum_{i=0}^{N} w_i(t) x_i(t)\right)\right]^2 \qquad (2)$$

(1) In the training stage, the values of the output vector are equal to the known values of y_0. Learning consists in giving the neuron input the values of $x(\tau)$, for which we know the correct values of the output data $d(\tau)$, called standard signals. The set of input signals and their corresponding standard signals is called the training sequence. After entering the input value, the output signal of the neuron is calculated. The weights are modified in such a way as to minimize the error between the standard signal and the neuron output. The learning algorithm is as follows [13]: We randomly select the initial weights of neurons.
(2) We give the training vector x to the neuron's input.
(3) The output value of the perceptron is calculated according to Formula (1).
(4) The output value $y(\tau)$ is compared with the reference value of the training chain $d(t)$.
(5) The weights are changed according to the following relationships
   a. $y(x(t)) \neq d(x(t)) \Rightarrow W_i(t+1) = w_i(t) + \eta(d - f(s))f'(s)x_i$;
   b. $y(x(t)) = d(x(t)) \Rightarrow w_e(t+1) = w_i(t)$.
(6) We return to point 2.

The algorithm is repeated for all input vectors included in the training sequence until the output error is smaller than the adopted tolerance.

2.2. Existing Methods of Damage Recognition

From the state of the literature, fault recognition in diagnosing internal combustion engines (ICEs) can be carried out using advisory systems, image analysis, and ANNs [14–18]. Probabilistic and deterministic state recognition methods have been used [16,18]. Lewicki and Hill [16] presented a method for diagnosing a marine ICE based on distance from a benchmark using metric recognition methods.

ANNs have also recently been used to diagnose defects in marine ICEs [19–23]. However, they rarely dealt with the classification of damage to the injection apparatus [2,15,17,24–27], which the co-author has been dealing with for many years.

Zoltowski and Cempel [3 chapter 22] used an ANN to diagnose technical objects and the control of an ICE [16]. ANNs have been used to diagnose selected inabilities of a marine ICE [27] and to classify the condition of the injection apparatus (He and Rutland [25],

Monieta and Choromanski [17]). The authors of the [27] paper limited themselves to measurements of pressure waveforms in the high-pressure fuel pipe. They attempted to classify them into different classes based on similarity without going into the state of the injection apparatus in the image-type model.

Pawletko [27] attempted to determine the usefulness of NNs for diagnosing selected states of a marine engine. The author verified an experimental diagnostic algorithm on a supercharged engine simulator. The diagnostics of the marine engine used the quantification of the parameters of the structure and the identification of the states that can be simulated in the computer program. Some marine engine inadequacies remained challenging to diagnose accurately and quantitatively due to occurrence and other failures. This author used advisory systems and experimental database research to extract diagnostic rules automatically. Acquiring knowledge from specialists was inefficient and time-consuming.

The research presented in Klimkiewicz's [2] study involved compression-ignition engine (CIE) vehicles equipped with distributor pumps. As input variables were selected, which were symptoms, measurements, and checkups, based on which experts determined the damage to the injector apparatus were the input variables and the engine one output variable. The best classification quality was obtained using probabilistic networks. An indicator of the suitability of individual variables for the classification performed by the network was the quotient of the network error obtained without using a given variable to the baseline error. The author did not make a quantitative assessment of the accuracy of correct classifications since only cases where the experience of the workers did not allow for the damage to be detected quickly were studied. The observed damage included the incorrect order of installation of injection lines, gasoline content in diesel fuel, low compression pressure, obstructed exhaust system, and fuel system airing. However, the impact of all significant and graded damage was not demonstrated.

Klimkiewicz [7] attempted to optimize the process of detecting defective components in injection pumps using artificial intelligence. A neural model based on probabilistic NNs was used. With the built model's help, the depth indicator's value for locating damaged components could be reduced. However, there was more than pump damage and unmeasurable factors.

Brzozowski and Nowakowski [28] presented the application of ANNs in a computational model of the working cycle of a CIE. To evaluate the usefulness of the method, the task of selecting the values of control parameters to achieve a reduction in the content of nitrogen oxides in the exhaust gas was solved as an optimization task. The model's equations were solved for parameter values obtained as the response of NNs to variable control parameters, including changes in emissions of harmful compounds and exhaust smoke.

Data generated during engine operation, using a diagnostic model based on ML, can help users of marine ICEs correctly identify types of damage and take prompt action to avoid serious accidents according to Xu et al. [20]. A multi-model fusion system was proposed based on the principle of evidence inference based on an ANN model. A method for determining the dependability of evidence by using the precision and stability of tested models was presented. A genetic algorithm optimized the validity to improve the efficiency of the merger arrangement. The suggested system was verified against a set of actual specimens taken from the operating ship's CIEs.

In their paper, Logan et al. [29] presented new developments in damage diagnosis on intelligent software representatives. The studies aimed to design a real-time agent capable of actual continuous ML using an ANN. A combustion engine simulator that modeled regular and abnormal operations was used to develop the controller's learning system and test results.

An intelligent sensor validation and online fault diagnosis technique were proposed and studied for a 6-cylinder turbocharged self-ignition engine (SIE) [30]. A singular autoassociative 3-layer ANN was taught to test the precision of the gauged quantities. A forward coupling SSN was trained to classify and recognize abnormal and fitful engine behavior over various operating conditions for online fault detection. The proposed technique was also equipped with an activated online learning mechanism. Effective state monitoring technique efforts have been implemented to diagnose SIE component damage. The vibration and acoustic emission signals were used and analyzed in the time, and frequency domain, performing various signal acquisition and analysis methods [31]. The elaborated methods (later NN) have accurately detected the scuffing states on engine components.

The complicated systems and trying operating conditions in ship's SIEs are susceptible to damage, leading to enormous significant economic losses. The enormous complexity of engines and sensor networks results in an increasing set of technical state monitoring information that often exceeds the capabilities of damage-finding systems. Wang et al. [32] presented a deep learning network structure for intelligent health monitoring SIEs using deep and ensemble learning. The authors have improved the optimizer, which uses the adaptive learning factors to speed up the network's learning and prevent detention at the local optimum.

The article by Cheliotis et al. [19] presents developed diagnostics for marine vessels based on operational data and damage probabilities. It was shown that complex physics-based models could detect the primary cause of damage by using "black-box neural networks" [19]. This study unites machine learning using damage finding, exponential-weighted average control plots, and "Bayesian diagnostic networks" to study the rate of damage growth increments. For validation, a "case study" of a major ship SIE was studied for the states of the air radiator and the compressed exchanges of the working medium subsystem [19]. It was found that any anomalous deflection in the main propulsion engine's outlet temperature was mainly due to exchanges working media system damage.

Wang et al. [32] applied the Bayesian analysis for the marine SIE's performance prognostics, uncertainty inferences, and results using probability distributions. The Bayesian NN model was studied for state diagnosis, while the "Bayesian logistic regression model" quantified the operation from the startup to damage of the ship CIE. The authors showed that the proposed approach was superior to the other methods used to be applied as an online technical state diagnosis for ship SIE forecasting.

Garczynska et al. [33] used engineering measurements on labile faces and docks, vessels, size assessment of marine units, studies of vessel hull distortions, dynamic surveying of units, and coordinate transformations with low uncertainty minor. The author's aim was dimensional control of vessels, platforms, and offshore objects. This method used ML for triaxial (3D) conversions. The suggested method was verified based on laboratory and field data. Zhu [34] investigated marine SIE state monitoring using an ANN. The vibration signals from the engine were measured and analyzed using the time series method. The assessment of valve lash changes and individual engine cylinders' relative loads was described. Diagnostic symptoms of the time waveforms of chatter signals received on the ICE were utilized to develop a proper ANN through the "backpropagation algorithm" for detecting damage [34].

Computer-simulated graded relations of injector technical state features—injector space pressure waveform—were subjected to a learning process by Monieta and Choromanski [17] using an ANN. After the learning process, an attempt was made to recognize simulated different technical states of fitness and unfitness with single and combined failures.

Monitoring the process of a ship's state is important because the parameters significantly impact the accuracy and reliability of the elaborated prognoses. Parameters of PANAMAX container ship main engine cylinder were clustered using the ANN [20]. The "case study results" [20] demonstrated the NNs' ability to assess the technical state of the main propulsion marine engine using parameter clusters. The results obtained were extended to diagnosing, genesis, and prognosticating of the vessel's diesel engine.

Zhang et al. [23] advertised diagnostics for SIE damage using a "quantum genetic algorithm" (QGA) to optimize the PNN model. A diagnosis model was elaborated using a PNN that considers a ship's SIE decomposition and exploitation characteristics. The parameter of the probabilistic network model was optimized using the QGA. The simulation showed that the ship SIE damage diagnosis process, using a "probabilistic neural network" optimized with the QGA, had significant merits in diagnosis correctness of 94.4% and the diagnosis performance, which was 40% higher than that of the BP NN model.

The diagnostics combined with expert analysis by the Wärtsilä service company was delivered as part of an existing agreement. The solution uses artificial intelligence methods and advanced rule-based diagnostics to continuously monitor operational data of marine SIE deviations and anomalous equipment behavior. One threat might be if customers are not distrustful of paying for the innovative service [35].

Recent years have seen the growth of ML in marine diesel engine diagnostics. Class imbalances were defects in parameters from the ship's CIEs as this significantly degraded the speed. Wang et al. [36] proposed a novel method for damage diagnosis using a graph convolutional network and probability to solve unbalanced classification. It is shown that the ordered data set of marine self-ignition engines and the used model can correct the classification quality in the data sets.

Information about the relative load is needed to optimize the parameter adjustments during the exploitation of an ICE. Shahid et al. [37] studied relative engine load classification using measured signals depending on the degree of crankshaft rotation. An ANN was trained using processed data for five grades of relative load and classified with an accuracy of 99.4% using a support vector machine.

Machinery malfunctions are inevitable on ship vessels, but their occurrence can be predicted by implementing "predictive maintenance". This article suggests a new method for damage localization by testing the "learning potential" of recorded voyage data [38]. A ridge polynomial regression model, with an R^2 test score of 0.96, can detect developing damage by monitoring the main drive engine outlet gas temperature and flushing air pressure. The rapid identification of propagating damage was a complement to diagnosing and planning repair work.

In another article, the search for crack-oriented injector nozzle analysis methods for analyzing vibration acceleration signals using multiple domains, color analysis, and ML were used for classification using ANNs [39]. Experimental investigations have shown the feasibility of early detection of damage development processes in CIEs, especially using decimation and time-domain window analysis.

Kowalski et al. [40] proposed intelligent damage diagnosis for ship's CIEs. They used an automatic engine fault detection system using ML. The engine-generated signals were monitored and used as input for pattern classification. A 15-grade task was solved by binarization, in which each extreme ML was trained in two classes. Furthermore, the output codes were used to reconstruct the original multi-class task. The results showed that the suggested approach provided high classification quality and short running time compared with many other methods. Diagnosing the condition of ICE subassemblies is a difficult task, and damage to them increases operating costs. Therefore, techniques have been implemented to diagnose the state of engine components. Ramteke et al. [41] presented the application of vibration and acoustic emission signals to diagnose seizures in CIE subassemblies. Analyses were conducted using fast and short-time Fourier analysis in the time, frequency, and time–frequency domains. The authors investigated ANN models for damage prediction and classification and proved that Fourier analysis was better for diagnosis, with classification efficiency reaching 100% accuracy.

Adamkewicz and Nikonchuk applied ML to the diagnosis of MDEs turbochargers [42]. The value of the coefficient of determination confirmed a good fit for the trend of compressor condition changes.

Zhao et al. [43] proposed a method of improved wavelet packet frequency and convolutional NN for valve timing diagnosis. The authors developed time–frequency distribution matrices and frequency bands of wavelet packet distribution. The results showed the effectiveness in diagnosing valve clearance of compression ignition engines.

The inadequacies in the state of knowledge have prompted the development of a reliable method for diagnosing injection equipment using ANNs, the main goal of which is to build a model between the diagnostic symptoms and technical states that is complex (with many independent variables) and nonlinear. In reviewing the state of the art, the application of NNs in developing and diagnostics of the fuel injection subsystem of ICEs, especially marine engines, needs to be improved.

3. Materials and Methods

3.1. Test Method

The measuring stand to investigate injectors was localized in separate accommodations in the Injection Apparatus Laboratory of the Maritime University of Szczecin. This stand consisted of a testing table and elements of marine self-ignition AL20/24 engine injection subsystems.

The measuring system consisted of three subsystems (Figure 2):
- A test table;
- A system of processing signals;
- A system of digital processing and analysis.

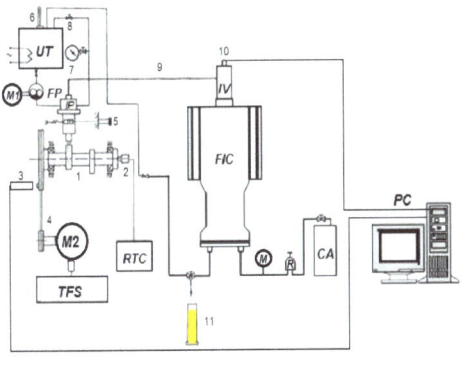

(a) (b)

Figure 2. Diagram of the built modernized injector test stand (**a**) and view of the test stand (**b**): 1—camshaft, 2—rotation sensor, 3—camshaft position photo-optical sensor, 4—belt transmission, 5—fuel setting depth gauge, 6—thermometer, 7—pressure gauge, 8—throttle valve, 9—injection pipe, 10—eddy current displacement sensor, 11—measuring cup, *FP*—booster fuel pump, *RTC*—revolution and time counter, *M1*—the electric motor for pump drive, *M2*—the electric motor for the camshaft drive, *CP*—compressed air tank *IP*—injection pump, *IV*—injection valve, *TFS*—thyristor supply system, *PC*—computer, *R*—pneumatic reducer, *FIC*—fuel injection chamber.

The mechanical part of this stand created a camshaft driven by a DC electric motor with a belt transmission. The calibration oil was supplied from the tank with the heater by gear pump and pressed by fuel filter to the injection pump.

Diagnostic signals were analyzed using the computer program signal analysis system (SAS). A preliminary process by selection in time, low-pass filtering, and synchronous averaging have been applied, and quantitative and qualitative analyses of spectra in the time and frequency domain have been applied [1,44]. This test stand was constantly developed and modernized.

The signal acquisition and analysis program allowed for analyzing a sequence of diagnostic signals originally presented in the cascaded form [1]. A cascade is a three-dimensional graph showing amplitudes as a function of time and sequence numbers. The functional, point and dimensionless estimates can be determined in this analysis option and summarized in the table. Functional estimates are the amplitude distribution function and probability function.

The mechanical part of this stand was formed by a camshaft driven by a *DC* electric motor through a belt transmission. Calibration oil was fed from a heater tank through a gear pump and pumped through a fuel filter to the injection pump. The tests were conducted for a conventional injection subsystem.

To identify the technical state, it is necessary to know the types and extent of wear and tear and the damage they have caused [6]. This study used simulation experiments, which have some advantages over numerical modeling. In the experimental studies undertaken in the past, there were significant limitations in that the components of the injection apparatus—new as well as one in service—have state features with random and widely varying values. Therefore, many manufactured samples of geometric and flow tests that met the requirements were selected [6]. Computer simulations were also conducted, but simplifying assumptions had to be made [45]. This made it possible to reduce the cost and time of testing and to analyze "isolated" damage, which in actual conditions of ICEs could encounter many difficulties.

The article formulates the following hypothesis: It is possible and expedient to use ML to classify and identify injectors' technical states and develop a diagnostic symptom—the technical state model of a marine engine injector.

The micro- and macrostructure of fuel injection determine the combustion process. One element that determines the injection's macrostructure is fuel distribution over time, called the injection course. The injection course is influenced by the design features of the injection subsystem [6], the fuel parameters, and the injection frequency [45].

3.2. Experimentally Simulated Technical States of the Injector

In this work, the effects of the control quantities and technical state of the following injector input quantities were studied:

- Injection pump supply pressure p_f from 0.15 to 0.30 MPa;
- Camshaft speed n from 150 to 450 rpm;
- Fuel temperature at the injector inlet $t_p = 20-50$ °C;
- Fuel dose control strip setting from 20 to 100%;
- Changes in the tension of the spring, causing changes in the opening pressure of the injector needle in the range $P_o = 15–30$ MPa (the strength of the spring tension to achieve the required injector opening pressure);
- Changes in the number of expensive holes in the injector nozzle $i = 4$ to 7;
- Changes in the diameter of the spray holes $d = 0.23$ to 0.28 mm;
- Changes in the average clearance at the leading part of the body and needle of the injector L_a;
- Changing the maximum needle lift h_{max};
- Changes in the nozzle needle cone angle $\alpha = 59$ to 61 °C.

The sum of the forces acting on the injector element $\sum F = F_1 + F_2 + F_3$ includes the force caused by the action of pressure p on the A component's surface $F_1 = pA$, spring preload force $F_2 = P_o$, and spring deflection Δh force with k_{sw} stiffness $F_3 = k_{sw}\Delta h$. Such values were dictated by the fact that these are essential input quantities for the conditions and state features, and by choice for the experimental plan [1,6]. In the first attempt to develop a network for solving a specific task, all variables that may be relevant should be included, as later at the design stage, this set is recommended to be reduced.

The state of the injection pump was investigated in a separate test. In this research, the state of the injection pump was assumed to be invariant. The supply pressure of the injection pump, $p_f = 0.15$ MPa, was the central value. Furthermore, the maximum inlet pressure to the injection pump could be 0.4 MPa. Thus, the following values of changes in fuel pressure before the injection pump $p_f = 0.1; 0.15, 0.2, 0.25,$ and 0.3 MPa were established.

The fuel setting at which fuel injection does not occur is 20%. The maximum (100%) fuel setting was taken as the central value. The effect of fuel setting N on needle displacement waveform values h depending on time τ at constant injector opening pressure, is shown in Figure 3.

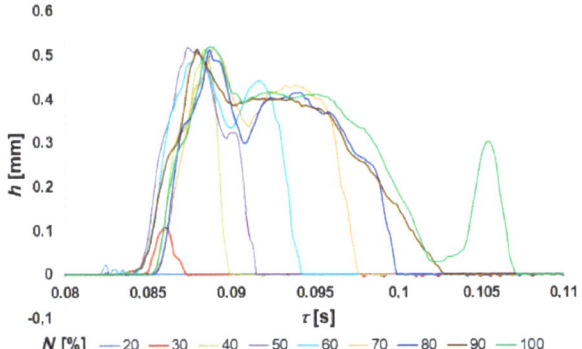

Figure 3. The effect of fuel setting N on the values of the needle displacement waveform h depends on the time τ at a constant injector opening pressure $P_o = 25$ MPa.

When the aforementioned research began and the injectors were purchased, the injector opening pressure of the AL20/24 type engine should be $25^{+0.5}$ MPa. It represented a central value. As a result of the relaxation of the injector spring, the opening pressure changes (usually decreases). Therefore, a gradation of opening pressure every 5 MPa was adopted, from $P_o = 15$ MPa to $P_o = 30$ MPa. Changes in the opening pressure of the injector (derating) were simulated by changing the spring tension. This was achieved by screwing the adjustment screw into the injector body or unscrewing it. Two new 7 × 0.26 R (standard) injector nozzles were used for these tests, and the results were presented in the paper [6]. The effect of the injector opening pressure control on the time waveforms of vibration displacement signals is shown in Figure 4.

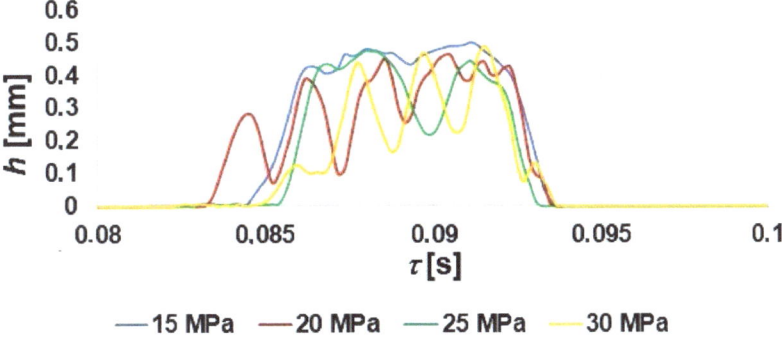

Figure 4. The effect of injector tension force P_o on the values of the nozzle lift waveform h as a function of time τ.

It can be observed from Figures 3 and 4 that there were ripples in the nozzle displacement waveforms, especially for larger fuel settings.

3.3. Data Preparation for the Neural Network Model

The artificial NN model's proper interpretation of injector failures depends on its ability to recognize characteristic symptoms of unfitness. For this purpose, sets consisting of different variants of single changes in state characteristics were developed while keeping constant the values in the state of unfitness of the others. However, the effect of combined damage when examining the nozzles of injectors in operation was also considered. In systems using ANNs, the more measurements, the more necessary data to train and validate the network.

The first important decision in building the network is determining which variables should be included and which cases should be collected. The displacement of the injector needle depends on the following input quantities that define the test conditions:

$$h = f(n, N, p_f, t_p) \tag{3}$$

where n is camshaft speed, N is fuel setting, p_f is the supply pressure of the injection pump, ρ is fuel density depending on the fuel type and the fuel temperature at the inlet t_p.

Fuel density depends on its temperature and pressure [6]. In experimental studies, it was assumed to be constant and determined at the inlet to the injection pump. In the experimental studies of the waveform of the injector nozzle needle displacement, the waveform of pressure in the combustion chamber was assumed constant, typically treated as continuous in the works. However, it was possible to supply compressed air at different pressures.

The steering values (fuel setting and rotational speed) were kept constant, and the state features were changed. Based on operational tests, the most common load was determined [1,46]. It was assumed that for unfitness, the listed state features receive a value of "0", while when they fall within the tolerance field of the suitability state, they receive a value of "1". Two sets of damage to the displacement waveforms of the nozzle needle were prepared. In reality, more features determine the condition of the injector nozzle, and the time waveforms of the pressure inside the nozzle sack are inadequate. However, in this research, they were assumed to be constant. These quantities appear in the model of the waveform of the nozzle needle displacement during fuel injection [45].

The input magnitudes determine the output magnitudes (simple and functional amplitude estimates), and thus the technical state of the injector S_t:

$$S_t = f(h_{rms}, h_{aver}, h_{peak}, K, C, I, B) \tag{4}$$

where h_{rms} is the root-mean-square values, h_{aver} is an average value, h_p is the peak value, K is the form factor, C is the peak factor, I is the impulsivity factor, and B is a dose of injected fuel.

Amplitude estimates are defined by the following measures [3]:
- Root-mean-square values:

$$\tilde{h}(\theta) = \left[\frac{1}{T}\int_0^T h^2(\tau,\theta)d\tau\right]^{\frac{1}{2}} = h_{rms}(\theta) \tag{5}$$

- Average value:

$$\overline{h}(\theta) = \frac{1}{T}\int_0^T |h(\tau,\theta)|d\tau = h_{aver}(\theta) \tag{6}$$

- Positive peak value:

$$\hat{h}(\theta) = max[+h(\tau,\theta)] = h_{peak}(\theta) \tag{7}$$

for $0 \leq \tau \leq T$.

Characterizing the quoted amplitude measures in terms of their sensitivity to the waveform of the process $h(\tau)$, it can be shown that each behaves differently. The *rms* value contains information about the energy of the signal under study, the average value represents the amplitude modulus of the signal, and the peak value contains information about the maximum values of the signal amplitudes.

The given definitions of amplitudes are instants of the stability and ergodic of the random processing $h(\tau)$ [3]. Thus, the selected dimensionless estimates of vibration processes, calculated from point measures according to the following relationships, were further used:
- Shape factor:

$$K = \frac{\tilde{h}}{\bar{\bar{h}}} \qquad (8)$$

- Peak factor:

$$C = \frac{\hat{h}}{\tilde{h}} \qquad (9)$$

- Impulsivity factor:

$$I = \frac{\hat{h}}{\bar{\bar{h}}} \qquad (10)$$

The advantage of dimensionless discriminants is their qualitative nature, which gives versatility for various diagnostic signals.

Analysis of the signals in the frequency domain showed that the most useful bandwidth was the frequency range of 0–125 Hz (Figure 5). These were polyharmonic signals with frequencies multiples of the rotational frequency nf_r. The amplitudes of the components of the spectra were read from the envelope.

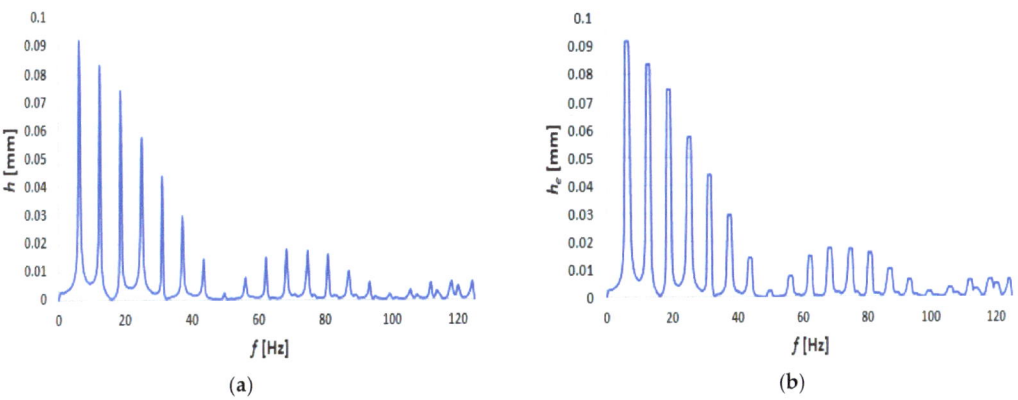

Figure 5. Example spectrum for fuel setting $N = 58\%$ and injector nozzle with a hole diameter $d = 0.28$ mm (**a**) and spectrum envelope (**b**).

Figure 5 shows that this was a polyharmonic line spectrum (discrete). Such a spectrum is represented by "spectral lines" with lengths proportional to the values of the H_i amplitudes of the individual components with frequencies f_o, $2f_o$, $3f_o$, ..., nf_o. Other dimensions of signal analysis are described in the article [1].

Traditional frequency analysis is unsuitable for observing nonstationary signals' properties with time-varying parameters. Slight changes in amplitudes and frequencies over time at the test stand outside the engine were caused by thermal processes. Thus, analyses were conducted using combined time–frequency representations of the signals. These analyses were performed as time–frequency (Figure 6) and time–scale [1] representations.

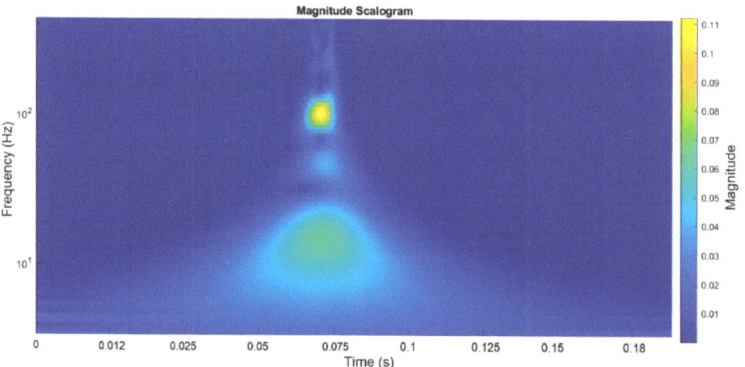

Figure 6. Time–frequency analysis of the needle vibration displacement signal for an injector opening pressure of 30 MPa and at a fuel setting of 58%.

The time–frequency maps, apart from interesting visualization, did not yield significant differentiation using image color analysis [5].

3.4. Estimating Maximum Relative Errors of Diagnoses

The maximum errors made in the study of the application of the vibration displacement signal for diagnosing marine engine injectors were evaluated based on the most significant reading error, depending on the accuracy class of the measuring instruments and the size range of the measured quantities. The resulting maximum relative mistake of fuel injection parameters $\delta(F_i)$ was determined from the relation:

$$\delta(F_i) = 100\% \sqrt{\left(\frac{\Delta t_p}{t_p}\right)^2 + \left(\frac{\Delta p_f}{p_f}\right)^2 + \left(\frac{\Delta N}{N}\right)^2 + \left(\frac{\Delta n}{n}\right)^2 + \left(\frac{\Delta P_o}{P_o}\right)^2} \qquad (11)$$

where Δt_p is the absolute error of supply fuel temperature measurement, Δt_p is the absolute error of supply fuel pressure measurement, ΔN is the absolute error of fuel setting, Δn is the absolute error of camshaft speed measurement, and ΔP_o is the absolute error of injector opening pressure.

The values of relative maximum errors of individual processing elements of diagnostic signals in Figure 2 are shown in the article [44].

The total errors of the vibration displacement signal processing track and the subsequent determination of individual diagnostic parameters were qualitatively estimated. The relative maximum error of individual diagnostic signal measurement track S_d:

$$\delta(S_d) = 100\% \sqrt{\left(\delta(F_i)^2 + \delta(P_e)^2\right)} \qquad (12)$$

where $d(P_e)$ is the relative maximum errors of individual processing elements of diagnostic signals.

The absolute root-mean-square error of the arithmetic mean of the diagnostic parameter, which was strongly influenced by the number of repeated measurements, can have the form:

$$\sigma_s = \frac{\sigma}{\sqrt{n}} = \sqrt{\frac{1}{n(n-1)} \sum_{i=1}^{n}(D_{pi} - D_{pa})^2} \qquad (13)$$

where n is a number of measurements, σ is the standard deviation, D_{pi} is i-th measurement result of the diagnostic parameter, and D_{pa} is the average value of the diagnostic parameter.

The measurement method is burdened with systematic and random errors; they should be summed up according to the formula for the total error σ_t:

$$\sigma_t(P_{dc}) = \sqrt{[\sigma_s(P_d)]^2 + \frac{1}{3}[\sigma(D_p)]^2} \qquad (14)$$

Formula (14) is valid for absolute and relative errors and the results can be used for final presentations of diagnostic symptoms. The summed relative total errors of the point amplitude estimates and the amplitude components of the envelope spectra of firing pin displacement signals did not exceed ±3.6%.

4. Results

4.1. Neural Network Learning Results

Table 1 identifies the studied technical states S_t and, thus, the suitability of individual diagnostic parameters. Meeting very high criteria by the diagnostic symptoms made selecting the most useful among the diagnostic parameters possible.

Table 1. Examples of the input and output quantities used in the tests and the resulting technical states S_t for the fuel setting N.

Name	$\tilde{h}$	$\overline{h}$	$\hat{h}$	P_o	i	d	L_a	h_{max}	α	n	N	p_f	t_p	B	S_t
Unit	mm	mm	mm	MPa	–	mm	mm	mm	°	rpm	%	MPa	°C	mm³	–
7 × 0.23R	0.0850	0.0185	0.5414	25	7	0.23	0.0045	0.465	60.50	375	58	0.15	20	48.4	0
7 × 0.25R	0.0818	0.0172	0.511	25	7	0.25	0.0035	0.456	60.80	375	58	0.15	20	52.9	0
7 × 0.26R	0.0700	0.0162	0.485	25	7	0.26	0.0035	0.463	60.97	375	58	0.15	20	53.0	1
7 × 0.28R	0.0589	0.0130	0.506	25	7	0.28	0.0040	0.465	60.98	375	58	0.15	20	50.7	0

Dimensionless estimates (8–10), resulting from point estimates (5–7), were discarded at the data preparation stage for learning.

Table 1 tells us that the results of the amplitude estimates were written as quantive outputs of the diagnostic signals used. The steering quantities and the state features were also reported as input quantities. The technical state for the cases shown was stored in the binary system as 1 for a suitability state and 0 for an unfit state.

At the outset of designing a network for solving a given task, one should include all variables that may be relevant so that the set can then be reduced. The rule recommends that the number of cases should be ten times the number of connections occurring in the network [47].

The first included only single faults, i.e., it included the following condition features:

- (0,1,1,1,1,1,1,1,1,1,1,1,1,1,1)—inappropriate opening pressure of the injector;
- (1,0,1,1, 1,1,1,1,1,1,1,1,1,1,1)—number of patented holes less than 7;
- (1,1,0,1, 1,1,1,1,1,1,1,1,1,1,1)—diameter of the pinholes not within the tolerance field;
- (1,1,1,1,0,1,1,1,1,1,1,1,1,1,1,1)—improper clearance in the guiding part of the body and the nozzle needle;
- (1,1,0,1, 1,1,1,1,1,1,1,1,1,1,1)—needle cone angle not within the tolerance field;
- (1,1,1,1,1, 1,1,0,1,1,1,1,1,1,1)—incorrect camshaft rotational speed;
- (1,1,1,1, 1,1,1,1,0,1,1,1,1,1)—incorrect fuel setting;
- (1,1,1,1, 1,1,1,1,1,0,1,1)—incorrect injection pump supply pressure;
- (1,1,1,1,1, 1,1,1,1,1,1,0,1)—improper fuel temperature;
- (1,1,1,1,1, 1,1,1,1,1,1,1,0)—incorrect dose of fuel pumped through the nozzle.

The second set additionally consisted of combined technical states. That is, its scope included the example operated injector nozzles: (1,1,1,0, 1,1,1,1,1,0).

Figure 7a shows the architecture of an example of a raw MLP network, and Figure 7b shows the network with descriptions of inputs, layers, and outputs. The network architectures were made in the Statistica 7 computer program.

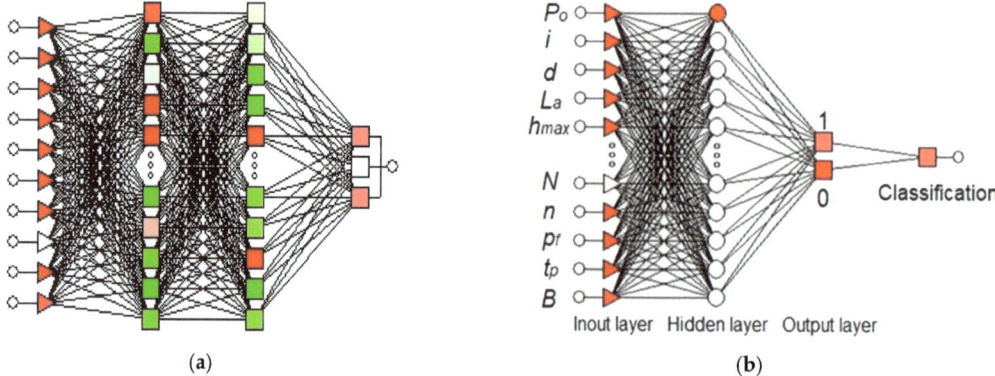

Figure 7. The architecture of an example raw MLP network (**a**) and with descriptions of inputs, layers, and outputs for GRNN network (**b**).

The network response mandates additional interpretation in classification networks to establish recognition based on which output layer neuron presents the highest signal parameter value.

In the diagnostic study, a large amount of data was collected, describing the diagnostic objects' inputs, states, and outputs. These data were initially classified by assigning technical state classes to them. Experimental data collected are declarative knowledge carriers subjected to ML in the Statistica 13.3 and MATLAB R2022a environment. All benchmark data and graded changes of single state characteristics and quantities defining test conditions were selected as a learning set, accounting for 63.26%.

ML in classification is the issue of determining to which set of technical state categories newly observed data belong. Assignment to a category was made based on previously entered training data whose membership in the class has been determined.

The injector was at full ability and not unfit for graded changes in fuel temperature and injection pump supply pressure. All data from new injectors were taken as validation data (18.37%). The test set for fuel setting 58 and 100% accounted for 18.37%. The global sensitivity measure shown in Table 2 assesses the data's variability. The MLP 14-5-2 network contains 14 input neurons, 5 to 13 neurons in the hidden ply, and two in the output ply.

Table 2. A global measure of the sensitivity of input quantities for the sample: learning, test, validation.

Item	Network Name	n	L_a	α	B	h_{rms}	P_o	N	i	p_f	h_{aver}	h_{peak}	h_{max}
1	MLP 14-5-2	1.051	1.293	1.248	1.136	1.848	1.161	1.084	1.315	1.090	1.145	0.996	1.103
2	MLP 14-13-2	2.819	1.748	1.845	1.754	1.473	1.274	1.549	1.191	1.218	1.195	1.173	1.100
3	MLP 14-11-2	2.314	1.963	1.656	1.551	1.313	1.265	1.332	1.160	1.129	1.216	1.170	1.125
4	MLP 14-6-2	2.077	1.698	1.880	1.679	1.605	1.634	1.436	1.655	1.394	1.196	1.359	1.046
5	MLP 14-8-2	1.844	1.272	1.228	1.260	1.077	1.127	1.146	1.102	1.101	1.024	1.059	1.074
6	MLP 14-7-2	1.877	1.742	1.603	1.560	1.094	1.306	1.104	1.171	1.189	1.125	1.083	1.068
	Average	1.997	1.619	1.577	1.490	1.402	1.295	1.275	1.266	1.187	1.150	1.140	1.086

The susceptibility of diagnostic parameter values was be determined by the sensitivity coefficient s_c defined by the formula:

$$s_c = \frac{y_{imax} - y_{imin}}{\bar{y}} \qquad (15)$$

where y_{imax} is the maximum quality of the measured diagnostic parameter, y_{imin}—the minimum quality of the measured diagnostic parameter, $\bar{y}$ is the average value of a given diagnostic parameter in a group of observations.

It can be observed from Table 2 that the highest sensitivity is the camshaft speed n, and the lowest sensitivity is the temperature of the fuel supply t_p for these data. Sensitivity analysis informs the utility of the input variables, that is, the practical usefulness.

4.2. Analysis and Interpretation of Test Results

The purpose of the classification is to assign the case to one of the predefined classes [47,48]. Each of the predicted classes corresponds to one neuron in the output layer. The outputs for these data are two-state in nature. The information is presented in the form of classification statistics. The summary results are recorded in Table 3.

Table 3. Summary of active networks.

Network ID	Network Name	Learning Quality	Testing Quality	Validation Quality	Error Function
1	MLP 14-5-2	92.06349	82.35294	100.0000	SOS
2	MLP 14-13-2	93.65079	88.23529	100.0000	Entropy
3	MLP 14-11-2	93.65079	94.11765	100.0000	SOS
4	MLP 14-6-2	90.47619	94.11765	94.1176	SOS
5	MLP 14-8-2	88.88889	94.11765	94.1176	Entropy
6	MLP 14-7-2	90.47619	94.11765	94.1176	Entropy

The best network turned out to be No. 3. From Table 3, there is a high quality of learning, testing, and validation (generally above 90%), and for individual networks, even 100%. However, many incorrect classifications exist, up to 6 for the MLP 14-6-2 network (Table 4). This high proportion of mistakes may be due to the correlation of diagnostic parameters with measured technical states. They form very accurate relationships ($k > 0.900$), accurate relationships ($k > 0.700$), and only in a few cases, significant relationships ($k > 0.500$). The minor confusion in the matrix in the classification summary was found in the MLP 14-11-2 network (Table 4).

Another measure of the quality of the network as a classifier is the receiver operating characteristic curve (ROC), which shows the cumulative specificity (false positives) on one axis and sensitivity (true positives) on the other. A classifier groups a set of features into decision groups so that all data located in this subspace corresponds to a distinguished group (class). The boundary values in these sets are called decision-marker subsurfaces. A decision boundary is an area in the feature space where the corresponding criterion functions (class membership functions) have the same value.

Table 4. Fragment of the confusion matrix in the classification summary for the sample: learning, test, validation.

Network ID	Network Name	Classification Summary	State 0	State 1	State All
1	MLP 14-5-2	Correct (%)	95.65217	82.14286	91.75258
		Incorrect (%)	4.34783	17.85714	8.24742
2	MLP 14-13-2	Correct (%)	97.10145	85.71429	93.81443
		Incorrect (%)	2.89855	14.28571	6.18557
3	MLP 14-11-2	Correct (%)	95.65217	92.85714	94.84536
		Incorrect (%)	4.34783	7.14286	5.15464

The closer the area under the curve (AUC) is to unity, the better the classifier's quality. As can be observed from the ROC and AUC values (Figure 8), the results obtained by the MLP network are excellent and slightly worse for the RBF.

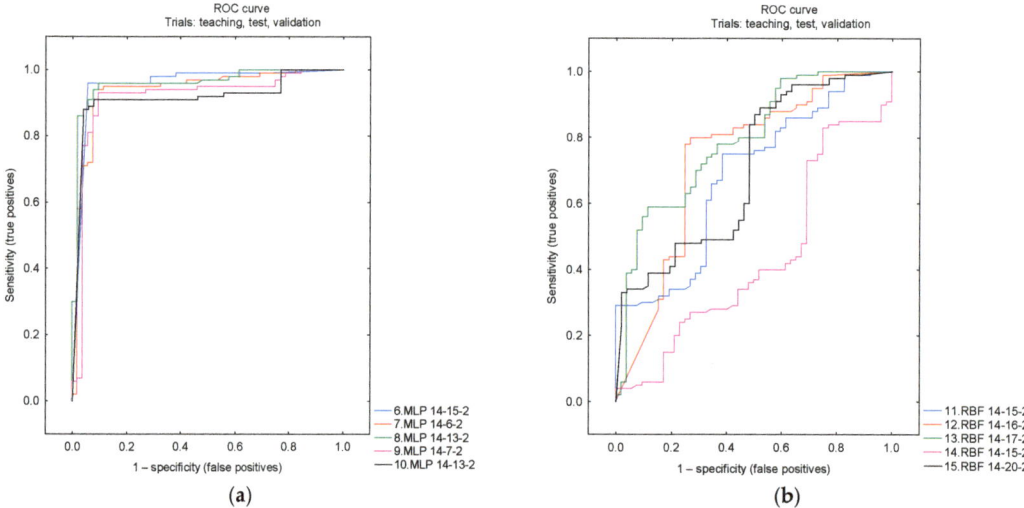

Figure 8. Receiver operating characteristics curve (ROC) for five MLP (**a**) and RBF (**b**) networks receiver.

Determining the area under the ROC graph from the allegation is the area for the ideal model (the area of a square with side 1). The area under ROC (AUROC) denotes the probability that the predictive model under study is higher (the value of the correlated element of the positive class than the random element of the accepted class). Table 5 shows the areas and ROC thresholds for learning, testing, and validation for the top six networks, from which it can be seen that the additional common one is MLP 14-13-2 network two.

From the presented results of the experimental study of the nozzle needle displacement signals of the execution of classification using the ANN in the Statistica 13 environment, the suitability for the state recognition is evident. Tests were also carried out for more measurement data by increasing the cases with frequency domain analysis for the first three components. Tests were also carried out for other signals [17] in various analysis domains (including wavelet transforms) and using the MATLAB R2022a computer program, and similar satisfactory results were obtained. The classification learner was used to train models that classify data. It belongs to the group of supervised learning. This application allows to select features, examine data, train models, and determine validation and testing.

Table 5. ROC plots and thresholds for learning, testing, and validation for the top six networks.

Network ID	Area and ROC Threshold for Sampling: Teaching, Testing, Validation					
	MLP 14-5-2	MLP 14-13-2	MLP 14-11-2	MLP 14-6-2	MLP 14-8-2	MLP 14-7-2
1	0.932712	0.972050	0.966874	0.966874	0.965839	0.965321
2	0.530628	0.682748	0.551377	0.551377	0.434712	0.482753

Figure 9a shows the different models used in the classification and their corresponding validation accuracies in MATLAB 2022a. The highest validation accuracies were obtained from models 4.24 of the Subspace KNN (k-nearest neighbor) = 95.0%. The algorithm was also trained using standard hyperparameters, and optimization methods were used. Figure 9b shows the minimum classification error for model 5.

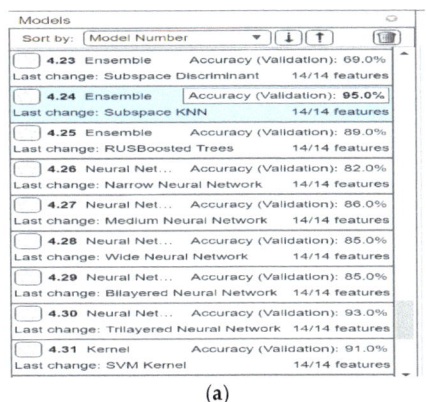
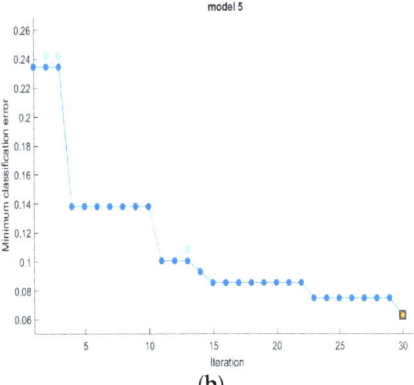

(a) (b)

Figure 9. The different models used in classification and their corresponding validation accuracies in MATLAB (**a**) and minimum classification error for model 5 (**b**).

The Statistica computer program in the automatic NN application has several options for learning MLP networks. These are Broyden–Fletcher–Goldfarb–Shanno (BFGS), coupled gradients, and the fastest gradient algorithm. The methods mentioned are optimization methods, and in this article, all MLP networks have the BFGS algorithm. Similar capabilities were in Matlab 2022a.

Figure 10a shows the classification learner matrix for the tree-layer NN. The matrix shows the percentage of observations and the proportion of correctly classified data (blue diagonal), and the cream-colored diagonal represents incorrectly classified data. Quality measures of a binary classifier are the "true positive rate" (*TPR*) to the "false positive rate" (*FPR*) [37,46]. Specificity (True Negative Rate) is the probability of detecting a passable state provided that there is one:

$$TNR = \frac{TN}{TN + EP} \tag{16}$$

where *TN* is negative cases the classifier detects as negative, and *FP* is negative cases falsely detected as positive. The false positive rate (false alarm) is the ratio of the number of cases falsely classified as an unfit condition to all cases:

$$FA = FPR = \frac{EP}{TN + EP} \tag{17}$$

Meanwhile, Figure 10b shows the *ROC* curve, which gives the ratio of correctly classified observations to incorrect ones for different classification thresholds.

The *TPR* (true positive rates) matrix represents the percentage of correctly classified observations for fitness class 1, and the *FNR* (false negative rates) represents the percentage of incorrectly classified observations for abnormal class 0. Running multiple variants of the tests for Statistica 13 and MATLAB R2022a programs and other signal analysis domains will help select the optimal solution.

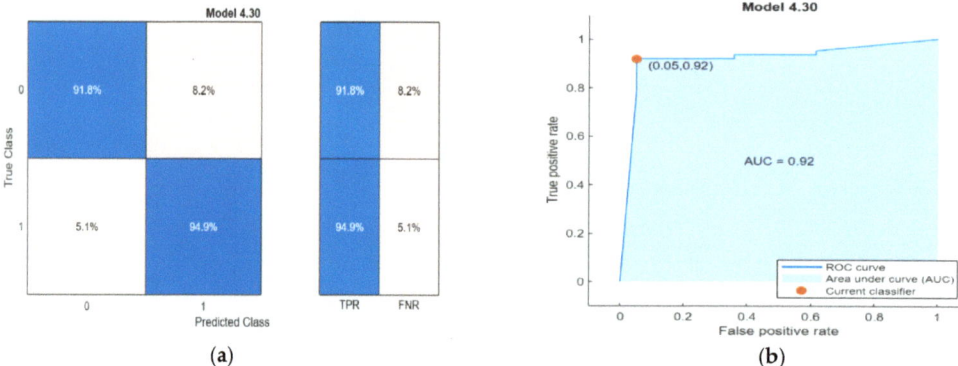

Figure 10. Example classification learner (**a**) and classification matrix *ROC* curve 10 (**b**) for the 4.30 model of the three-layer NN for parameters in the amplitude and frequency domain in MATLAB software.

5. Discussion

Some authors have studied the effect of operating conditions of the facility, and here, data have been collected over the entire range of variation in the parameters determining the conditions.

Successful results encourage further research using pressure signals in the fuel injection subsystems, in-cylinders of marine ICEs, and speed and vibration acceleration. In addition to analyses in the time, amplitude, and frequency domains, further time–frequency analyses, wavelet transforms, and other advanced analyses of diagnostic signals are being conducted.

The authors' contribution is also the execution of the research in the broad area of planned experimental studies. The diagnostic parameters used in the amplitude and frequency domain were related to the conducted measurements of the geometric and flow features of the technical state. Thus, deterministic models were obtained. Classifications carried out using NNs and two computer programs confirmed the nonrandomness of good validation and testing qualities.

Future research should also use diagnostic parameters recorded on marine ICEs under operating conditions. Operational tests have been carried out under natural conditions on more than 100 marine engines driving generators under marine ship conditions using vibration displacement signals and other working and residual processes.

6. Conclusions

NNs help identify typical faults in diagnosing marine engine injectors using displacement vibration signals. The study analyzed 10,000 types of networks, and MLP networks were selected from which the best one can be identified. Satisfactory quality of learning, validation, and testing by close and very close relationships with the values of diagnostic parameters in the studied ranges of variability were obtained. Test quality reached more than 90% using selected NNs of Statistica and MATLAB environments, and for the MLP networks, validation quality was 100%.

The experimental data of learning, validation, and testing sets in manual and random ways influenced the results. Higher qualities were obtained for random selection. The contribution of the present study to state of the art is the inclusion of several properties of the state features measured experimentally for model and randomly worn nozzles.

In further work, efforts should be made to reduce errors in classifying technical states by selecting insensitive data and matching hyperparameters when learning NNs.

7. Patents

Monieta, J. Method and device for diagnosing injectors. Patent of the Patent Office of the Republic of Poland WUP 09/08 2008, No. 199362B1, pp. 1–6.

Author Contributions: Conceptualization, J.M. and L.K.; methodology, J.M.; software, J.M. and L.K.; validation, J.M. and L.K.; formal analysis, J.M. and L.K.; investigation, J.M.; resources, J.M.; data curation, J.M.; writing—original draft preparation, J.M. and L.K.; writing—review and editing, J.M.; visualization, J.M.; supervision, J.M.; project administration, J.M.; funding acquisition, J.M. and L.K. All authors have read and agreed to the published version of the manuscript.

Funding: Ministry of Science and Higher Education of Poland. Award Number: 1/S/IESO/2014.

Data Availability Statement: Not applicable.

Acknowledgments: The authors thank the H. Cegielski Metal Industry Plant in Poznań for manufacturing research injector nozzles with graded technical states.

Conflicts of Interest: The authors declare no conflict of interest.

Nomenclature

ANN	artificial neural networks
B	fuel injection dose
d	diameter of the pinholes
f	frequency
F	force
h	needle lift
h_{aver}	average value of needle lift
H_i	values of the amplitudes of the individual components
h_{max}	the maximum needle lift
h_p	peak value of needle lift
h_{rms}	root-mean-square values of needle lift
I	number of patented holes
ICE	internal combustion engine
L_a	clearance in the guiding part of the body and the nozzle needle
ML	machine learning
MLP	multilayer perceptron
n	camshaft rotational speed
NN	neural network
N	fuel setting
p	pressure
p_f	injection pump supply pressure
P_o	opening pressure of the injector
SIE	self-ignition engine
S_t	technical state
t_p	fuel temperature at the inlet
a	needle cone angle
τ	time

References

1. Monieta, J. Selection of diagnostic symptoms and injection subsystems of marine reciprocating internal combustion engines. *Appl. Sci.* **2019**, *9*, 1540. [CrossRef]
2. Klimkiewicz, M. Application of neural networks in the diagnostics of the fuel apparatus of compression-ignition engines. *Agricult. Eng.* **2005**, *8*, 153–160.
3. Żółtowski, B.; Cempel, C. Elements of the Theory of Technical Diagnostics. In *Engineering of Diagnostics Machines*; Society of Technical Diagnostics: Bydgoszcz, Poland, 2004; pp. 17–525.
4. Marine Engine IMO Tier II and Tier III Programme. Engineering for Future—Since 1758. MAN Diesel & Turbo. 2015. Available online: https://www.man-es.com/marine/products/planning-tools-and-downloads/marine-engine-programme (accessed on 16 January 2023).

5. Monieta, J. Application of image color analysis for the assessment of injector nozzle deposits in internal combustion engines. *SAE Int. J. Fuels Lubr.* **2022**, *2*, 1–12. [CrossRef]
6. Monieta, J.; Łukomski, M. Methods and means of estimation of technical state features of the marine diesel engines injector nozzles type Sulzer 6AL20/24. *Scient. J. Mar. Univ. Szczecin* **2005**, *5*, 383–392. Available online: https://wydawnictwo_am_szczecin_plzeszyty-naukowepobierzcategory19-zn-5-77-omiuo-2005 (accessed on 15 March 2023).
7. Klimkiewicz, M. Neural model for localization of damage to injection pumps. *Agricult. Eng.* **2008**, *1*, 159–164.
8. Pantelelis, N.G.; Kanarachos, A.E.; Gotzias, N.D.; Papandreou, N.; Gu, F. Combining vibrations and acoustics for the fault detection of marine diesel engines using neural networks and wavelets. In Proceedings of the 14th International Congress on Condition Monitoring and Diagnostic Engineering Management (COMADEM 2001), Manchester, UK, 4–6 September 2001; pp. 649–656.
9. Haukin, S. *Neural Networks, A Comprehensive Foundation*; Macmillan College Publishing Company: New York, NY, USA, 2000.
10. Kecman, V. *Support Vector Machines, Neural Networks and Fuzzy Logic Models*; MIT Press: Cambridge, MA, USA, 2001.
11. Osowski, S. *Neural Networks for Information Processing*; Publishing House of the Warsaw University of Technology: Warsaw, Poland, 2006.
12. Schoellkopf, B.S.; Smola, A.J. *Learning with Kernels*; MIT Press: Cambridge, MA, USA, 2002.
13. Rutkowski, L. *Methods and Techniques of Artificial Intelligence*; National Scientific Publishing House: Warsaw, Poland, 2012.
14. Brzeżański, M.; Golomb, P. Application of the neural network method for the analysis of particulate emissions in the exhaust of CI and SI engines. *Combust. Eng.* **2009**, *SC1*, 331–337.
15. Jankowski, M.; Kwidziński, M. *Application of Neural Network for Automatic Classification of Injection Apparatus*; Congress of Technical Diagnostics: Gdansk, Poland, 1996; pp. 312–318.
16. Lewicki, P.; Hill, T. *Statistics Methods and Applications Book*; StatSoft: Tulsa, OK, USA, 2011.
17. Monieta, J.; Choromański, T. The application of artificial intelligence to identification of typical damages of marine diesel engines injectors. *Diagnostyka* **2008**, *4*, 139–144.
18. Sitnik, L. Neural networks in diesel engine control. In Proceedings of the PTNSS Congress—2005, The Development of Combustion Engines, Szczyrk, Poland, 25–28 September 2005; pp. 1–15.
19. Cheliotis, M.; Lazakis, I.; Cheliotis, A. Bayesian and machine learning-based fault detection and diagnostics for marine applications. *Offshor. Struct.* **2022**, *1*, 1–13. [CrossRef]
20. Raptodimos, Y.; Lazakis, I. Using artificial neural network-self-organising map for data clustering of marine engine condition monitoring applications. *Ship. Offshor. Struct.* **2018**, *13*, 649–656. [CrossRef]
21. Wang, R.H.; Chen, H.; Guan, C. Random convolutional neural network structure: An intelligent health monitoring scheme for diesel engines. *Measurement* **2021**, *171*, 108786. [CrossRef]
22. Xu, X.; Zao, Z.; Xu, X.; Yang, J.; Chang, L.; Yan, X.; Wang, G. Machine learning-based wear fault diagnosis for marine diesel engine by fusing multiple data-driven models. *Knowl. Based Syst.* **2020**, *190*, 105324. [CrossRef]
23. Zhang, J.; Zhang, D.; Shen, G.; Yang, J. Research on fault diagnosis of marine diesel engine based on probabilistic neural network optimized by quantum genetic algorithm. In Proceedings of the 2021 IEEE 4th International Conference on Automation, Electronics and Electrical Engineering (AUTEEE), Shenyang, China, 19–21 November 2021; pp. 729–733.
24. Abramek, K.; Osipowicz, T.; Mozga, Ł. The use of neural network algorithms for modeling injection doses of modern fuel injectors. *Combust. Eng.* **2021**, *185*, 10–14. [CrossRef]
25. He, Y.; Rutland, C.J. Application of artificial neural networks in engine modeling. *Int. J. Eng. Res.* **2004**, *5*, 281–296. [CrossRef]
26. Pawletko, R. The Use of neural networks for the faults classification of a marine diesel engine fuel injection system. *Glob. J. Eng. Educ.* **2005**, *9*, 143–148.
27. Pawletko, R. Using a neural network to diagnose selected malfunctions of a marine engine. *Diagnostyka* **2002**, *27*, 43–47.
28. Brzozowski, K.; Nowakowski, J. Application of artificial neural network to identify the working cycle model of compression ignition engine. In Proceedings of the First International Congress on Combustion Engines. PTNSS Congress—2005. The Development of Combustion Engines, Szczyrk, Poland, 25–28 September 2005; pp. 1–9.
29. Logan, K.; Inozu, B.; Roy, P.; Hetet, J.F.; Chesse, P.; Tauzia, X. Real-time marine diesel engine simulation for fault diagnosis. *Mar. Technol. Snam. News* **2002**, *39*, 21–28. [CrossRef]
30. Mesbahi, E. An intelligent sensor validation and fault diagnostic technique for diesel engines. *J. Dyn. Sys. Measur. Contr. Trans. ASME* **2001**, *123*, 141–144. [CrossRef]
31. Ramteke, S.M.; Chelladurai, H.; Amarnath, M. Diagnosis of liner scuffing fault of a diesel engine via vibration and acoustic emission analysis. *J. Vib. Eng. Technol.* **2019**, *8*, 815–833. [CrossRef]
32. Wang, R.H.; Chen, H.; Guan, C. A Bayesian inference-based approach for performance prognostics towards uncertainty quantification and its applications on the marine diesel engine. *ISA Trans.* **2021**, *118*, 159–173. [CrossRef]
33. Garczynska, I.; Tomczak, A.; Stepien, G.; Kasyk, L.; Slaczka, W.; Kogut, T. Applicability of machine learning for vessel dimension survey with a minimum number of common points. *Appl. Sci.* **2022**, *12*, 3453. [CrossRef]
34. Zhu, J.Y. Marine Diesel engine condition monitoring by use of bp neural network. In Proceedings of the International Multi Conference of Engineers and Computer Scientists Vol II IMECS 2009, Hong Kong, 18–20 March 2009; pp. 1–4.
35. Klinkmann, J.E. Wärtsilä Asset Diagnostic Configurator. Bachelor's Thesis, Novia University of Applied Science, Vaasa, Finland, 2019.

36. Wang, R.H.; Chen, H.; Guan, C. DPGCN model: A novel fault diagnosis method for marine diesel engines based on imbalanced datasets. *IEEE Trans. Instrum. Meas.* **2023**, *72*, 3504011. [CrossRef]
37. Shahid, S.M.; Ko, S.; Kwon, S. Real-Time Classification of Diesel Marine Engine Loads Using Machine Learning. *Sensors* **2019**, *19*, 3172. [CrossRef] [PubMed]
38. Cheliotis, M.; Lazakis, I.; Theotokatos, G. Machine learning and data-driven fault detection for ship systems operations. *Ocean Eng.* **2020**, *216*, 107968. [CrossRef]
39. Monieta, J. Diagnosing cracks in the injector nozzles of marine internal combustion engines during operation using vibration symptoms. *Appl. Sci.* **2023**, *13*, 9599. [CrossRef]
40. Kowalski, J.; Krawczyk, B.; Woźniak, M. Fault diagnosis of marine 4-stroke diesel engines using a one-vs-one extreme learning ensemble. *Eng. Appl. Artif. Intell.* **2017**, *57*, 134–141. [CrossRef]
41. Ramteke, S.M.; Chelladurai, H.; Amarnath, M. Diagnosis and Classification of Diesel Engine Components Faults Using Time-Frequency and Machine Learning Approach. *J. Vib. Eng.* **2022**, *10*, 175–192. [CrossRef]
42. Adamkiewicz, A.; Nikonczuk, P. An attempt at applying machine learning in diagnosing marine ship engine turbochargers. *Mainten. Reliab.* **2022**, *24*, 795–804. [CrossRef]
43. Zhao, H.; Mao, Z.; Chen, K.; Jiangn, Z. An intelligent fault diagnosis method for a diesel engine valve based on improved wavelet packet-Mel frequency and convolutional neural network. In Proceedings of the International Conference on Sensing, Diagnostics, Prognostics, and Control (SDPC) 2019, Beijing, China, 15–17 August 2019; pp. 354–359. [CrossRef]
44. Monieta, J. Method and a device for testing the friction force in precision pairs of injection apparatus of the self-ignition engines. *Energies* **2022**, *15*, 6898. [CrossRef]
45. Monieta, J.; Kasyk, L. Optimization of design and technology of injector nozzles in terms of minimizing energy losses on friction in compression ignition engines. *Appl. Sci.* **2021**, *11*, 7341. [CrossRef]
46. StatSoft, Inc. *Electronic Statistic Textbook*; StatSoft: Tulsa, OK, USA, 2013; pp. 1984–2011. ISBN 13: 9781884233593.
47. Tadeusiewicz, R.; Szaleniec, M. *Lexicon of Neural Networks*; Publisher of the Project Science Foundation: Wrocław, Poland, 2015.
48. Çelebi, K.; Uludamar, E.; Tosun, E.; Yıldızhan, Ş.; Aydın, K.; Özcanlı, M. Experimental and artificial neural network approach of noise and vibration characteristic of an unmodified diesel engine fuelled with conventional diesel, and biodiesel blends with natural gas addition. *Fuel* **2017**, *197*, 159–173. [CrossRef]

Disclaimer/Publisher's Note: The statements, opinions and data contained in all publications are solely those of the individual author(s) and contributor(s) and not of MDPI and/or the editor(s). MDPI and/or the editor(s) disclaim responsibility for any injury to people or property resulting from any ideas, methods, instructions or products referred to in the content.

Brief Report

On-Board Fuel Consumption Meter Field Testing Results

Peter Tapak [1,*], Michal Kocur [1] and Juraj Matej [2]

[1] Faculty of Electrical Engineering and Information Technology, Slovak University of Technology in Bratislava, Ilkovicova 3, 812 19 Bratislava, Slovakia; michal.kocur@stuba.sk

[2] TESTEK, a.s., Authorized Technical Service for Technical Inspections of Vehicles, P.O. Box 42, 820 17 Bratislava, Slovakia; juraj.matej@testek.sk

* Correspondence: peter.tapak@stuba.sk

Abstract: This paper aims to investigate and compare the fuel consumption data obtained from on-board fuel consumption meters (OBFCMs) from approximately 1000 vehicles through field testing. Furthermore, this research aims not only to compare the OBFCM readings but also to juxtapose them against the fuel consumption specifications provided by the respective vehicle manufacturers. To collect data, a cost-effective on-board diagnostics (OBD) reader and a user-friendly mobile app were employed, providing an accessible and efficient method for fuel consumption analysis. Field testing involved a diverse range of vehicles, covering various makes, models, and years of production. The OBCFM readings were recorded over a 9-month period, probably capturing a wide range of driving conditions and patterns. In order to ensure the reliability of the OBCFM readings, the fuel consumption measurements obtained from the manufacturers specifications were utilized as a reference benchmark. Preliminary data analysis indicates that there are noticeable variations in the fuel consumption data obtained from the OBCFM and the manufacturer specifications. These differences can be attributed to various factors. The novelty of the presented data lies in using a new feature implemented in EU cars since 2019. The study capitalizes on this feature, allowing for the collection of data from a broad spectrum of vehicles throughout the country under genuine driving conditions.

Keywords: on-board fuel consumption meter; on-board diagnostics; vehicle inspection; hybrid; fuel consumption gap

Citation: Tapak, P.; Kocur, M.; Matej, J. On-Board Fuel Consumption Meter Field Testing Results. *Energies* **2023**, *16*, 6861. https://doi.org/10.3390/en16196861

Academic Editors: Kazimierz Lejda, Artur Jaworski and Maksymilian Mądziel

Received: 22 August 2023
Revised: 21 September 2023
Accepted: 25 September 2023
Published: 28 September 2023

Copyright: © 2023 by the authors. Licensee MDPI, Basel, Switzerland. This article is an open access article distributed under the terms and conditions of the Creative Commons Attribution (CC BY) license (https:// creativecommons.org/licenses/by/ 4.0/).

1. Introduction

The transportation sector witnessed a substantial growth rate in recent decades and represented approximately 29% of global final energy consumption in 2019 according to [1], stating also that among all transportation modes, road transportation dominates as the most used, representing approximately three-quarters of both traffic demand and carbon emissions. The accurate measurement and monitoring of fuel consumption in vehicles are crucial for various purposes, including energy efficiency assessment, emission control, and policy formulation. On-board fuel consumption meters have emerged as a promising technology for providing real-time fuel consumption data, enabling drivers, researchers, and policymakers to make informed decisions via regulation [2] and regulation implementation [3]. This article aims to contribute to the existing body of knowledge and promote the adoption of OBFCMs as an effective tool for sustainable transportation.

Transportation accounts for a significant portion of energy consumption and carbon emissions in Europe, and requires the development and implementation of effective measures to improve energy efficiency and reduce environmental impacts. Fuel consumption is a key factor in determining the energy efficiency of vehicles, and the accurate measurement and monitoring of this parameter are essential to assess the performance of different vehicle types and optimize energy consumption. OBFCMs are devices integrated into vehicles that enable the reading of vehicle fuel-consumption-related data via OBD; the particular values are described in Section 2.

The European Union (EU) has recognized the importance of monitoring fuel consumption (FC) and has implemented regulatory directives to standardize the use of OBFCMs. The EU Transport in Figures Statistical Pocketbook [4] highlights the significance of fuel consumption reduction in achieving sustainable transportation goals. Regulation [5] emphasizes the approval and market surveillance of motor vehicles, including the implementation of systems such as OBFCMs. Starting in 2021, all newly registered passenger cars within the EU are required to have an on-board fuel and/or energy consumption monitoring (OBFCM) device installed. This device records the amount of fuel or energy consumed by the vehicle, as outlined in a regulation from 2019. According to Regulation 2021/392 (established in 2021), manufacturers are obligated to gather real-world data for new vehicles registered after 1 January 2021. These data can be obtained through direct transfers from the vehicle or during service and repair visits.

Since May 2023, EU member states have been mandated to collect real-world fuel and energy consumption data when vehicles undergo roadworthiness tests, known as periodic technical inspections (PTIs). These data must then be reported to the European Commission on a yearly basis [3].

Accuracy is one of the crucial factors in the measurement area. The study [6] analyzed the precision of the measurement of OBFCM fuel consumption simulating real-world conditions. It shows that, for most vehicles, OBFCM fuel consumption was measured with accuracy ±5% (on-road and lab) and that, for most vehicles, OBFCM distance was measured with accuracy ±1.5% (on-road and lab). It is important to note that the accuracy improves for monitoring over longer periods. The data in this paper are obtained only from vehicles from M1 category, for which the accuracy in measuring the fuel consumed is also set by regulation to ±5%.

One can find studies that compare real fuel consumption with the real-world fuel consumption, such as [7]. One of the primary challenges concerning exhaust emissions and energy consumption tests for cars relates to the discrepancies between laboratory-type approval tests and real road tests [8–10]. Typically, the results obtained on the road [11,12] exceed those from laboratory tests [13–16].

Several factors contribute to these discrepancies, including variations in driving cycles [17–21] and traffic resistance encountered during real driving [22–25]. Additionally, the car's performance in urban traffic conditions is heavily influenced by the road infrastructure, such as the type of intersections [26,27], energy recovery possibilities during braking [28], and the driver's driving style [29,30].

OBD, short for on-board diagnostics, serves as an integrated self-diagnostic system integrated within vehicles. The system includes an OBD-II socket and a 16-pin connector typically located near the driver's wheel, under the dashboard. Its origin traces back to California, where the California Air Resources Board (CARB) enforced stringent emission control regulations and raised awareness regarding emissions control.

Following CARB directives, OBD became mandatory in all cars manufactured after 1991, ensuring the monitoring and regulation of harmful gas emissions [31]. The significance of OBD-II in the automotive industry lies in its ability to log crucial vehicle parameters, allowing for the analysis and diagnostics of various vehicle systems. Additionally, it facilitates comprehensive fleet management and route optimization [31].

OBD-II, short for on-board diagnostics version 2, operates on the CAN (controller area network) protocol. This protocol allows vehicles to connect with OBD scanners and scanning tools externally. These scanners, which can be connected to PCs or laptops, provide users with access to valuable information from the vehicle's onboard computer. This information includes driving speeds, engine speeds, coolant levels, coolant temperature, emission control, idle time of the engine, and other essential engine-related data [32].

As mentioned in the introduction, according to [3], the OBFCM data must be read from vehicles as part of PTI. In 1935, the Slovak Republic, which was a part of Czechoslovakia at the time, made its initial effort to implement mandatory periodic technical inspections (PTIs). Today, nearly all vehicles in Slovakia are obligated to undergo these inspections. The article primarily focuses on category M1 vehicles, which are passenger cars with four wheels. Such vehicles must undergo the PTI within four years from their first registration and subsequently every two years thereafter.

However, certain vehicles have specific requirements for PTI. Those used for medical emergency services or mining rescue services, equipped with priority driving rights, along with ambulance vehicles, vehicles used for gas facility maintenance, and vehicles used for taxi services are required to undergo PTI annually. Therefore, taking into account also the date from which the OBFCM device became mandatory in the new cars, most of the vehicle data presented in this paper are from the group of vehicles required to undergo PTI annually.

It is worth noting that PTI intervals are similar among the member states of the European Union, and Slovakia follows this trend.

Furthermore, in Slovakia, emission checks are carried out as a separate and distinct inspection procedure.

The article presents the results obtained by field testing the OBFCM reading feature of the mSTK mobile app available at [33]. The field testing started in May 2022. Nevertheless, this application has been utilized in PTIs in Slovak Republic for OBD reading since 2021; in other words, it has been mandatory to use this app for OBD reading at all PTIs in Slovak Republic since 2021. The OBFCM data are obtained from the car via the OBD port using a standard diagnostic session by an OBD request defined in [34], so implementing this feature was an expansion of the well-established feature of the already existing app. The OBFCM device is mandatory for cars in the EU with type approval from 2019. Therefore, there are not many cars that have this device ready for the PTI inspection; usually, these are cars that are being used as a taxi or emergency vehicles, etc. However, over the 9-month period, more than 1000 cars with the OBFCM device have been successfully read at PTI service stations in the Slovak Republic. The goal of this paper is to present the data obtained during OBFCM reading field tests in an engaging and informative way for the reader, such as fuel consumption and reading success, while adhering to all applicable data protection laws.

In Section 2, we will present how and which OBFCM data were obtained in this study, and we will also provide information on the dataset. In Section 3, the gathered data are presented in several ways. In Section 4, the statistical properties of the data are analyzed and extreme values are linked to particular vehicles. Also, the fuel consumption gap is analyzed for hybrid vehicles and internal combustion vehicles. The yearly statistics are provided regarding OBFCM reading success rate. In Section 5, the results are summarized.

2. Materials and Methods

The OBD reading during the PTIs in Slovakia is made by using a smartphone app and a low-end OBD reader. The utilization of this process, together with its benefits and results, was described in [35]. The mSTK Android mobile application was developed in Java programming language in partnership with TESTEK a.s., a joint stock company providing vehicle inspection services for the Slovak Ministry of Transportation.

The data required by the European Commission via [3] to be collected and submitted starting on 20 May 2023 are the following :

- Total distance traveled (lifetime);
- Total fuel consumed (lifetime);
- Total distance traveled in charge-depleting operation with engine off (lifetime);
- Total distance traveled in charge-depleting operation with engine running (lifetime);
- Total distance traveled in driver-selectable charge-increasing operation (lifetime);
- Total fuel consumed in charge-depleting operation (lifetime);
- Total fuel consumed in driver-selectable charge-increasing operation (lifetime);
- Total grid energy into the battery (lifetime).

Among the other features described in [35], the application reads the OBFCM data via the ELM327 OBD reader [36], providing a very cost-effective solution.

The major benefit lies in using this cost-effective yet straightforward, and already proven to be reliable, technology. The OBFCM reading is an expansion of the already existing application by new OBD requests for the OBFCM reading. The only barrier is that it does not use freely available OBD requests. However, the OBD requests for OBFCM are standardized in [34], so the reading process is the same for each vehicle that has an OBFCM device implemented. After the successful read, the data are sent to national database. Every vehicle of the M_1 category has been requested for OBFCM data during PTIs during the 9-month field testing period [37].

The necessary OBFCM data are regarded as personal information, thus affording the vehicle owner the choice to opt out of sharing OBFCM readings. Additionally, a switch is incorporated within the mobile application's reading activity screen, allowing PTI inspectors to deactivate this function while still retrieving other essential OBD data like diagnostic trouble codes. This design ensures that owners have control over their data privacy while enabling the collection of crucial vehicle information. The mobile application refrains from retaining any data on a local basis; instead, it promptly transmits the information to the national PTI database. The OBFCM data are not shown on the display to the inspector either. This approach serves as a safeguard, ensuring that the personal data of car owners remain secure.

3. Results

Figure 1 presents chronological readings in three graphs to give the reader an overview on how the data are distributed over time. Panel (a) displays fuel consumption data obtained through the OBFCM reading (depicted by the red line) along with the manufacturer's declared fuel consumption. Panel (b) illustrates the corresponding odometer values, also gathered from the OBFCM reading request. Lastly, panel (c) showcases the difference between the declared and measured fuel consumption in [L/100 km] (in blue) and relative to the one declared by the manufacturer in [%] (in red). Figure 2 represents the same data as Figure 1, but the data are sorted by the difference in fuel consumption, with the average value of 1.9466 [L/100 km]. The study [38] states that the average fuel consumption gap can be accurately estimated with a sample of less than 0.05% of the fleet. The total number of passenger vehicles registered in Slovakia is 2,627,983 according to [39], where 0.05% of total cars yields 1314 cars. The article presents a dataset containing a substantial total of 1100 car records. Although this figure falls 27% below the established threshold, it remains noteworthy and interesting to the readers. Although the size of the dataset does not reach the threshold, it is still substantial enough to pique the interest of those engaging with the study.

The total number of vehicles in this paper is 1100, and only 57 readings, approximately 5.2%, were below the manufacturer declared values. The highest gap was 14.6531 [L/100 km] and the lowest gap was −3.221 [L/100 km]. In terms of relative gap, the maximum was 966.1867%, while the minimum was −45.3823%. The average relative fuel consumption gap in the study was 35.0474%. Figure 3 in panel (a) shows the comparison of fuel consumption declared by the manufacturer vs. the fuel consumption obtained via the OBFCM. One can see the hybrid vehicles data mostly in the lower left corner, where almost all of them are above the linear approximation. This indicates that most of the hybrid vehicles presented have a higher fuel consumption than what is declared by the manufacturer. In panel (b) of Figure 3, one can find the relative fuel consumption difference over the manufacturer-declared fuel consumption. It is obvious that the presented hybrid cars have the most relative difference in real and declared fuel consumption.

The results of the study suggest that there is a significant amount of variation in the manufacturer's declared fuel consumption and the measured values. This is shown by the linear regression results presented in Figure 3a, which have a mean value of $\mu = 1.94656$ and a standard deviation of $\sigma = 1.5452$.

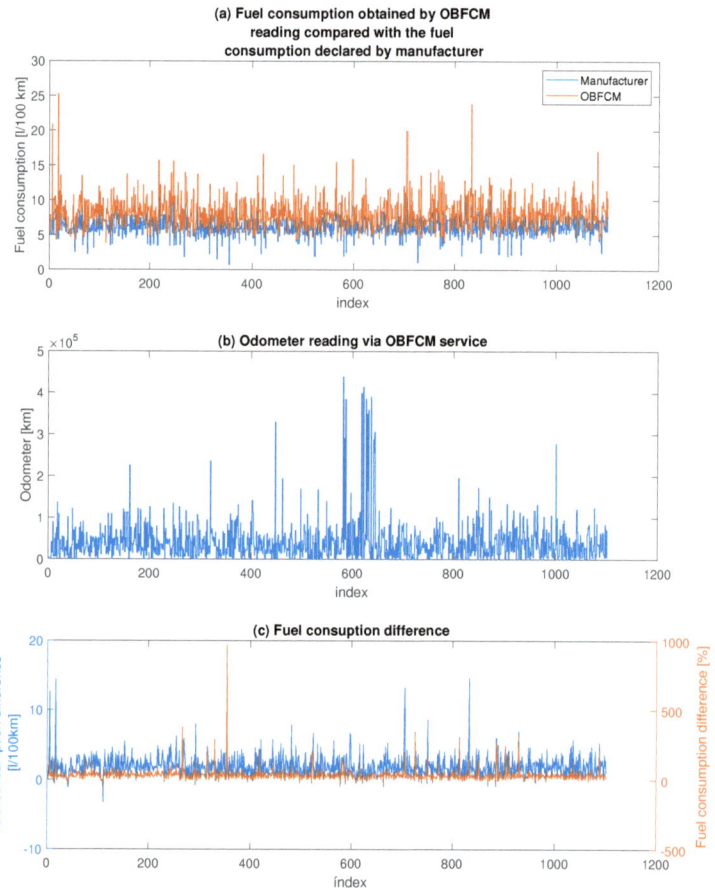

Figure 1. Fuel consumption obtained by OBFCM reading compared with the fuel consumption declared by manufacturer displayed in chronological order: (**a**) fuel consumption comparison, (**b**) odometer reading, (**c**) fuel consumption difference.

Figure 4 in panel (a) shows a fuel difference histogram fitted by normal distribution with mean value $\mu = 1.94656$ and standard deviation $\sigma = 1.5452$. Panel (b) of Figure 4 presents a difference histogram fitted by logistic distribution with mean $\mu = 1.81604$ and scale parameter $\sigma = 0.773935$. The bottom part of Figure 4 in panel (c) shows a relative fuel difference histogram fitted by normal distribution with mean value $\mu = 35.0474$ and standard deviation $\sigma = 45.0191$, and panel (d) presents a relative fuel consumption difference histogram fitted by logistic distribution with mean $\mu = 30.3732$ and scale parameter $\sigma = 14.758$.

Only 24 of the data presented were from vehicles that had hybrid propulsion. This corresponds to the low penetration of electric cars in the Slovak Republic. The mean gap in fuel consumption with the declared value of the manufacturers of these hybrid vehicles was 4.1261 [L/100 km]. However, only one hybrid vehicle recorded a lower fuel consumption than the manufacturer declared, as shown in the histogram in Figure 5.

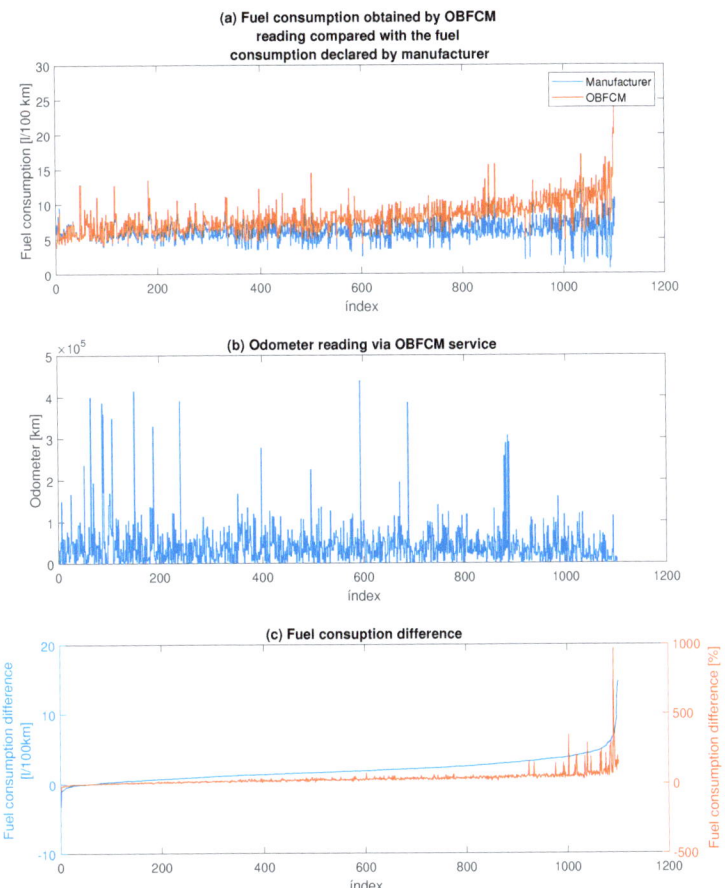

Figure 2. Fuel consumption obtained by OBFCM reading compared with the fuel consumption declared by manufacturer sorted by fuel consumption difference: (**a**) fuel consumption comparison, (**b**) odometer reading, (**c**) fuel consumption difference.

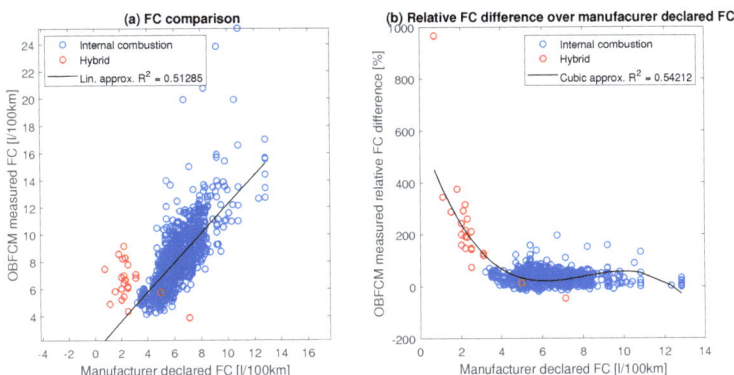

Figure 3. Fuel consumption (FC) comparison: (**a**) manufacturer-declared FC vs. OBFCM-measured FC, (**b**) relative FC difference obtained via OBFCM data over manufacturer-declared FC.

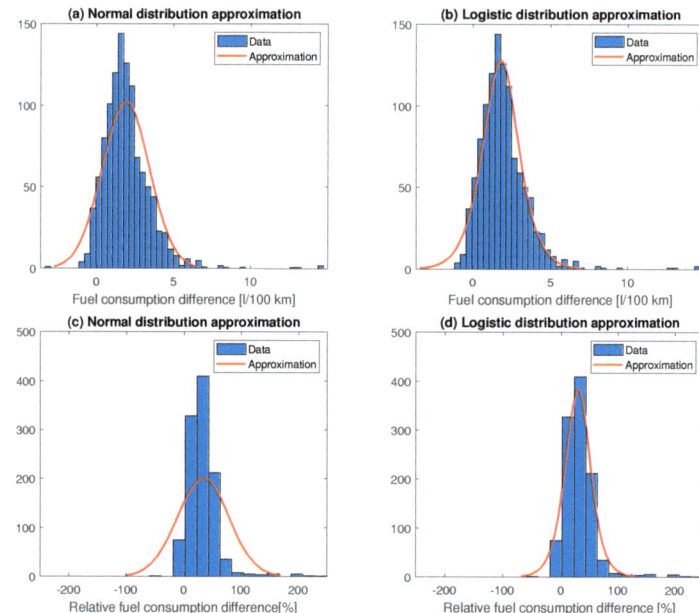

Figure 4. Fuel consumption difference histogram approximations: (**a**) normal distribution, (**b**) logistic distribution; relative fuel consumption difference histogram approximations: (**c**) normal distribution, (**d**) logistic distribution.

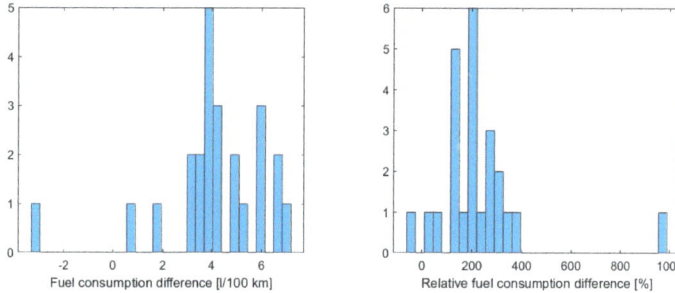

Figure 5. Fuel consumption gap histogram—hybrid vehicles.

4. Discussion

The highest fuel consumption gap amongst hybrid vehicles was 6.945 [L/100 km], which is about 50% of the average difference in all vehicles group. However, the mean fuel consumption difference amongst hybrids corresponds to 215% of the average of the all-vehicles group's mean fuel consumption difference. Table 1 summarizes the comparison of these two groups of vehicles included in our study. The results of laboratory and road tests of vehicles can differ significantly due to a number of factors, including the driving cycle and road load. These parameters also affect the hybrid drive control, which can switch between the combustion engine and electric mode [40].

Most studies of vehicle emissions and fuel consumption use type approval cycles, such as the NEDC, WLTC, or FTP. These cycles are designed to be repeatable and fair, but they do not always reflect real-world driving conditions. For example, they often do not include stop-and-go traffic or high speeds.

As a result, the results of laboratory tests can be significantly different from the results of road tests. This is because the actual driving cycles on the road are unrepeatable and can vary depending on the driver, the traffic, and the weather [8].

Table 1. Fuel consumption difference comparison.

Propulsion		IC	Hybrid	All
Number of vehicles		1076	24	1100
Odometer [km]	Mean	41,848	47,190	41,965
Fuel consumption gap [L/100 km]	Min	−1.09	−3.22	−3.22
	Mean	1.90	4.13	1.95
	Max	14.65	6.95	14.65
	SD	1.49	2.16	1.54
Relative fuel consumption gap [%]	Min	−18.71	−45.38	−45.38
	Mean	30.77	227.00	35.05
	Max	197.30	966.19	966.19
	SD	22.21	185.67	45.0191

The maximal relative fuel gap from Table 1 corresponds to almost nine times the fuel consumption declared by the manufacturer. The corresponding car was a plug-in hybrid sport utility vehicle (SUV) with an approximately 2.0 L diesel engine. The manufacturer-declared fuel consumption is 0.7 L/100 km. The odometer of the car reads 29,467.5 km. The data read from the OBFCM of this particular vehicle are in Table 2.

Table 2. OBFCM data (lifetime)—highest fuel consumption—hybrid.

Total distance traveled	29,467.5 km
Total fuel consumed	2199.25 L
Total distance traveled in charge-depleting operation with engine off	5021.4 km
Total distance traveled in charge-depleting operation with engine running	2794.9 km
Total distance traveled in driver-selectable charge-increasing operation	331.4 km
Total fuel consumed in charge-depleting operation	289.11 L
Total fuel consumed in driver-selectable charge-increasing operation	39.65 L
Total grid energy into the battery	1245.5 kWh

The only hybrid car with a negative relative fuel difference corresponding to almost one half of fuel consumption declared by the manufacturer was a hybrid small family estate with an approximately 1.6 L petrol engine. The manufacturer-declared fuel consumption is 7.1 L/100 km. The data read from the OBFCM of this particular vehicle are in Table 3.

Table 3. OBFCM data (lifetime)—lowest fuel consumption difference—hybrid.

Total distance traveled	31,872.0 km
Total fuel consumed	1235.95 L
Total distance traveled in charge-depleting operation with engine off	15,395.0 km
Total distance traveled in charge-depleting operation with engine running	649.0 km
Total distance traveled in driver-selectable charge-increasing operation	2546.0 km
Total fuel consumed in charge-depleting operation	218.28 L
Total fuel consumed in driver-selectable charge-increasing operation	856.54 L
Total grid energy into the battery	3546.30 kWh

The car with highest fuel consumption gap in this paper is a 3.0 L petrol sporty compact saloon, with a declared fuel consumption of 9.2 L/100 km. The data read from the OBFCM of this particular vehicle are in Table 4.

Table 4. OBFCM data (lifetime)—highest fuel consumption gap.

Total distance traveled	11,190.0 km
Total fuel consumed	2669.16 L

As can be seen, the data contain several cars with an extremely high relative fuel consumption gap. In Figure 6, the fuel gap is represented without vehicles, where the relative fuel gap is outside the range of $[-99; 99]\%$. Removing these 33 vehicles also shifts the overall statistical values presented in Table 5, showing a 0.1357 L/100 km decrease in the mean fuel consumption gap, corresponding to -5.4855%; in other words, the new mean relative fuel consumption gap in the dataset is 29.5619%.

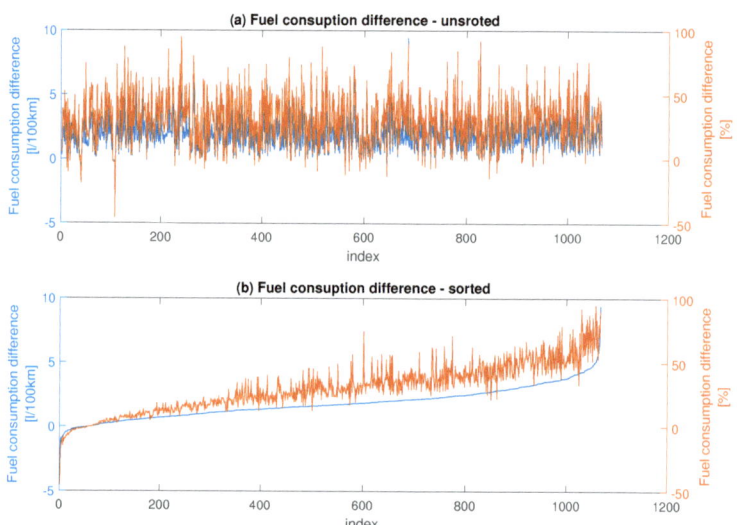

Figure 6. Fuel consumption obtained by OBFCM reading compared with the fuel consumption declared by manufacturer without marginal cases: (**a**) sorted chronologically, (**b**) sorted by fuel consumption difference.

Table 5. Fuel consumption (FC) gap comparison.

Dataset		Relative FC below 99%	All	Δ
Number of vehicles		1064	1100	−36
Odometer [km]	Mean	42,300	41,965	335
FC gap [L/100 km]	Min	−3.22	−3.22	0
	Mean	1.81	1.95	−0.14
	Max	9.42	14.65	−5.23
	SD	1.25	1.54	−0.29
Relative FC gap [%]	Min	−45.38	−45.38	0
	Mean	29.56	35.05	−5.49
	Max	94.78	966.19	−868.41
	SD	19.37	45.02	−25.65

Another interesting statistic for the reader can be the yearly percentage of successful OBFCM readings, which is represented in Table 6. It shows that some of the cars have been equipped with the OBFCM device much earlier than required by the authorities; however, the percentage of successful readings before 2014 is quite low, lower than 50%. The overall percentage of data validity is approximately 76%. In the remaining 24%, the data showed the fuel or energy consumption gap by several orders of magnitude; these data were not taken into account in the paper, and only 1100 records of 1434 were considered in the results presented. The resulting fuel consumption gap around 35% corresponds to similar research based on users' feedback data collection [41] or laboratory-measurement-based tests such as [42].

Table 6. OBFCM reading per year of registration.

Year of Registration	Number of OBFCM Readings	Number of Valid OBFCM Readings	% of Valid OBFCM Readings
2009	2	0	0.00
2012	13	6	46.15
2014	2	2	100.00
2015	3	3	100.00
2016	38	38	100.00
2017	47	43	91.49
2018	343	208	60.64
2019	410	296	72.20
2020	95	83	87.37
2021	208	182	87.50
2022	229	201	87.77
2023	44	30	68.18
SUM	1434	1100	76.71

In Table 7, the statistics are grouped by the displacement size of the engine. Still, in all groups, the OBFCM-obtained fuel consumption is more than 20% above the nominal one. Figure 7 presents the fuel consumption and relative fuel consumption difference represented by the groups from Table 7. One can see that the vehicles with the largest relative fuel consumption gap are with engine sizes [1000, 2000) ccm. The lowest mean relative fuel consumption gap can be found in the group of vehicles with engine sizes from [2000, 2500) ccm.

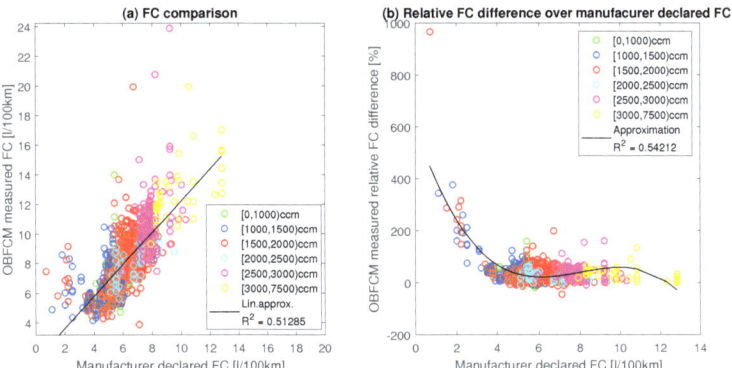

Figure 7. Fuel consumption (FC) comparison grouped by engine displacement: (**a**) manufacturer-declared FC vs. OBFCM-measured FC, (**b**) relative FC difference obtained via OBFCM data over manufacturer-declared FC.

Table 7. Fuel gap per engine displacement

Engine [ccm]	N	Odometer [km]	ΔFC [L/100 km]				Relative FC [%]			
			Min	Mean	Max	SD	Min	Mean	Max	SD
[0, 1000)	81	18,014.81	−1.09	1.55	8.58	1.35	−18.71	29.14	158.83	25.29
[1000, 1500)	331	32,132.12	−0.60	1.70	6.77	1.14	−10.13	38.12	376.08	42.81
[1500, 2000)	442	53,018.17	−3.22	1.89	13.22	1.50	−45.38	35.52	966.19	57.11
[2000, 2500)	80	44,480.50	−0.74	1.47	3.55	0.82	−9.17	22.00	64.50	13.71
[2500, 3000)	129	44,301.30	−0.84	2.94	14.65	1.97	−10.16	37.60	159.27	23.93
[3000, 7500)	37	36,743.84	−0.07	3.28	14.37	2.67	−0.52	34.26	133.08	26.08

In Table 8, the statistics are grouped by the year of registration. Vehicles from the interval [2009–2017] are grouped due to the low number of cars per year in that period, as can be seen from Table 6. However, in all groups, the fuel consumption obtained from the OBFCM is more than 30% above the nominal one. Figure 8 presents the fuel consumption and relative fuel consumption difference represented by the groups in Table 8. There is no significant correlation between the fuel consumption gap and the year of vehicle registration, with a correlation coefficient of −0.0393.

There is no significant correlation (correlation coefficient of −0.0393) between the fuel consumption gap and the year of vehicle registration.

Table 8. Fuel gap per years of registration.

Year	N	Odometer [km]	ΔFC [L/100 km]				Relative FC [%]			
			Min	Mean	Max	SD	Min	Mean	Max	SD
[2009–2017)	44	42,971.80	−0.19	1.74	4.62	1.09	−2.59	30.09	87.19	19.24
[2017–2018)	39	37,569.89	−0.74	1.85	6.26	1.45	−7.81	32.14	100.97	24.09
[2018–2019)	272	35,059.42	−3.22	2.15	14.65	1.72	−45.38	35.86	260.65	33.11
[2019–2020)	314	41,780.65	−0.46	1.96	13.22	1.51	−9.31	35.96	376.08	40.10
[2020–2021)	66	39,887.26	−0.88	2.19	9.42	1.67	−14.80	36.39	192.65	29.89
[2021–2022)	156	42,782.55	−1.09	1.72	14.37	1.66	−18.71	37.27	966.19	84.75
[2022–2023)	164	41,832.56	−0.60	1.81	6.70	1.28	−10.13	31.07	217.06	27.17
[2023–2024)	45	88,518.72	−0.54	1.88	6.13	1.31	−9.12	35.99	291.73	45.72

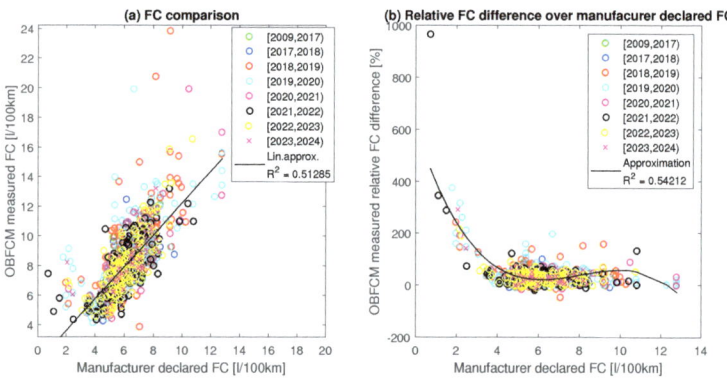

Figure 8. Fuel consumption (FC) comparison grouped by year of registrations: (a) manufacturer-declared FC vs. OBFCM-measured FC, (b) relative FC difference obtained via OBFCM data over manufacturer-declared FC.

The analysis of relevant data from 1100 vehicles shows the 35% gap between the fuel consumption declared by the manufacturers and the fuel consumption obtained by the OBFCM, which is similar to consumer-feedback-based studies [41,42].

The summary of comparisons with the fuel gap from other studies can be found in Table 9. The datasets denoted as TC2018, TC2019, TC2020, SM2018, and SM2019 are from a feedback-based study [38]. The dataset denoted as CH2016 is based on customer feedback in China from 2016. Detailed information about the values in the dataset denoted as E2020 can be found in the study [14], which is based on one-year measurements using a single vehicle with 20 drivers.

Table 9. Fuel gap comparison with other studies.

Dataset	Relative FC [%]	Difference with This Study [%]
TC2018	40.00	5.00
TC2019	36.00	1.00
SM2018	27.00	−8.00
SM2019	24.00	−11.00
SM2020	33.00	−2.00
CH2016	25.10	−9.90
E2020	29.00	−6.00

5. Conclusions

This paper highlights the results of an extensive field testing of OBFCM data collection from all PTI stations in the Slovak Republic, which took place from August 2022 to April 2023. Mobile applications with low-cost OBD readers were successfully implemented for this purpose.

One of the goals was to analyze the reading success rate of OBFCM devices, which presents the readiness of the devices. Approximately 1 million vehicles were tested at PTI stations during this period, and approximately 1500 had an OBFCM device. Of these, only 1100 provided consistent data, which were taken into account in the further data analysis.

The study presented a fuel consumption gap of 35%, which is similar to other studies discussed in Section 4. No correlation between the year of registration and fuel gap has been found in the data obtained by this study.

It is worth noting that the data primarily originated from vehicles with specific purposes, such as taxis and emergency vehicles, potentially influencing driving styles due to the specialized nature of these vehicles' operations. However, it is important to consider that all vehicles were subjected to genuine real-world conditions under the ownership of their respective drivers, with no synthetic testing employed. This brief report summarizes the findings of a study, but does not draw any definitive conclusions. Further research is needed to confirm the results.

Author Contributions: Conceptualization, P.T.; data curation, P.T. and J.M.; formal analysis, P.T., M.K. and J.M.; methodology, P.T., M.K. and J.M.; software, P.T. and M.K.; validation, P.T., M.K. and J.M.; writing—original draft, P.T.; writing—review and editing, P.T., M.K. and J.M. All authors have read and agreed to the published version of the manuscript.

Funding: This research was funded by the Slovak Grant Agency, grant KEGA 039STU-4-2021 and by Slovak Research and Development Agency, grant No. APVV-21-0125.

Data Availability Statement: The data presented in this study are available on request from the corresponding author. The data are not publicly available due to possible violation of personal data privacy.

Conflicts of Interest: The authors declare no conflict of interest.

Abbreviations

The following abbreviations are used in this manuscript:

OBD	On-board diagnostics
DTC	Diagnostic trouble code
PTI	Periodical technical inspection
OBFCM	On-board fuel consumption meter
CARB	California Air Resources Board
FC	Fuel consumption
EU	European union
SUV	Sport utility vehicle
SD	Standard deviation

References

1. Wu, R.; Zhang, S.; Yang, L.; Wu, X.; Yang, Z.; Zhang, X.; Zhang, D.; Huang, J.; Wu, Y. Corporate average fuel consumption evaluation and non-compliance disaggregation based on real-world data. *Appl. Energy* **2023**, *347*, 121353. [CrossRef]
2. Commission Regulation (EU) 2018/1832. 2018. Available online: https://eur-lex.europa.eu/legal-content/GA/TXT/?uri=CELEX%3A32018R1832 (accessed on 29 June 2023).
3. Implementing Regulation (EU) 2021/392—Data Protection and the CO_2 Data Transfer. 2021. Available online: https://eur-lex.europa.eu/legal-content/EN/TXT/PDF/?uri=CELEX:32021R0392&rid=1 (accessed on 29 June 2023).
4. EU Transport in Figures Statistical Pocketbook. 2019. Available online: https://transport.ec.europa.eu/facts-funding/studies-data/eu-transport-figures-statistical-pocketbook/statistical-pocketbook-2019_en (accessed on 29 June 2023).
5. Regulation (EU) 2018/858 of the European Parliament and of the Council of 30 May 2018. 2018. Available online: https://eur-lex.europa.eu/legal-content/EN/TXT/?uri=CELEX (accessed on 29 June 2023).
6. Pavlovic, J.; Fontaras, G.; Broekaert, S.; Ciuffo, B.; Ktistakis, M.; Grigoratos, T. How accurately can we measure vehicle fuel consumption in real world operation? *Transp. Res. Part Transp. Environ.* **2021**, *90*, 102666. [CrossRef]
7. SÁnchez, P.; Ndiaye, A.; Martín-Cejas, R. Plug-In Hybrid Electric Vehicles (PHEVS): A possible perverse effect generated by Environmental Policies. *Int. J. Transp. Dev. Integr.* **2019**, *3*, 259–270. [CrossRef]
8. Fontaras, G.; Zacharof, N.G.; Ciuffo, B. Fuel consumption and CO_2 emissions from passenger cars in Europe—Laboratory versus real-world emissions. *Prog. Energy Combust. Sci.* **2017**, *60*, 97–131. [CrossRef]
9. Merkisz, J.; Pielecha, J.; Bielaczyc, P.; Woodburn, J.; Szalek, A. A comparison of tailpipe gaseous emissions from the RDE and WLTP test procedures on a hybrid passenger car. Technical report, SAE Technical Paper. In Proceedings of the SAE Powertrains, Fuels & Lubricants Meeting, Krakow, Poland, 22–24 September 2020.
10. Riemersma, I.; Mock, P. *Too Low to Be True? How to Measure Fuel Consumption and CO_2 Emissions of Plug-In Hybrid Vehicles, Today and in the Future*; Technical Report; The International Council on Clean Transportation: Washington, DC, USA; 2017.
11. García-Contreras, R.; Soriano, J.A.; Fernández-Yáñez, P.; Sánchez-Rodríguez, L.; Mata, C.; Gómez, A.; Armas, O.; Cárdenas, M.D. Impact of regulated pollutant emissions of Euro 6d-Temp light-duty diesel vehicles under real driving conditions. *J. Clean. Prod.* **2021**, *286*, 124927. [CrossRef]
12. Triantafyllopoulos, G.; Katsaounis, D.; Karamitros, D.; Ntziachristos, L.; Samaras, Z. Experimental assessment of the potential to decrease diesel NOx emissions beyond minimum requirements for Euro 6 Real Drive Emissions (RDE) compliance. *Sci. Total Environ.* **2018**, *618*, 1400–1407. [CrossRef]
13. Tansini, A.; Pavlovic, J.; Fontaras, G. Quantifying the real-world CO_2 emissions and energy consumption of modern plug-in hybrid vehicles. *J. Clean. Prod.* **2022**, *362*, 132191. [CrossRef]
14. Pavlovic, J.; Fontaras, G.; Ktistakis, M.; Anagnostopoulos, K.; Komnos, D.; Ciuffo, B.; Clairotte, M.; Valverde, V. Understanding the origins and variability of the fuel consumption gap: Lessons learned from laboratory tests and a real-driving campaign. *Environ. Sci. Eur.* **2020**, *32*, 53. [CrossRef]
15. Lejda, K.; Jaworski, A.; Mądziel, M.; Balawender, K.; Ustrzycki, A.; Savostin-Kosiak, D. Assessment of Petrol and Natural Gas Vehicle Carbon Oxides Emissions in the Laboratory and On-Road Tests. *Energies* **2021**, *14*, 1631. [CrossRef]
16. Tietge, U.; Zacharof, N.; Mock, P.; Franco, V.; German, J.; Bandivadekar, A.; Lambrecht, U. *From Laboratory to Road: A 2015 Update of Official and "Real-World" Fuel Consumption and CO_2 Values for Passenger Cars in Europe*; ICCT: Washington, DC, USA, 2015.
17. Kadijk, G.; Verbeek, R.; Smokers, R.; Spreen, J.; Patuleia, A.; van Ras, M. *Supporting Analysis Regarding Test Procedure Flexibilities and Technology Deployment for Review of the Light Duty Vehicle CO_2 Regulations*; European Commission: Brussel, Belgium, 2012. [CrossRef]
18. Fontaras, G.; Ciuffo, B.; Zacharof, N.; Tsiakmakis, S.; Marotta, A.; Pavlovic, J.; Anagnostopoulos, K. The difference between reported and real-world CO_2 emissions: How much improvement can be expected by WLTP introduction? *Transp. Res. Procedia* **2017**, *25*, 3933–3943. [CrossRef]
19. Pavlovic, J.; Marotta, A.; Ciuffo, B. CO_2 emissions and energy demands of vehicles tested under the NEDC and the new WLTP type approval test procedures. *Appl. Energy* **2016**, *177*, 661–670. [CrossRef]

20. Tsiakmakis, S.; Fontaras, G.; Cubito, C.; Pavlovic, J.; Anagnostopoulos, K.; Ciuffo, B. *From NEDC to WLTP: Effect on the Type-Approval CO_2 Emissions of Light-Duty Vehicles*; Publications Office of the European Union: Luxembourg, 2017; Volume 50.
21. Liu, X.; Zhao, F.; Hao, H.; Chen, K.; Liu, Z.; Babiker, H.; Amer, A.A. From NEDC to WLTP: Effect on the Energy Consumption, NEV Credits, and Subsidies Policies of PHEV in the Chinese Market. *Sustainability* **2020**, *12*, 5747. [CrossRef]
22. Jaworski, A.; Mądziel, M.; Lew, K.; Campisi, T.; Woś, P.; Kuszewski, H.; Wojewoda, P.; Ustrzycki, A.; Balawender, K.; Jakubowski, M. Evaluation of the Effect of Chassis Dynamometer Load Setting on CO_2 Emissions and Energy Demand of a Full Hybrid Vehicle. *Energies* **2022**, *15*, 122. [CrossRef]
23. Jaworski, A.; Mądziel, M.; Kuszewski, H.; Lejda, K.; Jaremcio, M.; Balawender, K.; Jakubowski, M.; Wos, P.; Lew, K. The impact of driving resistances on the emission of exhaust pollutants from vehicles with the spark ignition engine fuelled with petrol and LPG. Technical report, SAE Technical Paper. In Proceedings of the SAE Powertrains, Fuels & Lubricants Meeting, Krakow, Poland, 22–24 September 2020 .
24. Kadijk, G.; Ligterink, N. Road Load Determination of Passenger Cars. 2012. Available online: https://www.tno.nl/media/1971/road_load_determination_passenger_cars_tno_r10237.pdf (accessed on 3 August 2023).
25. Kühlwein, J. The Impact of Official Versus Real-World Road Loads on CO_2 Emissions and Fuel Consumption of European Passenger Cars. The International Council on a Clean Transportation. 2016. Available online: https://theicct.org/sites/default/files/publications/ICCT_Coastdowns-EU_201605.pdf (accessed on 4 August 2023).
26. Jaworski, A.; Mądziel, M.; Lejda, K. Creating an emission model based on portable emission measurement system for the purpose of a roundabout. *Environ. Sci. Pollut. Res.* **2019**, *26*, 21641–21654. [CrossRef]
27. Mądziel, M.; Campisi, T.; Jaworski, A.; Kuszewski, H.; Woś, P. Assessing Vehicle Emissions from a Multi-Lane to Turbo Roundabout Conversion Using a Microsimulation Tool. *Energies* **2021**, *14*, 4399. [CrossRef]
28. Szumska, E.M.; Jurecki, R. The Analysis of Energy Recovered during the Braking of an Electric Vehicle in Different Driving Conditions. *Energies* **2022**, *15*, 9369. [CrossRef]
29. Szumska, E.M.; Jurecki, R. The Effect of Aggressive Driving on Vehicle Parameters. *Energies* **2020**, *13*, 6675. [CrossRef]
30. Kurtyka, K.; Pielecha, J. The evaluation of exhaust emission in RDE tests including dynamic driving conditions. *Transp. Res. Procedia* **2019**, *40*, 338–345 .
31. X 431 EURO LINK. 2023. Available online: https://www.csselectronics.com/pages/obd2-explained-simple-intro (accessed on 14 June 2023).
32. Meenakshi.; Nandal, R.; Awasthi, N. OBD-II and Big Data: A Powerful Combination to Solve the Issues of Automobile Care. In Proceedings of the Computational Methods and Data Engineering, Proceedings of the ICMDE 2020, Bangkok, Thailand, 26–28 February 2020; Singh, V., Asari, V.K., Kumar, S., Patel, R.B., Eds.; Springer: Singapore, 2020; pp. 177–189.
33. mSTK. 2020. Available online: https://play.google.com/store/apps/details?id=sk.deletech.uitk&authuser=0 (accessed on 1 January 2010).
34. Society of Automobile Engineers, E/E Diagnostic Test Modes J1979_201702. 2017. Available online: https://www.sae.org/standards/content/j1979_201702/ (accessed on 4 May 2023).
35. Tapak, P.; Kocur, M.; Rabek, M.; Matej, J. Periodical Vehicle Inspections with Smart Technology. *Appl. Sci.* **2023**, *13*, 7241. [CrossRef]
36. ELM 327. 2017. Available online: https://www.elmelectronics.com/wp-content/uploads/2017/01/ELM327DS.pdf (accessed on 15 July 2021).
37. Vehicle Categories. 2021. Available online: https://ec.europa.eu/growth/sectors/automotive/vehicle-categories_nn (accessed on 15 July 2021).
38. Ktistakis, M.A.; Pavlovic, J.; Fontaras, G. Developing an optimal sampling design to monitor the vehicle fuel consumption gap. *Sci. Total Environ.* **2022**, *832*, 154943. [CrossRef]
39. Number of Registered Vehicles in SLovak Republic—Ministry of Interior. 2023. Available online: https://www.minv.sk/?celkovy-pocet-evidovanych-vozidiel-v-sr&subor=487941 (accessed on 9 August 2023).
40. Jaworski, A.; Kuszewski, H.; Lew, K.; Wojewoda, P.; Balawender, K.; Woś, P.; Longwic, R.; Boichenko, S. Assessment of the Effect of Road Load on Energy Consumption and Exhaust Emissions of a Hybrid Vehicle in an Urban Road Driving Cycle–Comparison of Road and Chassis Dynamometer Tests. *Energies* **2023**, *16*, 5723. [CrossRef]
41. Auto-Abc, Real Fuel Consumption. 2023. Available online: https://www.auto-abc.eu/info/real-fuel-consumption (accessed on 9 August 2023).
42. ADAC. Ecotest. 2023. Available online: https://www.adac.de/rund-ums-fahrzeug/autokatalog/ecotest/ecotest-ranking-sauberste-autos/ (accessed on 9 August 2023).

Disclaimer/Publisher's Note: The statements, opinions and data contained in all publications are solely those of the individual author(s) and contributor(s) and not of MDPI and/or the editor(s). MDPI and/or the editor(s) disclaim responsibility for any injury to people or property resulting from any ideas, methods, instructions or products referred to in the content.

Review

Vehicle Emission Models and Traffic Simulators: A Review

Maksymilian Mądziel

Faculty of Mechanical Engineering and Aeronautics, Rzeszow University of Technology, 35-959 Rzeszow, Poland; mmadziel@prz.edu.pl

Abstract: Accurate estimations and assessments of vehicle emissions can support decision-making processes. Current emission estimation tools involve several calculation methods that provide estimates of the exhaust components that result from driving on urban arterial roads. This is an important consideration, as the emissions generated have a direct impact on the health of pedestrians near the roads. In recent years, there has been an increase in the use of emission models, especially in combination with traffic simulator models. This is because it is very difficult to obtain an actual measurement of road emissions for all vehicles travelling along the analysed road section. This paper concerns a review of selected traffic simulations and the estimation of exhaust gas components models. The models presented have been aggregated into a group with respect to their scale of accuracy as micro, meso, and macro. This paper also presents an overview of selected works that combine both traffic and emission models. The presented literature review also emphasises the proper calibration process of simulation models as the most important factor in obtaining accurate estimates. This work also contains information and recommendations on modelling that may be helpful in selecting appropriate emission estimation tools to support decision-making processes for, e.g., road managers.

Keywords: vehicles; emission modelling; traffic simulation; air pollution; emission compounds

1. Introduction

Air pollution is a major component of the risk to human health and the state of the environment [1]. Outdoor pollution is responsible for approximately 1.2 million annual deaths [2]. Road transport, which is widely considered the only major source of pollution in urban areas, contributes to this [3]. Therefore, great effort is being made to reduce the exhaust emissions from transport. As a result, many new vehicle designs and new fuel components are being implemented, as well as strengthening traffic management in cities to maximize traffic flow [4,5]. Vehicle emissions have become a key issue for the global community in recent years. The number of vehicles on the road is constantly increasing, compounding the problem, and making road transport a major contributor to pollution [6,7]. The main components of vehicle exhaust emissions include carbon monoxide (CO), nitrogen oxides (NO_x), total hydrocarbons (THC), particulate matter (PM), and greenhouse gases such as carbon dioxide (CO_2) [8–10]. However, it should be borne in mind that the future share of electric vehicles will increase, so that the proportion of pollutants in the road environment will decrease dramatically, which will not make their impact on human health any less significant.

Vehicle exhaust emissions depend on many factors. Estimating emissions is challenging and depends, among other things, on the scale at which vehicle emissions are measured [11,12]. Exhaust emissions in terms of modelling depend, among other things, on the vehicle category and also on the traffic flow of a certain group of vehicles. These emissions depend on a number of factors, e.g., the specification of the fuel used, the emission control technology, and the environmental and operational conditions of the vehicle (driving style, contribution of cold-start emissions, etc.) [13,14]. Vehicle emission models are used to accurately estimate emissions from road transport. It is important to understand the connection between vehicle emissions and the environment in order to develop, among other things, strategies that can minimize the impact of exhaust emissions on the environment. For this purpose, vehicle emission models and vehicle traffic simulation models are used.

Vehicle emission models estimate the amount of exhaust emissions based on various input data. These data are differentiated according to the scale of the model used [15,16]. Examples of input data can be detailed vehicle data, for example, engine specifications, driving conditions, and fuel properties [17,18]. Traffic simulations can be a certain complement to emission models. Traffic simulation allows for a virtual representation of real-world traffic scenarios, allowing for the analysis of vehicle interactions and their impact on the immediate environment and the health of pedestrians travelling on arterial roads [19–21]. To generate a certain set of vehicle traffic data, simulators allow valuable information to be provided on the magnitude of vehicle emissions versus, for example, air quality to inform decision makers in controlling air pollution [22,23].

Models to estimate vehicle emissions are becoming increasingly sophisticated and extensive [24]. This is undoubtedly related to the increasing number of vehicles and the diversity of pollutants, modifications, and fuel types used [25]. For emission models, it is as important to distinguish between models for different engine thermal states, e.g., cold start, hot running, etc. The first emission models, which were developed in the 1980s and 1990s, allowed the prediction of emissions from the warmed-up state of a vehicle's engine for only a few major emission components based on certified test data. Modern emission models allow the estimation of all known exhaust compounds, regulated and unregulated, estimate fuel consumption, and are based on data from actual road tests under different driving conditions [26]. The first emission models were based solely on data from a dozen vehicles, while the current emission models use data from thousands of different vehicles [27]. Over the years, the approach has also changed from the classic modelling for driving mode (acceleration, deceleration, idle, cruise) to also include other variables, such as data from vehicle engines: engine load, speed, engine temperature, air-to-fuel ratio, etc [28,29].

This paper is a literature synthesis of the current state-of-the-art in the field of exhaust emission models and traffic simulation. The topics described could be helpful to those involved in modelling vehicle traffic and modelling environmental aspects more widely. This work is divided into three main parts. The first presents the characteristics of the described topic based on a Web of Science core collection search. This is followed by a description of vehicles, the main components of exhaust gases, a description of selected emission models and traffic simulators, examples of their use, and trends, and further developments in the field. The work also considers the problem of calibrating vehicle traffic summation models for a more accurate estimation of vehicle emissions.

The scope of this work also includes a comparison of exhaust emission models with selected parameters. Such an overview of the work could be helpful for those seeking information on the complexity and usefulness of emission models and their potential use for traffic simulation purposes. The numerous examples and comparisons could also be useful to decision makers in traffic management. Based on the review, recommendations on the usefulness of the selected models were also developed in terms of their potential use for traffic simulation and environmental analysis of the impact of road transport on exhaust emissions for use by transport decision-makers.

To show the contribution of this article to the development of the topic, Table 1 presents the contribution of this article compared to other review papers dealing with similar research topics. The comparative papers dealt with the topics of emission model reviews, traffic simulation models, the combination of the two topics, and the description of additional things related to the topic, such as the characterisation of vehicle exhaust components and the essence of model calibration.

Table 1. Authors contribution table.

Author(s)	Emission Models Review	Traffic Simulators Models Review	Pollutants Emission Description	Calibration Problem Description
Boulter et al. [30]	√ (instantaneous emission models)			
Smit et al. [31]	√			
Faris et al. [32]	√		√	√
Demir et al. [33]	√ (review of 6 emission models)			
Gokhale, et al. [34]	√			
Wang et al. [35]	√			
Pel et al. [36]			√	
Nguyen et al. [37]		√ (agent based)		
Ejericto et al. [38]		√		
Liu et al. [39]		√		
Alghamdi et al. [40]		√		
Mubasher et al. [41]		√		
Forehead et al. [42]	√			
Algers et al. [43]		√ (micro)		
Madi [44]		√	√	
Franco et al. [45]	√ (emission factors)			
This paper	√	√	√	√

2. Methodology and Characteristics of the Literature Review

The general structure of this paper is shown in Figure 1. The literature review was based on prior identification of the main keywords of the topic under analysis. This analysis was based on the Web of Science Core Collection search engine. Web of Science Core Collection is a database of scholarly literature that provides access to some of the world's most influential scientific, technical, and medical research. It is widely considered one of the most comprehensive and authoritative sources of research information, and it is widely used by researchers [46,47].

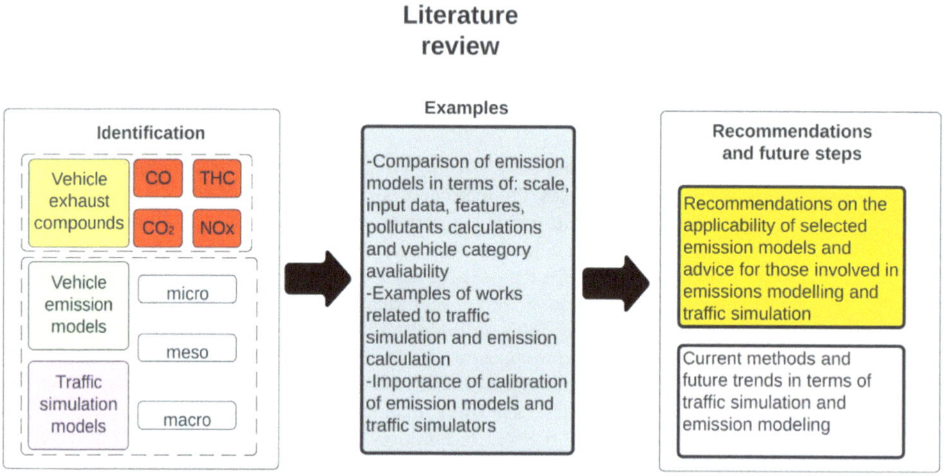

Figure 1. General structure of this work.

The general research methodology for the literature analysis of the topic described was based on a preliminary identification of the main tools that are used in the context of emissions modelling and traffic modelling. Subsequently, this paper describes selected emission models and traffic models, provides examples of the use of these tools, and identifies directions for further work.

The detailed description of the literature review methodology carried out on the topic described first dealt with identification. This identification consisted of the following:

- Analysis of vehicle exhaust components and modelling methods resulting from the method of data collection: chassis dynamometer data, road data using the PEMS system;
- Analysis of selected micro-, meso-, and macro-scale exhaust emission models, their characteristics, and a presentation of their main features;
- Analysis of selected micro, meso, and macro traffic simulation models and their description;

The second step of the literature analysis consisted of providing examples for:

- Combining the use of emission models and traffic simulation models with their brief characteristics;
- Calibration of the simulation model and emission models as the main determinant of the results obtained.

In a third step, this paper describes recommendations for the applicability of the described emission models in combination with traffic simulation models. Future trends in the development of future simulation models are also described.

The following keywords were used to identify the main keywords linked to the topic in the Web of Science Core collection search: emission model and traffic simulation, and vehicle—1250 results were retrieved (information obtained for 12 April 2023). No time frame was specified for the search. Figure 2 presents a keyword map based on the data recovered from the database and the VOSviewer software. From the keyword map, we can see which keywords occur most frequently for the topic described, for example. Based on this, we can see, e.g., keywords such as VT-Micro, car following model, VISSIM, COPERT, SUMO, etc. This makes it possible to pre-classify the most common emission models and simulators used in research. For such analyses, we can additionally search for new papers with the addition of a new keyword related to, for example, the emission model. This way, it is additionally possible to determine the exact number of papers that use, among other things, the analysed tool. For example, with the additional keyword COPERT, we can find at least 155 work results. On this basis, a set of emission models and traffic simulation software were selected, which are described in the following section.

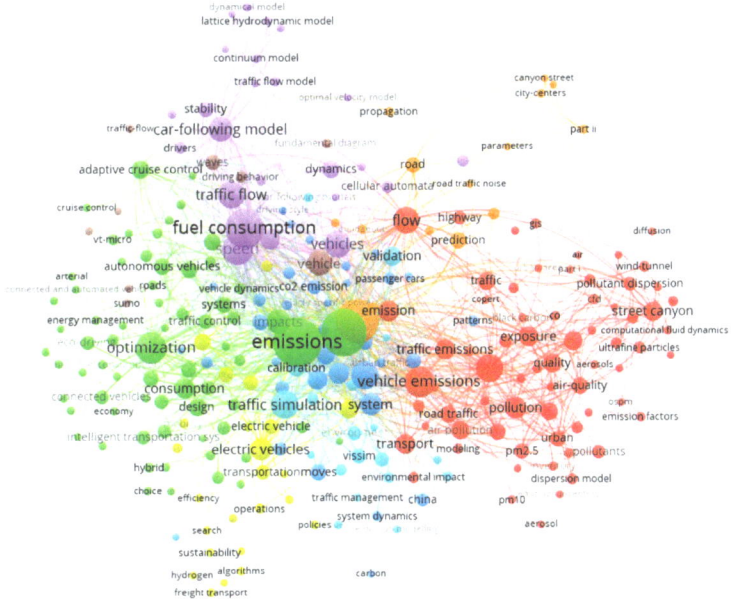

Figure 2. VOSviewer connection map of keywords related to the described topic, based on data from Web of Science core collection.

Based on data from the Web of Science Core collection, a world map was also generated showing the frequency of papers in the searched article database (Figure 3).

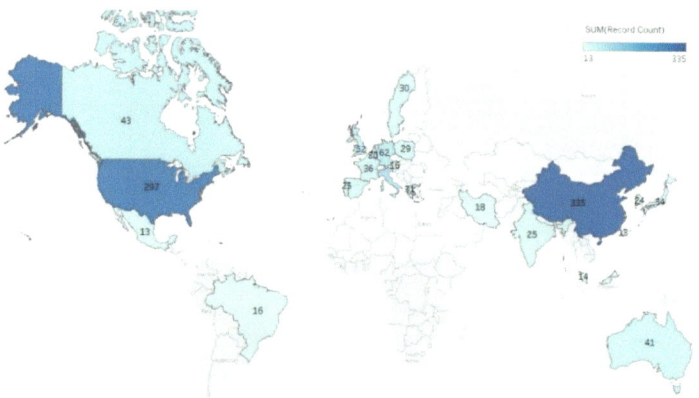

Figure 3. World map of papers related to the described topic.

Figure 4 shows the number of records for the entries for the topics studied. From it we can see an upward trend for published works. The first records analysed date from 1997, while the most recent are from 2023.

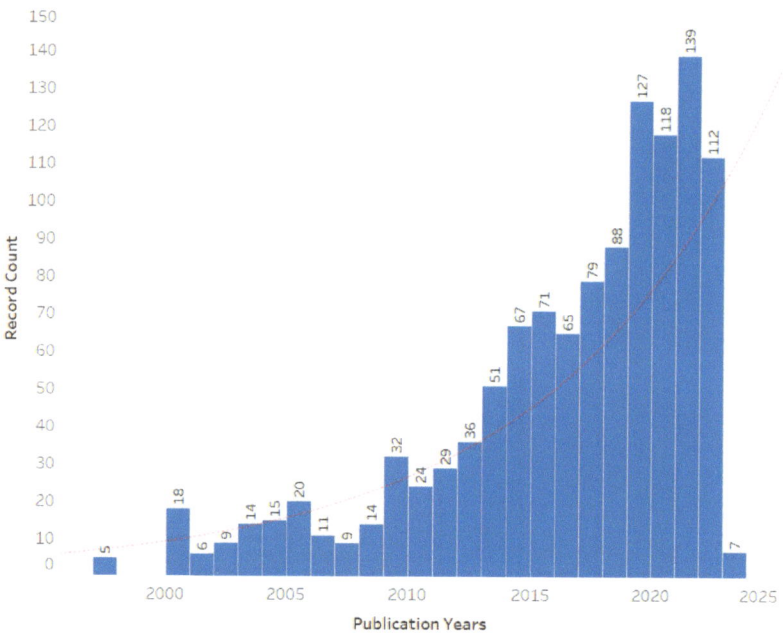

Figure 4. Number of publications in relation to years related to the described topic.

The content presented in this paper in the field of emission and traffic simulation models can be useful particularly in the following aspects:
- This work presents an up-to-date overview of emission models indicating their calculation capabilities for a selected range of model inputs and outputs, which will speed up the process of selecting a suitable tool for those involved in modelling;

- The overview of the selected vehicle traffic simulation models presented can be useful for selecting a suitable tool with mapping vehicle traffic for different scales of accuracy;
- The link between the topic of emission models and the topic of traffic simulation models gives some insight into the capabilities of the software for different accuracy scales, which can also speed up the selection and correct use of a given tool for emission estimation and traffic simulation;
- This work presents the essence of the calibration of emissions and traffic simulation models in order to obtain results close to real values, which gives information for people modelling vehicle traffic and emissions, whose parameters for which the models need to be calibrated and checked against real values, e.g., traffic volume or speed profile;
- From the whole work, the most important recommendations for emissions and traffic modelling have been selected;
- This work shows future trends in the modification of existing emission and traffic models as well as the development of completely new models.

3. Emission from Vehicles—Literature Review

3.1. Main Components of Vehicle Exhaust

Conventional internal combustion engines, as well as hybrid systems, are the main sources of exhaust pollutants in urban areas [48,49]. These emissions result from the combustion of both gaseous and liquid fuels. The main groups of exhaust emission components considering their formation and source include [50–52]:

- High-temperature combustion products such as nitrogen oxide (NO_x);
- Products of incomplete combustion, including particulate matter (PM), carbon monoxide (CO), and hydrocarbons (THC);
- Combustion products from waste fuels, including heavy metals and sulphur oxides (Sox);
- Products from other sources, e.g., volatile organic compounds (VOC);
- Products of total combustion that generate the greenhouse effect (CO_2).

PM can be classified into three types of categories: PM_1, these are particles not exceeding 1 µm in diameter; $PM_{2.5}$, particles not exceeding 2.5 µm in diameter; and PM_{10}, particles not exceeding 10 µm in diameter (containing carbon particles) [53]. Motor vehicles are an important source of PM_1 and $PM_{2.5}$ emissions. For different engines, the particle emissions are different. Diesel engines emit mainly 20–130 µm, while petrol engines emit 20–60 µm [54,55].

In addition to emissions resulting from combustion processes, particulate matter is also produced as a result of the wear and friction of materials, mainly brake friction linings and vehicle tyres [56,57].

The harmful components of the exhaust gases contained in vehicle exhaust negatively affect human health and the state of the surrounding environment [58–61]:

- Carbon monoxide—The poisonous effect is due to the reaction of its combined with hemoglobin and metalloproteins; internal organs such as the heart and central nervous system are damaged; at low concentrations of gas, loss of consciousness occurs; CO shares in the inhaled air already at the level of 0.02% have a negative effect on human health, and life-threatening concentrations of 0.1% cause.
- Short-term exposure of nitrogen oxides to high concentrations leads to lung edema and death, lower concentrations cause the so-called silo disease; in addition to negative direct effects on human health, nitrogen oxides are the main cause of photochemical smog (Los Angeles-type smog), which appears over cities during hot, sunny weather.
- Carbon dioxide—one of the greenhouse gases.
- Hydrocarbons—one of the components of smog; they irritate the conjunctiva, cause allergies, and have strong carcinogenic effects.
- Particulate matter—breathing air polluted with $PM_{2.5}$ can lead to atherosclerosis, complications of pregnancy, and respiratory diseases; PM_{10} can contain toxic sub-

stances such as polycyclic aromatic hydrocarbons (e.g., benzo/a/pyrene), dioxins and furans, which are carcinogenic; the limit level for the average concentration is 50 µg/m^3 and must not exceed more than 35 days per year; particulate matter is part of London smog, which mainly occurs in the months of November to February during temperature inversions.

In view of the increasing emission of exhaust pollutants, coordinated action is being taken in the European Union to reduce the harmful effects of vehicle exhaust. One of the key decisions to reduce exhaust emissions was the introduction of EURO standards [62]. Currently, the regulations cover NO$_x$, THC, CO, and PM emissions [63]. The New European Driving Cycle (NEDC) test, which was carried out under controlled laboratory conditions to ensure the reproducibility and comparability of the results obtained, was used to determine motor vehicle emissions until September 2017 [64,65]. However, it was found that the NEDC test often did not reflect the conditions actually experienced when driving on public roads [66,67]. To improve the emissions and fuel consumption that occur when driving under road conditions, a new WLTP (World Harmonised Light Duty Procedure) driving test procedure was created [68]. With the new procedure, in addition to the determination of emissions during controlled driving tests with a chassis dynamometer, real driving emission (RDE) road tests are also carried out using mobile emission measurement systems—PEMS (Portable Emissions Measurement System) [69–71].

Based on the described dynamometer tests, initially using data from measurement bags in which car exhaust was collected, the first calculation models for exhaust emissions were developed [72]. As the first aggregations of emissions data were for bagged emissions, the models only allowed the calculation of emission factors. Subsequent emission tests involved a modal analysis of the exhaust gas, which allows for a detailed analysis of the emissions of individual exhaust components during the entire dynamometer test. With the development of technology, today, the latest emission models are based on road tests that use on-board exhaust emission measurement systems (PEMS) to record the data [73,74].

3.2. Overview of Selected Exhaust Emission Models

Emission models are divided according to their precision scale into macro (regional, national area), meso (local area), and micro (areas of a dedicated part of a city, intersection road sections) [75].

We can divide emission models into two categories [76,77]:
- Based on parameters: average speed, vehicle type, etc.
- Based on traffic parameters such as acceleration, deceleration, idle, and continuous driving.

Another division of emission modes is based on [78,79]:
- Models that require the uploading of speed profiles;
- Models that generate the speed profiles themselves as part of the emissions modelling process;
- Models that include uploaded velocity profiles.

3.2.1. Macroscopic Emission Models

Macroscopic models are mainly based on the parameter of the average driving speed in the analysed section of the road [80]. They are based on the relation presented in Equation (1).

$$F = A + \left(\frac{B}{V}\right) + C \cdot V + D \cdot V^2 \tag{1}$$

where:

A, B, C, D—coefficients selected according to the type of vehicle and road;
V—average driving speed (km/h);
F—fuel consumption (l/100 km).

In terms of modelling, fuel consumption is highly correlated with CO_2 emissions, compared to CO, THC, and NO_x [81].

The macro model is used to estimate fuel consumption and road traffic's environmental impact. They can determine the total energy consumption of projects and road infrastructure development strategies and assess the impact of greenhouse gas emissions on the study area. Some environmental impacts are local, regional, or global and can be short- and long-term. Macro-scale emission models allow transport impacts to be determined on a large regional scale (regional transport corridors) [82,83].

An example of software that includes macro-scale emission calculation models is the Operat FB package. It is used to model the dispersion of pollutants in ambient air from point, line, and surface sources [84]. It has a "car" module, which makes it possible to calculate the emissions from car traffic on roads. Pollutant emission results are calculated according to the EMEP/Corinair B710 and B760 methodology [85]. This involves calculating the hot emissions from the exhaust of the vehicle engine, the cold emissions arising at the start of engine operation, and the evaporative emissions, the sources of which include changes in the volume of fuel vapour in the vehicle tanks and the cooling of the tank after the vehicle engine has stopped. Brake pads and abrasion can also be calculated [86]. The Corinair model distinguishes more than 200 vehicle categories, allocated to six groups: passenger vehicles, vans, lorries, buses, mopeds, and motorbikes. A further criterion for classification is the engine capacity of the vehicle and the compliance with a given emission standard. To calculate emissions, it is necessary to specify the share of specific vehicles in a given section of the road, specify their speed, and indicate whether the trip was made on an urban, suburban, or motorway road section [87].

Another model to calculate macro-scale emissions is COPERT. It is based on European data using: kilometrage, vehicle structure, driving speed, humidity, and air temperature. Emission factors are calculated for the following vehicle categories [88,89]:

- Passenger cars;
- Vans (<3.5 t);
- Lorries (>3.5 t);
- Motorbikes and mopeds.

In the COPERT model, emissions are calculated based on Equation (2):

$$E_i = \Sigma_j \left[\Sigma_m \left(FC_{j,m} \cdot EF_{i,j,m} \right) \right] \tag{2}$$

where:

E_i—emission of exhaust component I (g);
$FC_{j,m}$—fuel consumption for vehicle category j, using fuel m (kg);
$eF_{i,j,m}$—fuel consumption emission factor of component i for vehicle category j and fuel m (g/kg).

We can consider petrol, diesel, LPG, and CNG as fuels in this model [90].

For US emissions data, the MOBILE model is used, which was developed by the US EPA [91]. This model includes data from vehicle manufacturers collected by CARB and the EPA. Emissions are calculated for a given class of post-vehicle traffic conditions for assumed traffic conditions. Emission calculations take into account model calibration factors such as operating conditions, vehicle specifications, and the surrounding environment [92,93].

Another example is the model for calculating emissions developed by the European Union, called Artemis. Calculate the total emissions of a given traffic component as a product of a given emission factor and the amount of vehicle traffic activity. Unlike other models, emission factors are calculated as a result of the kinematics of vehicle movement in the modelled area. This approach to calculating emissions is contained in three steps [94,95]:

- Identification of cycles by the kinematics of vehicle motion;
- Selection of an appropriate cycle representing a specific group, which determines the corresponding emission factors for road activities;

- Determination of correction factors to determine reference emission factors.

The current approach to macro-scale emission models used for the determination of harmful exhaust components is based on two calculation steps. The first is the selection of a set of emission factors that determine the emissions for a given traffic condition, while the second stage is the assessment of vehicle activity in the area under study. Emissions are calculated by multiplying these two stages. However, this methodology has two main disadvantages:

- Inaccuracy of emission results—most of the data included in the macro models are based on measured emissions data from engine dynamometers, while the driving cycles that are used do not correctly represent real-world driving (e.g., they take into account frequent rapid acceleration or braking); in addition, they may not be fully reliable, as some car manufacturers have been found to have manipulated emissions results.
- Inadequate characterisation of current driver behaviour—current methods of determining emission factors are based on the average driving performance over a predetermined driving cycle used for certifying vehicles with emission standards; recent modifications to the introduction of the new driving cycle have rendered older generation cycles obsolete.

3.2.2. Microscopic Emission Models

Micro-scale models need a large amount of data related to the measurements of vehicle parameters such as acceleration and speed, as well as road parameters such as terrain gradient and position coordinates [96]. Micro-scale models usually have a recording resolution of 1 s. To date, several micro-scale models have been developed that are based both on vehicle exploration parameters, such as engine power, and vehicle speed [97].

A classification of micro-scale models that takes into account the type of data used for input includes [98–101]:

- Based on speed profile: Enviver Versit+, VT, RoundaboutEM;
- Based on vehicle parameters, such as power: CSIRO, CMEM, VT-CPFM, LPGemission;
- Combining the above methods.

The power-based CMEM model was developed in 2006. Emission processes fall into different categories that correspond to physical phenomena associated with vehicle operations. Each component of the exhaust gas is modelled analytically and includes parameters specific to its generation process [102]. In developing these models, emissions were measured both directly in the engine and at the exhaust outlet of the vehicle under study. In total, more than 300 vehicles were tested under laboratory conditions for three driving cycles [103]. Developing this type of model requires a large amount of data. To calculate real-time emissions, many physical variables must be collected, and vehicle speed profiles must be entered.

Another model based on engine power is CSIRO. This model predicts fuel consumption, THC, and CO emissions very well, in relation to brake and road test data [104].

The power equation used in this model is shown as follows:

$$Zt = Zd + Zr + Za + Ze + Zm \qquad (3)$$

where:

Zt—vehicle power;
Zd—vehicle power to overcome internal losses; $Zd = 2.36 \cdot 10^{-7} \, V^2 M$;
Zr—vehicle power to overcome rolling resistance; $Zr = (3.72 \cdot 10^{-5} \, V + 3.09 \cdot 10^{-8} \cdot V^2) M$;
Za—vehicle power to overcome air resistance; $Za = 1.29 \cdot 10^{-5} \, CdAV^3$;
Ze—power needed to overcome the inertia force and climbing resistance;
$Ze = 2.78 \cdot 10^{-4} \, (a + g\sin\theta)MV$.
Zm—power lost to power the vehicle's accessories;
M—vehicle mass (kg);

V—vehicle speed (km/h);
a—vehicle acceleration (m/s²);
Cd—vehicle drag coefficient;
A—vehicle frontal area (m²);
θ—road gradient (%).

The emission rate (Em) is calculated from Equation (4).

$$Em = \alpha' + \beta' Z_t \qquad (4)$$

where:

α'—emissions at idle;
β'—emission per unit power output of the vehicle.

Equation (4) reflects the emissions when the engine is idle (or fuel consumption) when the useful power is equal to 0. This method can provide adequate estimates of emissions for micro-scale emission models. However, these results are based on the statistical average of power, fuel consumption, and emissions. In micro-modelling, changes in the characteristics of the vehicle (e.g., age and catalytic converters) must be taken into account.

However, sometimes parameters such as the vehicle drag value are not available. The MOVES emission model uses the VSP to determine the amount of time a vehicle spends in each operating mode bin. This model uses vehicle-specific power (VSP), defined as the output of the engine per vehicle unit and expressed as a function of vehicle speed, road gradient, and acceleration [105]:

$$VSP = V \cdot 1.1 \cdot a + 9.81 \cdot \theta \cdot 0.132\, V + 0.000302 \cdot V^3 \qquad (5)$$

where:

VSP—vehicle specific power (kW/t);
V—vehicle speed (km/h);
a—vehicle acceleration (m/s²);
θ—road gradient (%).

The proposed model of Ahn et al. [106] is based on the relationship between emissions and speed. The results show that exhaust emissions and fuel consumption increase as the speed increases, even when the vehicle accelerates. This phenomenon cannot be described in a model based on vehicle power [107]. The model is based on linear acceleration and speed regression and is relatively compatible with raw data ($R^2 > 0.92$). Regression results are sensitive to the test methods used, that is, the cycles of driving and selected vehicles represent the composition of the fleet. The model is based on Equation (6):

$$EM = \sum_{i=0}^{3} \sum_{j=0}^{3} k_{i,j}^e \cdot V^i \cdot A^j \qquad (6)$$

where:

EM—instantaneous fuel consumption and emissions of harmful exhaust components;
$k_{i,j}^e$—coefficients of the regression model for MOE, both for speed i and for acceleration j;
V^i—velocity (m/s);
A^j—acceleration (m/s²).

The model described in [108] contains six variable values. The emission functions for each vehicle provide on-the-fly results with speed and acceleration inputs, using nonlinear multiple linear regressions. The model was calibrated using emissions data from 25 vehicles (6 buses, 2 trucks, and 17 cars). As a result of the fact that the data from the tests obtained revealed a clear distinction in emissions for the acceleration and braking phases, the entire data set was divided into three elements, namely: acceleration, greater than 0.5 m/s²; deceleration less than −0.5 m/s², constant speed driving, and idling. The model can be used to estimate NO$_x$ emissions, volatile organic compounds (VOCs), CO$_2$, and PM. The model has the form shown in Equation (7):

$$E_n(t) = \max[E_0, f_1 + f_2V_n(t) + f_3V_n(t)^2 + f_4A_n(t) + f_5A_n(t)^2 + f_6V_n(t)\,A_n(t)] \quad (7)$$

where:

$V_n(t)$—vehicle speed n at time t;
$A_n(t)$—acceleration of vehicle n at time t;
$E_0(t)$—lower emission limit for a specific emission type of an exhaust constituent [g/s];
f_1-f_6—emission constants determined by regression analysis for a given type of vehicle and exhaust constituents.

As part of Enviver, the VERSIT+ speed profile-based emissions model includes a variable related to the vehicle's driving cycle. From the speed profiles obtained in VISSIM, emission factors (g/km) can be estimated for different classes of vehicles [109]. VERSIT+ offers 246 emissions models for each category and toxic exhaust component. The speed profiles used in the model represent actual road conditions, as opposed to those derived from the New European Driving Cycle (NEDC) [110]. The emission factors ($EF_{j,k,l}$) were obtained by multiple linear regression to find the empirical relationship between emission rates, speed profiles, and dynamic variables [111]. Road transportation exhaust emissions (g/h) for specific exhaust components in one or more road sections are calculated from Equation (8):

$$TEj = \sum k, m \left(EF_{j,k,l} \cdot TV_{k,m} \cdot L_m \right) \quad (8)$$

where:

$EF_{j,k,l}$—average emission factor (g/km);
j—emission component;
k—vehicle class;
l—speed profile;
$TV_{k,m}$—volume of—road traffic (vehicles/h);
m—road section;
L_m—length of the road section (km).

Another software package in which the emissions of harmful exhaust components can be calculated on the micro scale is the AVL simulation package, in particular, Cruise, which uses Matlab/Simulink for modelling and simulation processes [112,113]. In addition to estimating model emissions, it is also possible to calculate vehicle fuel consumption. This package is designed to model any vehicle powertrain configuration (including EVs, fuel cells, and HEVs) [114]. To estimate the emissions, it is necessary to determine the trajectory using an uploaded speed profile. For speed profiles, there is a free choice, depending on whether we estimate vehicle emissions for a driving test, for example, WLTC, FTP-75, NEDC, or for a real-world driving trip [115]. The modelling approach in the AVL simulation package includes detailed vehicle specifications and projections for selected chassis components, for which we have the freedom to model shape [116].

3.2.3. Comparison of Selected Emission Models

A summary of the selected emission models in terms of their scale, the input data needed for the calculation, and the main model-specific features is presented below in the form of Table 2. Information on the availability of an update is also provided, which may be helpful for those wishing to make up-to-date estimates of vehicle emissions. The update feature is also relevant in the context of new vehicle designs that are approved for road use, while they need further testing of actual road emissions to provide valuable data for updating emission models. Updating emission models will also be a challenge in the context of the upcoming EURO7 emission standards, which will be a new emissions standard and will revolutionise the subject of vehicle emissions.

Table 2. Selected emission models with their characteristics regarding input data, scale of detail, selected features, and information on availability of updates.

Model	Scale	Input Data	Features	Updates Information	Source
COPERT	Macro	Vehicle category, number of vehicles, weather conditions, load, average speed, distance travelled, etc.	The wide availability of vehicle types and emission components studied.	Continuously updated, current version: 5.6.	[117,118]
MOVES	Macro	Vehicle category, number of vehicles, weather conditions, load, average speed, distance travelled, etc.	Ability to calculate emissions for a large number of exhaust components, including: HC, CO, CO_2, NO_x, CH_4, N_2O, and PM.	Continuously updated, current version: MOVES3.	[119–121]
PHEM	Micro	Among other things, the speed profiles of the vehicles tested.	Accuracy of emission estimation for the entire route, wide range of engine types and test vehicles, time resolution 1 Hz.	Continuously updated.	[122,123]
CMEM	Micro/macro	Among other things, the speed profiles of the vehicles tested.	In addition to application at the micro scale, it is also possible to estimate emissions at the macro scale, making the model versatile	Currently without update support.	[124]
Versit+/Enviver	Micro	Speed and acceleration profiles of vehicles.	Automatic generation of emission maps, full support for selected traffic simulation models, e.g., VISSIM.	Currently without update support.	[125,126]
VT-Micro	micro	Speed and acceleration profiles of vehicles.	The ability to calculate continuous emissions along the route and fuel consumption for the exhaust gases: CO_2, NO_x, CO, and THC	Currently without update support.	[127,128]
ESTM BOSH	micro	Speed and acceleration profiles of vehicles.	The possibility of creating emission maps within the scope of the VISSIM software, which allows very precise localisation of areas of increased concentrations of exhaust constituents.	Continuously updated.	[129]
EMPA	micro	Speed and acceleration profiles of vehicles.	Possibility to calculate emissions for LDVs only.	Currently without update support.	[130,131]
EMFAC	macro	e.g., average vehicle speed, type structure of vehicles, vehicle load, ambient conditions: temperature, humidity, etc.	Ability to calculate emissions for a number of indicators: THC, CO, NO_x, PM, SO_x and CO_2.	Last update in 2021.	[132,133]
MODEM	micro	Speed and acceleration profiles of vehicles.	Continuous emission estimation; no emission estimation possible for heavy duty vehicles.	Currently without update support.	[134,135]
HBEFA	macro	e.g., average vehicle speeds, type structure of vehicles.	Estimation of emission factors for vehicles of different categories: PC, LDV, HGV, urban buses, and motorbikes.	Last update in 2022.	[136–138]

The selected emission models presented above, together with their characteristics, mainly apply to the micro- and macro-scale. For the given example micro- and macro-scales, Table 2 also gives examples of the input data required for the model emissions calculation. For example, for macro-scale models, it is necessary to determine the average speed of the vehicles, their type of structure, the fuel used, the vehicle load, and, for example, the percentage of urban, extra-urban, and motorway driving [139]. For the macro input data, data on ambient conditions such as minimum, maximum, and average temperatures for the whole year for the given months as well as the prevailing atmospheric humidity are often also required [140]. For the micro scale, a vehicle speed profile recorded at a frequency of at least 1 Hz is usually the basic parameter required. Such input data can be obtained from the vehicle's GPS and OBDII system, or, for example, can be obtained as simulation data. Most simulation software, such as VISSIM or Matlab Simulink, allow the generation of vehicle parameter data from the simulation, and these data are then loaded into micro-scale models. Sometimes, for the specifics of some micro-scale models, it is necessary to adapt the input file for the appropriate coding for the emission model, e.g., when uploading speed profiles to Enviver Versit+, it is necessary to adapt the input data and save them in .fzp format.

Tables 3 and 4 show the characteristics of the selected parameters for selected emission models. Table 2 deals with the estimation of the emission for a selected component of the exhaust gas and shows which model offers the calculation possibilities for a given emission factor. Table 3, on the other hand, shows the emission calculation possibilities for a given vehicle category for selected emission models.

Table 3. Presentation of emission calculation options for selected vehicle exhaust components for selected emission models.

Pollutant	Emission Model				
	PHEM (Micro)	COPERT5 (Macro)	HBEFA (Macro)	Versit+ (Micro)	MODEM (Micro)
CO_2	No	Yes	Yes	Yes	Yes
CO	Yes	Yes	Yes	Yes	Yes
THC	Yes	Yes	Yes	Yes	Yes
PM	Yes	Yes	Yes	Yes	Yes
NO_x	Yes	Yes	Yes	Yes	Yes
CH_4	No	Yes	Yes	No	No
Benzene	No	Yes	Yes	No	No
Toluene	No	No	Yes	No	No
Xylene	No	No	Yes	No	No
NO_2	No	Yes	Yes	No	No
N_2O	No	Yes	Yes	No	No
1,3-butadiene	No	Yes	No	No	No
SO_2	No	No	Yes	No	No
Fuel consumption	Yes	Yes	Yes	No	Yes

Table 4. Presentation of emission calculation options for selected vehicle categories for selected emission models.

Vehicle Category	Emission Model				
	PHEM (Micro)	COPERT5 (Macro)	HBEFA (Macro)	Versit+ (Micro)	MODEM (Micro)
Passenger cars	Yes	Yes	Yes	Yes	Yes
LGV	No	Yes	Yes	No	No
HGV	Yes	Yes	Yes	No	No
Urban bus	Yes	Yes	Yes	No	No
Coach	Yes	Yes	Yes	No	No
Motorcycles	No	Yes	Yes	No	No

Based on Table 3, it can be observed that the lower the detail of the model (macro-scale models), theoretically they allow emission calculations for a larger number of exhaust components. Micro-scale models are characterised by lower emission estimates. This is due to the fact that micro-scale models allow for the generation of detailed second-by-second emission profiles, which are generally difficult to obtain when building an emission model. Macro-scale models allow for either averaged estimates or emission factors, which is simpler to model, as it allows for a larger margin of error.

Figure 5 shows the overall combination of the emission models for different scales for different transport models of vehicle traffic simulations. From it, it can be seen for which example applications and input data the emissions of vehicle exhaust components can be estimated. However, an unambiguous link between a given emission model and a given type of vehicle traffic simulation is not clear-cut, as will be shown later in the presentation of case studies of other works.

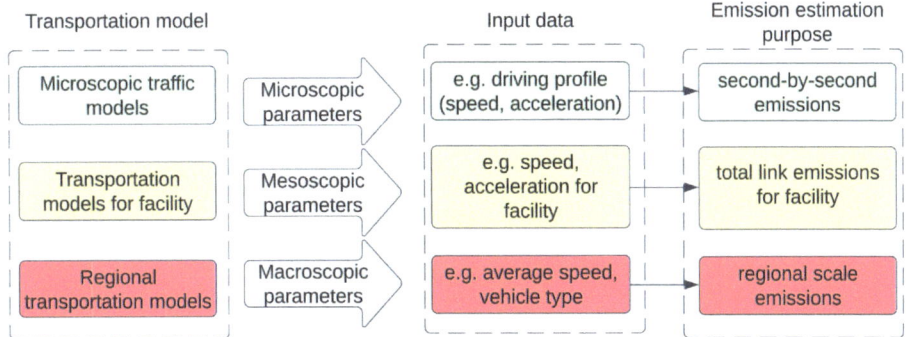

Figure 5. Representation of the scale of the emission model to use for vehicle traffic models in relation to different scales of detail and input data.

A list of websites to access the simulation software is in the Appendix A.

3.3. Traffic Simulation Tools and Exhaust Emissions

Traffic simulation techniques have been used since the middle of the 20th century. The intensive development of information technology, including numerical computer techniques coupled with high-powered computers, has contributed significantly to the development of modern computational procedures used in traffic engineering. These procedures take into account an increasing group of factors that affect the complex phenomenon of traffic [141]. Three levels of analysis have emerged: network—macro, local—micro, and intermediate—meso. Macroscopic models assess vehicle traffic at a high level of aggregation as traffic flow (the number of vehicles passing a given point in an hour) without focussing on its components (vehicles) [142]. Microscopic models describe the behaviour of the units that make up the traffic stream, as well as the interactions that occur between them [143]. Mesoscopic models, on the other hand, are characterised by an intermediate level of detail. They describe individual vehicles, but do not take into account the relationships between them [144].

Macroscopic models are used to estimate the emission of exhaust pollutants on the basis of the average speed and kilometres travelled by a specific group of vehicles. Microscopic models, on the other hand, simulate the movement of individual vehicles included in traffic streams, giving them specific speeds and accelerations, creating very large amounts of data when recording these situations. These data are successively used by additional applications to calculate vehicle emissions [145,146].

One example of traffic microsimulation software is TRANSIMS. It is a system that enables the analysis of transport systems on a regional scale. It contains a number of modules, among which we can highlight the Traffic Microsimulator. It is based on the theory of cellular automata (CA) system and uses the Nagel–Schreckenberg model [147]. The main feature of CA-based models is temporal and spatial discretisation. For this reason, each path in the network is divided into small cells of equal size. Each cell can be in one of two states: occupied by a vehicle or empty. Due to this simplification, compared to other micro-scale models, the model is efficient and can be applied practically to simulate large and complex regional and even national road networks [148]. TRANSIMS is also equipped with an additional module that is used to evaluate vehicle emissions from the model. It allows the estimation of THC, CO, and NO$_x$ [149].

The software for traffic microsimulation is SUMO (Simulation for Urban Mobility). It is a system developed by the German Aerospace Centre (DLR). It includes the Krauss leader driving model, an extension of the Gipps model, and the Krajzewicz lane change model [150,151]. The system allows traffic simulations for different types of vehicles and different intersections with or without traffic lights, and for different road networks with more than 10,000 roads. SUMO also allows dynamic traffic distribution procedures and graphical visualisation of traffic in 2D [152]. It also allows for the estimation of exhaust emissions based on the HBEFA and PHEM models, which provide results for CO_2, CO, THC, NO$_x$, PM, and vehicle fuel consumption [153].

The most popular software for traffic microsimulation is VISSIM. It is a system developed by the German company PTV. It uses the psychophysiological model of Wiedemann's driver behaviour [154,155] to model driving behind the leader. This software implements a rule-based model of lane change based on Wiedemann's work [156]. The VISSIM software does not follow the classical approach of modelling a road network with a dedicated graph consisting of vertices (nodes) and edges (segments). Instead, the road network is built using sections connected to each other by connectors. This solution makes it possible to model almost any road system. VISSIM is not only characterised by a high degree of accuracy in modelling the geometry and parameters of the road network, but also enables a precise representation of vehicle traffic. In addition to motor vehicles, the program can simulate pedestrians, cyclists, motorbikes, and rail vehicles. Dynamic traffic distribution can also be used in the program to iterate the driver learning process [157]. Two-dimensional and three-dimensional simulations can be visualized using the software. The leader driving model used in the programme also describes the boundaries between the distinguished states, which are [158,159]:

- Free driving;
- Approaching;
- Dependent driving;
- Braking.

In the software, in addition to the 1974 Wiedemann model, there is also the 1999 model, which is more extensive (with a greater number of calibration parameters). In order to calculate vehicle emissions, the software can use the VERSIT+ emissions model developed by the Dutch Organisation for Applied Scientific Research (TNO) and also the integrated ESTM BOSH emissions model. [160,161].

3.4. Simulation of Vehicle Traffic and Emission Models—Examples

The emission models described above are used in vehicle movement simulation studies. The use of these models is noted in many cases for VISSIM and SUMO. The models represent the movement of vehicles on a micro scale, i.e., in addition to measuring parameters such as speed and vehicle acceleration, they measure any relationships that occur between vehicles appearing in the simulation. A similar use is for meso-scale models, except that the interaction between vehicles is then excluded. The typical use of these data is to use, among other things, the previously mentioned parameters for the computation of micro-scale emission models. However, it is possible to aggregate these data to some extent, i.e., to determine, for example, the average speed parameter or the distance travelled for a given vehicle or group of vehicles. As a result of such operations, it is also possible to use, e.g., emission models on a macro scale, for the input data, in addition to general parameters of the studied fleet, such as the number of vehicles, their type, size, and fulfilment of the emission class, need, e.g., vehicle operation data, such as, e.g., speed or vehicle load. Table 5 below shows a cross section of the work on vehicle traffic simulation and the calculation models used to assess emissions.

Table 5. Examples of work using traffic simulations and emission models.

Traffic Simulator Model (Scale)	Emission Model	Description	Source
Dynamic Traffic Assignment (meso)	Integrated microscopic estimation model (MOVES)	The study proposed an alternative approach for a mesoscopic traffic simulation model with an integrated microscopic emission model. The study used a postprocessing procedure to generate detailed vehicle trajectories.	[162]
VISSIM (micro)	MOVES2010a	Emission prediction with the use of VISSIM road model and emission model MOVES of vehicles on a limited-access motorway.	[163]
VISUM/AIMSUN Next (macro)	HERMESv3/PHEMLight	The study presents a macroscopic traffic emission model for Barcelona. The developed system is used to, e.g., quantify the hourly level of NO_x and PM10 emissions.	[164]
AIMSUN (micro)	Cruise/COPERT	The studies demonstrate a new approach for the estimation of fuel consumption of vehicles with different levels of congestion in Turin city.	[165]
EMME 4.0	COPERT–Chile	Traffic simulation model used for environmental impact modelling for the South American city, Bogota.	[166]
VISSIM (micro)	COPERT	The work describes the computing of CO_2 and NO_x based on average speed as the input variable. A real-world study describing intercity corridors with many alternative routes are presented.	[167]
VISSIM (micro)	VSP	Two scenarios based on real-world data networks were analysed in VISSIM in the context of CO, HC, and NO_x emissions. The used emission model was VSP-calibrated using the collected field emission data.	[168]
VML (macro)	HERMESv3	The work describes the changes in the air quality at street level for different traffic management strategies in Barcelona.	[169]
VISSIM (micro)	Enviver Pro	The work presents a case study for the city of The Hague during the morning peak in 2040. Several SAV market penetration scenarios were tested. Emission analysis concerns CO_2, NO_x, and PM10.	[170]
VISSIM (micro)	Enviver Pro	The article addresses the use of different rights of way at four-arm intersections. The emission analysis is related to the comparison of NO_x and PM10 for the assumed VISSIM road models.	[171]
SUMO (micro)	GT-Suite	This work is the combination of traffic flow and vehicle simulations to investigate the vehicle performance. In the simulation of SUMO, the authors used the real-world elevation profile and rolling resistance factor.	[172]
SUMO (micro)	PHEMlight	The work describes the adoption of smooth driving habits on emissions in large-scale urban networks. The dataset contains a total of 4156 urban trips from 100 drivers.	[173]
NFD (meso)	Integrated microscopic estimation model	The network-wide fundamental diagram (NFD) and microscopic emission models are estimated using mesoscopic traffic simulation tools at different scales for various traffic compositions for 13 simulated scenarios.	[174]
MATSim (macro)	MOVES/R-Line	The simulation is performed in MATSim to simulate individual vehicle emissions.	[175]
VISUM (macro)	Integrated emission model	Estimation of CO_2 and NO_x emissions based on average speed as input variables for the scenarios with shared vehicle (SV), automated vehicle (AV), and electric vehicles (EV).	[176]
DTA (macro)	Integrated emission model	The work presents a macroscopic dynamic modelling framework in an urban city that has numerous central business districts to assess the emissions. The computation of emissions includes NO_x, VOC, CO_2, and PM in urban areas.	[177]
VISSIM (micro)	Enviver/RoundaboutEM	The work assumed the analysis of different roundabout scenarios and the impact of the traffic flow on them on the emissions of CO_2, CO, and THC.	[160]

The large cross section of vehicle traffic and emission models used shows the wide applicability of such tools for estimating vehicle emissions into the environment. Table 1 shows a selection of works that combine the topics of emission models with vehicle traffic simulation models for the years 2015–2023. From this, it can be seen that the development of models at the micro- or macro-scale does not always come down to using a model that corresponds to the scale of the simulation. Examples of the number of papers that use a particular type of software are as follows; based on the Web of Science core collection for a search of article abstracts where the key search term was simulation, emission, and VISSIM for the years 2008–2023, 120 results were found. Using the same keywords, but for VISSIM software, six papers were found. On the other hand, 73 results were found for the SUMO traffic simulation tool.

Figure 6 shows the classification and grouping for the simulation models of the delineation for micro-, meso-, and macro-scales. Some traffic simulation tools, for example, are located in two clusters; for example, for VISSIM, the simulation can be carried out at two levels of detail: micro and meso. The respective clusters of traffic simulators are combined with groups of emission models. For traffic simulators on the micro-scale, the connection to the emission model naturally occurs not only for the micro-scale emission models, but also for the macro-scale models. The reason for this is that we can generate instantaneous speed and acceleration data from these models, which can then be aggregated and used to calculate, for example, the average speed of a vehicle. Such a parameter, together with traffic volume information, already allows such micro-scale simulation models to be used to calculate macro-scale emissions.

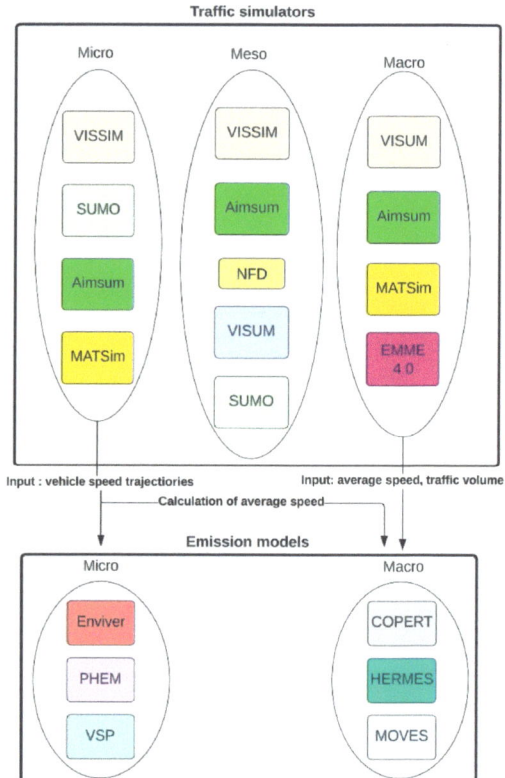

Figure 6. Classification of the main groups of traffic simulators models and emission models with the relation between them.

3.5. Importance of Traffic Simulator Calibration in the Context of Emission Calculation

An important aspect in the context of emission and traffic simulation models is their appropriate calibration [178]. A previously performed analysis of papers in the Web of

Science Core collection database confirmed this statement, while the calibration itself was found to be in the closest proximity to the main keyword emission. The calibration of various parameters in simulation models is crucial to achieve sufficiently good results for the application under analysis. The calibration aspect mainly concerns micro-scale models, where we have many vehicle movement parameters that affect their dynamics, and consequently the generated emissions. These parameters are strictly according to the particular driving model that is used in the application [179]. Examples include parameters such as minimum spacing between vehicles, the time it takes to join the traffic stream, acceleration from 80 km/h, and any statistical distribution of speed and acceleration [180,181]. In VISSIM, the parameters related to vehicle speed and acceleration are defined as the desired speed and acceleration and are functions that we can freely shape based on real data. If we do not perform this step of model calibration, the standard values of the model parameters provided by the software manufacturer are only applicable for specific conditions, which are not described in the software manual files. The model calibration should be performed not only for the national scale, but also for the regional scale. For example, the driving style of drivers in Tokyo, London, or Paris is different from that of drivers in smaller cities such as Enna, Görlitz, and Rzeszow [182,183]. The number of calibration variables in simulation models is usually large, which makes the process complicated for the normal user. Sometimes the calibration carried out is also wrong, because, for the vehicle traffic example, for simulation models, users are guided by the calibration of, e.g., the traffic volume parameter, so that the simulation is as close as possible to the real one, which is not efficient considering aspects of vehicle emissions [184].

For micro-scale simulation models, vehicle trajectories, vehicle speeds, and acceleration should be taken into account in particular [185]. The software-generated results of these trajectories should then be as close as possible to the real ones. Although some macro traffic characteristics, such as queues and travel times, are the results of the simulation, more detailed characteristics, such as speed and acceleration profiles, are not guaranteed to be realistic enough to predict emissions. To obtain an accurate estimate of emissions, calibration must be carried out from the perspective of real vehicle trajectories. In case studies, the default parameter settings of, e.g., the VISSIM and AIMSUN microsimulation software did not produce realistic trajectory results [186–188]. It is clear that the default parameter settings do not produce an effective result for estimating emissions.

An overview of the work in which traffic simulation models were calibrated to estimate emissions from computational models is presented in Table 6.

Table 6. Selected papers describing the essence of calibrating traffic simulation models and exhaust emission models.

Traffic Simulator Model	Emission Model	Calibration Description	Source
VISSIM	VERSIT+	This paper describes the process of calibrating a microscopic model for an intersection in Rotterdam. The selected pairs of parameters for calibration that have the greatest influence on the calculation of emissions are: desired speed distribution, desired deceleration and acceleration functions, time headway, threshold for entering the state following, following variation, oscillation acceleration, acceleration to standstill. Failure to carry out this process would result in emission underestimates for CO_2 of 9.3%, for NO_x of 13%, and for PM_{10} of 0.3%.	[189]
Paramics	MOBILE6	This paper proposes an intermediate component model that provides a better estimation of vehicle speeds on sections. The model was developed using multimodel linear regression and was then calibrated using a traffic microscopic simulation model. This calibration was mainly based on the vehicle speed parameter of the road sections under study.	[190]

Table 6. Cont.

Traffic Simulator Model	Emission Model	Calibration Description	Source
VISSIM	PHEM	This work involves combining traffic simulation models with emission models to assess the impact of traffic on environmental factors. Based on the actual results, a traffic model calibration was carried out for the acceleration and desired speed parameters. It was shown that not carrying out the calibration step results in an underestimation of fuel consumption by about 12%, NO_x emissions by 19%, CO by 11%, HC by 29%, PM by 17%, PN by 16%, and NO by 25%.	[191]
NGSIM	Instantaneous vehicle emission model TU Graz	This work assumes the use of 90 vehicle trajectories along with nine collected in the NGSIM software database. This work uses two car-following models for which the trajectories were generated: Newell and Gipps. The models were calibrated using standard goodness-of-fit indicators. The focus of this study was on the analysis of fuel consumption, NO_x, and PM emissions.	[192]
VISSIM/NGSIM	MOVES	The calibration process is based on different compositions of the measure of effect (MOE) (calibration levels), which contain aggregated traffic volume data in order to identify those variables that affect the micro-scale emission estimates. First, five more detailed measurement calibration levels are defined, important calibration levels are identified, and finally reliable calibration levels are selected based on the available traffic data. The influence of vehicle type composition (light and heavy vehicles) on the estimated emissions is also assessed in a well-calibrated simulation.	[193]
AIMSUN	VSP	This paper describes the calibration of a micro-scale emission model based on real data from the PEMS system. This calibration relates to the embedded emissions model in the AIMSUN software. The road data refers to 35 vehicles. The recommendation of this work is to integrate the emission data from the AIMSUN software with the MOVES emission models.	[194]
VISSIM	COPERT SL/R-LINE	The aim of this study was to analyse vehicle emissions as a result of reducing vehicle traffic and increasing pedestrian traffic in the city of Madrid. Modelling was done on a macro scale, while it was important to validate and calibrate the traffic volume obtained on the studied streets with that of the VISUM programme. This aspect had the greatest impact on the CO, NO_x, and PM parameters studied. The model was checked with the GEH.	[195]
VISSIM	Enviver/RoundaboutEM	In this study, microsimulation models of vehicle traffic were calibrated for the parameters: time headway, following variation, oscillation acceleration, acceleration from standstill, threshold for entering the state following, and time gap and minimum headway for yield sign. In addition to this, parameters such as speed and acceleration desired for the actual journey through the roundabout under study were calibrated.	[160]
VISSIM	VSP/EMEP/EEA	This work deals with the evaluation of emissions from autonomous vehicles for their different traffic shares. In this work, the driving model behind the Wiedemann 99 leader was used. The parameters that were calibrated were acceleration at 80 km/h, standstill acceleration, oscillation, acceleration, following threshold, headway time, and standstill distance. For the lane changing model, the minimum headway and safety distance reduction factor were calibrated. With these assumptions, CO_2, CO, NO_x, and HC emissions were analysed in the paper.	[196]
VISSIM	VERSIT+	This paper describes the process of calibrating the car-following model to reflect the actual driver behaviour for eco-driving style. The parameters of the car-following model that were modified were: desired acceleration from standstill at 80 km/h, oscillation during acceleration, influence of distance on speed oscillation, positive speed difference during the following process, start of deceleration process, distance difference, headway time, and standstill distance. The emission parameters studied were CO_2 and NO_x.	[197]
VISSIM	VSP	This work concerns a micro-scale simulation model with a signal controller, for which light duty vehicles were equipped with a cooperative vehicle infrastructure system. The following parameters were calibrated in VISSIM: maximum look-ahead distance, average standstill distance, the number of observed, vehicle waiting time before, diffusion, additive part of safety distance, and desired speed. After the calibration process, emissions were measured for the parameters CO_2, NO_x, HC, and CO.	[198]

4. Future Steps in Emission Modelling and Traffic Simulation and Recommendations

The subject of emissions modelling, as well as traffic modelling has developed rapidly in recent years. Initially, emission models were created on the basis of chassis dynamometer tests and the collection of emission data during their execution. Subsequent years and the technological development of emission measurement equipment made it possible to collect emission data while driving in real traffic conditions. This was made possible by PEMS equipment [199]. Many research institutes around the world now carry out vehicle emission tests in this way, generating large amounts of data that can then be used to develop new and more accurate emission models through processing [200,201].

In terms of vehicle traffic modelling, calibration of the traffic model is also an important topic in the work described earlier. The calibration of the traffic models for further analysis is of great importance because this process has a major impact on the results obtained from the exhaust emissions from simulated vehicles [202,203]. Traffic simulation calibration is the process of adjusting the simulation parameters in relation to the actual observed traffic conditions. This is crucial to ensure that accurate and reliable results are obtained. This process is usually concerned with adjusting parameters such as vehicle speed and acceleration, but also includes parameters related to, for example, vehicle spacing or determined issues of priority. Failure to carry out this process accurately, or omitting it altogether, can result in differences in exhaust emissions, differences of up to several dozen percent [204]. In the context of emission calculations, acceleration and vehicle speed are the most important simulation parameters, particularly at the micro scale.

Vehicle emission models need to be continuously developed and improved as more and more vehicles are put on the road every year. New vehicle designs require on-road emissions testing, and the data collected is used to create new emissions models and update the current database. This applies to almost any scale of emission models. New hybrid vehicles are another challenge facing the aspect of emission modelling [205]. These vehicles do not always generate emissions while driving, which means that micro-scale models in particular need to be sensitive to such emission characteristics. To prepare, for example, an emissions map, modern emission models must show precisely where there are, for example, increased emissions [206]. The development of new emission models is also influenced by the emergence of new fuel mixtures for road transport [207–209]. This is linked to the emergence of hydrogen as a fuel and of biofuels, which can be produced from renewable fuels [210–213]. In recent years, there has been an increase in the use of, for example, artificial intelligence computing methods so that newly developed models can reflect the emissions of hybrid vehicles [214,215].

There is also great potential in the use of data banks where information is collected, for example, for the management of a road section. Large amounts of data are generated for the main arterial roads in the city, which we can describe as big data sets. These data can also be processed in the context of accurately calibrating traffic simulation models, as well as being used as input data for emission models. The increased volume and detail of data also allows models to be developed with improved spatial resolution [216]. The development of more advanced mapping and geospatial technologies is leading to a greater focus on high-resolution emission models that take into account the unique characteristics of specific locations [217]. This is particularly relevant for traffic and emission simulations in urban areas. The increased accuracy of such models allows for better analysis of the spread of emissions near roads, pavements, cycleways, as well as in pedestrian crossing areas.

The directions for further work for both emission models and traffic simulation models are concerning electric vehicles. The traffic characteristics of these types of vehicles are somewhat different from those of conventional vehicles, i.e., they have different torque characteristics, resulting in a slightly different acceleration distribution of acceleration [218–220]. In the context of traffic simulation models, this state of affairs requires the additional calibration of vehicle movements with this type of vehicle. Regarding the modelling of emissions from EVs, it is important to consider the topic of energy efficiency, as the converted emissions generated by an EV depend on the power source used to charge this type of vehicle. This is mainly the case in countries where non-renewable energy, such as coal deposits, is used for energy production, where we can speak of non-direct emissions.

Recommendations on the Applicability of Emission Models and Traffic Simulation Models

Based on the information from the review of the work presented above, some recommendations can be made to those who are or will be involved in the field of emissions modelling and traffic simulation. Selected recommendations for modelling are as follows:

- For micro-scale models, it is often necessary to collect data at a frequency of at least 1 Hz. Such data for road tests, e.g., for the speed parameter, can be obtained using GPS and the OBDII interface; traffic models such as, e.g., VISSIM can also aggregate simulation data, which can then be exported to emission models such as Versit+.
- The combination of emission models and simulation models is not always at the same scale, and it is often possible to find work that uses micro-scale traffic models while using macro-scale emission models for emission estimations.
- For micro-scale traffic models, it is possible to aggregate the results of, for example, all vehicle trips into the average speed parameter, which can be used as an input for the macro-scale emission models.
- For both micro and macro scales, it is necessary to calibrate the traffic simulation models, as this has a significant impact on the emissions calculation.
- At the micro scale, calibration concerns parameters related to vehicle dynamics such as acceleration and desired speed, but also other parameters such as vehicle lateral distance, among others.
- Calibration is necessary because the emission models do not contain universal data that are characteristic of the whole world, and emission estimates are to a large extent influenced by the driving style of the driver, which varies according to the size and characteristics of the region.
- Standard settings of traffic simulation software, for example, VISSIM and Aimsum, often do not include actual vehicle trajectories.

5. Conclusions

This paper provides an overview of the emission models used for different traffic simulation methods for environmental analyses. The use of emission models with a combination of traffic simulation models is essential for transportation management and planning. This approach allows for a better understanding of the emissions generated by vehicles and their impact on air and environmental quality. In the context of emissions and traffic modelling, accurate calibration is necessary so that emission results can be reliably obtained. Simulation models are used as a tool to generate travel data from the input data of emission models. A common use of simulation is as a decision support tool, and the same is true for driving simulators, which can help decision makers in the road transport sector. The right choice of tools is crucial for an adequate understanding of the problem under analysis. The variety of tools available on the market, both in terms of emission models and traffic simulation models, is a challenge for those who wish to use them to calculate emissions from road transport sources. This paper also draws attention to the important aspect of calibrating traffic simulation models to obtain as reliable vehicle emission results as possible. For example, in the context of micro-scale models, it is important to obtain vehicle speed and acceleration profiles that are as close to the actual trajectories.

Future topics related to emissions and traffic modelling include better access to databases, which are created by road operators, and which can be used to calibrate models, and the use of modern computing techniques based on artificial intelligence methods. Vehicle emission models and traffic simulators should be integrated with other models and data sources, such as land use and energy models, to provide a complete and more accurate picture of the vehicle emissions generated. The shortcomings of this study are related to the choice of papers and models to be analysed, while the rationale for this is that it was guided by the popularity of the solutions used and the scope of this work had to be limited. Another shortcoming of this study is the broader description of topics related to artificial intelligence and support for modelling of vehicle traffic and emissions, while the shortcomings mentioned will also be the scope of new and other works.

This main conclusions from the literature review are the following:

- In order to select suitable emission calculations and vehicle movement modelling tools, it is important to know their various capabilities in advance.
- It is necessary to carry out a calibration process for the traffic model, e.g., for micro-scale modelling, it is necessary to calibrate the vehicle acceleration function and the vehicle speed.
- It is possible to use, e.g., micro-scale traffic models and use the emission model for the macro scale, but an input data set must be prepared beforehand, and these data must be validated accordingly.
- Standard data generated by models are often a generalisation and assume some standard data of the software manufacturer. It is necessary to adapt these data and validate them to the local, regional, and national conditions under analysis.

Funding: This research received no external funding.

Data Availability Statement: Not applicable.

Conflicts of Interest: The author declares no conflict of interest.

Appendix A

Table A1. A list of access pages for selected emission models and traffic simulation models.

Type of Model	Name of the Model	Source
Emission model (macro)	COPERT	[221]
Emission model (macro)	MOBILE	[222]
Emission model (macro)	HBEFA	[223]
Emission model (micro)	CMEM	[224]
Emission model (micro)	ESTM BOSH	[225]
Emission model (micro)	EMPA	[226]
Traffic simulator model (micro)	VISSIM	[227]
Traffic simulator model (macro)	AIMSUN	[228]
Traffic simulator model (micro)	SUMO	[229]
Traffic simulator model (macro)	VISUM	[230]
Traffic simulator model (macro)	MATSim	[231]
Traffic simulator model (macro)	EMME 4.0	[232]

References

1. Jin, Y.; Andersson, H.; Zhang, S. Air Pollution Control Policies in China: A Retrospective and Prospects. *Int. J. Environ. Res. Public Health* **2016**, *13*, 1219. [CrossRef] [PubMed]
2. Bo, M.; Salizzoni, P.; Clerico, M.; Buccolieri, R. Assessment of Indoor-Outdoor Particulate Matter Air Pollution: A Review. *Atmosphere* **2017**, *8*, 136. [CrossRef]
3. Inkinen, T.; Hämäläinen, E. Reviewing Truck Logistics: Solutions for Achieving Low Emission Road Freight Transport. *Sustainability* **2020**, *12*, 6714. [CrossRef]
4. Kuszewski, H.; Jaworski, A.; Mądziel, M. Lubricity of Ethanol–Diesel Fuel Blends—Study with the Four-Ball Machine Method. *Materials* **2021**, *14*, 2492. [CrossRef] [PubMed]
5. Nellore, K.; Hancke, G.P. A survey on urban traffic management system using wireless sensor networks. *Sensors* **2016**, *16*, 157. [CrossRef]
6. Vidhi, R.; Shrivastava, P. A Review of Electric Vehicle Lifecycle Emissions and Policy Recommendations to Increase EV Penetration in India. *Energies* **2018**, *11*, 483. [CrossRef]
7. Wang, L.; Zhang, F.; Pilot, E.; Yu, J.; Nie, C.; Holdaway, J.; Yang, L.; Li, Y.; Wang, W.; Vardoulakis, S.; et al. Taking Action on Air Pollution Control in the Beijing-Tianjin-Hebei (BTH) Region: Progress, Challenges and Opportunities. *Int. J. Environ. Res. Public Health* **2018**, *15*, 306. [CrossRef]
8. Liu, H.; Wang, X.; Zhang, D.; Dong, F.; Liu, X.; Yang, Y.; Huang, H.; Wang, Y.; Wang, Q.; Zheng, Z. Investigation on Blending Effects of Gasoline Fuel with N-Butanol, DMF, and Ethanol on the Fuel Consumption and Harmful Emissions in a GDI Vehicle. *Energies* **2019**, *12*, 1845. [CrossRef]
9. Ziółkowski, A.; Fuć, P.; Lijewski, P.; Jagielski, A.; Bednarek, M.; Kusiak, W. Analysis of Exhaust Emissions from Heavy-Duty Vehicles on Different Applications. *Energies* **2022**, *15*, 7886. [CrossRef]

10. Jaworski, A.; Lejda, K.; Mądziel, M.; Ustrzycki, A. Assessment of the emission of harmful car exhaust components in real traffic conditions. *IOP Conf. Ser. Mater. Sci. Eng.* **2018**, *421*, 042031. [CrossRef]
11. Kawamoto, R.; Mochizuki, H.; Moriguchi, Y.; Nakano, T.; Motohashi, M.; Sakai, Y.; Inaba, A. Estimation of CO_2 Emissions of Internal Combustion Engine Vehicle and Battery Electric Vehicle Using LCA. *Sustainability* **2019**, *11*, 2690. [CrossRef]
12. Mazza, S.; Aiello, D.; Macario, A.; De Luca, P. Vehicular Emission: Estimate of Air Pollutants to Guide Local Political Choices. A Case Study. *Environments* **2020**, *7*, 37. [CrossRef]
13. Hu, H.; Lee, G.; Kim, J.H.; Shin, H. Estimating Micro-Level On-Road Vehicle Emissions Using the K-Means Clustering Method with GPS Big Data. *Electronics* **2020**, *9*, 2151. [CrossRef]
14. Kachba, Y.; Chiroli, D.M.D.G.; Belotti, J.T.; Antonini Alves, T.; de Souza Tadano, Y.; Siqueira, H. Artificial neural networks to estimate the influence of vehicular emission variables on morbidity and mortality in the largest metropolis in South America. *Sustainability* **2020**, *12*, 2621. [CrossRef]
15. Obaid, M.; Torok, A.; Ortega, J. A Comprehensive Emissions Model Combining Autonomous Vehicles with Park and Ride and Electric Vehicle Transportation Policies. *Sustainability* **2021**, *13*, 4653. [CrossRef]
16. Zhang, Y.; Zhou, R.; Peng, S.; Mao, H.; Yang, Z.; Andre, M.; Zhang, X. Development of Vehicle Emission Model Based on Real-Road Test and Driving Conditions in Tianjin, China. *Atmosphere* **2022**, *13*, 595. [CrossRef]
17. Hatzopoulou, M.; Miller, E.J.; Santos, B. Integrating vehicle emission modeling with activity-based travel demand modeling: Case study of the Greater Toronto, Canada, Area. *Transp. Res. Rec.* **2007**, *2011*, 29–39. [CrossRef]
18. Wang, L.; Chen, X.; Xia, Y.; Jiang, L.; Ye, J.; Hou, T.; Wang, L.; Zhang, Y.; Li, M.; Li, Z.; et al. Operational Data-Driven Intelligent Modelling and Visualization System for Real-World, On-Road Vehicle Emissions—A Case Study in Hangzhou City, China. *Sustainability* **2022**, *14*, 5434. [CrossRef]
19. Beza, A.D.; Maghrour Zefreh, M.; Torok, A. Impacts of different types of automated vehicles on traffic flow characteristics and emissions: A microscopic traffic simulation of different freeway segments. *Energies* **2022**, *15*, 6669. [CrossRef]
20. Plakolb, S.; Jäger, G.; Hofer, C.; Füllsack, M. Mesoscopic Urban-Traffic Simulation Based on Mobility Behavior to Calculate NOx Emissions Caused by Private Motorized Transport. *Atmosphere* **2019**, *10*, 293. [CrossRef]
21. Gupta, M.; Mohan, M.; Bhati, S. Assessment of Air Pollution Mitigation Measures on Secondary Pollutants PM_{10} and Ozone Using Chemical Transport Modelling over Megacity Delhi, India. *Urban Sci.* **2022**, *6*, 27. [CrossRef]
22. Liu, L.; Zhang, X.; Xu, W.; Liu, X.; Lu, X.; Wang, S.; Zhang, W.; Zhao, L. Ground Ammonia Concentrations over China Derived from Satellite and Atmospheric Transport Modeling. *Remote. Sens.* **2017**, *9*, 467. [CrossRef]
23. Al-Turki, M.; Jamal, A.; Al-Ahmadi, H.M.; Al-Sughaiyer, M.A.; Zahid, M. On the Potential Impacts of Smart Traffic Control for Delay, Fuel Energy Consumption, and Emissions: An NSGA-II-Based Optimization Case Study from Dhahran, Saudi Arabia. *Sustainability* **2020**, *12*, 7394. [CrossRef]
24. Maurer, R.; Kossioris, T.; Sterlepper, S.; Günther, M.; Pischinger, S. Achieving Zero-Impact Emissions with a Gasoline Passenger Car. *Atmosphere* **2023**, *14*, 313. [CrossRef]
25. Progiou, A.; Liora, N.; Sebos, I.; Chatzimichail, C.; Melas, D. Measures and Policies for Reducing PM Exceedances through the Use of Air Quality Modeling: The Case of Thessaloniki, Greece. *Sustainability* **2023**, *15*, 930. [CrossRef]
26. Koupal, J.; Beardsley, M.; Brzezinski, D.; Warila, J.; Faler, W. *US EPA's MOVES2010 Vehicle Emission Model: Overview and Considerations for International Application*; US Environmental Protection Agency, Office of Transportation and Air Quality: Ann Arbor, MI, USA, 2010. Available online: https://www.epa.gov/sites/default/files/2019-08/documents/paper137-tap2010.pdf (accessed on 4 May 2023).
27. Ahn, K. Microscopic Fuel Consumption and Emission Modeling. Ph.D. Thesis, Virginia Tech, Blacksburg, VA, USA, 1998.
28. Jamshidnejad, A.; Papamichail, I.; Papageorgiou, M.; De Schutter, B. A mesoscopic integrated urban traffic flow-emission model. *Transp. Res. Part C Emerg. Technol.* **2017**, *75*, 45–83. [CrossRef]
29. Kean, A.J.; Harley, R.A.; Kendall, G.R. Effects of vehicle speed and engine load on motor vehicle emissions. *Environ. Sci. Technol.* **2003**, *37*, 3739–3746. [CrossRef]
30. Young Park, J.; Noland, R.B.; Polak, J.W. Microscopic model of air pollutant concentrations: Comparison of simulated results with measured and macroscopic estimates. *Transp. Res. Rec.* **2001**, *1750*, 64–73. [CrossRef]
31. Ajtay, D.; Weilenmann, M.; Soltic, P. Towards accurate instantaneous emission models. *Atmos. Environ.* **2005**, *39*, 2443–2449. [CrossRef]
32. Faris, W.; Rakha, H.; Kafafy, R.I.; Idres, M.; Elmoselhy, S. Vehicle fuel consumption and emission modelling: An in-depth literature review. *Int. J. Veh. Syst. Model. Test.* **2011**, *6*, 318. [CrossRef]
33. Demir, E.; Bektaş, T.; Laporte, G. A comparative analysis of several vehicle emission models for road freight transportation. *Transp. Res. Part D Transp. Environ.* **2011**, *16*, 347–357. [CrossRef]
34. Gokhale, S.; Khare, M. A review of deterministic, stochastic and hybrid vehicular exhaust emission models. *Int. J. Transp. Manag.* **2004**, *2*, 59–74. [CrossRef]
35. Wang, H.; McGlinchy, I. Review of vehicle emission modelling and the issues for New Zealand. In Proceedings of the 32nd Australasian Transport Research Forum, Auckland, New Zealand, 29 September–1 October 2009.
36. Pel, A.J.; Bliemer, M.C.J.; Hoogendoorn, S.P. A review on travel behaviour modelling in dynamic traffic simulation models for evacuations. *Transportation* **2012**, *39*, 97–123. [CrossRef]

37. Nguyen, J.; Powers, S.T.; Urquhart, N.; Farrenkopf, T.; Guckert, M. An overview of agent-based traffic simulators. *Transp. Res. Interdiscip. Perspect.* **2021**, *12*, 100486. [CrossRef]
38. Ejercito, P.M.; Nebrija, K.G.E.; Feria, R.P.; Lara-Figueroa, L.L. Traffic simulation software review. In Proceedings of the 2017 8th International Conference on Information, Intelligence, Systems & Applications (IISA), Larnaca, Cyprus, 28–30 August 2017; pp. 1–4.
39. Liu, C.-J.; Liu, Z.; Chai, Y.-J.; Liu, T.-T. Review of Virtual Traffic Simulation and Its Applications. *J. Adv. Transp.* **2020**, *2020*, 8237649. [CrossRef]
40. Alghamdi, T.; Mostafi, S.; Abdelkader, G.; Elgazzar, K. A Comparative Study on Traffic Modeling Techniques for Predicting and Simulating Traffic Behavior. *Future Internet* **2022**, *14*, 294. [CrossRef]
41. Mubasher, M.M.; ul Qounain, J.S.W. Systematic literature review of vehicular traffic flow simulators. In Proceedings of the 2015 International Conference on Open Source Software Computing (OSSCOM), Amman, Jordan, 10–13 September 2015; pp. 1–6.
42. Forehead, H.; Huynh, N. Review of modelling air pollution from traffic at street-level-The state of the science. *Environ. Pollut.* **2018**, *241*, 775–786. [CrossRef]
43. Algers, S.; Bernauer, E.; Boero, M.; Breheret, L.; Di Taranto, C.; Dougherty, M.; Fox, K.; Gabard, J.F. Review of Micro-Simulation Models. Review Report of the SMARTEST Project. 1997. Available online: https://citeseerx.ist.psu.edu/document?repid=rep1&type=pdf&doi=f981a04dbf276f4298bdaab3b5acf43af6fcea5f (accessed on 4 May 2023).
44. Madi, M.Y. Investigating and calibrating the dynamics of vehicles in traffic micro-simulations models. *Transp. Res. Procedia* **2016**, *14*, 1782–1791. [CrossRef]
45. Franco, V.; Kousoulidou, M.; Muntean, M.; Ntziachristos, L.; Hausberger, S.; Dilara, P. Road vehicle emission factors development: A review. *Atmos. Environ.* **2013**, *70*, 84–97. [CrossRef]
46. Chen, B.; Shin, S. Bibliometric Analysis on Research Trend of Accidental Falls in Older Adults by Using Citespace—Focused on Web of Science Core Collection (2010–2020). *Int. J. Environ. Res. Public Health* **2021**, *18*, 1663. [CrossRef]
47. Barneo-Alcántara, M.; Díaz-Pérez, M.; Gómez-Galán, M.; Carreño-Ortega, Á.; Callejón-Ferre, Á.J. Musculoskeletal disorders in agriculture: A review from web of science core collection. *Agronomy* **2021**, *11*, 2017. [CrossRef]
48. Wallington, T.J.; Anderson, J.E.; Dolan, R.H.; Winkler, S.L. Vehicle Emissions and Urban Air Quality: 60 Years of Progress. *Atmosphere* **2022**, *13*, 650. [CrossRef]
49. Piracha, A.; Chaudhary, M.T. Urban Air Pollution, Urban Heat Island and Human Health: A Review of the Literature. *Sustainability* **2022**, *14*, 9234. [CrossRef]
50. Holnicki, P.; Nahorski, Z.; Kałuszko, A. Impact of Vehicle Fleet Modernization on the Traffic-Originated Air Pollution in an Urban Area—A Case Study. *Atmosphere* **2021**, *12*, 1581. [CrossRef]
51. Ko, S.; Park, J.; Kim, H.; Kang, G.; Lee, J.; Kim, J.; Lee, J. NOx Emissions from Euro 5 and Euro 6 Heavy-Duty Diesel Vehicles under Real Driving Conditions. *Energies* **2020**, *13*, 218. [CrossRef]
52. Hooftman, N.; Oliveira, L.; Messagie, M.; Coosemans, T.; Van Mierlo, J. Environmental Analysis of Petrol, Diesel and Electric Passenger Cars in a Belgian Urban Setting. *Energies* **2016**, *9*, 84. [CrossRef]
53. Iqbal, A.; Afroze, S.; Rahman, M. Vehicular PM Emissions and Urban Public Health Sustainability: A Probabilistic Analysis for Dhaka City. *Sustainability* **2020**, *12*, 6284. [CrossRef]
54. Penkała, M.; Ogrodnik, P.; Rogula-Kozłowska, W. Particulate Matter from the Road Surface Abrasion as a Problem of Non-Exhaust Emission Control. *Environments* **2018**, *5*, 9. [CrossRef]
55. Giechaskiel, B.; Forloni, F.; Carriero, M.; Baldini, G.; Castellano, P.; Vermeulen, R.; Kontses, D.; Fragkiadoulakis, P.; Samaras, Z.; Fontaras, G. Effect of Tampering on On-Road and Off-Road Diesel Vehicle Emissions. *Sustainability* **2022**, *14*, 6065. [CrossRef]
56. Kole, P.J.; Löhr, A.J.; Van Belleghem, F.G.A.J.; Ragas, A.M.J. Wear and Tear of Tyres: A Stealthy Source of Microplastics in the Environment. *Int. J. Environ. Res. Public Health* **2017**, *14*, 1265. [CrossRef]
57. Bessagnet, B.; Allemand, N.; Putaud, J.P.; Couvidat, F.; André, J.M.; Simpson, D.; Pisoni, E.; Murphy, B.N.; Thunis, P. Emissions of Carbonaceous Particulate Matter and Ultrafine Particles from Vehicles—A Scientific Review in a Cross-Cutting Context of Air Pollution and Climate Change. *Appl. Sci.* **2022**, *12*, 3623. [CrossRef] [PubMed]
58. Wang, L.; Zhong, B.; Vardoulakis, S.; Zhang, F.; Pilot, E.; Li, Y.; Yang, L.; Wang, W.; Krafft, T. Air Quality Strategies on Public Health and Health Equity in Europe—A Systematic Review. *Int. J. Environ. Res. Public Health* **2016**, *13*, 1196. [CrossRef]
59. Connerton, P.; Vicente de Assunção, J.; Maura de Miranda, R.; Dorothée Slovic, A.; José Pérez-Martínez, P.; Ribeiro, H. Air quality during COVID-19 in four megacities: Lessons and challenges for public health. *Int. J. Environ. Res. Public Health* **2020**, *17*, 5067. [CrossRef] [PubMed]
60. Sun, C.; Zhang, J.; Ma, Q.; Chen, Y. Human health and ecological risk assessment of 16 polycyclic aromatic hydro-carbons in drinking source water from a large mixed-use reservoir. *Int. J. Environ. Res. Public Health* **2015**, *12*, 13956–13969. [CrossRef] [PubMed]
61. Haque, M.S.; Singh, R.B. Air pollution and human health in Kolkata, India: A case study. *Climate* **2017**, *5*, 77. [CrossRef]
62. Selleri, T.; Melas, A.D.; Joshi, A.; Manara, D.; Perujo, A.; Suarez-Bertoa, R. An Overview of Lean Exhaust deNOx Aftertreatment Technologies and NOx Emission Regulations in the European Union. *Catalysts* **2021**, *11*, 404. [CrossRef]
63. Selleri, T.; Gioria, R.; Melas, A.D.; Giechaskiel, B.; Forloni, F.; Villafuerte, P.M.; Demuynck, J.; Bosteels, D.; Wilkes, T.; Simons, O.; et al. Measuring Emissions from a Demonstrator Heavy-Duty Diesel Vehicle under Real-World Conditions—Moving Forward to Euro VII. *Catalysts* **2022**, *12*, 184. [CrossRef]

64. Liu, X.; Zhao, F.; Hao, H.; Chen, K.; Liu, Z.; Babiker, H.; Amer, A.A. From NEDC to WLTP: Effect on the Energy Consumption, NEV Credits, and Subsidies Policies of PHEV in the Chinese Market. *Sustainability* **2020**, *12*, 5747. [CrossRef]
65. Lee, H.; Lee, K. Comparative Evaluation of the Effect of Vehicle Parameters on Fuel Consumption under NEDC and WLTP. *Energies* **2020**, *13*, 4245. [CrossRef]
66. Giakoumis, E.G.; Zachiotis, A.T. Investigation of a Diesel-Engined Vehicle's Performance and Emissions during the WLTC Driving Cycle—Comparison with the NEDC. *Energies* **2017**, *10*, 240. [CrossRef]
67. Kaya, T.; Kutlar, O.A.; Taskiran, O.O. Evaluation of the Effects of Biodiesel on Emissions and Performance by Comparing the Results of the New European Drive Cycle and Worldwide Harmonized Light Vehicles Test Cycle. *Energies* **2018**, *11*, 2814. [CrossRef]
68. Grigoratos, T.; Agudelo, C.; Grochowicz, J.; Gramstat, S.; Robere, M.; Perricone, G.; Sin, A.; Paulus, A.; Zessinger, M.; Hortet, A.; et al. Statistical Assessment and Temperature Study from the Interlaboratory Application of the WLTP–Brake Cycle. *Atmosphere* **2020**, *11*, 1309. [CrossRef]
69. Bodisco, T.; Zare, A. Practicalities and driving dynamics of a real driving emissions (RDE) Euro 6 regulation homologation test. *Energies* **2019**, *12*, 2306. [CrossRef]
70. Andrych-Zalewska, M.; Chlopek, Z.; Merkisz, J.; Pielecha, J. Comparison of Gasoline Engine Exhaust Emissions of a Passenger Car through the WLTC and RDE Type Approval Tests. *Energies* **2022**, *15*, 8157. [CrossRef]
71. Pielecha, J.; Skobiej, K.; Gis, M.; Gis, W. Particle Number Emission from Vehicles of Various Drives in the RDE Tests. *Energies* **2022**, *15*, 6471. [CrossRef]
72. Varella, R.A.; Giechaskiel, B.; Sousa, L.; Duarte, G. Comparison of Portable Emissions Measurement Systems (PEMS) with Laboratory Grade Equipment. *Appl. Sci.* **2018**, *8*, 1633. [CrossRef]
73. Giechaskiel, B.; Casadei, S.; Rossi, T.; Forloni, F.; Di Domenico, A. Measurements of the Emissions of a "Golden" Vehicle at Seven Laboratories with Portable Emission Measurement Systems (PEMS). *Sustainability* **2021**, *13*, 8762. [CrossRef]
74. Chen, J.; Li, Y.; Meng, Z.; Feng, X.; Wang, J.; Zhou, H.; Li, J.; Shi, J.; Chen, Q.; Shi, H.; et al. Study on Emission Characteristics and Emission Reduction Effect for Construction Machinery under Actual Operating Conditions Using a Portable Emission Measurement System (Pems). *Int. J. Environ. Res. Public Health* **2022**, *19*, 9546. [CrossRef]
75. Zhang, R.; Wang, Y.; Pang, Y.; Zhang, B.; Wei, Y.; Wang, M.; Zhu, R. A Deep Learning Micro-Scale Model to Estimate the CO_2 Emissions from Light-Duty Diesel Trucks Based on Real-World Driving. *Atmosphere* **2022**, *13*, 1466. [CrossRef]
76. Bifulco, G.N.; Galante, F.; Pariota, L.; Spena, M.R. A Linear Model for the Estimation of Fuel Consumption and the Impact Evaluation of Advanced Driving Assistance Systems. *Sustainability* **2015**, *7*, 14326–14343. [CrossRef]
77. Mądziel, M.; Campisi, T. Assessment of vehicle emissions at roundabouts: A comparative study of PEMS data and microscale emission model. *Arch. Transp.* **2022**, *63*, 35–51. [CrossRef]
78. Smit, R.; Ntziachristos, L.; Boulter, P. Validation of road vehicle and traffic emission models—A review and meta-analysis. *Atmos. Environ.* **2010**, *44*, 2943–2953. [CrossRef]
79. Davis, N.; Lents, J.; Osses, M.; Nikkila, N.; Barth, M. Development and Application of an International Vehicle Emissions Model. *Transp. Res. Rec. J. Transp. Res. Board* **2005**, *1939*, 156–165. [CrossRef]
80. De Nunzio, G.; Laraki, M.; Thibault, L. Road Traffic Dynamic Pollutant Emissions Estimation: From Macroscopic Road Information to Microscopic Environmental Impact. *Atmosphere* **2020**, *12*, 53. [CrossRef]
81. Rakha, H.A.; Ahn, K.; Moran, K.; Saerens, B.; Van Den Bulck, E. Virginia Tech Comprehensive Power-Based Fuel Consumption Model: Model development and testing. *Transp. Res. Part D Transp. Environ.* **2011**, *16*, 492–503. [CrossRef]
82. Obaid, M.; Torok, A. Macroscopic Traffic Simulation of Autonomous Vehicle Effects. *Vehicles* **2021**, *3*, 187–196. [CrossRef]
83. Schnieder, M.; Hinde, C.; West, A. Emission Estimation of On-Demand Meal Delivery Services Using a Macroscopic Simulation. *Int. J. Environ. Res. Public Health* **2022**, *19*, 11667. [CrossRef]
84. Sówka, I.; Pawnuk, M.; Miller, U.; Grzelka, A.; Wroniszewska, A.; Bezyk, Y. Assessment of the Odour Impact Range of a Selected Agricultural Processing Plant. *Sustainability* **2020**, *12*, 7289. [CrossRef]
85. Nowakowicz-Dębek, B.; Wlazło, Ł.; Szymula, A.; Ossowski, M.; Kasela, M.; Chmielowiec-Korzeniowska, A.; Bis-Wencel, H. Estimating Methane Emissions from a Dairy Farm Using a Computer Program. *Atmosphere* **2020**, *11*, 803. [CrossRef]
86. Wlazło, Ł.; Nowakowicz-Dębek, B.; Ossowski, M.; Stasińska, B.; Kułażyński, M. Estimation of ammonia emissions from a dairy farm using a computer program. *Carbon Manag.* **2020**, *11*, 195–201. [CrossRef]
87. De Blasiis, M.R.; Ferrante, C.; Palmieri, F.; Veraldi, V. Coupling Virtual Reality Simulator with Instantaneous Emission Model: A New Method for Estimating Road Traffic Emissions. *Sustainability* **2022**, *14*, 6793. [CrossRef]
88. Li, F.; Zhuang, J.; Cheng, X.; Li, M.; Wang, J.; Yan, Z. Investigation and Prediction of Heavy-Duty Diesel Passenger Bus Emissions in Hainan Using a COPERT Model. *Atmosphere* **2019**, *10*, 106. [CrossRef]
89. Jaworski, A.; Mądziel, M.; Kuszewski, H. Sustainable Public Transport Strategies—Decomposition of the Bus Fleet and Its Influence on the Decrease in Greenhouse Gas Emissions. *Energies* **2022**, *15*, 2238. [CrossRef]
90. Ali, M.; Kamal, M.D.; Tahir, A.; Atif, S. Fuel Consumption Monitoring through COPERT Model—A Case Study for Urban Sustainability. *Sustainability* **2021**, *13*, 11614. [CrossRef]
91. Weng, J.; Wang, R.; Wang, M.; Rong, J. Fuel Consumption and Vehicle Emission Models for Evaluating Environmental Impacts of the ETC System. *Sustainability* **2015**, *7*, 8934–8949. [CrossRef]

92. Dong, Y.; Xu, J.; Liu, X.; Gao, C.; Ru, H.; Duan, Z. Carbon Emissions and Expressway Traffic Flow Patterns in China. *Sustainability* **2019**, *11*, 2824. [CrossRef]
93. Hagan, R.; Markey, E.; Clancy, J.; Keating, M.; Donnelly, A.; O'Connor, D.J.; Morrison, L.; McGillicuddy, E.J. Non-Road Mobile Machinery Emissions and Regulations: A Review. *Air* **2022**, *1*, 14–36. [CrossRef]
94. Tucki, K. A Computer Tool for Modelling CO_2 Emissions in Driving Cycles for Spark Ignition Engines Powered by Biofuels. *Energies* **2021**, *14*, 1400. [CrossRef]
95. El-Sehiemy, R.; Hamida, M.A.; Elattar, E.; Shaheen, A.; Ginidi, A. Nonlinear Dynamic Model for Parameter Estimation of Li-Ion Batteries Using Supply–Demand Algorithm. *Energies* **2022**, *15*, 4556. [CrossRef]
96. Mao, F.; Li, Z.; Zhang, K. A Comparison of Carbon Dioxide Emissions between Battery Electric Buses and Conventional Diesel Buses. *Sustainability* **2021**, *13*, 5170. [CrossRef]
97. Mądziel, M. Liquified Petroleum Gas-Fuelled Vehicle CO_2 Emission Modelling Based on Portable Emission Measurement System, On-Board Diagnostics Data, and Gradient-Boosting Machine Learning. *Energies* **2023**, *16*, 2754. [CrossRef]
98. Mądziel, M.; Campisi, T.; Jaworski, A.; Tesoriere, G. The Development of Strategies to Reduce Exhaust Emissions from Passenger Cars in Rzeszow City—Poland. A Preliminary Assessment of the Results Produced by the Increase of E-Fleet. *Energies* **2021**, *14*, 1046. [CrossRef]
99. Yu, Q.; Lu, L.; Li, T.; Tu, R. Quantifying the Impact of Alternative Bus Stop Platforms on Vehicle Emissions and Individual Pollution Exposure at Bus Stops. *Int. J. Environ. Res. Public Health* **2022**, *19*, 6552. [CrossRef] [PubMed]
100. Guérette, E.-A.; Chang, L.T.-C.; Cope, M.E.; Duc, H.N.; Emmerson, K.M.; Monk, K.; Rayner, P.J.; Scorgie, Y.; Silver, J.D.; Simmons, J.; et al. Evaluation of Regional Air Quality Models over Sydney, Australia: Part 2, Comparison of $PM_{2.5}$ and Ozone. *Atmosphere* **2020**, *11*, 233. [CrossRef]
101. Smit, R.; McBroom, J. Use of microscopic simulation models to predict traffic emissions. *Road Transp. Res. A J. Aust. New Zealand Res. Pract.* **2009**, *18*, 49–54.
102. Scora, G.; Barth, M. *Comprehensive Modal Emissions Model (cmem), Version 3.01. User Guide. Centre for Environmental Research and Technology*; University of California: Riverside, CA, USA, 2006.
103. Chamberlin, R.; Swanson, B.; Talbot, E.; Dumont, J.; Pesci, S. Analysis of MOVES and CMEM for evaluating the emissions impact of an intersection control change. No. 11-0673. In Proceedings of the Transportation Research Board 90th Annual Meeting, Washington, DC, USA, 21–27 January 2011.
104. Smit, R.; Ormerod, R.; Bridge, I. Vehicle emission models and their application-Emission inventories. *Clean Air Environ. Qual.* **2002**, *36*, 30–34.
105. Perugu, H. Emission modelling of light-duty vehicles in India using the revamped VSP-based MOVES model: The case study of Hyderabad. *Transp. Res. Part D Transp. Environ.* **2019**, *68*, 150–163. [CrossRef]
106. Ahn, K.; Rakha, H.; Trani, A.; Van Aerde, M. Estimating vehicle fuel consumption and emissions based on instantaneous speed and acceleration levels. *J. Transp. Eng.* **2002**, *128*, 182–190. [CrossRef]
107. Rakha, H.; Ding, Y. Impact of Stops on Vehicle Fuel Consumption and Emissions. *J. Transp. Eng.* **2003**, *129*, 23–32. [CrossRef]
108. Panis, L.I.; Broekx, S.; Liu, R. Modelling instantaneous traffic emission and the influence of traffic speed limits. *Sci. Total Environ.* **2006**, *371*, 270–285. [CrossRef]
109. Dias, H.L.F.; Bertoncini, B.V.; Oliveira, M.L.M.D.; Cavalcante, F.S.Á.; Lima, E.P. Analysis of emission models integrated with traffic models for freight transportation study in urban areas. *Int. J. Environ. Technol. Manag.* **2017**, *20*, 60–77. [CrossRef]
110. Borge, R.; Quaassdorff, C.; Pérez, J.; de la Paz, D.; Lumbreras, J.; de Andrés, J.M.; Narros, A.; Rodríguez, E. Development of road traffic emission inventories for urban air quality modeling in Madrid (Spain). In Proceedings of the 21st USEPA International Emission Inventory Conference, San Diego, CA, USA, 13–16 April 2015.
111. Quaassdorff, C.; Borge, R.; Pérez, J.; Lumbreras, J.; de la Paz, D.; de Andrés, J.M. Microscale traffic simulation and emission estimation in a heavily trafficked roundabout in Madrid (Spain). *Sci. Total Environ.* **2016**, *566–567*, 416–427. [CrossRef] [PubMed]
112. Yang, Y.; Zhao, H.; Jiang, H. Drive Train Design and Modeling of a Parallel Diesel Hybrid Electric Bus Based on AVL/Cruise. *World Electr. Veh. J.* **2010**, *4*, 75–81. [CrossRef]
113. Ilimbetov, R.Y.; Popov, V.V.; Vozmilov, A.G. Comparative Analysis of "NGTU–Electro" Electric Car Movement Processes Modeling in MATLAB Simulink and AVL Cruise Software. *Procedia Eng.* **2015**, *129*, 879–885. [CrossRef]
114. Srinivasan, P. *Performance Fuel Economy and CO_2 Prediction of a Vehicle Using AVL Cruise Simulation Techniques*; (No. 2009-01-1862); SAE Technical Paper; SAE: Warrendale, PA, USA, 2009.
115. Cioroianu, C.C.; Marinescu, D.G.; Iorga, A.; Sibiceanu, A.R. Simulation of an electric vehicle model on the new WLTC test cycle using AVL CRUISE software. *IOP Conf. Ser. Mater. Sci. Eng.* **2017**, *252*, 012060. [CrossRef]
116. Wang, B.H.; Luo, Y.G. AVL cruise-based modeling and simulation of EQ6110 hybrid electric public bus. In Proceedings of the 2010 International Conference on Computer Application and System Modeling (ICCASM 2010), Taiyuan, China, 22–24 October 2010; Volume 7, pp. V7-252–V7-255.
117. O'Driscoll, R.; ApSimon, H.M.; Oxley, T.; Molden, N.; Stettler, M.E.; Thiyagarajah, A. A Portable Emissions Measurement System (PEMS) study of NOx and primary NO_2 emissions from Euro 6 diesel passenger cars and comparison with COPERT emission factors. *Atmos. Environ.* **2016**, *145*, 81–91. [CrossRef]
118. Information Related to COPERT Emission Model. Available online: https://www.emisia.com/utilities/copert/ (accessed on 10 April 2023).

119. Ozguven, E.E.; Ozbay, K.; Iyer, S. A simplified emissions estimation methodology based on MOVES to estimate vehicle emissions from transportation assignment and simulation models. In Proceedings of the 92nd Annual Meeting of the Transportation Research Board, Washington, DC, USA, 14–16 January 2013.
120. Information Related to MOVES Emission Model. Available online: https://www.epa.gov/moves (accessed on 10 April 2023).
121. Koupal, J.; Michaels, H.; Cumberworth, M.; Bailey, C.; Brzezinski, D. EPA's plan for MOVES: A comprehensive mobile source emissions model. In Proceedings of the 12th CRC On-Road Vehicle Emissions Workshop, San Diego, CA, USA, 15–17 April 2002; pp. 15–17.
122. Wyatt, D.W.; Li, H.; Tate, J.E. The impact of road grade on carbon dioxide (CO_2) emission of a passenger vehicle in real-world driving. *Transp. Res. Part D Transp. Environ.* **2014**, *32*, 160–170. [CrossRef]
123. Information Related to PHEM Emission Model. Available online: https://sumo.dlr.de/docs/Models/Emissions/PHEMlight.html (accessed on 10 April 2023).
124. Oswald, D.; Scora, G.; Williams, N.; Hao, P.; Barth, M. Evaluating the environmental impacts of connected and automated vehicles: Potential shortcomings of a binned-based emissions model. In Proceedings of the 2019 IEEE Intelligent Transportation Systems Conference (ITSC), Auckland, New Zealand, 27–30 October 2019; IEEE: Piscataway, NJ, USA, 2019; pp. 3639–3644.
125. De Coensel, B.; Can, A.; Degraeuwe, B.; De Vlieger, I.; Botteldooren, D. Effects of traffic signal coordination on noise and air pollutant emissions. *Environ. Model. Softw.* **2012**, *35*, 74–83. [CrossRef]
126. Ligterink, N.E.; De Lange, R.; Schoen, E. Refined vehicle and driving-behaviour dependencies in the VERSIT+ emission model. In Proceedings of the ETAPP Symposium, Toulouse, France, 2–4 June 2009; pp. 1–8.
127. Rakha, H.; Ahn, K.; Trani, A. Comparison of MOBILE5a, MOBILE6, VT-MICRO, and CMEM models for estimating hot-stabilized light-duty gasoline vehicle emissions. *Can. J. Civ. Eng.* **2003**, *30*, 1010–1021. [CrossRef]
128. Rakha, H.; Ahn, K.; Trani, A. Development of VT-Micro model for estimating hot stabilized light duty vehicle and truck emissions. *Transp. Res. Part D Transp. Environ.* **2004**, *9*, 49–74. [CrossRef]
129. Šarić, A.; Sulejmanović, S.; Albinović, S.; Pozder, M.; Ljevo, Ž. The Role of Intersection Geometry in Urban Air Pollution Management. *Sustainability* **2023**, *15*, 5234. [CrossRef]
130. Ajtay, D.; Weilenmann, M. Static and dynamic instantaneous emission modelling. *Int. J. Environ. Pollut.* **2004**, *22*, 226. [CrossRef]
131. Saharidis, G.K.; Konstantzos, G.E. Critical overview of emission calculation models in order to evaluate their potential use in estimation of Greenhouse Gas emissions from in port truck operations. *J. Clean. Prod.* **2018**, *185*, 1024–1031. [CrossRef]
132. Bai, S.; Eisinger, D.; Niemeier, D. MOVES vs. EMFAC: A comparison of greenhouse gas emissions using Los Angeles County. In Proceedings of the Transportation Research Board 88th Annual Meeting, Washington, DC, USA, 11–15 January 2009; pp. 09–0692.
133. Shah, S.D.; Johnson, K.C.; Miller, J.W.; Cocker, D.R., III. Emission rates of regulated pollutants from on-road heavy-duty diesel vehicles. *Atmos. Environ.* **2006**, *40*, 147–153. [CrossRef]
134. Boulter, P.G.; McCrae, I.S.; Barlow, T.J. *A Review of Instantaneous Emission Models for Road Vehicles*; TRL: Crowthorne, UK, 2007.
135. Esteves-Booth, A.; Muneer, T.; Kubie, J.; Kirby, H. A review of vehicular emission models and driving cycles. *Proc. Inst. Mech. Eng. Part C J. Mech. Eng. Sci.* **2002**, *216*, 777–797. [CrossRef]
136. Colberg, C.A.; Tona, B.; Stahel, W.A.; Meier, M.; Staehelin, J. Comparison of a road traffic emission model (HBEFA) with emissions derived from measurements in the Gubrist road tunnel, Switzerland. *Atmos. Environ.* **2005**, *39*, 4703–4714. [CrossRef]
137. Borge, R.; de Miguel, I.; de la Paz, D.; Lumbreras, J.; Pérez, J.; Rodríguez, E. Comparison of road traffic emission models in Madrid (Spain). *Atmos. Environ.* **2012**, *62*, 461–471. [CrossRef]
138. Boveroux, F.; Cassiers, S.; De Meyer, P.; Buekenhoudt, P.; Bergmans, B.; Idczak, F.; Jeanmart, H.; Verhelst, S.; Contino, F. Impact of mileage on particle number emission factors for EURO5 and EURO6 diesel passenger cars. *Atmos. Environ.* **2021**, *244*, 117975. [CrossRef]
139. Savostin-Kosiak, D.; Mądziel, M.; Jaworski, A.; Ivanushko, O.; Tsiuman, N.; Loboda, A. Establishing the regularities of correlation between ambient temperature and fuel consumption by city diesel buses. *East. Eur. J. Enterp. Technol.* **2020**, *6*, 23–32. [CrossRef]
140. Mądziel, M.; Campisi, T.; Jaworski, A.; Kuszewski, H.; Woś, P. Assessing Vehicle Emissions from a Multi-Lane to Turbo Roundabout Conversion Using a Microsimulation Tool. *Energies* **2021**, *14*, 4399. [CrossRef]
141. Khan, Z.H.; Imran, W.; Azeem, S.; Khattak, K.S.; Gulliver, T.A.; Aslam, M.S. A macroscopic traffic model based on driver reaction and traffic stimuli. *Appl. Sci.* **2019**, *9*, 2848. [CrossRef]
142. Lee, H.-W.; Kim, D. Macroscopic traffic models from microscopic car-following models. *Phys. Rev. E* **2001**, *64*, 056126. [CrossRef] [PubMed]
143. Krivda, V.; Petru, J.; Macha, D.; Novak, J. Use of Microsimulation Traffic Models as Means for Ensuring Public Transport Sustainability and Accessibility. *Sustainability* **2021**, *13*, 2709. [CrossRef]
144. Ferrara, A.; Sacone, S.; Siri, S. Microscopic and Mesoscopic Traffic Models. In *Freeway Traffic Modelling and Control*; Springer: Berlin/Heidelberg, Germany, 2018; pp. 113–143. [CrossRef]
145. Ma, X.; Jin, J.; Lei, W. Multi-criteria analysis of optimal signal plans using microscopic traffic models. *Transp. Res. Part D Transp. Environ.* **2014**, *32*, 1–14. [CrossRef]
146. Marsden, G.; Bell, M.; Reynolds, S. Towards a real-time microscopic emissions model. *Transp. Res. Part D Transp. Environ.* **2001**, *6*, 37–60. [CrossRef]

147. Kim, M.; Cho, G.-H. Influence of Evacuation Policy on Clearance Time under Large-Scale Chemical Accident: An Agent-Based Modeling. *Int. J. Environ. Res. Public Health* **2020**, *17*, 9442. [CrossRef] [PubMed]
148. Lee, K.S.; Eom, J.K.; Moon, D.S. Applications of TRANSIMS in transportation: A literature review. *Procedia Comput. Sci.* **2014**, *32*, 769–773. [CrossRef]
149. Li, M.; Luo, D.; Liu, B.; Zhang, X.; Liu, Z.; Li, M. Arterial coordination control optimization based on AM–BAND–PBAND model. *Sustainability* **2022**, *14*, 10065. [CrossRef]
150. Olaverri-Monreal, C.; Errea-Moreno, J.; Díaz-Álvarez, A.; Biurrun-Quel, C.; Serrano-Arriezu, L.; Kuba, M. Connection of the SUMO microscopic traffic simulator and the unity 3D game engine to evaluate V2X communication-based systems. *Sensors* **2018**, *18*, 4399. [CrossRef]
151. Cárdenas-Benítez, N.; Aquino-Santos, R.; Magaña-Espinoza, P.; Aguilar-Velazco, J.; Edwards-Block, A.; Cass, A.M. Traffic Congestion Detection System through Connected Vehicles and Big Data. *Sensors* **2016**, *16*, 599. [CrossRef]
152. Schweizer, J.; Poliziani, C.; Rupi, F.; Morgano, D.; Magi, M. Building a Large-Scale Micro-Simulation Transport Scenario Using Big Data. *ISPRS Int. J. Geo-Inf.* **2021**, *10*, 165. [CrossRef]
153. Kővári, B.; Szőke, L.; Bécsi, T.; Aradi, S.; Gáspár, P. Traffic signal control via reinforcement learning for reducing global vehicle emission. *Sustainability* **2021**, *13*, 11254. [CrossRef]
154. Weyland, C.M.; Baumann, M.V.; Buck, H.S.; Vortisch, P. Parameters influencing lane flow distribution on multilane freeways in PTV VISSIM. *Procedia Comput. Sci.* **2021**, *184*, 453–460. [CrossRef]
155. Zeidler, V.; Buck, H.S.; Kautzsch, L.; Vortisch, P.; Weyland, C.M. Simulation of autonomous vehicles based on Wiedemann's car following model in PTV vissim. In Proceedings of the 98th Annual Meeting of the Transportation Research Board (TRB), Washington, DC, USA, 13–17 January 2019; pp. 13–17.
156. Nalic, D.; Pandurevic, A.; Eichberger, A.; Fellendorf, M.; Rogic, B. Software Framework for Testing of Automated Driving Systems in the Traffic Environment of Vissim. *Energies* **2021**, *14*, 3135. [CrossRef]
157. Severino, A.; Pappalardo, G.; Curto, S.; Trubia, S.; Olayode, I.O. Safety Evaluation of Flower Roundabout Considering Autonomous Vehicles Operation. *Sustainability* **2021**, *13*, 10120. [CrossRef]
158. Campisi, T.; Mądziel, M.; Nikiforiadis, A.; Basbas, S.; Tesoriere, G. An Estimation of Emission Patterns from Vehicle Traffic Highlighting Decarbonisation Effects from Increased e-fleet in Areas Surrounding the City of Rzeszow (Poland). In *Computational Science and Its Applications—ICCSA 2021*; Lecture Notes in Computer, Science; Gervasi, O., Murgante, B., Mira, S., Garau, C., Blecic, I., Taniar, D., Apduhan, B.O., Rocha, A.M.A.C., Tarantino, E., Torre, C.M., Eds.; Springer: Cham, Switzerland, 2021; Volume 12953.
159. Shen, T.; Cummings, M.L. A microsimulation comparison for lane-merging driver behaviors. In Proceedings of the Human Factors and Ergonomics Society Annual Meeting, Atlanta, GA, USA, 10–14 October 2022; SAGE Publications: Los Angeles, CA, USA, 2022; Volume 66, No. 1. pp. 1433–1436.
160. Madziel, M.; Jaworski, A.; Savostin-Kosiak, D.; Lejda, K. The Impact of Exhaust Emission from Combustion Engines on the Environment: Modelling of Vehicle Movement at Roundabouts. *Int. J. Automot. Mech. Eng.* **2020**, *17*, 8360–8371. [CrossRef]
161. Ziemska, M. Exhaust Emissions and Fuel Consumption Analysis on the Example of an Increasing Number of HGVs in the Port City. *Sustainability* **2021**, *13*, 7428. [CrossRef]
162. Zhou, X.; Tanvir, S.; Lei, H.; Taylor, J.; Liu, B.; Rouphail, N.M.; Frey, H.C. Integrating a simplified emission estimation model and mesoscopic dynamic traffic simulator to efficiently evaluate emission impacts of traffic management strategies. *Transp. Res. Part D Transp. Environ.* **2015**, *37*, 123–136. [CrossRef]
163. Abou-Senna, H.; Radwan, E.; Westerlund, K.; Cooper, C.D. Using a traffic simulation model (VISSIM) with an emissions model (MOVES) to predict emissions from vehicles on a limited-access highway. *J. Air Waste Manag. Assoc.* **2013**, *63*, 819–831. [CrossRef]
164. Rodriguez-Rey, D.; Guevara, M.; Linares, M.P.; Casanovas, J.; Salmerón, J.; Soret, A.; Jorba, O.; Tena, C.; Pérez García-Pando, C. A coupled macroscopic traffic and pollutant emission modelling system for Barcelona. *Transp. Res. Part D Transp. Environ.* **2021**, *92*, 102725. [CrossRef]
165. Samaras, C.; Tsokolis, D.; Toffolo, S.; Magra, G.; Ntziachristos, L.; Samaras, Z. Enhancing average speed emission models to account for congestion impacts in traffic network link-based simulations. *Transp. Res. Part D Transp. Environ.* **2019**, *75*, 197–210. [CrossRef]
166. Mangones, S.C.; Jaramillo, P.; Fischbeck, P.; Rojas, N.Y. Development of a high-resolution traffic emission model: Lessons and key insights from the case of Bogotá, Colombia. *Environ. Pollut.* **2019**, *253*, 552–559. [CrossRef] [PubMed]
167. Macedo, E.; Tomás, R.; Fernandes, P.; Coelho, M.C.; Bandeira, J.M. Quantifying road traffic emissions embedded in a multi-objective traffic assignment model. *Transp. Res. Procedia* **2020**, *47*, 648–655. [CrossRef]
168. Fan, J.; Gao, K.; Xing, Y.; Lu, J. Evaluating the effects of one-way traffic management on different vehicle exhaust emissions using an integrated approach. *J. Adv. Transp.* **2019**, *2019*, 1–11. [CrossRef]
169. Rodriguez-Rey, D.; Guevara, M.; Linares, M.P.; Casanovas, J.; Armengol, J.M.; Benavides, J.; Soret, A.; Jorba, O.; Tena, C.; García-Pando, C.P. To what extent the traffic restriction policies applied in Barcelona city can improve its air quality? *Sci. Total Environ.* **2022**, *807*, 150743. [CrossRef]
170. Overtoom, I.; Correia, G.; Huang, Y.; Verbraeck, A. Assessing the impacts of shared autonomous vehicles on congestion and curb use: A traffic simulation study in The Hague, Netherlands. *Int. J. Transp. Sci. Technol.* **2020**, *9*, 195–206. [CrossRef]

171. Mądziel, M.; Campisi, T. Investigation of Vehicular Pollutant Emissions at 4-Arm Intersections for the Improvement of Integrated Actions in the Sustainable Urban Mobility Plans (SUMPs). *Sustainability* **2023**, *15*, 1860. [CrossRef]
172. Gao, J.; Chen, H.; Dave, K.; Chen, J.; Jia, D. Fuel economy and exhaust emissions of a diesel vehicle under real traffic conditions. *Energy Sci. Eng.* **2020**, *8*, 1781–1792. [CrossRef]
173. Adamidis, F.K.; Mantouka, E.G.; Vlahogianni, E.I. Effects of controlling aggressive driving behavior on network-wide traffic flow and emissions. *Int. J. Transp. Sci. Technol.* **2020**, *9*, 263–276. [CrossRef]
174. Saedi, R.; Verma, R.; Zockaie, A.; Ghamami, M.; Gates, T.J. Comparison of Support Vector and Non-Linear Regression Models for Estimating Large-Scale Vehicular Emissions, Incorporating Network-Wide Fundamental Diagram for Heterogeneous Vehicles. *Transp. Res. Rec. J. Transp. Res. Board* **2020**, *2674*, 70–84. [CrossRef]
175. Gräbe, R.J.; Joubert, J.W. Are we getting vehicle emissions estimation right? *Transp. Res. Part D Transp. Environ.* **2022**, *112*, 103477. [CrossRef]
176. Fernandes, P.; Bandeira, J.M.; Coelho, M.C. A macroscopic approach for assessing the environmental performance of shared, automated, electric mobility in an intercity corridor. *J. Intell. Transp. Syst.* **2021**, 1–17. [CrossRef]
177. Jiang, Y.; Ding, Z.; Zhou, J.; Wu, P.; Chen, B. Estimation of traffic emissions in a polycentric urban city based on a macroscopic approach. *Phys. A Stat. Mech. Its Appl.* **2022**, *602*, 127391. [CrossRef]
178. Kaths, H.; Keler, A.; Bogenberger, K. Calibrating the wiedemann 99 car-following model for bicycle traffic. *Sustainability* **2021**, *13*, 3487. [CrossRef]
179. Ziemska-Osuch, M.; Osuch, D. Modeling the assessment of intersections with traffic lights and the significance level of the number of pedestrians in microsimulation models based on the PTV Vissim tool. *Sustainability* **2022**, *14*, 8945. [CrossRef]
180. Azam, M.; Hassan, S.A.; Puan, O.C. Calibration methodologies of VISSIM-based microsimulation model for heterogeneous traffic conditions-a survey. *Adv. Transp. Stud.* **2023**, *59*, 123–146.
181. Li, S.; Xiang, Q.; Ma, Y.; Gu, X.; Li, H. Crash Risk Prediction Modeling Based on the Traffic Conflict Technique and a Microscopic Simulation for Freeway Interchange Merging Areas. *Int. J. Environ. Res. Public Health* **2016**, *13*, 1157. [CrossRef]
182. Meiring, G.A.M.; Myburgh, H.C. A review of intelligent driving style analysis systems and related artificial intelligence algorithms. *Sensors* **2015**, *15*, 30653–30682. [CrossRef]
183. Nai, W.; Yang, Z.; Wei, Y.; Sang, J.; Wang, J.; Wang, Z.; Mo, P. A Comprehensive Review of Driving Style Evaluation Approaches and Product Designs Applied to Vehicle Usage-Based Insurance. *Sustainability* **2022**, *14*, 7705. [CrossRef]
184. Paszkowski, J.; Herrmann, M.; Richter, M.; Szarata, A. Modelling the Effects of Traffic-Calming Introduction to Volume–Delay Functions and Traffic Assignment. *Energies* **2021**, *14*, 3726. [CrossRef]
185. Deluka Tibljaš, A.; Giuffrè, T.; Surdonja, S.; Trubia, S. Introduction of Autonomous Vehicles: Roundabouts Design and Safety Performance Evaluation. *Sustainability* **2018**, *10*, 1060. [CrossRef]
186. Dutta, M.; Ahmed, M.A. Calibration of VISSIM models at three-legged unsignalized intersections under mixed traffic conditions. *Adv. Transp. Stud.* **2019**, *48*, 31–46.
187. Wilmink, I.R.; Viti, F.; Baalen, J.V.; Li, M. Emission modelling at signalised intersections using microscopic models. In Proceedings of the 16th ITS World Congress and Exhibition on Intelligent Transport Systems and Services, Stockholm, Sweden, 21–25 September 2009.
188. Li, J.; Van Zuylen, H.; Chen, Y.; Viti, F.; Wilmink, I. Optimizing Traffic Control for Emission Reduction: The calibration of the simulation model. In Proceedings of the Mobil.TUM 2009—International Scientific Conference on Mobility and Transport-ITS for Larger Cities, Munich, Germany, 12–13 May 2009.
189. Jie, L.; Van Zuylen, H.; Chen, Y.; Viti, F.; Wilmink, I. Calibration of a microscopic simulation model for emission calculation. *Transp. Res. Part C Emerg. Technol.* **2013**, *31*, 172–184. [CrossRef]
190. Nesamani, K.; Chu, L.; McNally, M.G.; Jayakrishnan, R. Estimation of vehicular emissions by capturing traffic variations. *Atmos. Environ.* **2007**, *41*, 2996–3008. [CrossRef]
191. Hirschmann, K.; Zallinger, M.; Fellendorf, M.; Hausberger, S. A new method to calculate emissions with simulated traffic conditions. In Proceedings of the 13th International IEEE Conference on Intelligent Transportation Systems, Funchal, Portugal, 19–22 September 2010; pp. 33–38.
192. da Rocha, T.V.; Leclercq, L.; Montanino, M.; Parzani, C.; Punzo, V.; Ciuffo, B.; Villegas, D. Does traffic-related calibration of car-following models provide accurate estimations of vehicle emissions? *Transp. Res. Part D Transp. Environ.* **2015**, *34*, 267–280. [CrossRef]
193. Kim, J.; Kim, J.H.; Lee, G.; Shin, H.-J.; Park, J.H. Microscopic Traffic Simulation Calibration Level for Reliable Estimation of Vehicle Emissions. *J. Adv. Transp.* **2020**, *2020*, 4038305. [CrossRef]
194. Swidan, H. Integrating AIMSUN Micro Simulation Model with Portable Emissions Measurement System (PEMS): Calibration and Validation Case Study. Master's Thesis, North Carolina State University, Raleigh, NC, USA, 2011.
195. Sánchez, J.M.; Ortega, E.; Lopez-Lambas, M.E.; Martín, B. Evaluation of emissions in traffic reduction and pedestrianization scenarios in Madrid. *Transp. Res. Part D Transp. Environ.* **2021**, *100*, 103064. [CrossRef]
196. Tomás, R.F.; Fernandes, P.; Macedo, E.; Bandeira, J.M.; Coelho, M.C. Assessing the emission impacts of autonomous vehicles on metropolitan freeways. *Transp. Res. Procedia* **2020**, *47*, 617–624. [CrossRef]
197. Garcia, M.R.; Monzon, A.; Valdes, C.; Romana, M. Modeling different penetration rates of eco-driving in urban areas: Impacts on traffic flow and emissions. *Int. J. Sustain. Transp.* **2017**, *11*, 282–294. [CrossRef]

198. Liao, R.; Chen, X.; Yu, L.; Sun, X. Analysis of Emission Effects Related to Drivers' Compliance Rates for Cooperative Vehicle-Infrastructure System at Signalized Intersections. *Int. J. Environ. Res. Public Health* **2018**, *15*, 122. [CrossRef]
199. Zhai, Z.; Tu, R.; Xu, J.; Wang, A.; Hatzopoulou, M. Capturing the Variability in Instantaneous Vehicle Emissions Based on Field Test Data. *Atmosphere* **2020**, *11*, 765. [CrossRef]
200. Mądziel, M.; Jaworski, A.; Kuszewski, H.; Woś, P.; Campisi, T.; Lew, K. The Development of CO_2 Instantaneous Emission Model of Full Hybrid Vehicle with the Use of Machine Learning Techniques. *Energies* **2022**, *15*, 142. [CrossRef]
201. Wen, H.T.; Lu, J.H.; Jhang, D.S. Features Importance Analysis of Diesel Vehicles' NOx and CO_2 Emission Predictions in Real Road Driving Based on Gradient Boosting Regression Model. *Int. J. Environ. Res. Public Health* **2021**, *18*, 13044. [CrossRef]
202. Chen, C.; Zhao, X.; Liu, H.; Ren, G.; Zhang, Y.; Liu, X. Assessing the Influence of Adverse Weather on Traffic Flow Characteristics Using a Driving Simulator and VISSIM. *Sustainability* **2019**, *11*, 830. [CrossRef]
203. Al-Ahmadi, H.M.; Jamal, A.; Reza, I.; Assi, K.J.; Ahmed, S.A. Using Microscopic Simulation-Based Analysis to Model Driving Behavior: A Case Study of Khobar-Dammam in Saudi Arabia. *Sustainability* **2019**, *11*, 3018. [CrossRef]
204. Shao, Y.; Han, X.; Wu, H.; Claudel, C.G. Evaluating Signalization and Channelization Selections at Intersections Based on an Entropy Method. *Entropy* **2019**, *21*, 808. [CrossRef] [PubMed]
205. Pielecha, J.; Skobiej, K.; Kurtyka, K. Exhaust Emissions and Energy Consumption Analysis of Conventional, Hybrid, and Electric Vehicles in Real Driving Cycles. *Energies* **2020**, *13*, 6423. [CrossRef]
206. Asher, Z.D.; Galang, A.A.; Briggs, W.; Johnston, B.; Bradley, T.H.; Jathar, S. *Economic and Efficient Hybrid Vehicle Fuel Economy and Emissions Modeling Using an Artificial Neural Network*; (No. 2018-01-0315); SAE Technical Paper; SAE: Warrendale, PA, USA, 2018.
207. Bachar, R.K.; Bhuniya, S.; Ghosh, S.K.; Sarkar, B. Controllable Energy Consumption in a Sustainable Smart Manufacturing Model Considering Superior Service, Flexible Demand, and Partial Outsourcing. *Mathematics* **2022**, *10*, 4517. [CrossRef]
208. Mittal, M.; Sarkar, B. Stochastic behavior of exchange rate on an international supply chain under random energy price. *Math. Comput. Simul.* **2023**, *205*, 232–250. [CrossRef]
209. Guchhait, R.; Sarkar, B. Increasing Growth of Renewable Energy: A State of Art. *Energies* **2023**, *16*, 2665. [CrossRef]
210. Dash, S.K.; Chakraborty, S.; Roccotelli, M.; Sahu, U.K. Hydrogen Fuel for Future Mobility: Challenges and Future Aspects. *Sustainability* **2022**, *14*, 8285. [CrossRef]
211. Kuszewski, H.; Jaworski, A.; Mądziel, M.; Woś, P. The investigation of auto-ignition properties of 1-butanol–biodiesel blends under various temperatures conditions. *Fuel* **2023**, *346*, 128388. [CrossRef]
212. Sarkar, B.; Mridha, B.; Pareek, S. A sustainable smart multi-type biofuel manufacturing with the optimum energy utilization under flexible production. *J. Clean. Prod.* **2022**, *332*, 129869. [CrossRef]
213. Mridha, B.; Ramana, G.V.; Pareek, S.; Sarkar, B. An efficient sustainable smart approach to biofuel production with emphasizing the environmental and energy aspects. *Fuel* **2023**, *336*, 126896. [CrossRef]
214. Abdullah, A.M.; Usmani, R.S.A.; Pillai, T.R.; Marjani, M.; Hashem, I.A.T. An Optimized Artificial Neural Network Model using Genetic Algorithm for Prediction of Traffic Emission Concentrations. *Int. J. Adv. Comput. Sci. Appl.* **2021**, *12*, 794–803. [CrossRef]
215. Khurana, S.; Saxena, S.; Jain, S.; Dixit, A. Predictive modeling of engine emissions using machine learning: A review. *Mater. Today Proc.* **2021**, *38*, 280–284. [CrossRef]
216. Jing, B.; Wu, L.; Mao, H.; Gong, S.; He, J.; Zou, C.; Song, G.; Li, X.; Wu, Z. Development of a vehicle emission inventory with high temporal–spatial resolution based on NRT traffic data and its impact on air pollution in Beijing—Part 1: Development and evaluation of vehicle emission inventory. *Atmos. Chem. Phys.* **2016**, *16*, 3161–3170. [CrossRef]
217. Batterman, S.; Ganguly, R.; Harbin, P. High Resolution Spatial and Temporal Mapping of Traffic-Related Air Pollutants. *Int. J. Environ. Res. Public Health* **2015**, *12*, 3646–3666. [CrossRef]
218. Mądziel, M.; Campisi, T. Energy Consumption of Electric Vehicles: Analysis of Selected Parameters Based on Created Database. *Energies* **2023**, *16*, 1437. [CrossRef]
219. Amara-Ouali, Y.; Goude, Y.; Massart, P.; Poggi, J.-M.; Yan, H. A Review of Electric Vehicle Load Open Data and Models. *Energies* **2021**, *14*, 2233. [CrossRef]
220. Kubik, A.; Turoń, K.; Folęga, P.; Chen, F. CO_2 Emissions—Evidence from Internal Combustion and Electric Engine Vehicles from Car-Sharing Systems. *Energies* **2023**, *16*, 2185. [CrossRef]
221. COPERT Emission Model Download Site. Available online: https://www.emisia.com/utilities/copert/download/ (accessed on 4 May 2023).
222. MOBILE Emission Model. Available online: https://www.epa.gov/moves/latest-version-motor-vehicle-emission-simulator-moves (accessed on 4 May 2023).
223. HBEFA Emission Model. Available online: https://www.hbefa.net/e/index.html (accessed on 4 May 2023).
224. CMEM Emission Model. Available online: https://www.cert.ucr.edu/cmem (accessed on 4 May 2023).
225. ESTM BOSH Emission Model. Available online: https://company.ptvgroup.com/en/ptv-vissim-emissions-calculation-from-bosch (accessed on 4 May 2023).
226. EMPA Emission Model. Available online: https://www.empa.ch/web/s503/modelling-remote-sensing (accessed on 4 May 2023).
227. Vissim Download Site. Available online: https://www.myptv.com/en/mobility-software/ptv-vissim/demo (accessed on 4 May 2023).
228. Aimsun Download Site. Available online: https://www.aimsun.com/aimsun-next-downloads/ (accessed on 4 May 2023).

229. Sumo Download Site. Available online: https://www.eclipse.org/sumo/ (accessed on 4 May 2023).
230. Visum Download Site. Available online: https://your.visum.ptvgroup.com/vision-traffic-suite-students-en (accessed on 4 May 2023).
231. MATSim Download Site. Available online: https://matsim.org/downloads/ (accessed on 4 May 2023).
232. Emme 4.0 Download Site. Available online: https://www.inrosoftware.com/en/products/emme/ (accessed on 4 May 2023).

Disclaimer/Publisher's Note: The statements, opinions and data contained in all publications are solely those of the individual author(s) and contributor(s) and not of MDPI and/or the editor(s). MDPI and/or the editor(s) disclaim responsibility for any injury to people or property resulting from any ideas, methods, instructions or products referred to in the content.

MDPI AG
Grosspeteranlage 5
4052 Basel
Switzerland
Tel.: +41 61 683 77 34

Energies Editorial Office
E-mail: energies@mdpi.com
www.mdpi.com/journal/energies

Disclaimer/Publisher's Note: The title and front matter of this reprint are at the discretion of the Guest Editors. The publisher is not responsible for their content or any associated concerns. The statements, opinions and data contained in all individual articles are solely those of the individual Editors and contributors and not of MDPI. MDPI disclaims responsibility for any injury to people or property resulting from any ideas, methods, instructions or products referred to in the content.